Physiological and Morphological Adaptation and Evolution

World Anthropology

General Editor

SOL TAX

Patrons

CLAUDE LÉVI-STRAUSS
MARGARET MEAD†
LAILA SHUKRY EL HAMAMSY
M. N. SRINIVAS

MOUTON PUBLISHERS · THE HAGUE · PARIS · NEW YORK

Physiological and Morphological Adaptation and Evolution

Editor

WILLIAM A. STINI

MOUTON PUBLISHERS · THE HAGUE · PARIS · NEW YORK

ISBN 90-279-7710-0 (Mouton)
0-202-90095-9 (AVC Inc.)
Jacket photo by Cas Oorthuys
Cover and jacket design by Jurriaan Schrofer
Indexes by Society of Indexers, Great Britain
Phototypeset in V.I.P. Times by
Western Printing Services Ltd, Bristol
Printed in Great Britain

General Editor's Preface

Until the current generation, morphological differences in the human species were treated as "racial" variations. Measurements and comparisons of proportions and indices classified people into types and subgroups of humanity. In recent years such anatomical characterizations of populations have been almost completely replaced as our knowledge of variations has been broadened to include physiological, cytological, chemical, and genetic characters that are understood all together in the context of adaptation and evolution. These papers have been contributed by human biologists from all parts of the world — inspired by the opportunity of an unusual international Congress — and the present volume provides a unique sample of the new understanding derived from research on such variations.

Like most contemporaty sciences, anthropology is a product of the European tradition. Some argue that it is a product of colonialism, with one small and self-interested part of the species dominating the study of the whole. If we are to understand the species, our science needs substantial input from scholars who represent a variety of the world's cultures. It was a deliberate purpose of the IXth International Congress of Anthropological and Ethnological Sciences to provide impetus in this direction. The *World Anthropology* volumes, therefore, offer a first glimpse of a human science in which members from all societies have played an active role. Each of the books is designed to be self-contained; each is an attempt to update its particular sector of scientific knowledge and is written by specialists from all parts of the world. Each volume should be read and reviewed individually as a separate volume on its own given subject. The set as a whole will indicate what changes are in store for anthropology as scholars from the developing countries join in studying the species of which we are all a part.

The IXth Congress was planned from the beginning not only to include as many of the scholars from every part of the world as possible, but also with a view toward the eventual publication of the papers in high-quality volumes. At previous Congresses scholars were invited to bring papers which were then read out loud. They were necessarily limited in length; many were only summarized; there was little time for discussion; and the sparse discussion could only be in one language. The IXth Congress was an experiment aimed at changing this. Papers were written with the intention of exchanging them before the Congress, particularly in extensive pre-Congress sessions; they were not intended to be read aloud at the Congress, that time being devoted to discussions — discussions which were simultaneously and professionally translated into five languages. The method for eliciting the papers was structured to make as representative a sample as was allowable when scholarly creativity — hence self-selection — was critically important. Scholars were asked both to propose papers of their own to suggest topics for sessions of the Congress which they might edit into volumes. All were then informed of the suggestions and encouraged to re-think their own papers and the topics. The process, therefore, was a continuous one of feedback and exchange and it has continued to be so even after the Congress. The some two thousand papers comprising *World Anthropology* certainly then offer a substantial sample of world anthropology. It has been said that anthropology is at a turning point; if this is so, these volumes will be the historical direction-markers.

As might have been foreseen in the first post-colonial generation, the large majority of the Congress papers (82 percent) are the work of scholars identified with the industrialized world which fathered our traditional discipline and the institution of the Congress itself: Eastern Europe (15 percent); Western Europe (16 percent); North America (47 percent); Japan, South Africa, Australia, and New Zealand (4 percent). Only 18 percent of the papers are from developing areas: Africa (4 percent); Asia-Oceania (9 percent); Latin America (5 percent). Aside from the substantial representation from the U.S.S.R. and the nations of Eastern Europe, a significant difference between this corpus of written material and that of other Congresses is the addition of the large proportion of contributions from Africa, Asia, and Latin America. "Only 18 percent" is two to four times as great a proportion as that of other Congresses; moreover, 18 percent of 2,000 papers is 360 papers, 10 times the number of "Third World" papers presented at previous Congresses. In fact, these 360 papers are more than the total of *all* papers published after the last International Congress of Anthropological and Ethnological Sciences which was held in the United States (Philadelphia, 1956).

The significance of the increase is not simply quantitative. The input of scholars from areas which have until recently been no more than subject matter for anthropology represents both feedback and also long-awaited theoretical contributions from the perspectives of very different cultural, social, and historical traditions. Many who attended the IXth Congress were convinced that anthropology would not be the same in the future. The fact that the Xth Congress (India, 1978) was our first in the "Third World" may be symbolic of the change. Meanwhile, sober consideration of the present set of books will show how much, and just where and how, our discipline is being revolutionized.

Readers of the present volume will be particularly interested in a score of others in this series which deal with biological anthropology in long-term evolutionary perspectives and as human biologists discover and study the interactions of cultural and biological phenomena in the widest variety of geographical, ecological, and social situations.

Chicago, Illinois SOL TAX
January 9, 1979

Table of Contents

General Editor's Preface v

Introduction 1
by *William A. Stini*

SECTION ONE: PHENOTYPIC VARIATION AMONG HUMAN POPULATIONS

Skin Color Variability in the Middle East 7
by *E. Sunderland*

Factor Analysis of Anthropometric Data for Fifteen Race-Age-National Origin Specific Groups 19
by *Edmund Churchill, David Rabinow,* and *Peggy Erskine*

Certain Problems of Ethnic Anthropology in Southwest Asia in the Light of the Latest Research 45
by *M. G. Abdushelishvili*

Recent Anthropological Data on the Ammassalimiut Eskimo of East Greenland: Comparative Studies 55
by *Albert Ducros* and *Jacqueline Ducros*

Natural Selection and Physical Adaptation to a Biotype of Tropical Africa 69
by *Emile Crognier*

Territorial Morphofunctional Studies in the Soviet Union: The Problem of Adaptive Types 81
by *T. I. Alekseeva*

Adaptation of Morphology to Climate in Central Africa 89
by *Jean Hiernaux*

The Differential Geography of Races 97
by *V. P. Alexeev*

Blood Levels of Some Biochemical Components among Negroid Individuals with Hemoglobin A and AS 111
by *V. Correnti, G. Spedini, M. Cresta, F. Vecchi*, and *E. Capucci*

Genogeography of Blood Factors in the Transcaucasus 115
by *A. A. Voronov*

Morphological Peculiarities of the Body Frame in Different Ethnoterritorial Groups 131
by *N. S. Smirnova*

Effects of Isolation on Genetic Variability: Macaque Populations as Model Systems 139
by *Laura Newell Morris* and *Peter E. Nute*

SECTION TWO: GROWTH AND DEVELOPMENT

Introduction 163
by *William A. Stini*

Variations in Female Physique 165
by *O. G. Eiben*

Body Composition of Japanese Children: Factor Analysis of Development 177
by *S. Kanda*

Genetics of Human Morphological Characters 187
by *C. Susanne*

Lean Body Mass in Young, Adult, and Advanced Age Groups 201
by *Jana Pařízková*

Skeletal Maturation of Japanese–American Hybrids 207
by *Kunihiko Kimura*

Growth and Environment Studies in Czechoslovakia 213
by *Miroslav Prokopec*

Some Biosocial Determinants of the Growth, Health, and Nutritional Status of Papua New Guinean Preschool Children 233
by *L. A. Malcolm*

Processes of Human Growth in High Altitudes 243
by *N. N. Miklashevskaya, V. S. Solovyeva*, and *E. Z. Godina*

Research on Puberty in Haiti 255
by *J. B. Romain*

The Development of Gypsy Children in Changing Living Conditions 263
by *Jaroslav Suchý*

The Relationship Between Human Body Constitution, Longevity, Disease, and Causes of Death 273
by *Peter Râmneantzu*

One of the Factors of Brachycephalization over the Centuries: The Susceptibility of Dolichocephals to Tuberculosis 279
by *G. Olivier* and *M. E. de Castro e Almeida* with the collaboration of *H. Tissier*

A Study of Morphological Variables and Genetic Markers in Urban and Rural South African Negro Male Populations 287
by *Hertha de Villiers*

Physical Growth and the Social Environment: A West African Example 301
by *D. K. Fiawoo*

Progressive Decline in Stature in India: A Study of Sixty Population Groups 315
by *Pranab Ganguly*

A Study of Height, Weight, and Arm Girth Measurements of Punjabi Males 339
by *Raghbir Singh*

Nutrition and Pregnancy 345
by *John R. K. Robson*

The Age at Menarche in Venezuelan Students Residing in Metropolitan Caracas 349
by *Fritzi Kohn de Brief*

The Relevance of Nourishment to the Reproductive Cycle of the Female in India 363
by *Judit Katona-Apte*

An Anatomical View of Demographic Data: Biomass, Fat Mass, and Lean Body Mass of the United States and Mexican Human Populations 369
by *Gabriel Lasker* and *Henry Womack*

Individual Differences in Skeletal Muscle Fiber Composition 379
by *S. Inokuchi*

Adaptive Strategies of Human Populations under Nutritional Stress 387
by *William A. Stini*

SECTION THREE: EFFECTS OF CHROMOSOMAL VARIATION ON HUMAN DEVELOPMENT

Comments on the Effects of Chromosomal Variation on Human Development 411
by *Fathi Abdel-Hameed*

The XX Male and the XY Female in Humans 417
by *Fathi Abdel-Hameed* and *Linda M. Gluesing*

Hormonal Factors in Formative Processes 451
by *E. N. Khrisanfova*

Polyploidy in Humans and Other Vertebrates 459
by *Fathi Abdel-Hameed*

Biographical Notes 499

Index of Names 509

Index of Subjects 519

Introduction

WILLIAM A. STINI

The topic of human variation has long been a concern of anthropologists. The earliest discussions of the variety of sizes, shapes, and colors seen among the peoples of the world were by contemporary standards naive and, in many cases, unashamedly racist. Nevertheless, we may detect even in some of the earliest accounts of human variability the suspicion that in some way people were shaped by their surroundings. With the advent of evolutionary theory, the mechanisms through which a species adapted to its environment became an increasingly important area of interest. When students of human variation began to speculate on the manner in which some populations acquired characteristic constellations of traits instead of viewing them as static entities of an exotic nature, the ground was laid for the emergence of the science of human biology. When the application of genetic theory to evolutionary studies became widely accepted, leading to the recognition of the importance of the study of populations and their variability, the contribution of human biology to general biological theory, as well as to anthropology, became significant. The discovery of traits of known mode of inheritance in a large, widely distributed species made the study of human variability meaningful not only as description; it ultimately opened lines of inquiry which have done much to illuminate the means by which descent with modification occurs.

Of course, the study of human biology is still a young science. Though no longer solely descriptive, it is not a discipline that has generated new methods of its own. Indeed, students of human biology provide a cross section of methodological approaches which closely parallels that of the biological sciences in general. While experimentation on humans is not a tool of human biology, animal experiments have shed much light on processes at work in humans. As a result, human biologists find themselves working with many other species at a time when specialists in other areas of biology have been drawn to the study of humans. This is, of course, a healthy state of affairs and promises to enhance our understanding of ourselves to a degree not possible in an earlier, more parochial period in the study of man. This volume contains a sampling of work being done today under the heading of "human biology." As can be seen from a perusal of the Table of Contents, the scope of the investigations is broad, but interest in the variety of human types is alive and flourishing. Could it really be otherwise?

The study of human differences is a fascinating one. The interest increases as new lines of inquiry emerge. From the observation of differences in pigmentation and body size to the calculation of cranial indices, sitting-height indices, and constitutional types, to the discovery of clines in blood group frequencies, polymorphisms in serum proteins, variations in polypeptide sequences of hemoglobin molecules, and immunological specificities of human tissues, is a continuum illustrating the polytypic nature of our species. When one studies the great phenotypic variability of our species, one must be impressed by the fact that under a wide variety of stresses in many ecological habitats, humans have survived by making adaptations which have permitted them to remain interfertile. We are still *one* species. The occupation of many habitats is made possible by adjustments which are not the product of changes in genetic material but are instead affected by exploitation of great phenotypic plasticity. Added to this plasticity is the human capacity to modify the environment in ways inconceivable to other forms of life. Much of human adaptation involves behavioral rather than physiological responses to stress, making possible the retention of many genotypes in human populations which might otherwise disappear as part of the price of natural selection. Such modifications in the intensity of natural selection allow the retention and accumulation of genetic traits which may not "earn their keep" most of the time but may, in the long run, prove valuable to a species whose own activities promise to alter the environment in ways which defy prediction. Man is a most adaptable species. This observation may be translated into the assumption that human evolution may yet have much ground to cover, that the nature of our descendents is not easily predictable through knowledge of our ancestors, and that in its aggregate our species has the potential for many changes. It is one of the functions of human biology to identify, and quantify if possible, the factors which are intrinsic to man's adaptation — past, present, and, as far as possible, future. With this type of adaptation in mind, one can hardly view the study of human variation as an esoteric exercise.

Much of the variation between human populations may be traced to genetic differences. But a surprisingly large component of human variability directly reflects the action of the environment, leaving us to conclude that certain characteristics which were formerly viewed as "racial" are more amenable to change than had been thought. If we were to compare the body sizes of nineteenth century Europeans with those of contemporary Japanese, we would be less convinced that body size is a diagnostic racial trait than were our predecessors a century ago. Not only are we in a better position to make valid comparisons today, based on the increased means of investigation at our disposal, but in addition, the populations themselves have changed over a relatively short span of time. Much of the change is directly traceable to an altered environment which includes changes in diet and health care as well as the breakdown of many barriers to migration and admixture. Central to the changes in body size and proportion seen in many parts of the world is the manner in which humans grow, develop, and mature under varying conditions. It is now widely accepted that there is a genetically determined maximum body size in humans which is seldom if ever attained due to environmental intervention in the growth process. The more rigorous the environment, the greater the distance between actual and potential growth. Many aspects of adaptation are associated with alterations in the timing and pace of the growth process. Consequently, much of the current research in human variation is concerned with the way people grow and mature. In recognition of this trend in research, the present volume has been divided into three sections. The first includes papers describing phenotypic variation among human populations; the second contains a number of papers illustrating the importance of the human growth process in determining population characteristics; while the third section is a brief foray into the realm of genetic aberrations, the possible raw material of future human evolution. More emphasis on the third section would be justifiable, but other volumes in this series will make appropriate material available, allowing space in this volume for materials not included elsewhere.

Brief mention of another attribute of this volume is appropriate. The papers herein represent the research interests of human biologists from many countries. It is an encouraging sign that scientists from every continent, sharing an interest in the range of human variation, have contributed papers through their participation in the IXth International Congress of Anthropological and Ethnological Sciences. We have much to learn from each other. It seems particularly appropriate that the study of human variability should bring together researchers from so many zones of habitation of this wide-ranging species. That scientists from France, England, Belgium, Italy, Czechoslovakia, Hungary, Rumania, the Soviet Union, India, Japan, New Guinea, Africa, South and North America share so many interests should not be surprising in view of the intrinsic interest and value of the subject of human biology. That we find communication easy and pleasant raises the possibility that, as young as it is, the science of human biology will contribute significantly to worldwide appreciation of our shared interest in preserving our common heritage, the habitat in which our species evolved, that fragile, narrow slice of the universe we call the biosphere.

SECTION ONE

Phenotypic Variation Among Human Populations

Skin Color Variability in the Middle East

E. SUNDERLAND

In areas of high levels of ultraviolet radiation, the selective value of dark pigmentation has been postulated by various authors including Barnicot (1959) and Weiner et al. (1964). Broadly speaking, the geographical distribution of skin color suggests that inhabitants of tropical areas are more darkly pigmented than are the inhabitants of temperate regions. However, the evidence for selection in favor of a heavily pigmented skin in tropical circumstances is equivocal. Dark skin absorbs heat more readily but skin color is not a significant factor in promoting radiant heat loss. Moreover, the incidence of skin cancer in whites living in regions of high ultraviolet radiation, such as South Africa and Australia, is higher than in dark skinned individuals in the same regions. The tumors tend, however, to affect individuals during the post-reproductive periods of their lives.

Skin color data, particularly if obtained by using a reflectance spectrophotometer, are still inadequate for most parts of the world. In the Middle East there are very few studies: Hulse (1970) reports on the Habbanites, a South Arabian Jewish group now resident in Israel; Lourie (1970, personal communication) reports upon Kurdish and Yemenite Jews now resident in Israel; Sunderland (1967) reports on the Arab, Druze, and Chechen inhabitants of Azraq, an oasis in eastern Jordan. The purpose of the present study is to provide broad parameters for skin color data in the Middle East, within which the local studies might be more meaningfully interpreted.

THE SKIN COLOR DATA

The data were collected in Benghazi, Libya and in Beirut, Lebanon

The fieldwork was partly supported by a grant (No. 1942) from the Wenner–Gren Foundation for Anthropological Research, New York.

during the latter months of 1966 and the spring of 1967, from individuals whose parental origins were generally Middle Eastern, including Libya and Lebanon. In all, 671 males and 177 females participated in the survey, and the parents of each individual were born within one major region, usually one country, of the Middle East.

For the male sample, fourteen geographical "units," single countries or otherwise, were utilized in the study: Tripolitania, Cyrenaica, Fezzan (all in Libya), Pakistan, Lebanon, the Arabian Peninsula, Afghanistan/Iran, Iraq/Syria, Jordan, Palestine (the population studied was non-Jewish and the subjects gave Palestine, approximately present-day Israel excluding Sinai, as their parental birth place), Turkey, Sudan, Kenya, and Morocco. The number examined from any one region varied from sixteen or fewer in Fezzan, Pakistan, Sudan, Kenya, and Morocco (all of which are geographically marginal to the central Middle East area) to thirty-five or more in all the other regions, including ninety-two individuals from Cyrenaica, 135 from Tripolitania, and 142 from Lebanon.

Middle Eastern females, for social reasons, are far less likely to participate in genetic surveys than are males, and "sufficiently large" numbers were available from only seven of the fourteen geographical units listed above. The numbers obtained varied from a minimum of nine (Tripolitania) to a maximum of forty-one (Lebanon). Thus, seven units were distinguished as follows: Tripolitania, Cyrenaica, Pakistan, Lebanon, Iraq/Syria, Palestine, and Turkey.

METHOD

A portable EEL (Evans Electroselenium Ltd.) reflectance spectrophotometer was used to obtain values for the percentage of light of known, specified wave lengths reflected from the skin surfaces of the medial aspect of the upper arm and forehead, relative to the reflectance from a white, magnesium carbonate block for each subject. The instrument has nine Ilford filters (numbered 601–609) and of these, three were used in the study: 601, 605, and 609, corresponding to the dominant wave lengths 425mμ, 545mμ, and 685mμ, respectively. The subjects were mostly young people, many being students. The average age of the males was 21.6 years and that of the females 22.1 years.

RESULTS

Geographical Distribution

Table 1 summarizes the data which were collected for males and females in each of the geographical units, showing the medial aspect of the upper arm and the forehead in each case. It is evident from Table 1 that considerable regional differences occur in the mean percentage reflectance values and in the variances at each wave length for both males and females. For instance, the mean reflectance values at 425mμ (arm) for males ranges from 11.9 (Kenya) to 28.8 (Turkey) and the variances from 2.1 (Kenya) to 30.3 (Jordan). At 685mμ the male percentage reflectance means range from 32.4 (Kenya) to 59.3 (Turkey), and the variances from 5.4 (Pakistan) to 99.8 (Fezzan). Among females the arm values at 425mμ range from 21.9 (Pakistan) to 31.2 (Iraq/Syria), and the variances from 15.6 (Lebanon) to 37.7 (Cyrenaica). At 685mμ the female arm means range from 51.3 (Pakistan) to 60.2 (Iraq/Syria), and the variances from 4.6 (Turkey) to 72.6 (Cyrenaica).

The mean percentage reflectance values for the forehead (males) at 425mμ range from 11.0 (Sudan) to 21.4 (Turkey) and the variances from 0.4 (Kenya) to 21.5 (Jordan). At 685mμ the male forehead reflectance values range from 29.4 (Kenya) to 55.9 (Turkey), and the variances from 10.3 (Kenya) to 89.9 (Fezzan). For females, the forehead values at 425mμ range from 17.8 (Pakistan) to 24.6 (Iraq/Syria), and the variances from 9.2 (Turkey) to 21.8 (Palestine). At 685mμ the mean percentage reflectance values for the forehead (females) range from 46.6 (Pakistan) to 57.4 (Turkey), and the variances from 13.3 (Palestine) to 72.8 (Cyrenaica).

These examples and the tabulated data generally denote very considerable regional differences in mean reflectance values and in the variability shown by the variances. The analyses of variance were computed in turn for the mean percentage reflectance values for each wave length on the arm and on the forehead. In all, six comparisons were made for the male series and six for the female series; in each case the within-region variance was compared with the between-region variance. Table 2 summarizes the results which were obtained. Without exception very large variance ratio values were obtained for males and females, in each case significant at the 0.001 level of probability or less. There can be no doubt that the regional heterogeneity of the data collected is very appreciable.

It should be noted that no females were examined from Kenya, Sudan, and Fezzan, three regions with a very dark male population. If the females from these regions are (relatively) as dark as the males, then the absence of female data from those areas might be expected to minimize somewhat the female variance ratio values. Even so, there is highly significant female regional heterogeneity.

It is known that the additive quality of genic effect and the independence of nonheritable variations seem greatest with a logarithmic transformation of reflectance values at 425mμ, without a transformation at 545mμ, and with an anti-logarithmic transformation at 685mμ. Table 3

Table 1. Means and variances of reflectance values at 425mμ, 545mμ, and 685mμ for

	No.	Mean age (years)	Medial aspect of upper arm 425mμ Mean	Var.	545mμ Mean	Var.
Males						
Tripolitania	135	21.8	19.8 ± 0.37	18.5	27.4 ± 0.45	27.3
Cyrenaica	92	22.0	19.0 ± 0.41	15.7	26.6 ± 0.58	30.9
Fezzan	16	21.4	14.9 ± 0.98	15.2	20.7 ± 1.80	51.5
Pakistan	16	28.2	20.5 ± 0.57	5.1	26.9 ± 0.69	7.5
Lebanon	142	20.9	27.0 ± 0.41	23.9	34.4 ± 0.40	22.7
Arabian Peninsula	37	20.8	21.0 ± 0.58	12.5	28.0 ± 0.65	15.6
Afghanistan/Iran	38	21.2	24.2 ± 0.57	12.4	31.7 ± 0.61	14.1
Iraq/Syria	35	20.9	27.4 ± 0.76	20.3	34.1 ± 0.74	19.4
Jordan	46	21.6	23.7 ± 0.81	30.3	31.1 ± 0.79	28.7
Palestine	40	20.9	26.4 ± 0.80	25.4	33.6 ± 0.75	22.5
Turkey	37	21.2	28.8 ± 0.80	23.8	36.6 ± 0.65	15.7
Sudan	15	23.7	12.2 ± 0.66	6.5	14.7 ± 0.99	14.7
Kenya	9	23.8	11.9 ± 0.48	2.1	14.0 ± 0.67	4.0
Morocco	13	20.2	21.2 ± 0.92	11.1	28.1 ± 1.12	16.5
Total	671					
Females						
Tripolitania	9	21.9	24.7 ± 1.62	23.5	33.6 ± 2.02	36.9
Cyrenaica	36	24.9	24.3 ± 1.02	37.7	32.3 ± 1.36	66.2
Pakistan	35	26.0	21.9 ± 0.79	21.7	28.4 ± 0.83	24.1
Lebanon	41	19.4	28.7 ± 0.62	15.6	36.6 ± 0.64	16.9
Iraq/Syria	19	19.8	31.2 ± 1.33	33.4	39.1 ± 1.03	20.0
Palestine	19	19.3	29.6 ± 1.05	20.8	37.3 ± 0.87	14.3
Turkey	18	20.6	30.0 ± 1.24	27.7	38.0 ± 0.94	15.9
Total	177					

Key: Var. = variance

summarizes the logarithmic and anti-logarithmic transformations for upper arm and forehead readings indicated above.

Whichever type of reading, transformed or otherwise, is used, it is evident that there is an extremely dark population in Kenya, Sudan, and Fezzan and that there is a gradation northwards from those regions to very much lighter skinned groups in Palestine, Lebanon, Iraq/Syria, and, especially, Turkey. At any wave length this gradation is as clearly evident in the arm values as in those of the forehead.

The extent to which this regional variation is due to adaptations by long-localized populations with very different genetic constitutions, and the extent to which it is due to relatively recent differential admixtures with sub-Saharan people in the Mediterranean littoral of North Africa and mainland Southwest Asia, remains problematical. Nonetheless, the existence of this long-postulated gradation is quantitatively verified here, making the demonstrated regional heterogeneity for both males and females unquestionable.

Middle Eastern Populations

		Forehead					
685mμ Mean	Var.	425mμ Mean	Var.	545mμ Mean	Var.	685mμ Mean	Var.
53.6 ± 0.47	29.3	15.4 ± 0.24	7.9	19.9 ± 0.30	12.3	48.8 ± 0.55	40.3
53.0 ± 0.51	24.3	15.1 ± 0.25	5.5	19.4 ± 0.32	9.2	48.7 ± 0.48	21.3
44.0 ± 2.50	99.8	12.4 ± 0.65	6.8	15.2 ± 1.06	18.0	38.4 ± 2.37	89.9
52.0 ± 0.58	5.4	15.6 ± 0.41	2.7	19.5 ± 0.43	2.9	46.6 ± 0.86	11.8
58.0 ± 0.24	8.5	20.8 ± 0.26	9.3	25.6 ± 0.27	10.4	54.7 ± 0.36	18.3
52.5 ± 0.55	11.7	15.1 ± 0.40	5.8	19.4 ± 0.49	9.0	47.1 ± 0.79	22.9
55.7 ± 0.44	7.2	17.9 ± 0.52	10.4	22.8 ± 0.60	13.7	51.2 ± 0.88	29.6
57.4 ± 0.52	9.6	20.9 ± 0.58	11.8	24.9 ± 0.64	14.3	55.0 ± 0.71	17.8
54.1 ± 0.65	19.2	15.4 ± 0.68	21.5	20.4 ± 0.75	25.8	49.0 ± 1.00	46.1
57.6 ± 0.49	9.5	20.5 ± 0.49	9.4	25.5 ± 0.59	13.7	54.6 ± 0.58	13.5
59.3 ± 0.50	9.1	21.4 ± 0.55	11.3	26.0 ± 0.64	15.1	55.9 ± 0.67	16.5
35.5 ± 1.97	58.1	11.0 ± 0.27	1.1	11.7 ± 0.48	3.5	30.5 ± 1.65	40.8
32.4 ± 1.60	23.1	11.2 ± 0.21	0.4	11.5 ± 0.28	0.7	29.4 ± 1.07	10.3
53.3 ± 0.83	9.0	17.5 ± 0.63	5.1	22.3 ± 0.84	9.2	51.1 ± 1.31	22.4
55.2 ± 1.42	18.2	18.7 ± 1.13	11.4	24.3 ± 1.75	27.6	51.9 ± 1.66	24.8
54.0 ± 1.42	72.6	19.9 ± 0.77	21.3	24.4 ± 1.01	37.0	49.9 ± 1.42	72.8
51.3 ± 0.68	16.3	17.8 ± 0.52	9.6	21.8 ± 0.56	11.0	46.6 ± 0.81	23.2
58.4 ± 0.36	5.3	22.2 ± 0.59	14.3	27.7 ± 0.72	21.6	55.9 ± 0.67	18.3
60.2 ± 0.54	5.5	24.6 ± 1.00	19.1	30.1 ± 0.96	17.7	57.0 ± 1.03	20.0
58.8 ± 0.53	5.4	22.9 ± 1.07	21.8	28.3 ± 1.04	20.6	55.3 ± 0.84	13.3
59.0 ± 0.51	4.6	24.2 ± 0.71	9.2	30.1 ± 0.79	11.1	57.4 ± 1.01	18.4

A Comparison of Exposed (Forehead) and Unexposed (Medial Aspect of the Upper Arm) Areas of Skin

The percentage reflectance values obtained from the less exposed parts of the upper arm and from the more exposed forehead differ considerably for males and females at each wave length. The comparisons are summarized in Table 4. In many cases the differences between arm and forehead readings for males are significant at the 0.001 level of probability or less, but there are certain exceptions:

601 Fezzan (0.02–0.05), Sudan (>0.10), Kenya (>0.10), Morocco (0.01–0.001).

605 Fezzan (0.01–0.02), Sudan (0.01–0.02).

609 Fezzan (>0.10), Iraq/Syria (0.01–0.001), Sudan (0.05–0.10), Kenya (>0.10), Morocco (>0.10).

In many instances these areas contain very dark skinned populations, and the effect of greater exposure of the forehead, compared with the

Table 2. Geographical variation of male and female skin color in the Middle East: analyses of variance, variance ratios

Arm			Forehead		
601	605	609	601	605	609
Males (n = 671)					
46.57	25.00	62.06	53.18	52.15	53.78
Females (n = 177)					
12.63	13.88	13.66	10.71	5.18	15.73

These are all significant at the 0.001 level of probability.

Table 3. Transformed reflectance values: medial aspect of upper arm and forehead

	Upper arm				Forehead			
	log. 425		Antilog. 685		log. 425		Antilog. 685	
Population	Mean	s.e.	Mean	s.e.	Mean	s.e.	Mean	s.e.
Tripolitania	1.29	0.008	0.35	0.003	1.18	0.007	0.31	0.004
Cyrenaica	1.27	0.009	0.34	0.004	1.18	0.007	0.31	0.003
Fezzan	1.16	0.031	0.28	0.014	1.08	0.025	0.25	0.013
Pakistan	1.31	0.012	0.33	0.004	1.19	0.012	0.29	0.005
Lebanon	1.42	0.007	0.38	0.002	1.31	0.005	0.35	0.003
Arabian Peninsula	1.32	0.013	0.34	0.004	1.17	0.011	0.30	0.006
Afghanistan/Iran	1.38	0.010	0.36	0.004	1.25	0.013	0.33	0.007
Iraq/Syria	1.43	0.012	0.38	0.005	1.31	0.012	0.36	0.005
Jordan	1.36	0.015	0.35	0.005	1.17	0.019	0.31	0.007
Palestine	1.41	0.013	0.38	0.004	1.31	0.010	0.35	0.004
Turkey	1.45	0.012	0.39	0.005	1.32	0.012	0.36	0.006
Sudan	1.08	0.021	0.23	0.011	1.04	0.011	0.20	0.007
Kenya	1.08	0.016	0.21	0.007	1.05	0.008	0.19	0.003
Morocco	1.32	0.016	0.34	0.006	1.24	0.016	0.33	0.009

Key: s.e. = standard error.

arm, appears to be minimal. Generally speaking, the same comments apply to the female data.

Thus, it is usually true that the differences between arm and forehead readings are least noticeable in darker skinned populations. Moreover, the differences between arm and forehead reflectance values are least clear at the red end of the spectrum. Yet it is here that reflectance values, transformed to anti-logarithmic equivalents, have the closest direct relationship to melanin concentrations. Since the magnitude of the differences in mean reflectance values from exposed and unexposed areas of skin is demonstrably considerable, it is important that in all skin color

Table 4. Percentage reflectance values: medial aspect of upper arm compared with forehead

Population	*n*	*t*	601	*t*	605	*t*	609
Males							
Tripolitania	135	11.34	<0.001	17.26	<0.001	10.69	<0.001
Cyrenaica	92	8.14	<0.001	10.92	<0.001	6.10	<0.001
Fezzan	16	2.15	0.02–0.05	2.66	0.01–0.02	1.61	>0.10
Pakistan	16	6.99	<0.001	9.20	<0.001	5.86	<0.001
Lebanon	142	12.81	<0.001	18.29	<0.001	7.59	<0.001
Arabian Pen.	37	8.39	<0.001	10.54	<0.001	5.59	<0.001
Afghanistan/Iran	38	8.15	<0.001	10.46	<0.001	4.58	<0.001
Iraq/Syria	35	6.81	<0.001	9.36	<0.001	2.72	0.01–0.001
Jordan	46	7.82	<0.001	9.84	<0.001	4.27	<0.001
Palestine	40	6.30	<0.001	8.49	<0.001	3.93	<0.001
Turkey	37	7.61	<0.001	11.61	<0.001	4.09	<0.001
Sudan	15	1.70	>0.10	2.71	0.01–0.02	1.95	0.05–0.10
Kenya	9	1.34	>0.10	6.70	<0.001	1.59	>0.10
Morocco	13	3.32	0.01–0.001	4.13	<0.001	1.43	>0.10
Females							
Tripolitania	9	1.44	>0.10	1.64	>0.10	0.71	>0.10
Cyrenaica	36	3.43	0.01–0.001	3.01	0.01–0.001	2.04	0.02–0.05
Pakistan	35	4.35	<0.001	6.73	<0.001	4.43	<0.001
Lebanon	41	7.61	<0.001	9.20	<0.001	3.31	0.01–0.001
Iraq/Syria	19	3.97	<0.001	6.41	<0.001	2.77	0.01–0.001
Palestine	19	4.47	<0.001	6.62	<0.001	3.51	<0.001
Turkey	18	4.05	<0.001	6.45	<0.001	1.41	>0.10

studies the site where the skin reflectance values were obtained must be specified.

A Comparison of Male and Female Skin Color

Differences between males and females were considered separately for each region to avoid introducing the geographical element already demonstrated to be present in the overall variability. Seven of the geographical units contained large enough samples of both males and females for the purposes of statistical comparisons: Tripolitania, Cyrenaica, Pakistan, Lebanon, Iraq/Syria, Palestine, and Turkey. T-tests were made at 601, 605, and 609 for each country for both the arm reflectance and the forehead. The results are summarized in Table 5.

The females have higher mean reflectance values in all cases (see Table 1), except for the following:

Arm 685mμ. (609) Pakistan. Male mean is higher.
Arm 685mμ. (609) Turkey. Male mean is higher.
Forehead 685mμ. (609) Pakistan. Male and female means are equal.

Thus, in forty-two comparisons, the females have the higher mean

Table 5. T-tests comparing males with females

	Arm readings						Forehead readings					
Populations	t	A 601	t	605	t	609	t	F 601	t	605	t	609
Tripolitania	3.81	<0.001	4.06	<0.001	1.86	0.05–0.10	3.77	<0.001	4.06	<0.001	2.03	0.02–0.05
Cyrenaica	6.12	<0.001	4.59	<0.001	0.82	>0.10	7.65	<0.001	3.37	<0.001	1.02	>0.10
Pakistan	1.13	>0.10	1.16	>0.10	0.59	>0.10	2.61	0.02–0.01	2.61	0.02–0.01	–	–
Lebanon	2.03	0.02–0.05	2.71	0.01–0.001	0.79	>0.10	2.43	0.01–0.02	3.27	0.01–0.001	1.58	>0.10
Iraq/Syria	2.70	0.01–0.001	3.97	<0.001	3.37	0.01–0.001	3.51	<0.001	4.67	<0.001	1.61	>0.10
Palestine	2.37	0.02–0.05	3.05	0.01–0.001	1.47	>0.10	2.37	0.02–0.05	2.51	0.01–0.02	0.72	>0.10
Turkey	0.84	>0.10	1.25	>0.10	0.38	>0.10	2.99	0.01–0.001	3.90	<0.001	1.22	>0.10

percentage reflectance values in all cases except those listed. The results of the comparisons may be summarized as follows:

Arm. 601 and 605: Significant male-female differences are demonstrated for each country, except for Pakistan and Turkey ($P = 0.05$ or less in each case and, in some instances, less than 0.001).
609: Male-female differences do not occur in any of the seven geographical units other than Iraq/Syria where the P value is 0.01–0.001.

Forehead. 601 and 605: Significant male-female differences are demonstrated for each country ($P=0.05$ or less in each case and again, in some instances, less than 0.001).
609: Male-female differences do not occur in any of the seven areas other than Tripolitania where the P value is 0.02–0.05.

Thus, while at 601 and 605 for both the arm and the forehead, marked male-female differences are demonstrable, this is not generally the case at the red end of the spectrum.

Religious Groupings

Many distinct religious groupings occur in Middle Eastern countries. It is likely that these have maintained a certain reproductive isolation over a long period of time and that each group might be expected to exhibit certain physical differences.

In the present study, sizeable numbers of Muslims and Christians from certain countries were examined. Very small numbers of Druze, Jews, and Bahais were included in the total regional samples, but because of their small numbers they were excluded from a consideration of skin color differences among the religious groupings.

The comparison of Muslims and Christians was done separately country by country for males and females to eliminate the geographical and sex variations already indicated. The results are summarized in Table 6.

Males. Comparisons of Muslims and Christians were feasible for Lebanon (52 Muslims, 80 Christians), Iraq/Syria (16 Muslims, 16 Christians), Jordan (10 Muslims, 6 Christians), Palestine (17 Muslims, 23 Christians), and Turkey (9 Muslims, 28 Christians).

Females. Comparisons of Muslims and Christians were feasible for Lebanon (14 Muslims, 21 Christians), Iraq/Syria (14 Muslims, 4 Christians), and Palestine (5 Muslims, 13 Christians).

Generally the magnitude of the Muslim-Christian "differences" are much smaller than are those between the arm and forehead or between the males and females. Moreover, they are not all systematic. For male Muslims and Christians there are no significant differences, ($P = 0.05$ or less) for Lebanon, Iraq/Syria, Jordan, Palestine, or Turkey. The sole exception is the Jordan forehead readings at 605 where the P value is 0.02. For female Muslims and Christians there are no significant differences ($P = 0.05$ or less) for Lebanon, Iraq/Syria, or Palestine, with the exception of the Iraq/Syria arm readings at 609 where the P value is 0.02–0.05.

One may usefully summarize the skin color results as follows:

a. Pronounced regional or geographical heterogeneity was demonstrated for both males and females, with a very dark population in Kenya, Sudan, and Fezzan, graduating through intermediate populations to very much lighter groups in Turkey and contiguous Asian countries.

b. The differences between the forehead and arm percentage reflectance readings are highly statistically significant in most cases, the forehead being the darker.

c. In almost all the comparisons of males and females the latter have statistically significantly higher mean values than the males. They are least clearly distinguishable at 609.

d. The "differences" between Christians and Muslims are of a smaller order than the others and are, in most instances, not statistically significant. Moreover, there are less systematic differences than in the other features compared.

Table 6. Skin color: percentage reflectance values, Muslims compared with Christians

Males	*n*	Mean age (years)	Arm 601 Mean	Arm 605 Mean	Arm 609 Mean	Forehead 601 Mean	Forehead 605 Mean	Forehead 609 Mean
Lebanese Muslims	52	20.7	26.7	34.4	57.7	20.4	25.6	53.9
Lebanese Christians	80	21.1	27.1	34.3	58.1	21.1	25.7	55.2
t			0.45	0.11	0.73	1.29	0.17	1.68
P			>0.10	>0.10	>0.10	>0.10	>0.10	0.05–0.10
Iraqi/Syrian Muslims	16	21.3	27.0	34.0	56.8	21.0	25.4	55.4
Iraqi/Syrian Christians	16	20.4	28.1	34.4	58.2	20.6	24.2	54.2
t			0.65	0.25	1.27	0.31	0.87	0.79
P			>0.10	>0.10	>0.10	>0.10	>0.10	>0.10
Jordanian Muslims	10	20.8	27.1	34.1	57.4	19.1	23.0	53.6
Jordanian Christians	6	19.7	28.7	34.9	57.8	21.9	27.5	56.5
t			0.62	0.33	0.27	1.76	2.62	1.52
P			>0.10	>0.10	>0.10	0.10	0.02	>0.10
Palestinian Muslims	17	21.1	25.2	32.6	57.2	19.6	24.6	53.5
Palestinian Christians	23	20.7	27.4	34.4	57.9	21.1	26.1	55.4
t			1.38	1.19	0.72	1.57	1.28	1.66
P			>0.10	>0.10	>0.10	>0.10	>0.10	>0.10
Turkish Muslims	9	20.6	29.5	37.0	60.3	21.6	25.5	55.3
Turkish Christians	28	21.4	28.6	36.5	59.0	21.4	26.2	56.2
t			0.47	0.31	1.12	0.16	0.47	0.57
P			>0.10	>0.10	>0.10	>0.10	>0.10	>0.10

Females	*n*	Mean age (years)	Arm 601 Mean	Arm 605 Mean	Arm 609 Mean	Forehead 601 Mean	Forehead 605 Mean	Forehead 609 Mean
Lebanese Muslims	14	19.6	28.9	36.9	59.0	22.0	28.3	56.3
Lebanese Christians	21	19.1	29.2	36.7	58.3	21.9	26.9	55.3
t			0.23	0.14	0.87	0.09	0.84	0.66
P			>0.10	>0.10	>0.10	>0.10	>0.10	>0.10
Iraqi/Syrian Muslims	14	19.9	31.1	39.0	60.0	24.7	30.0	57.8
Iraqi/Syrian Christians	4	19.3	32.9	41.2	62.4	25.8	32.0	56.5
t			0.55	0.92	2.34	0.46	0.86	0.55
P			>0.10	>0.10	0.02–0.05	>0.10	>0.10	>0.10
Palestinian Muslims	5	18.4	29.1	37.2	59.2	25.2	30.1	56.9
Palestinian Christians	13	19.7	29.9	37.3	58.7	21.6	27.5	54.4
t			0.32	0.04	0.38	1.52	1.06	1.35
P			>0.10	>0.10	>0.10	>0.10	>0.10	>0.10

COMPARATIVE DATA

As stated above, few studies of skin color in localized Middle Eastern communities exist. However, the data presented in Table 7 appear to be generally in agreement with the Middle Eastern material presented in this paper (Table 1). Percentage reflectance readings at filter 609 relate very closely to the concentration of melanin in the epidermis. The samples in Table 7 are arranged in order of increasing mean percentage reflectance values at 609 and therefore in order of decreasing melanin concentration. The Arab and Habbanite male values are very similar and somewhat darker than other similar series, the Druze and the Yemenites. The latter in turn are darker than a third pair of samples, the Kurds and the Chechen. All three female samples are lighter than the lightest male series. The Habbanite and Yemenite samples are very similar to one another while both are much darker than the Kurdish females.

Table 7. Middle Eastern populations skin reflectance values: medial aspect of the upper arm

Population	*n*	601 Mean	s.d.	605 Mean	s.d.	609 Mean	s.d.
Males							
Arabs (Sunderland)	30	21.6	4.60	29.3	4.99	52.2	3.95
Habbanites (Hulse)	—	19.74	3.00	—	—	52.32	4.12
Druze (Sunderland)	42	22.5	5.13	30.0	4.72	52.8	4.37
Yemenites (Lourie)	32/33	18.36	2.25	26.50	2.67	53.15	2.81
Kurds (Lourie)	52	21.11	4.21	29.19	5.32	54.89	5.88
Chechen (Sunderland)	23	27.4	4.70	33.6	5.04	55.0	3.96
Females							
Habbanites (Hulse)	—	22.87	3.89	—	—	56.65	3.99
Yemenites (Lourie)	38	21.61	4.77	31.36	4.86	56.94	4.84
Kurds (Lourie)	39	24.19	3.53	33.73	3.52	60.02	3.10

Key: s.d. = standard deviation

CONCLUSION

Most authorities concur that the paramount selective factor in the determination of skin color is the intensity of ultraviolet light, but quantitative data on this parameter in the Middle East are virtually non-existent. In general terms however, climatic differences within the Middle East might well provide sufficient variation in selective pressures upon the populations to account for the observed skin color variability. Genetic differences in the area are, moreover, explicable in terms of population movements and consequent hybridization.

REFERENCES

BARNICOT, N. A.
1959 Climatic factors in the evolution of human populations. Cold Spring Harbor Symposium on Quantitative Biology 24:115–129.

HULSE, F. S.
1970 "Skin colour among the Yemenite Jews of the isolate from Habban," in *Proceedings of the VIII Congress of Anthropological and Ethnological Sciences*, 226–228. Tokyo.

SUNDERLAND, E.
1967 The skin colour of the people of Azraq, eastern Jordan, *Human Biology* 39:65–70.

WEINER, J. S., G. A. HARRISON, R. SINGER, R. HARRIS, W. JOPP
1964 Skin colour in Southern Africa. *Human Biology* 36:294–307.

Factor Analysis of Anthropometric Data for Fifteen Race-Age-National Origin Specific Groups

EDMUND CHURCHILL, DAVID RABINOW, and PEGGY ERSKINE

One may speculate that factor analysis reached its peak as a research tool in the field of physical anthropometry some ten or fifteen years ago. Up to that time, many anthropologists were impressed by the vast amount of work, both theoretical and computational, that psychologists had invested in developing and using this technique. More than a few undoubtedly also were impressed by the intensity of the internecine battles among the factor analytic psychologists. Surely, here was a methodology which had much to contribute to every field of research, certainly to physical anthropology. However, the problems posed by the substantial and presumably complicated computations involved in factor analysis kept anthropologists from rushing to use it. But we all believed that the then youthful computer would soon reach adulthood and be able to do these analyses without effort. Then factor analyses would really roll.

The computer matured, and soon there were a variety of program libraries, such as the BIMD and BMD (see Dixon 1971) that demanded nothing more than some data and a few control cards as the price of factor analysis. But, if one is to judge by the contents of the *American Journal of Physical Anthropology* and *Human Biology*, the anticipated factor analyses, by and large, did not take place. It is quite possible that these analyses were carried out, that the results were scanned and deemed to contribute little or nothing, worthy of no more than burial in the "never-to-be-published" file.

If our last hypothesis is correct, and we think it is, it is unfortunate for

The research reported in this paper was supported in part by the Aerospace Medical Research Laboratories, Aerospace Medical Division, Air Force System Command, Wright-Patterson Air Force Base, Ohio, under contract AF 33615-72-C-1006, with Webb Associates, Yellow Springs, Ohio. Further reproduction is authorized to satisfy the United States government.

two reasons. First, if a technique as important as factor analysis apparently was once thought to be is killed off and buried, it deserves something of a decent funeral — one paper, at least, showing that a massive effort to squeeze useful answers out of the technique got us nowhere. Secondly, if we do not publish evidence showing that a particular methodology is a useless one, we tempt successive waves of students to discover its uselessness anew, wasting time and effort which would better have been spent elsewhere.

Two things prompted this study. One was the availability of great quantities of anthropometric data; i.e., the Aerospace Medical Research Laboratory's data bank collection of about twenty large-scale surveys (Churchill, Churchill, and Kikta 1977), from which it would be possible to extract data for a massive attempt at factor analysis. The second was the apparent interest in the use of factor analysis in comparative studies. Vandenberg (1968) contrasted factor analyses of Dutch and American women; Choi and Trotter (1970) compared analyses of all four combinations of black–white and male–female fetal skeletons; Rhoads (1970) studied differences between analyses of several American groups with that of Solomon Island men. Since the data used here would make possible a fair number of sex-race-age-social status comparisons, a contribution could be made to this body of literature.

The design of this study was primarily as follows:

1. A set of forty measurements was selected for use. Most of these had been taken in all the surveys we planned to use. Computations were sometimes necessary for a few variables since they had not measured directly. For example, elbow–wrist length was derived several times by subtracting hand length from elbow–finger tip length. In a few other cases, slight substitutions (knee height, standing for knee height, sitting, for example) were made.

2. An initial study of how consistent factor analysis results are when obtained from essentially identical data was carried out by running factor analyses on ten groups of 400 men selected (without duplication) from a particular survey sample.

3. To minimize the effect that using samples of different sizes might have on the comparisons of the analyses of the various groups, analyses were based on groups of no more than 400 subjects. (The number 400 was arrived at by multiplying forty, the number of variables, by a factor of ten.) Limiting the sample size to this number also facilitated obtaining samples that were often fairly homogeneous in terms of age and socioeconomic background as indicated by military status.

4. The BMD factor analysis program (BMD03M) was used on a CDC 6600 computer. The long word length of this machine provided a level of computational accuracy almost equivalent to double precision calculations on most standard size computers.

5. Analyses were carried out on the following fifteen samples:
   a. young United States Air Force (USAF) flying personnel;
   b. older USAF flying personnel;
   c. black USAF basic trainees;
   d. white USAF basic trainees;
   e. older men from the same survey as groups c and d;
   f. black women (mostly trainees) of the USAF;
   g. eighteen- and nineteen-year-old white USAF women;
   h. white USAF women thirty years old and older;
   i. young privates in Turkish military service;
   j. young privates in Greek military service;
   k. young privates in Italian military service;
   l. German flying personnel;
   m. young Vietnamese soldiers;
   n. Korean flying personnel;
   o. Iranian military personnel.

6. The primary reason for using military data in this study was their availability. Comparable data suitable for factor analyses based on civilian samples, providing the contrasts which we wished to study, simply do not exist. In this country, the only large-scale anthropometric survey of adult civilians was one of women made more than three decades ago (O'Brien and Shelton 1941). Although military samples have certain biases automatically built into them, it is possible for the most part to make judgments as to the nature of these biases. On the other hand, civilian surveys similar to the one just mentioned, based on a haphazard selection of volunteers, also have their biases; but it is difficult to determine what they are. Finally, what little evidence exists suggests that there is likely to be considerable agreement between military and civilian data. The National Health Examination survey gathered a small amount of anthropometric data based on probability samples of civilian men and women (Stoudt et al. 1965). Comparison of these data with those from the USAF surveys from which samples A and B, and F, G, and H, were obtained show close agreement except that the data from the military samples have somewhat narrow ranges.

DATA SOURCES

The Consistency Study

Data for this study were taken from a survey of flying personnel made in 1950 (Hertzberg, Daniels, and Churchill 1954). Ten samples were obtained by starting, for each sample, with one of the first ten men in the total series and then taking every tenth man. There is no cogent reason to suppose that substantial background differences existed among these ten samples. All subjects were on active flying status: 61 percent were officers; 15 percent aviation cadets; 24 percent enlisted, men and noncommissioned officers; and almost all (99.3 percent) were Caucasians. Minor adjustments have been made to the data for this survey since publication, and small differences exist between the statistics given here for the total sample and the previously published values (Table 1).

The results of the factor analyses performed on the ten samples of 400 and the total sample of 4,000 are presented in Table 2. Two conventions were followed in preparing this table. First, "appearance" indicates, for a particular factor and a particular sample, the order in which this factor occurred. Thus, for example, the head breadth factor (the one we have designated as factor 3) was actually the seventh factor to appear for sample 5. Second, factor loadings are listed only for those variables that had a loading in excess of 0.316 for at least one sample. Loadings less than this value, when reported, are listed in parentheses. While the choice of 0.316 ($=\sqrt{0.10}$) as the criterion for inclusion in the table was quite arbitrary, it is doubtful that smaller loadings have any real significance.

Any useful evaluation of these results must be somewhat subjective, both because of the overall complexity of the results of these and all other factor analyses, from a mathematical point of view, and because meaningful differences rarely have genuinely objective definitions.

The writer's subjective judgment is that the results appear to be good since the same conclusions probably would be drawn from the analyses of each of the ten subsamples and the total sample series of 4,000. Samples of 400, with only forty variables, seem adequate for analysis. Ten factors with significant loadings appear in each analysis, and these factors seem virtually identical for all samples. For at least the first five factors, they appear almost always in the same sequence.

These factors are:

1. A long bone factor. This is, as Thurstone (1946, 1947) observed several decades ago, not a stature factor. The highest loading in all cases is crotch height with knee height more often than not having the second highest loading. All the heights, waist–back (surface distance from waist level to cervicale), hand length and foot length have substantial weightings. Perhaps as interesting is the list of measurements which meet our criterion of a 0.316 loading some, but not all, of the time: weight, and three rather boney breadths — hip, hand, and foot. For all analyses, this long bone factor appeared first.

2. A weight–circumferential factor. Weight and buttock circumference consistently have the highest weightings, with the other three torso circumferences (shoulder, chest and waist, and thigh and biceps circumferences) not far behind. Neck and calf circumference, and shoulder and

hip breadth also have weightings of about 0.7 or above. Ankle and wrist circumferences and face (bizygomatic) breadth also have more than minimum weightings for all samples. Foot breadth and head circumference have low weightings but above 0.316 for most, but not all, samples. This factor was the second factor for all samples.

3. A head breadth factor. Note that while the first two factors usually had loadings of 0.90 or more, the maximum loading for the head breadth is always somewhere in the 0.7–0.8 range. In ten of the eleven analyses, this factor occurred third; the other time it was seventh.

4. The sitting height factor. Sitting height and eye height, sitting alternately share first and second highest loading. There are also weak but significant loadings for stature and waist back and, sometimes, for shoulder height. This factor always occurred fourth.

5. The hand–foot factor. By now, the factors are beginning to get a bit vague. The total sample weightings suggest the general pattern: hand breadth, 0.57; foot breadth, 0.47; wrist circumference, 0.46; hand length, 0.41; foot length, 0.34; ankle circumference, 0.30. Only for the first two variables are these their highest weightings. Oddly enough, despite the low weightings, this factor appeared as the fifth factor in all cases.

6. The head length factor. There were only two significant weightings: head length in the 0.7–0.8 range and head circumference in the 0.6–0.7 range. Seven of the eleven analyses yield this as the sixth factor; once it was the third and three times, the seventh factor. The weightings do not seem to be related to the order in which the factor occurred.

7. A sort of lower-leg circumference factor, with weighting for calf and ankle circumference of about 0.5, lower than either had on factor 2. This factor's appearance ranged from sixth to thirteenth.

8. Face length and interpupillary factors. This most peculiar factor demonstrates that a minor factor can perhaps be an artifact of a measurement technique. Face length (menton–sellion length) and interpupillary distance should be somewhat independent since this is almost always the case with two head measurements made in perpendicular planes. On the appearance of this factor, an earlier publication (Churchill and Truett 1957) was checked and a correlation of 0.37 was found between face length and interpupillary distance (plus virtually zero correlations between face length and both interocular breadth and biocular breadths). It would appear, then, that the factor analyses are restating something which is a peculiarity of the data, rather than something concocted by the factor analysis process. No explanation can be given for this. If this factor is not sufficiently strange already, the appearance of hand breadth with a significant weighting in the case of three samples should surely make it so. Eight times this factor occurred eighth, once seventh and twice ninth.

Table 1. Mean values in consistency study (weight in kilograms)

	1	2	3	4	5	6	7	8	9	10	Total
1 Weight	74.4	74.4	74.2	74.9	73.6	73.7	74.4	74.4	74.3	74.1	74.3
2 Stature	175.6	175.9	175.3	175.6	175.2	175.3	175.8	175.7	175.7	175.5	175.6
3 Acromial height	143.6	143.9	143.4	143.8	143.0	143.3	143.8	143.6	143.7	143.4	143.5
4 Waist height	106.8	107.0	106.7	106.8	106.4	106.7	107.0	106.9	106.8	106.7	106.8
5 Crotch height	83.3	83.6	83.2	83.4	83.1	83.2	83.6	83.7	83.5	83.3	83.4
6 Sitting height	91.4	91.2	91.1	91.4	91.2	91.1	91.3	91.5	91.3	91.4	91.3
7 Eye height, sitting	80.1	79.8	79.8	80.0	79.9	79.8	80.0	80.1	80.0	80.1	80.0
8 Knee height, sitting	55.0	55.2	55.0	55.1	54.9	54.9	55.1	55.1	55.1	54.9	55.0
9 Popliteal height	43.1	43.3	43.0	43.1	42.9	43.0	43.2	43.2	43.2	43.0	43.1
10 Buttock–knee length	59.9	60.2	59.9	60.1	59.8	59.9	60.1	60.0	60.1	60.0	60.0
11 Shoulder–elbow length	36.3	36.5	36.3	36.3	36.3	36.3	36.4	36.5	36.4	36.3	36.4
12 Elbow–wrist length	28.9	28.9	28.8	28.9	28.8	28.8	28.9	29.0	28.9	28.8	28.9
13 Sleeve inseam	50.1	50.5	50.2	50.4	50.1	50.2	50.2	50.6	50.3	50.2	50.3
14 Sleeve length	85.2	85.6	85.3	85.3	85.2	85.3	85.4	85.6	85.3	85.3	85.3
15 Thumb-tip reach	81.9	82.4	82.0	82.1	81.6	82.0	82.1	82.3	82.2	82.0	82.0
16 Neck circumference	37.9	38.0	38.1	38.1	37.9	38.0	37.9	38.0	37.9	37.9	38.0
17 Shoulder circumference	115.0	114.9	115.2	115.3	114.6	114.8	115.2	114.7	114.9	114.8	114.9
18 Chest circumference	98.6	98.5	98.3	98.7	98.1	98.3	98.8	98.5	98.6	98.4	98.5
19 Waist circumference	81.1	81.5	81.5	81.7	80.6	80.8	81.4	81.4	81.3	80.9	81.2
20 Buttock circumference	95.9	95.9	96.2	96.5	95.6	95.6	95.9	96.1	96.0	95.9	96.0
21 Upper thigh circumference	56.9	56.8	56.9	57.2	56.6	56.6	56.9	56.9	56.9	56.7	56.8

Table 1. *(continued)*

22 Calf circumference	36.5	36.6	36.5	36.7	36.4	36.5	36.6	36.5	36.5	36.6	36.5
23 Ankle circumference	22.5	22.5	22.5	22.7	22.5	22.6	22.6	22.5	22.6	22.6	22.6
24 Biceps circumference, flexed	32.5	32.3	32.4	32.6	32.3	32.4	32.5	32.3	32.3	32.4	32.4
25 Wrist circumference	17.4	17.4	17.4	17.4	17.3	17.4	17.4	17.4	17.4	17.4	17.4
26 Shoulder breadth	45.5	45.5	45.5	45.5	45.3	45.4	45.4	45.3	45.4	45.3	45.4
27 Interscye	49.7	49.7	49.8	49.8	49.7	49.8	49.8	49.9	49.8	49.7	49.8
28 Hip breadth	33.4	33.5	33.5	33.5	33.4	33.3	33.5	33.5	33.6	33.5	33.5
29 Waist–back	45.0	45.0	44.9	45.0	44.8	44.9	45.0	45.0	45.1	45.0	45.0
30 Hand length	19.0	19.1	19.0	19.0	19.0	19.0	19.0	19.1	19.1	19.0	19.0
31 Hand breadth	8.8	8.8	8.8	8.8	8.8	8.8	8.8	8.8	8.8	8.8	8.8
32 Foot length	26.7	26.7	26.6	26.7	26.7	26.6	26.7	26.7	26.7	26.6	26.7
33 Foot breadth	9.7	9.7	9.6	9.6	9.6	9.7	9.7	9.6	9.6	9.7	9.7
34 Head circumference	57.1	57.1	57.0	57.0	57.0	57.1	57.1	56.9	57.0	57.0	57.0
35 Head length	19.7	19.7	19.7	19.7	19.7	19.7	19.7	19.7	19.7	19.7	19.7
36 Head breadth	15.4	15.4	15.4	15.4	15.4	15.4	15.4	15.4	15.4	15.4	15.4
37 Head height	12.9	13.0	13.0	12.9	12.9	13.0	13.0	13.0	13.0	13.0	13.0
38 Face breadth	14.1	14.1	14.1	14.1	14.1	14.1	14.1	14.1	14.1	14.1	14.1
39 Face length	11.8	11.8	11.8	11.7	11.8	11.8	11.8	11.8	11.7	11.8	11.8
40 Interpupillary breadth	6.3	6.3	6.3	6.3	6.3	6.3	6.3	6.3	6.3	6.3	6.3
41 Age	28.0	27.8	28.2	27.8	27.8	27.9	28.2	27.9	27.8	27.7	27.9
n	400	400	400	400	400	400	400	400	400	400	400

Table 2. Results of factor analyses on ten samples and the total survey

	Loadings of factor 1										
Sample	1	2	3	4	5	6	7	8	9	10	Total
Appearance	1	1	1	1	1	1	1	1	1	1	
1 Weight	(27)	34	33	(28)	32	34	(29)	(27)	(28)	(30)	(30)
2 Stature	88	87	89	86	84	88	88	86	87	86	87
3 Acromial height	87	88	88	87	85	89	89	86	87	87	87
4 Waist height	90	91	90	90	89	92	91	89	89	87	89
5 Crotch height	92	93	93	92	92	93	94	92	94	93	92
6 Sitting height	47	43	47	37	38	49	46	38	51	40	44
7 Eye height, sitting	46	40	41	34	36	42	44	36	47	36	40
8 Knee height, sitting	92	90	91	89	88	91	92	91	91	89	90
9 Popliteal height	89	90	89	86	85	90	88	89	89	89	88
10 Buttock–knee length	82	84	83	80	83	81	85	83	82	82	82
11 Shoulder–elbow length	81	82	81	83	80	86	80	83	82	82	82
12 Elbow–wrist length	76	80	80	82	79	83	83	82	83	80	81
13 Sleeve inseam	73	74	74	76	74	77	74	75	74	73	75
14 Sleeve length	75	77	76	76	74	79	75	75	75	78	76
15 Thumb-tip reach	67	68	66	74	68	69	67	69	72	68	69
28 Hip breadth	34	37	34	(29)	37	35	37	(30)	34	32	34
29 Waist–back	44	42	51	38	38	44	40	35	41	43	42
30 Hand length	64	62	60	61	61	68	69	64	62	58	63
31 Hand breadth	34	33	(22)	(27)	33	37	35	(26)	34	(27)	(31)
32 Foot length	64	66	61	64	62	67	72	67	70	64	66
33 Foot breadth	(28)	40	(28)	(30)	(31)	37	33	(28)	(31)	(24)	(31)

Table 2. (*continued*)

	Loadings of factor 2										
Sample	1	2	3	4	5	6	7	8	9	10	Total
Appearance	2	2	2	2	2	2	2	2	2	2	2
1 Weight	90	89	88	92	90	90	91	91	89	90	90
10 Buttock–knee length	34	34	31	38	38	40	(30)	(30)	35	36	35
14 Sleeve length	(28)	(29)	(28)	(26)	34	(29)	(31)	(29)	(24)	(23)	(28)
16 Neck circumference	69	69	71	69	66	67	68	65	71	67	68
17 Shoulder circumference	87	88	89	87	90	89	89	87	88	85	88
18 Chest circumference	86	86	88	87	88	87	88	87	88	85	87
19 Waist circumference	87	87	89	87	88	88	87	88	88	87	87
20 Buttock circumference	90	89	89	92	89	90	91	89	89	90	90
21 Upper thigh circumference	84	88	88	90	86	88	86	87	87	87	87
22 Calf circumference	65	73	74	76	74	75	67	66	68	71	71
23 Ankle circumference	53	58	56	64	57	65	53	49	53	60	57
24 Biceps circumference, flexed	86	87	85	85	87	86	85	84	87	86	86
25 Wrist circumference	60	60	58	60	64	58	60	54	56	61	59
26 Shoulder breadth	83	83	83	81	83	82	82	82	81	80	82
27 Interscye	45	49	52	51	51	53	54	54	58	48	52
28 Hip breadth	71	71	71	74	71	72	70	67	74	73	71
31 Hand breadth	(26)	(20)	(28)	36	(26)	(31)	(28)	(21)	(27)	33	(28)
33 Foot breadth	32	(30)	(25)	36	33	34	34	(25)	(29)	33	32
34 Head circumference	36	35	35	36	(30)	32	(29)	(30)	32	34	33
38 Face breadth	37	43	36	36	43	40	33	39	37	40	38

	Loadings of factor 3										
Sample	1	2	3	4	5	6	7	8	9	10	Total
Appearance	3	3	3	3	7	3	3	3	3	3	3
34 Head circumference	52	45	42	49	41	52	55	52	46	42	47
36 Head breadth	78	76	76	78	71	79	79	74	76	72	75
38 Face breadth	72	64	66	63	63	66	67	63	67	60	64

Table 2. *(continued)*

	Loadings of factor 4										
Sample	1	2	3	4	5	6	7	8	9	10	Total
Appearance	4	4	4	4	4	4	4	4	4	4	4
2 Stature	37	38	36	40	41	36	40	41	40	37	39
3 Acromial height	(30)	(31)	(31)	32	34	(29)	32	34	34	(28)	32
6 Sitting height	81	81	80	86	84	79	83	83	78	83	82
7 Eye height, sitting	78	82	81	85	83	81	81	83	79	82	81
29 Waist–back	37	32	34	36	39	35	33	32	41	34	36
	Loadings of factor 5										
Sample	1	2	3	4	5	6	7	8	9	10	Total
Appearance	5	5	5	5	5	5	5	5	5	5	5
23 Ankle circumference	(23)	(28)	37	32	(16)	(18)	32	(21)	(21)	47	(30)
25 Wrist circumference	41	49	49	49	40	40	49	42	45	45	46
30 Hand length	41	37	38	38	33	(25)	37	45	44	45	41
31 Hand breadth	63	61	63	59	62	48	61	56	56	53	57
32 Foot length	32	32	32	33	(24)	(15)	32	39	37	43	34
33 Foot breadth	50	45	52	48	46	36	50	46	43	58	47
	Loadings of factor 6										
Sample	1	2	3	4	5	6	7	8	9	10	Total
Appearance	7	6	6	6	3	6	6	7	7	6	6
34 Head circumference	64	68	65	65	68	63	58	64	69	66	65
35 Head length	76	79	75	76	79	76	69	76	79	75	75

Table 2. (*continued*)

	Loadings of factor 7										
Sample	1	2	3	4	5	6	7	8	9	10	Total
Appearance	13	7	7	8	6	7	7	6	6	10	7
22 Calf circumference	53	43	39	40	47	37	49	57	51	38	46
23 Ankle circumference	57	50	46	41	54	48	53	60	58	37	49
	Loadings of factor 8										
Sample	1	2	3	4	5	6	7	8	9	10	Total
Appearance	8	9	8	9	8	8	8	8	8	7	8
31 Hand breadth	(21)	(22)	(21)	35	(29)	39	(15)	(30)	(25)	32	(28)
39 Face length	52	54	56	49	55	51	52	52	53	52	50
40 Interpupillary breadth	48	53	57	55	59	50	56	56	54	47	51
	Loadings of factor 9										
Sample	1	2	3	4	5	6	7	8	9	10	Total
Appearance	6	8	10	11	10	12	10	10	11	9	12
37 Head height	42	41	44	43	51	39	50	43	43	44	36
	Loadings of factor 10										
Sample	1	2	3	4	5	6	7	8	9	10	Total
Appearance	12	10	11	12	12	10	11	9	10	11	11
29 Waist–back	49	59	47	45	50	52	50	59	43	50	47

9. A head height factor. Unlike the preceding factor, head height appeared all by itself. It occurred, variously, as the sixth, eighth, ninth, tenth, eleventh, and twelfth factors.

10. A waist–back factor. Only waist–back has a significant weighting here; a low value (0.5) but higher than the weightings for waist–back on factors 1 and 4.

No other factors appeared for any of the samples. Many additional factors would undoubtedly have shown up if, for example, the entire set of 132 measurements made in this survey had been included in the factor analyses.

The Comparative Factor Analyses

The data for Groups A and B come from the survey conducted in 1967 of 2,500 rated and student pilots and navigators in the USAF (Churchill, Churchill, and Kikta 1977). All subjects were officers and almost all college graduates; 98 percent were Caucasians. The two groups consisted, in effect, of the 400 youngest and the 400 oldest subjects. Group A included all twenty-one and twenty-two-year-olds plus 205 twenty-three-year-olds; Group B included fifty-five thirty-six-year-olds plus all older subjects.

The data for Groups C, D, and E come from an unpublished survey of USAF personnel made in 1965. Most of the subjects in this survey were young men undergoing basic training; only a few of the non-trainees were flying personnel. The trainees were almost uniformly high-school graduates. Group C contained 383 basic trainees who classified themselves as Negro. Group D included the first 400 basic trainees who classified themselves as white. (This group includes those who might now be classified as Chicano.) Group E included all white survey subjects thirty-two years old or older.

Groups F, G, and H are the only samples of women in the study and were measured in 1968 in a survey described by Clauser, Tucker, and McConville (1972). Group F contained all subjects who classified themselves as Negro; Group G, the eighteen- and nineteen-year-olds who classified themselves as "white"; and Group H, the 230 survey subjects who were white and at least thirty years old.

Groups I, J, and K are based on data collected in a survey of military personnel in Turkey, Greece, and Italy, conducted by a team of United States anthropologists operating under the North Atlantic Treaty Organization sponsorship during 1960–1961 (Hertzberg, Churchill, and Dupertuis 1963). The samples used here were selected to consist of young enlisted men, most of whom were serving in the army. Each sample appeared to be rather homogeneous ethnically. Group I contained all

Turkish army and navy privates; Group J, Greek privates; and Group K, Italian privates.

Group L included the first 400 subjects measured in a recent survey of German military pilots (Grunhofer and Kroh 1975).

Group M contained the first 400 subjects aged twenty, twenty-one, or twenty-two years in a survey of Vietnam military personnel conducted under the direction of Robert White (1964).

Group N included all the subjects in a survey of Korean flying personnel made by the Korean Air Force following the pattern of the USAF survey which yielded the data for the present consistency study (Kay 1961).

Sample sizes and mean values for each of the forty variables, and age, are given in Table 3. The results of the fifteen factor analyses are summarized in Tables 4 and 5.

These results include a variety of consistencies and inconsistencies. Table 4 presents most of the consistencies. As for the samples of the consistency study, there are clearly two major factors for each of the fifteen samples. The long bone and the weight-circumference factors appeared first and second respectively in all cases and had weightings generally much the same as those of the consistency study. Factor 3 (whose point of appearance ranged from third to ninth) is factor 6 of the consistency study. Factor 4, which appeared third or fourth, is the sitting height factor (also factor 4) of the consistency study.

Only these first four factors are clearly defined for all samples. We have, however, included three additional factors in Table 4.

Factor 5 is the same hand–foot factor as factor 5 of the consistency study. However, this factor did not appear in the samples of black women and Vietnamese. Factor 6 is, for many of the samples, quite similar to the head breadth factor of the consistency study, but is hardly the same factor for all fifteen analyses. Finally, the seventh factor, a calf and ankle circumference, is well defined and similar to the seventh factor of the consistency study for ten of the samples, but either marginal or missing for the other five samples.

All the loadings above 0.316 not shown in Table 4 are listed in Table 5, grouped by sample. Factor 7 means the seventh factor to appear; other such terms should be interpreted similarly.

In attempting to interpret these results, it may be useful to keep in mind several points. First, while measuring techniques differed from survey to survey and these differences probably are reflected in differences in the factor analysis, there are several sets of samples measured by the same measuring terms using the same techniques. These are: samples A and B; samples C, D, and E; samples F, G, and H; and samples I, J, and K. Samples A and B were measured more or less at the same time, as were samples C and D and samples F, G, and H. Most individuals in sample E

Table 3. Mean values in comparative study

	A	B	C	D	E	F	G	H	I	J	K	L	M	N	O
1 Weight	77.5	80.8	68.5	69.1	77.5	56.6	57.5	61.7	64.0	66.4	68.7	75.0	51.0	63.0	60.8
2 Stature	177.4	177.2	174.9	175.0	175.8	161.4	162.3	162.5	168.6	170.6	170.1	176.3	161.1	168.6	168.7
3 Acromial height	144.9	145.4	142.8	142.7	144.4	131.8	132.1	132.5	137.2	138.7	138.4	146.3	131.6	138.6	137.9
4 Waist height	106.5	106.3	107.0	105.2	105.0	101.1	100.4	100.4	100.0	100.5	101.3	106.2	93.3	103.9	101.0
5 Crotch height	85.4	84.5	86.0	83.0	81.8	76.8	74.8	73.7	78.1	80.2	80.7	83.1	74.5	75.2	78.2
6 Sitting height	93.3	93.0	88.2	91.3	91.5	83.0	85.7	86.1	89.2	90.2	89.6	91.4	85.3	90.8	88.9
7 Eye height, sitting	80.9	80.8	77.1	79.9	80.0	71.0	73.7	74.4	76.7	78.0	77.9	79.9	73.2	79.0	76.7
8 Knee height, sitting	55.7	55.8	56.6	55.2	55.2	42.8	42.1	41.9	52.6	52.4	53.6	54.5	48.4	50.1	51.6
9 Popliteal height	43.9	43.3	46.2	44.5	44.0	41.8	41.0	40.7	42.0	40.5	40.4	43.6	41.4	40.7	40.2
10 Buttock–knee length	60.4	60.7	61.9	60.1	60.4	59.0	57.4	57.7	57.3	57.4	58.1	60.2	52.8	53.8	55.7
11 Shoulder–elbow length	35.8	36.1	36.8	36.3	37.2	31.5	30.9	31.1	35.2	35.4	35.7	37.0	35.1	35.4	35.4
12 Elbow–wrist length	30.0	30.0	30.0	28.2	28.8	24.4	23.5	23.4	27.7	27.9	28.0	28.8	27.0	26.2	27.4
13 Sleeve inseam	48.5	48.4	51.1	49.5	48.4	45.7	44.5	43.6	45.1	46.0	44.5	51.2	44.6	45.1	46.4
14 Sleeve length	90.5	91.0	89.4	87.6	88.3	80.6	79.5	80.0	83.8	84.8	85.3	87.8	77.7	81.8	82.6
15 Thumb-tip reach	79.7	81.0	81.1	78.1	78.7	75.7	73.9	74.8	76.3	76.0	75.3	80.8	81.3	69.3	84.1
16 Neck circumference	37.9	38.9	36.3	36.2	38.6	33.5	33.9	34.3	35.5	36.3	37.4	38.3	33.5	36.5	36.3
17 Shoulder circumference	116.9	118.5	111.2	111.5	116.6	99.4	100.2	103.3	108.4	109.8	111.7	116.2	100.6	110.0	108.0
18 Chest circumference	96.7	101.3	90.1	91.8	99.7	82.2	84.0	87.8	90.0	91.7	93.7	96.4	80.8	90.5	88.6
19 Waist circumference	84.9	90.4	74.9	78.0	87.0	66.4	67.1	70.4	79.0	79.4	80.7	84.1	66.6	75.1	74.7
20 Buttock circumference	97.9	99.5	90.5	92.5	97.7	92.2[1]	93.3[1]	96.7[1]	90.9	93.5	94.4	97.1	82.5	89.4	91.1
21 Upper thigh circumference	59.1	58.7	54.3	54.2	56.7	54.7	55.3	57.0	52.4	54.3	53.8	56.1	47.5	53.2	51.6
22 Calf circumference	37.3	37.4	35.9	36.2	36.9	33.4	34.4	34.4	35.4	35.9	36.3	37.3	32.6	35.3	34.1

Table 3. (*continued*)

23 Ankle circumference	22.5	22.5	22.4	22.7	22.5	20.8	21.4	21.0	22.5	22.6	22.7	22.9	20.3	21.6	21.6
24 Biceps circumference, flexed	32.5	32.9	30.7	30.4	32.8	26.5	26.6	28.5	29.3	30.3	30.7	32.5	24.0	30.0	28.1
25 Wrist circumference	17.4	17.8	17.3	17.1	17.4	15.1	15.0	15.1	17.1	16.9	17.6	18.1	14.9	16.6	16.3
26 Shoulder breadth	47.9	48.7	45.8	45.6	47.7	41.4	41.8	43.1	44.2	44.1	45.7	46.7	41.0	44.7	42.3
27 Interscye[2]	38.4	39.7	37.2	36.8	38.7	34.6	34.8	36.2	38.1	36.6	39.3	43.6	35.5	41.0	38.0
28 Hip breadth	35.0	35.6	32.8	33.9	35.1	33.9	34.9	36.0	32.4	33.5	33.9	35.6	28.9	31.1	32.2
29 Waist–back	46.4	47.0	43.5	45.7	47.4	40.0	40.7	40.5	44.6	45.3	45.3	45.5	46.3	45.3	46.1
30 Hand length	19.1	19.2	20.4	19.5	19.6	19.2	18.2	18.5	19.0	18.9	19.0	19.0	17.7	18.1	18.6
31 Hand breadth	8.8	9.0	9.0	8.9	8.9	7.7	7.5	7.7	8.8	8.9	9.1	8.7	8.0	8.4	8.6
32 Foot length	27.0	27.1	27.5	26.7	26.9	24.8	24.0	24.1	26.4	26.4	26.5	26.4	23.9	24.8	26.2
33 Foot breadth	9.8	9.7	10.0	9.8	9.8	9.1	8.9	8.8	10.4	10.2	10.4	10.0	9.3	9.6	10.1
34 Head circumference	57.4	57.8	56.2	55.9	57.3	55.9	54.7	55.1	54.8	55.6	56.4	57.2	54.2	55.9	55.5
35 Head length	19.9	20.0	19.8	19.5	19.9	18.7	18.3	18.5	18.4	18.9	19.2	19.2	18.1	18.4	18.8
36 Head breadth	15.5	15.7	15.1	15.3	15.6	14.4	14.4	14.8	15.7	15.4	15.6	15.7	14.9	15.5	15.1
37 Head height	13.5	13.3	13.0	13.2	13.0	12.9	12.7	12.8	13.1	13.5	13.0	13.1	12.3	13.0	12.4
38 Face breadth	14.1	14.4	13.8	13.9	14.3	12.9	12.8	13.0	14.2	14.1	14.3	13.5	12.6	13.9	12.1
39 Face length	12.0	12.2	12.2	11.8	11.9	10.7	10.5	10.8	12.2	11.8	11.9	12.2	11.3	12.4	12.1
40 Interpupillary breadth	6.2	6.3	6.5	6.2	6.4	10.2[3]	10.0[3]	10.4[3]	6.3[4]	6.3	6.4	6.3	6.3	5.9	6.1
41 Age	23.0	40.8	19.1	19.3	37.5	22.2	19.1	38.1	21.0	21.8	22.4	N.A.	21.5	28.1	26.2
n	400	400	383	400	400	146	400	230	344	400	398	400	400	263	400

Key: [1] 7″ below waist.
[2] Techniques for measuring interscye differed substantially from survey to survey.
[3] Bigonial breadth.
[4] Mean of interocular and biocular breadths.

Table 4. Loadings of factors one through seven

	Loadings of factor 1														
Sample	A	B	C	D	E	F	G	H	I	J	K	L	M	N	O
Appearance	1	1	1	1	1	1	1	1	1	1	1	1	1	1	1
1 Weight	32	(30)	32	(28)	32	32	36	32	38	37	36	38	37	(26)	33
2 Stature	86	84	87	88	83	85	80	85	87	85	84	90	82	84	86
3 Acromial height	86	85	90	90	84	89	82	86	89	86	86	91	82	87	88
4 Waist height	93	92	94	94	92	90	90	91	93	93	92	94	85	83	92
5 Crotch height	95	92	97	96	95	93	92	92	90	96	94	94	88	90	88
6 Sitting height	45	40	38	47	38	40	39	49	52	39	34	49	40	33	41
7 Eye height, sitting	42	40	38	44	34	38	33	44	47	35	36	46	(29)	(27)	42
8 Knee height, sitting	92	90	93	89	87	85	82	88	89	88	88	90	78	86	89
9 Popliteal height	91	91	94	91	89	77	80	82	82	92	90	88	67	78	89
10 Buttock–knee length	76	78	80	80	83	78	74	79	88	84	81	84	81	75	78
11 Shoulder–elbow height	81	84	85	88	84	76	79	83	87	87	86	73	80	80	80
12 Elbow–wrist length	82	84	83	83	80	74	77	77	82	82	78	74	74	78	78
13 Sleeve inseam	81	88	89	88	85	81	87	89	87	86	86	68	76	80	79
14 Sleeve length	75	82	77	74	78	75	81	82	79	80	80	69	73	63	66
15 Thumb-tip reach	70	76	76	74	76	68	70	71	76	80	76	69	80	65	67
25 Wrist circumference	(18)	(19)	(29)	(26)	(24)	(19)	(30)	(28)	34	(29)	(24)	(18)	(16)	(24)	(26)
28 Hip breadth	(24)	(28)	(22)	(27)	(29)	(13)	(18)	(20)	46	36	35	(20)	41	(27)	36
29 Waist–back	42	(31)	42	(30)	(23)	(28)	(30)	35	33	33	37	35	34	(21)	40
30 Hand length	53	65	63	58	56	58	49	56	67	65	55	64	64	68	62
31 Hand breadth	(24)	33	(26)	34	(27)	33	(26)	(27)	(31)	36	(28)	(30)	(29)	(28)	(28)
32 Foot length	57	64	68	61	61	62	57	64	67	66	54	64	62	60	62
33 Foot breadth	(27)	34	(25)	(30)	(29)	(16)	(26)	(31)	(26)	35	(24)	35	(21)	(25)	(17)

Table 4. (*continued*)

Sample	Loadings of factor 2 A	B	C	D	E	F	G	H	I	J	K	L	M	N	O
Appearance	2	2	2	2	2	2	2	2	2	2	2	2	2	2	2
1 Weight	90	89	90	92	89	86	83	87	86	85	85	85	82	92	89
10 Buttock–knee length	44	33	43	48	39	43	45	36	(27)	38	43	(31)	(24)	37	(30)
14 Sleeve length	34	(26)	32	39	(29)	(30)	(24)	(25)	32	(30)	32	(26)	32	(27)	(29)
16 Neck circumference	65	70	70	75	70	64	37	56	79	64	67	71	68	71	70
17 Shoulder circumference	82	84	85	90	78	89	84	90	89	81	77	85	77	46	90
18 Chest circumference	87	85	86	90	82	90	81	90	85	81	79	87	73	83	87
19 Waist circumference	90	86	87	91	84	80	80	89	79	80	80	88	71	86	83
20 Buttock circumference	93	87	93	92	90	83	85	86	84	87	88	85	78	87	82
21 Upper thigh circumference	93	85	95	94	89	83	81	75	84	90	90	77	76	74	84
22 Calf circumference	76	70	80	83	77	66	60	60	76	68	67	52	73	78	79
23 Ankle circumference	60	56	57	68	60	49	36	41	58	52	51	37	63	68	60
24 Biceps circumference, flexed	83	82	82	90	81	85	83	85	85	80	83	74	81	80	84
25 Wrist circumference	43	44	46	53	46	60	43	51	60	46	45	54	68	65	59
26 Shoulder breadth	78	83	78	86	75	85	81	89	84	72	68	75	59	78	75
27 Interscye	38	44	53	57	37	60	51	63	50	35	32	44	49	34	38
28 Hip breadth	80	67	83	83	71	62	75	63	57	71	70	74	51	66	61
31 Hand breadth	(27)	(25)	(27)	33	(21)	(26)	(23)	(19)	53	(30)	(26)	(25)	39	35	39
33 Foot breadth	33	(22)	32	40	(27)	44	(21)	(13)	40	37	34	(22)	37	(30)	(30)
34 Head circumference	(23)	(24)	(23)	(28)	(22)	(26)	(19)	(15)	37	(24)	(22)	(21)	(24)	39	35
38 Face breadth	(26)	37	39	41	32	(27)	(16)	(26)	42	42	34	(20)	39	(27)	(30)

Table 4. *(continued)*

	Loadings of factor 3														
Sample	A	B	C	D	E	F	G	H	I	J	K	L	M	N	O
Appearance	3	3	3	3	3	4	5	8	5	3	6	9	3	4	6
34 Head circumference	82	76	81	81	78	65	74	70	61	72	68	53	74	71	71
35 Head length	90	87	89	92	87	71	78	78	83	89	86	65	80	81	83

	Loadings of factor 4														
Sample	A	B	C	D	E	F	G	H	I	J	K	L	M	N	O
Appearance	4	4	4	4	4	3	3	4	4	4	4	4	4	3	4
2 Stature	42	43	40	39	48	46	49	44	38	42	43	35	46	38	44
3 Acromial height	33	37	(30)	(30)	41	32	41	38	(29)	35	35	(26)	38	32	35
6 Sitting height	82	82	86	82	86	82	84	79	74	83	87	80	80	83	81
7 Eye height, sitting	83	82	87	82	88	81	85	79	77	83	85	79	80	84	78
29 Waist–back	60	58	54	60	56	58	65	57	53	48	56	40	40	(31)	41

	Loadings of factor 5														
Sample	A	B	C	D	E	F	G	H	I	J	K	L	M	N	O
Appearance	6	5	5	6	5		7	5	6	6	5	6		9	7
22 Calf circumference	(25)	(31)	(25)	(22)	(24)		(17)	(12)	(09)	(15)	38	(12)		(22)	(14)
23 Ankle circumference	42	48	41	34	38		(26)	(27)	(27)	(29)	52	(24)		(19)	(23)
25 Wrist circumference	51	58	55	53	55		45	52	51	53	60	52		(28)	(24)
30 Hand length	45	46	45	44	51		40	52	33	45	42	(29)		(21)	(22)
31 Hand breadth	64	62	68	59	61		56	64	56	62	68	60		58	49
32 Foot length	45	42	33	41	44		47	39	(23)	43	47	(28)		(25)	(28)
33 Foot breadth	50	50	50	42	51		51	52	40	33	64	38		50	50

Table 4. (*continued*)

	Loadings of factor 6														
Sample	A	B	C	D	E	F	G	H	I	J	K	L	M	N	O
Appearance	5	6	7	5	7	5	4	7	3	5	3	3	8	5	3
34 Head circumference	37	41	42	34	47	(07)	(29)	(11)	61	52	58	41	34	34	43
36 Head breadth	72	79	79	70	81	40	61	39	85	79	84	51	58	65	78
38 Face breadth	69	70	57	72	64	73	75	66	69	72	72	(18)	(26)	(23)	(30)
40 Interpupillary breadth	40	(31)	(26)	47	(25)	64	57	60	36	48	40	55	(11)	(02)	(26)

	Loadings of factor 7														
Sample	A	B	C	D	E	F	G	H	I	J	K	L	M	N	O
Appearance	11	11	10		13		10	10	7	10	8	5		14	10
22 Calf circumference	43	46	34		41		59	56	49	51	44	60		32	(31)
23 Ankle circumference	43	42	44		41		63	63	50	50	38	63		(24)	44

Table 5. A history of all loadings in excess of 0.316 not shown in Table 4

	Appearance 1	2	3	4	5	6	7	8	9	10	11	12	13	14	15
Sample A															
26 Shoulder breadth							40								
27 Interscye							55								
30 Hand length								33							
36 Head breadth									33						
37 Head height			38						51						
39 Face length														35	
Sample B															
10 Buttock-knee length														32	
20 Buttock circumference							32								
27 Interscye								51							
28 Hip breadth							46								
29 Waist–back												34			
37 Head height										53					
39 Face length												34			
Sample C															
17 Shoulder circumference						32									
26 Shoulder breadth						35									
27 Interscye						49									
37 Head height									57						
38 Face breadth												35			
40 Interpupillary breadth												48			

Table 5. *(continued)*

	Appearance 1	2	3	4	5	6	7	8	9	10	11	12	13	14	15
Sample D															
27 Interscye							42								
36 Head breadth								39							
37 Head height								49							
Sample E															
17 Shoulder circumference						51									
26 Shoulder breadth						51									
27 Interscye						57									
37 Head height							32								34
39 Face length									45						
40 Interpupillary breadth											35				
Sample F															
13 Sleeve inseam									39						
14 Sleeve length									34						
16 Neck circumference							33								
22 Calf circumference								52							
23 Ankle circumference								62							
25 Wrist circumference								38							
27 Interscye													43		
29 Waist–back										45					
31 Hand breadth							37								
32 Foot length								33							
33 Foot breadth								40							
36 Head breadth						46									
37 Head height						60									
39 Face length							53								

Table 5. (*continued*)

	Appearance 1	2	3	4	5	6	7	8	9	10	11	12	13	14	15
Sample G															
16 Neck circumference									35						
17 Shoulder circumference						33									
27 Interscye						47									
37 Head height								48							
Sample H															
16 Neck circumference														40	
20 Buttock circumference						37									
21 Upper thigh circumference						45									
27 Interscye													39		
28 Hip breadth						56									
34 Head circumference			41												
36 Head breadth			60												
37 Head height			56												
39 Face length											43				
Sample I															
28 Hip breadth										37					
37 Head height												46			
39 Face length											40				
40 Interpupillary breadth									47						
Sample J															
17 Shoulder circumference							37								
26 Shoulder breadth							41								
27 Interscye							58								
28 Hip breadth									−35						
37 Head height											49				
39 Face length												42			

	Appearance 1	2	3	4	5	6	7	8	9	10	11	12	13	14	15
Sample K															
17 Shoulder circumference							46								
26 Shoulder breadth							44								
27 Interscye							55								
29 Waist–back										39					
30 Hand length													37		
32 Foot length													34		
37 Head height											48				
39 Face length					33										
40 Interpupillary breadth														42	
Sample L															
21 Upper thigh circumference					36										
26 Shoulder breadth								33							
27 Interscye										37					
34 Head circumference							36								
37 Head height							54								
38 Face breadth											58				
39 Face length													49		
Sample M															
8 Knee height, sitting							39								
9 Popliteal height							47								
17 Shoulder circumference													34		
18 Chest circumference													35		
27 Interscye													44		
28 Hip breadth												45			
29 Waist–back											39				
30 Hand length															42
32 Foot length															33
33 Foot breadth					45										
37 Head height						57									
38 Face breadth										37					
39 Face length									45						
40 Interpupillary breadth										44					

Table 5. (*continued*)

	Appearance 1	2	3	4	5	6	7	8	9	10	11	12	13	14	15
Sample N															
14 Sleeve length								34							
15 Thumb-tip reach												36			
27 Interscye								44							
29 Waist–back							52								
30 Hand length											34				
32 Foot length											32				
37 Head height										55					
38 Face breadth						51									
39 Face length													36		
Sample O															
12 Elbow-wrist length											36				
20 Buttock circumference								37							
27 Interscye									47						
28 Hip breadth								42							
30 Hand length											53				
32 Foot length											35				
39 Face length												45			

were measured about two months earlier than samples C and D. Sample J was measured three months after sample I and sample K three months later.

Secondly, the contrasts of A–B, D–E, and G–H presumably should be affected only by age. The contrasts C–D and F–G are black–white contrasts; the contrasts C–F, D–G, and H–E are male–female contrasts. It could be hypothesized that samples I, J, and K should provide results falling in some sort of a sequence, as do the mean values for the data in these samples.

However, while there are differences among the results, we see no particular pattern to these differences. We surely see no patterns which we can associate with any rational *a priori* hypotheses as to what differences should occur; nor do we see patterns that suggest hypotheses that a more resourceful person might have concocted *a priori.*

CONCLUSIONS

The major hypothesis to which these results seem to lend support is the one stated at the beginning of this study: that the difficulties of drawing useful conclusions from factor analyses of anthropometric data have been a major consideration in restricting the publication of such analyses. We are presenting these results, however, in the hope that they may be useful both as an indication of what is obtained when one factor-analyzes anthropometric data and as raw material for further analysis.

REFERENCES

CHOI, SUNG C., MILDRED TROTTER
1970 A statistical study of the multivariate structure and race-sex differences of American white and Negro fetal skeletons. *American Journal of Physical Anthropology* 33:307–312.

CHURCHILL, EDMUND, THOMAS CHURCHILL, PAUL KIKTA
1977 *The AMRL anthropometric data bank library: volumes I—V.* AMRL Technical Report 77–01. Wright-Patterson Air Force Base, Ohio: Aerospace Medical Research Laboratory.

CHURCHILL, EDMUND, BRUCE TRUETT
1957 *Metrical relations among dimensions of the head and face.* WADC Technical Report 56–621. Wright-Patterson Air Force Base, Ohio: Wright Air Development Center.

CLAUSER, C. E., PEARL E. TUCKER, JOHN T. McCONVILLE
1972 *Anthropometry of air force women.* AMRL Technical Report 70–5. Wright-Patterson Air Force Base, Ohio: Aerospace Medical Research Laboratory.

DIXON, W. J., *editor*
1971 *BMD (BioMedical Computer Programs).* Berkeley: University of California.

GRUNHOFER, H. J., G. KROH
1975 *A review of Anthropometric data of German Air Force and United States Air Force flying personnel 1967–1968*. AGARD-AG-205 Neuilly sur Seine: Advisory Group for Aerospace Research and Development.
HERTZBERG, H. T. E., EDMUND CHURCHILL, C. W. DUPERTUIS
1963 *Anthropometric survey of Turkey, Greece, and Italy*. New York: Macmillan.
HERTZBERG, H. T. E., G. S. DANIELS, E. CHURCHILL
1954 *Anthropometry of flying personnel 1950*. WADC Technical Report 52–321. Wright-Patterson Air Force Base, Ohio: Wright Air Development Center.
KAY, WON CHUEL
1961 Anthropometry of the ROKAF pilots. *Korean Journal of Aviation Medicine* 9:61–113. (Abstract and data summaries in English)
O'BRIEN, RUTH, WILLIAM C. SHELTON
1941 *Women's measurements for garment and pattern construction*. U.S. Department of Agriculture Miscellaneous Publication 454. Washington, D.C.: U.S. Government Printing Office.
RHOADS, JOHN G.
1970 "Factors of physique in American and Solomon Island males." Paper presented at the American Association of Physical Anthropologists annual meeting, Boston.
STOUDT, HOWARD W., ALBERT DAMON, ROSS McFARLAND, JEAN ROBERTS
1965 *Weight, height, and selected body dimensions of adults, United States, 1960–1962*. Public Health Service Publication 1,000, series 11, No. 8. Washington, D.C.: U.S. Government Printing Office.
THURSTONE, L. L.
1946 Factor analysis and body types. *Psychometrika* 11:15–21.
1947 Factorial analysis of body measurements. *American Journal of Physical Anthropology*, n.s. 5:15–28.
VANDENBERG, STEVEN G.
1968 A factor analysis of garment measures of Dutch women. *Human Biology* 40:295–313.
WHITE, ROBERT M.
1964 *Anthropometric survey of the armed forces of the Republic of Vietnam*. Natick, Mass.: United States Army Natick Laboratories.

Certain Problems of Ethnic Anthropology in Southwest Asia in the Light of the Latest Research

M. G. ABDUSHELISHVILI

Our paper delivered at the VIIIth International Congress of Ethnographic and Anthropological Sciences in Tokyo (Abdushelishvili 1969) dealt with certain problems concerning the anthropological genesis of the aboriginal population of the Caucasus. The report was based on years of study of all the ethnic and territorial groups of the present-day population of the Caucasus, as well as the paleoanthropological series referring to nearly all the chronological periods, beginning from the Aeneolithic up to the present. We noted at the congress that the problems emerging in connection with the study of the genesis of the ancient and modern anthropological types of the Caucasus extend far beyond and are closely interwoven with the problems of the historical anthropology of the peoples of Asia Minor and central and southwestern Asia. This definition of the problem naturally led to the need for a more attentive examination of the anthropological material in the above-mentioned areas, the result of which is the current paper. This is a continuation of, and to a certain degree an extension of, the problems posed earlier hypothetically. There are, in turn, a great number of similar problems in the ethnic history of the peoples of Asia Minor and central and southwestern Asia. A comparison of anthropological materials from all these territories can shed light on many issues concerning the genesis of the peoples of southwestern Asia. Such a comparison is a matter for the future. It requires further research by scientists of different countries and in different fields. Significantly, of special value are methodical and methodological comparisons of materials, since only unified studies can lead to the solution of issues concerning peoples so far removed ethnically, linguistically, historically, and territorially. In such cases it is much easier to explain the differences between authors (both in bias and method) than the similarities.

In 1971 we had the happy experience of visiting India, which is a long distance away from the Caucasus, to make a direct study of local groups. The comparison of these data with those studied earlier in the Caucasus provided the basis for this paper. But first of all, a few words about the problems emerging in connection with the study of the physical anthropology of the population of the Caucasus.

As we have noted repeatedly, all the aboriginal population of the Caucasus belonged to the southern branch of the race which has lately become known as the Indo-European race. This race is extensive and inhabits the area roughly between the forty-fourth and twentieth parallels, from India to the Atlantic Ocean. It is divided into three large subraces: the Indo–Pamir, Mediterranean–Balkan, and Asian Minor. The Mediterranean–Balkan group of types, which is of the general Europoid type, is characterized by wavy hair, strongly developed tertiary hair, a narrow and sharply profiled face, a moderately projecting, narrow nose, and somewhat thickish lips. This group dominates the populations of southern Europe and North Africa. The Indo–Pamir group of types is very similar morphologically to the Mediterranean–Balkan group, differing from it by its somewhat darker pigmentation. This group is predominant among certain peoples of central Asia (the Tajiks, Uzbeks, Turkmen), Afghanistan, Iran, and northern India. Between the Indo–Pamir and the Balkan–Mediterranean groups is the group of types known as the Asian Minor. Morphologically, this group is characterized by its strongly developed tertiary hair, slant of the forehead, clearly expressed superciliary area, and sharply projecting nose of medium width, with salient nasal bone and very high bridge. This type is spread through the Transcaucasus, Anatolia, Mesopotamia, Syria, Palestine, and Arabia.

All three groups of the Indo-Mediterranean race are encountered among the contemporary population of the Caucasus, though with certain specific features of their own. The Caspian type clearly corresponds to the Indo–Pamir group of types; the Pontic corresponds to the Mediterranean–Balkan; while the Asian Minor is the most widespread type among the present-day population of the Caucasus. In turn, these three basic groups are represented in the Caucasus by different local types: the Caspian group by the West Caspian type, the Pontic by the Adighe type, and the Asian Minor by the Assyroid, Armenoid, Iberian, Caucasian, and Kolkhic types. The Caspian types in the Caucasus dominate among the Azerbaijan, Kurdic, and certain Daghestan groups, as well as the Persian Azerbaijan group and those of the Caspian coast areas of Iran. This type is characterized by its dark pigmentation, low cranial index, narrow face, moderately projecting cheekbones, smaller average height, and moderately projecting nasal bone. The Asian Minor type dominates among the Assyrian, Armenian, Georgian, West Azerbaijan, and certain West Daghestan groups. This type is generally characterized by its dark hair, smaller average height, high cranial index, relatively wide face, projecting curve of the nasal bone, highly developed tertiary hair, etc. The Adighe type exists among the aboriginal groups of the northwestern Caucasus. These are, first and foremost, the Kabardinians, the Adighes, the Circassians, the Abazins, the Ingushes, and, to a certain degree, certain West Georgian groups. The characteristic features of this type are: horizontal position of the tip of the nose, high percentage of cases of folds on the upper eye lid, straight nasal bone, lower than average cranial index, narrow forehead and face, thin beard, and relatively light pigmentation.

Thus, the Mediterranean–Balkan group of types, spreading over the territory of southern Europe, North Africa, and the entire Mediterranean area (through the Adighe type of the Pontic race), extends to the northwest Caucasus. The Indo–Pamir group of types, spreading to central and southern Asia through the West Caspian type of the Caspian race, extends to the eastern Transcaucasus. The Asian Minor group of types, extending over the Caucasus and more southern regions, is concentrated between the above-mentioned two types. All these types, represented in the aboriginal population of the Caucasus, despite the clear-cut distinctions between them, differ much less from each other than from the territorially more remote representatives of the basic branches of the South Europoid race. Therefore, it is possible, to a certain degree, to speak of the Caucasian community of the South Europoid or Indo-Mediterranean race, the roots of which apparently emerged in ancient times. Today there can be no doubt that the Caucasus is the only territory with a concentration of the local varieties of all three basic branches of the Indo–Mediterranean race. This fact, no doubt, requires an explanation. It could be the result of a chance overlap of the distributions of the three branches on the narrow Caucasian isthmus of Eurasia, and subsequent mixture among them. An alternative possibility is local formation of the specific features of these different types in the territory of the Caucasus itself, and the spread further west, east, and south of South Europoid forms already differentiated (but not to the full extent, since the full differentiation of the racial specific features of these three branches apparently took place later in the territories where they now live, in different climatic, geographical, and social conditions).

Thus, the Caucasus is the territory where all the basic branches of the Indo-Mediterranean race can be found. Naturally, the question arises of the autochthonic character of these racial types. First of all, a few words about the Asian Minor which is most widespread among the Caucasian population. There can be no doubt that the Asian Minor type has extremely ancient roots. This type (Bunak 1927) was known from ancient times and was widely spread in all parts of the Tigris and Euphrates area, from its southern part to the rivers' upper reaches, and from the western tip of the Anatolia peninsula to the inland regions of Iran. Bunak stressed that no other racial elements are to be found in the most ancient populations of Asia Minor. On the other hand, in all paleoanthropological materials from Asia Minor, dolichocranic forms clearly dominate. These forms are rather close to each other, and their regular epochal transformation led to the formation of the modern Asian Minor types. As Debets (1951) justly noted the data on the present spread of the Asian Minor group makes it possible to state that the group's formation took place approximately on the same territory over which it is currently spread. Thus, the formation of the Asian Minor types is definitely linked to Asia Minor, where dolichocephalic forms clearly dominated. We feel that there is no reason to seek the ancient representatives of the Asian Minor type solely in the brachycephalic series in Asia Minor. First, one must remember the general and natural process of brachycephalization, which does not in the least rule out the possibility of affinity between the present-day brachycephals and the ancient ones. On the contrary, it supports such an affinity. On the other hand, one should not forget the observations made by Oshanin (1957), that the complex of features characteristic for Armenoids is sometimes combined with the long shape of the skull. We absolutely agree with Debets (1951), who concluded that "the importance of the cephalic index for classifying the Europoid types of Asia Minor has a considerable less effect than is usually supposed. One cannot regard as Armenoid all the brachycephalic skulls." Thus, there are no grounds for seeking exclusively brachycephalic ancestors for the present-day representatives of the Asian Minor type, although the possibility of such cannot be ruled out. If this is the case, it is not difficult to find in the Mediterranean Dolichoids of Asia Minor skull shapes that are rather close to the Asian Minor in face skeletal structure, particularly of the nose. This leads us to recognize that the main Asian Minor substrata, the northern-most variety of which is represented in the ancient population of the Caucasus, formed the basis for the further formation of the main anthropological types of the present-day population of the Caucasus. It is worth noting the regionality associated with genetic changes in some anthropological types in the Caucasus, which determined the deviation from the "classical" Asian Minor shapes, both in ancient and present-day representatives of the Asian Minor types in the Caucasus. Despite these morphological differences one can discern common features in the processes which took place in Asia Minor and the Caucasus in forming the local types of the Asian Minor race. Thus, paleoanthropological materials provide a basis for regarding the Caucasus as a territory where the shaping of the Asian Minor race was taking place, a race that cannot be regarded as coming from elsewhere.

And now, a few words about another Caucasian branch of the South Europoid race, the so-called Pontic type of the Mediterranean–Balkan race. The Pontic type in the population of the northern Caucasus was first identified by Bunak (1932) who related it to the Mediterranean branch of the Europoid race in the absence of admixture with the northern branch of Europoids (Bunak 1956). Debets (1958) determined the place of the Pontic type among the European groups of anthropological types. In his chart the Pontic type is shown as a transitional form resulting from mixture between the main subdivisions of the Europoid race. Another point of view on the shaping of the Pontic type was expressed by Alexeev (1974), who made the following main points: "The Pontic type was formed in the framework of the Mediterranean branch of the Europoid race, without the participation of an admixture from the northern branch. The basic process of forming that type is the gracilization of the massive, protomorphic type. The autochthonic origin of the Pontic type is evidence of the local sources of the ethnogenesis of the peoples representing this type." One must agree with Bunak and Alexeev that the Pontic race belongs to the southern branch of Europoids, without admixture of northern elements. We contend that Alexeev is also right when he states that the peoples related to the Pontic type are descendants of the most ancient population of the northern Caucasus. As to the emergence of the Pontic type as a result of the gracilization of the ancient, massive type of Europoid race which, according to the writer, provided the basis for the emergence of the Caucasian type, we disagree (Abdushelishvili 1964). Therefore we shall here merely note our contention that the emergence of a Pontic, in this case Adighe type, results from the development of the local Caucasian paleoanthropological types whose roots can be clearly traced in the ancient craniological series in the territory of the Caucasus.

Finally, the third main branch of the Indo-Mediterranean race, the Caspian type, is related to the Indo–Pamir group of types. The paleoanthropological materials from the territory where the West Caspian and Asian Minor types are settled in present times indicate the great similarity between them: everywhere we find that the long-headed, narrow-faced, and rather gracile type prevails. Analogical forms beginning from Aeneolithic time have also been known outside the Caucasus in the territory of Asia Minor and central and southwest Asia, which undoubtedly indicates the wide-spread distribution of these forms. In the paleoanthropological materials from the eastern Caucasus, there are no data indicating the emergence of new elements differing from the local ones. On the contrary, everything points to the continuity of the population which, from time immemorial, has resided in the above-mentioned territory. Therefore, the paleoanthropological materials provide, as we see it, no reason for claiming that the Caspian type was formed in the process of populating the Caucasus from the southeast (Debets 1958). Yet the West Caspian type has certain features that distinguish it from the others in the Caucasus. It should be noted that these distinctive features become gradually less pronounced at more remote chronological periods, an indication of the homogeneity of Caucasian anthropological types as presently differentiated. The distinguishing features of the West Caspian types are: greater cephalic length, smaller cephalic breadth, lower cephalic index, lower nose, lower frequency of protruding forms of nasal bone, more pronounced cheekbones, lower nose bridge, more pronounced upper lip, less hair, and somewhat darker pigmentation. In search of analogical forms we turned to the data collected by Debets (1967) in Afghanistan in 1963–1967. Comparison with these indicates that in Afghanistan the predominant groups differ from the Caucasian types in the same direction as the West Caspian, but to a much greater degree. Thus, for example, the cephalic breadth that Debets studied in the Afghan groups appeared less in the West Caspian groups; the cephalic index lower, the face narrower and lower, the nose lower, the frequency of protruding forms of the nose lower, the cheekbones more expressed, the nose bridge lower, the hair less developed, and the pigmentation darker.

The search for more distant parallels repeatedly led us to groups residing in the northwestern part of India. The materials of Soviet anthropologists (Cheboksarov and Zubov n.d.) and their Indian counterparts (Guha 1935) point to the existence of analogical forms in northwestern India. However, since the research methods are different, it is difficult to establish the degree to which the distinctive features are expressed. We mentioned previously that in 1971 we visited India and studied several groups in the Harian state and the union territory of Delhi (Table 1). As a result, valuable data were obtained bearing on the problem of ethnic anthropology in southwestern India. The value of these materials, certainly, lies in their similarity to the materials of the Caucasus, central Asia, and Afghanistan. In the materials we studied in India the above-mentioned differentiating features are expressed to the greatest degree: the head length is more than in the Afghan groups; the breadth less; the cephalic index is very low, nowhere exceeding the limits of the dolichocephalic categories; the face is narrower, the nose lower, the percentage of the protruding forms of the nasal bone approximately the same; the cheekbones are more pronounced, the nose bridge lower, the profile of the upper lip more protruding; the hair coverage on the body is expressed just as in the Afghan groups, the beard is thinner, the pigmentation darker, etc.

Thus, at present it is possible to reconstruct the ethnogeographic changes in the differentiating features of the Pontic, Asian Minor, and Caspian types. This differentiation, through a gradually growing trend of distinctions, leads from the Caucasus to northwestern India. This pattern of differentiation can hardly be attributed to chance, particularly since it was traced on nearly ideal uniform material. It is no doubt an indication of the ancient genetic continuity of the population of southwestern Asia.

Table 1. The average data of the Indian groups studied in 1971

Anthropological indications	Groups studied				
	Rajputs	Julahs (Chamars)	Gujars	Ahirs	Jats
Number of observations	92	101	100	102	103
1. Body length	168.02	165.31	170.00	168.93	169.16
2. Head length	195.98	192.47	193.19	192.76	192.89
3. Head breadth	143.16	139.73	141.53	141.25	143.09
4. Head index	73.19	72.27	73.22	73.38	73.70
5. Skull diameter	137.59	133.35	134.33	134.18	136.12
6. Morphological height of the head	122.82	119.57	121.35	119.60	120.69
7. Physiological height of the head	179.67	176.99	177.05	176.04	178.48
8. Nose height from lower eyebrows	53.07	51.95	53.96	52.24	52.83
9. Nose width	38.03	37.05	36.39	37.25	36.57
10. General profile of nasal bone (%)	22.83	25.74	42.00	35.29	32.04
11. Thickness of both lips	19.97	20.06	19.26	19.07	19.58
12. Forehead incline	2.33	2.52	2.64	2.65	2.69
13. Horizontal face profile	2.74	2.61	2.87	2.79	2.63
14. Cheek-bone protrusion	1.76	1.67	1.56	1.53	1.50
15. Height of nose bridge	2.14	2.13	2.33	2.30	2.22
16. Width of eye slit	2.06	2.08	2.03	2.19	2.09
17. Incline of eye slit	2.14	2.33	2.15	2.08	2.24
18. Fold of upper eyelid (%)	67.78	53.47	51.00	73.53	59.22
19. Eye colour	1.96	1.90	1.76	1.89	1.91
20. Hair colour (point II)	5.65	7.25	7.81	8.70	8.06
21. Skin colour	1.57	1.82	1.41	1.49	1.48
22. Nostril incline	2.41	2.24	2.46	2.38	2.52
23. Position of nose base	1.52	1.74	1.90	1.77	1.84
24. Height of nose wings	2.05	2.05	2.00	2.04	2.11
25. Degree of expression of wing fissures	2.18	2.40	2.22	2.19	2.20
26. Projection of nose wings	2.40	2.19	2.14	2.29	2.35
27. Height of skin part of upper lip	1.87	1.78	2.00	1.81	1.82
28. Profile of upper lip	1.87	1.78	2.00	1.81	1.82
29. Length of beard (from 25 years on)	3.17	2.80	3.72	3.14	2.95
30. Size of eyebrows	2.30	2.09	2.27	2.26	2.25
31. Size of chest hair (from 25 years on)	2.55	1.46	2.56	2.31	2.02

However, the direction of this continuity is not yet clear. The only point that seems indisputable is that the Caspian type in the Caucasus is not the result of the penetration of southern forms since the roots of this type are clearly present in the paleoanthropological materials of the Caucasus.

Unfortunately, our material does not enable us to decide to which specific Europoid form we should relate the Europoid component traced in India. The Indo-Aryans who settled in the territory of northern India at the end of the eleventh and the beginning of the tenth century B.C. were not the first Europoids in India. To find a final solution to the question, it is necessary to study over Europoid groups in India, drawing extensively on paleoanthropological data and using the same unified method that was applied in 1971 to study the Hindi language groups in the Harian state and the Union territory of Delhi. One cannot but recognize that genetic generalizations based on physical distinctions alone are insufficient. Therefore, both in India and in the Caucasus, the research program has included rather extensive genetic research. Unfortunately, these data are still under laboratory examination and cannot yet serve as the basis for comparison with the data of physical anthropology. The issues we raise now are offered for the purpose of posing problems, rather than solving them. They must serve as a basis for further research into the problems of ethnic anthropology of southwestern Asia, research that can open new areas in the study of the ethnogenesis of the populations under consideration, as well as in the study of race genesis in general, and of Mediterranean race genesis in particular.

REFERENCES

ABDUSHELISHVILI, M. G.

1964 *Antropologiia drevnego i sovremennogo naseleniia Gruzii* [Anthropology of the ancient and modern populations of Georgia]. Tbilisi.

1969 The genesis of the aboriginal population of the Caucasus in the light of anthropological data, in *Proceedings of the VIII International Congress of Anthropological and Ethnological Sciences,* volume 1. Tokyo: Kyoto.

ALEXEEV, V. P.

1974 *Proiskhozdenie norodov Kavkaza* [Origin of Caucasian peoples]. Moscow: Nauka.

BUNAK, V. V.

1927 *Crania armenica.* Moscow.

1932 Neues Material zur Ausscheiderung anthropologischer Typen unter der Bevölkerung Osteuropas. *Zeitschrift für Morphologie und Anthropologie* 30(3).

1956 Antropologicheskie issledovaniia v iuzhnoĭ Belorussii [Anthropological research in southern Belorussia]. *Antropologicheskiĭ sbornik I, Trudy Instituta ètnografii Akademii nauk SSSR, novaia seriia* 33.

CHEBOKSAROV, N. N., A. A. ZUBOV.

n.d. Otchët ob antropologicheskikh issledovaniiakh v Indii [Account of anthropological research in India]. Unpublished manuscript.

DEBETS, G. F.

1951 Zaseleniia Iûzhnoĭ i Peredneĭ Azii po dannym antropologii [The settlement of South Asia and northern Asia Minor according to anthro-

pological data]. *Trudy Instituta ėtnografii Akademii nauk SSSR, novaia seriia* 16.
1958 Opyt graficheskogo izobrazheniia genealogicheskoĭ klassifikatsii chelovecheskikh ras [Attempt at a graphic depiction of the genealogical classification of the races of mankind]. *Sovetskaia etnografiia* 4.
1967 Antropologicheskie issledovaniia v Afganistane [Anthropological research in Afghanistan]. *Sovetskaia ėtnografiia* 4.

GUHA, B. S.
1935 Racial affinities of the peoples of India. *Census of India* 1(3).

OSHANIN, L. V.
1957 *Antropologicheskiĭ sostav naseleniia Sredneĭ Azii i ėtnogenez eë narodov* [The anthropological composition of the population of Central Asia and the ethnogenesis of its peoples]. Erevan.

Recent Anthropological Data on the Ammassalimiut Eskimo of East Greenland: Comparative Studies

ALBERT DUCROS and JACQUELINE DUCROS

At present, 2,500 persons live on the east coast of Greenland in the district of Angmagssalik (65° 30′ north latitude), and 420 descended from a small hunters' colony, founded in 1925, live at Scoresbysund (70° 25′ north latitude). Participating in research in anthropological and sociological biology conducted by the Center of Anthropological Research (Museum of Man, Paris) one of us has collected several new anthropological data on this accultured, little-intermixed group, descended from 413 subjects discovered by Holm in 1884, whose genealogies are known (Gessain 1960).

SKIN FOLDS AND BODY FAT

Along with their height and weight, the thickness of three skinfolds was measured on eighty-five adult men from twenty to forty-nine years of age (five subjects from fifty to fifty-five years of age were included). The sample was examined at Angmagssalik. It is made up of a few hunters and, primarily, of salaried workers (sailors and dock and road workers). We used a Harpenden[1] caliper to measure the tricipital, the subscapular, and the suprailiac[2] folds, on the left side. The results are shown in Table 1.

Results

1. The analysis of variance on the transformed data (logarithmic transformation) shows that there is no significant difference between the three

[1] Harpenden is a registered trademark.
[2] This is directly above the iliac crest in the vertical prolongation of the extremity of the twelfth rib; the fold is faintly slanted from top to bottom and from the outside inward.

age groups at probability = 0.05 for a group of this size. A similar result was obtained for height and weight. These Eskimo have the thickest panniculus adiposus at the level of the trunk and the thinnest on the arm.

2. If we regard the values of the skinfold thicknesses of these Eskimo with comparable data (gathered with the same technique) from other authors (Figures 1 and 2), we see that the stability that continues with age is not peculiar to the Eskimo group. It may be found in other populations: South African blacks, Zulu (Slome et al. 1960), Kabyles (Bourlière and Parot 1962), Australians (Abbie 1967), and Sara (Crognier 1969); but it is not found in two samples of city populations with a higher standard of living: Americans (Brozek and Kinzey 1960) and Parisians (Bourlière and Parot 1962).

Not having the mean of each fold for each sample to compare, it is impossible to say whether the Eskimo present a special localization of the greatest thickness of the panniculus adiposus. But it should be noted that their averages are weak. An estimate of the body density and of the percentage of total fat (by equations established in a Japanese group by Nagamine and Suzuki in 1964) from the tricipital and subscapular folds (Damon and Goldman 1964) was made for the class aged twenty to twenty-nine and compared in Table 2 to results obtained, by densitometric methods, in other populations.

Although the smallest in height, the Eskimo are relatively the heaviest. Their great weight is due to the importance of the lean mass. Their

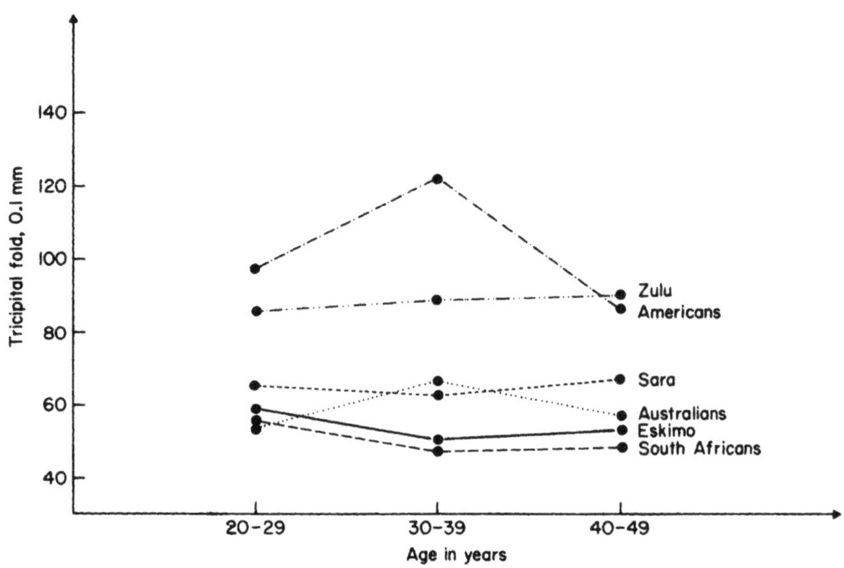

Figure 1. Comparison of tricipital fold thickness in Ammassalimiut Eskimo and other male populations

Table 1. Height, weight, and skinfolds of Ammassalimiut Eskimo

Age group Number in sample	20–29 years 29		30–39 years 34		40–50 years 22		Test F	Total sample 85	
	m	s.d.	m	s.d.	m	s.d.		m	s.d.
Age	24.20	2.57	34.00	2.80	44.00	3.60	–	33.00	8.17
Height (cm)	164.15	6.37	165.27	5.74	163.29	4.20	0.83	164.49	5.59
Weight (kg)	66.39	6.14	66.47	6.91	65.02	7.80	0.33	66.07	6.81
Tricipital fold									
absolute fold (0.1 mm)	58.72	15.20	50.20	13.92	53.17	16.18	–	53.87	15.21
100 × log. absolute fold	175.53	10.65	168.67	10.61	170.71	12.65	2.92	171.54	11.59
Subscapular fold									
absolute fold (0.1 mm)	79.72	23.78	75.23	29.35	81.87	32.64	–	78.52	28.37
100 × log. absolute fold	188.64	10.98	185.35	13.21	188.78	14.12	0.68	187.37	12.87
Suprailiac fold									
absolute fold (0.1 mm)	69.89	23.34	60.97	16.81	59.30	17.77	–	63.53	19.77
100 × log. absolute fold	182.18	14.37	177.03	11.16	175.58	12.08	1.99	178.38	12.87

Note: Here, for $P = 0.05$, the limit of significance is F = 3.11.
Key: m = mean. s.d. = standard deviation.

estimated body density is close to that of subjects of twenty years, or of sportsmen (Le Bideau [1959] finds 1.080 for young trained French sportsmen).

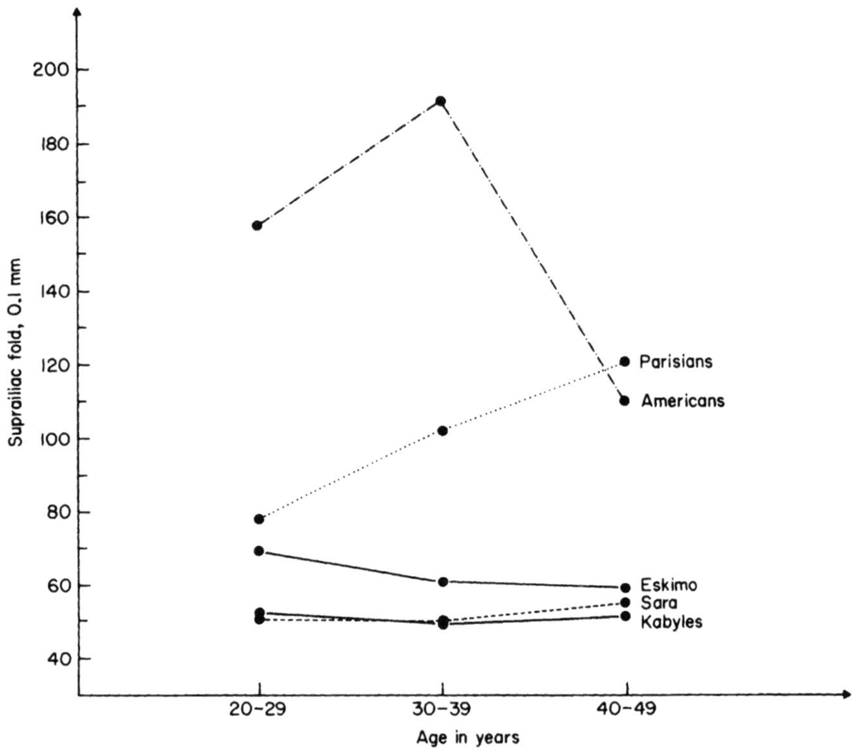

Figure 2. Comparison of suprailiac fold thickness in Ammassalimiut Eskimo and other male populations

Discussion

Without entering into a lengthy discussion on this subject (Ducros 1971), we note that the thickness of subcutaneous fat depends on factors of nutrition and also of exercise (which could explain, to a certain degree, the stability which continues with age, shown in the samples from subjects who are physically active during their entire life), but it also depends upon genetic factors (Osborne and De George 1959). The effective protection against cold which is assured by the adipose covering has been mentioned by many authors and demonstrated in several experiments. One could thus expect that, by natural selection, arctic populations would have a greater thickness of cutaneous fat than populations living in other

Table 2. Body density and adiposity in Ammassalimiut Eskimo and other groups aged twenty to twenty-nine

Number in sample	Eskimo 29		Japanese 96		Americans 88		Belgians 50	
	m	s.d.	m	s.d.	m	s.d.	m	s.d.
Age	24.10	2.57	22.10	–	22.00	–	19 yrs. 11 m.	–
Height (cm)	164.15	6.37	167.20	5.80	172.40	13.40	174.00	
Weight (kg)	66.39	6.14	58.90	7.70	68.30	11.10	67.87	7.30
Relation weight/height	40.44	–	35.20	–	39.60	–	39.00	–
Skinfolds (0.1 mm)								
Tricipital	58.70	15.20	80.00	44.10	121.00	46.00	–	–
Subscapular	79.70	23.70	109.00	56.40	122.00	50.00	–	–
Suprailiac	69.90	23.30	87.00	57.40	164.00	88.00	–	–
Density	1.0752	0.0040	1.0694	0.0147	1.0678	0.0120	1.0752	0.004
Lean mass	60.13	5.43	52.13	5.81	60.01	–	61.46	6.41

Source: Eskimo: Ducros (this study).
Japanese: Nagamine and Suzuki (1964).
Americans: Pascale et al. (1956).
Belgians: Crenier (1966).

Key: m = mean. s.d. = standard deviation.

climates. Rodahl (1957) and Laughlin (1966), however, have already emphasized that the Eskimo are rather thin.

Protection against cold can be obtained, on the one hand, through cultural techniques and, on the other hand, through various physiological processes: the heightening of the basal metabolism, in relation to the richness in protein of the diet (Rodahl 1952) or of the vasomotor reactions, such as an increase in blood flow to the extremities. Of course, all the Ammassalimiut do not have identical occupation and ways of life (some are salaried workers, others are hunters). These differences could go hand in hand with the differences in thickness of body fat that research now in progress could reveal.

With regard to adaptation to cold, it seems that the panniculus adiposus does not play a sufficiently essential role, or is not sufficiently independent of nutrition, to give credence to the belief that selection operates so that arctic populations possess a greater quantity of body fat than do human groups living in other climates.

SKIN REFLECTOMETRY

The skin reflectance was measured in 102 male Ammassalimiut at the following wavelengths: 425, 545, and 685 nm (using filters 601, 605, and 609 of the Evans Electric Ltd. spectroreflectometer).

Results

1. The principal characteristics of distributions are shown in Table 3. For our group, the curves of the reflectance at the arm do not deviate significantly from the norm for filters 601 and 605. For filter 609, the distribution shows an excess of values lower than the mean. The curves of reflectance at the forehead deviate from the norm, especially with regard to skewness. The differences between the values observed at the forehead and those observed at the arm are explained by the facial sunburn, which was very marked in several subjects, especially hunters.

2. The correlations between the different values of reflectance at the arm were calculated and compared to those obtained in a European sample (Leguèbe 1961):

	r: 102 Eskimos	r: 145 Belgians	t (not significant)
Arm 601–605	0.87	0.82	1.34
601–609	0.69	0.76	1.13
605–609	0.76	–	–

These correlations are high and close in the two samples compared. The regularity of distributions and the form of the scatter diagrams do not

Table 3. Reflectance in 102 male Ammassalimiut: tests of normality, skewness, and kurtosis

	Filter 601		605		609	
	Forehead	Arm	Forehead	Arm	Forehead	Arm
Mean	11.22	22.41	16.11	29.91	43.56	55.65
Standard deviation	3.85	4.23	4.49	3.88	5.82	4.42
Chi square	15.77	3.90	13.37	4.95	22.83[a]	12.65
Degrees of freedom	9	9	9	8	13	10
Skewness						
g_1	0.923	0.395	0.886	−0.340	0.596	−0.674
sg_1	0.239	0.239	0.239	0.239	0.239	0.239
t	3.86[b]	1.65	3.70[b]	1.42	2.49[a]	2.82[b]
Kurtosis						
g_2	1.055	0.198	0.085	−0.158	−0.026	−0.012
sg_2	0.474	0.474	0.474	0.474	0.474	0.474
t	2.22[a]	0.41	0.18	0.33	0.05	0.02

[a] $P<0.05$.
[b] $P<0.01$.

suggest the existence of a curvilinear relation (see also Huizinga 1965), as was the case in the interpopulational comparisons (Weiner et al. 1964).

3. Compared to the reflectance of other populations, that of the Ammassalimiut Eskimo is close to that which Hulse (1967) observed in a sample from the Japanese middle class; but the variance is significantly higher for the 609 filter:

	605		609	
	m	s.d.	m	s.d.
102 Eskimo	29.91	3.88	55.65	4.42
140 Japanese	27.36	4.48	52.32	3.31
F	1.33	(not significant)	1.78	(P = 0.05)

TOTAL RIDGE COUNT AND FINGER PATTERNS AT SCORESBYSUND

In 1971, one of us collected the prints (500 fingers) of a sample of the last generation born in the little colony of Scoresbysund. Fifty children of both sexes from seven to thirteen years of age were considered. There is no concentration of related persons in the group, although, because of the origin of the population, certain subjects have common ancestors (Robert 1971).

It is in this colony that Abel (1934) made the first ridge count of an Eskimo population. Because of the family composition of his sample, his results have not always been taken into consideration (Cummins and Midlo 1943). This sample brought together the founders of the colony: brothers and sisters with their immediate ascendants and descendants.

Results

The distribution of patterns according to hands, fingers, and sex, in Table 4, conforms to the usual distribution; the feminine sample, however, presents a greater frequency of whorls than the masculine sample, a fact which has already been pointed out for other Eskimo and Asiatic groups but which could show the accidents of sampling, rather than a particularity of sexual dimorphism.

The various results published on the Eskimo are not suitable for statistical exploitation (Ducros and Ducros 1972). Our sample presents fewer whorls than that of Abel (whose index of pattern intensity of men and women together is 17.1), and the nature of the patterns seems closer to an Angmagssalik sample examined in 1936 (Gessain 1959); see Figure 3 for a comparison.

Table 4. Detailed sequence of finger patterns and detailed ridge count for twenty-five male and twenty-five female Ammassalimiut children

Finger patterns (percent)	Boys				Girls				Ridge count	Boys		Girls	
	Arch	Ulnar loop	Radial loop	Whorl	Arch	Ulnar loop	Radial loop	Whorl		m	s.d.	m	s.d.
Right hand									Right hand				
I	0	12.0	0	88.0	0	20.0	0	80.0	I	24.32	4.38	20.95	5.47
II	4.0	32.0	8.0	56.0	12.0	24.0	4.0	60.0	II	16.56	6.66	15.72	7.38
III	4.0	48.0	0	48.0	0	56.0	4.0	40.0	III	17.20	5.24	17.60	6.15
IV	0	16.0	8.0	76.0	0	20.0	0	80.0	IV	20.60	5.36	19.72	5.72
V	4.0	64.0	0	32.0	4.0	40.0	0	56.0	V	14.92	6.04	15.80	5.06
Total right	2.4	34.4	3.2	60.0	3.2	32.0	1.6	63.2	Total right	93.60	21.06	89.79	24.56
Left hand									Left hand				
I	0	28.0	0	72.0	0	20.0	0	80.0	I	21.80	4.09	17.92	5.19
II	0	48.0	16.0	36.0	16.0	20.0	4.0	60.0	II	15.23	4.74	15.16	8.21
III	4.0	52.0	0	44.0	4.0	44.0	0	52.0	III	17.04	5.32	16.84	6.44
IV	0	28.0	4.0	68.0	4.0	24.0	0	72.0	IV	20.52	4.72	17.60	6.93
V	4.0	80.0	0	16.0	0	52.0	0	48.0	V	14.28	5.43	15.12	4.66
Total left	1.6	47.2	4.0	47.2	4.8	32.0	0.8	62.4	Total left	88.87	18.61	82.64	27.57
Total both hands	2.0	40.8	3.6	53.6	4.0	32.0	1.2	62.8	Total ridge count	182.47	38.80	172.43	51.28
Pattern intensity index	m = 15.16 s.d. = 3.41				m = 15.88 s.d. = 4.05								

Key: m = mean. s.d. = standard deviation.

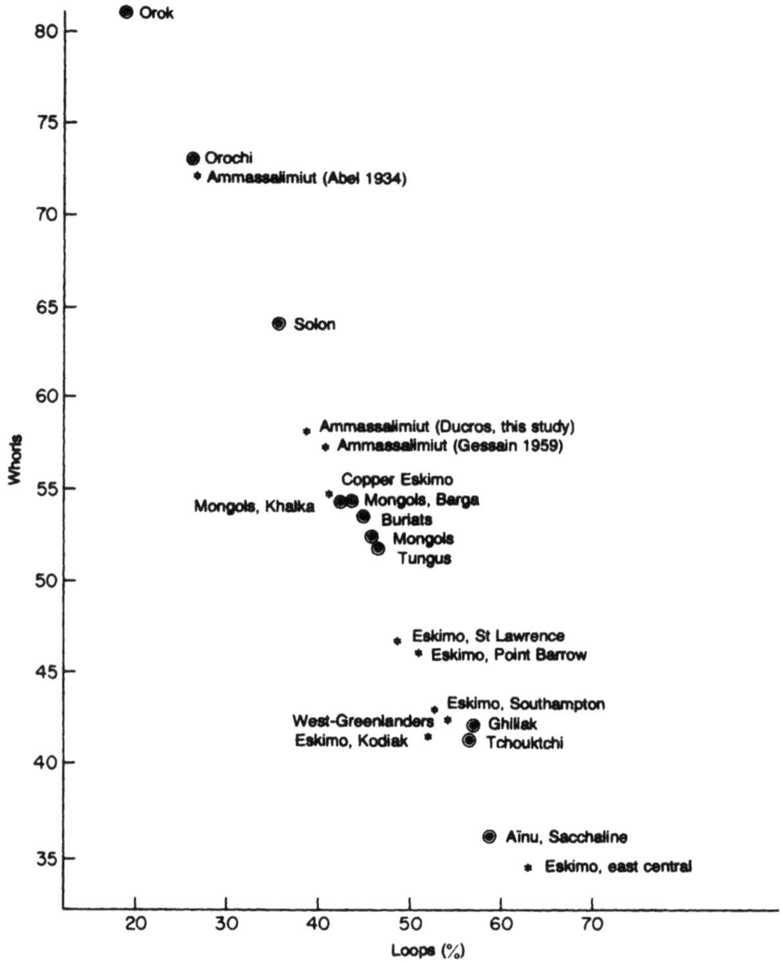

Figure 3. Frequency of whorls and loops among Eskimo and north Asiatic populations (both sexes): * = Eskimo or hybrids; ● = north Asiatic

Compared to other populations, this sample has a higher frequency of whorls than all other Eskimo samples, but still higher frequencies have been found among small samples of north Asiatic populations: Orok, Orochi, and Solons (Ducros and Ducros 1972).

The ridge count of our masculine and feminine sample is significantly smaller than that of the population examined by Abel, which we have calculated according to the author's published individual data:

Ammassalimiut (Ducros)			Ammassalimiut (Abel)				
	m	s.d.		m	s.d.	*t*	*P*
Men (25)	182.5	38.8	Men (29)	209.5	37.2	2.59	0.02
Women (25)	172.4	51.3	Women (30)	205.5	39.3	2.63	0.02

Except for that of Abel, the ridge count in our Eskimo sample is the highest ever observed in the entire world (Ducros and Ducros 1972).

Discussion and Conclusions

Given the conditions of the colony formation, the higher values of whorls and the higher ridge counts which we have observed could be considered as an example of the "founder effect"; this sample, however, does not seem fundamentally different, in the nature of its patterns, from the mother population of Angmagssalik examined in 1936 (Gessain 1959), whose total ridge count must also be higher. It could reasonably be concluded that the Eskimo from the eastern coast of Greenland occupy, with regard to dermatoglyphics, a particular place within the Eskimo population. They are closest to the known north Mongoloid populations. Of course, in small groups whose sizes fluctuate, one can always attribute their peculiarities to known genetic phenomena such as genetic drift, especially for characteristics which would have few determining factors. It is plausible, however, that this group, isolated for a long time on the east coast of Greenland and only recently discovered, possesses a former characteristic of arctic populations which was modified, essentially by admixture, in those Canadian and American Eskimo groups which have for some time been more open to outside influences.

REFERENCES

ABBIE, A. A.
1967 Skinfold thickness in Australian aborigines. *Archaeology and Physical Anthropology in Oceania* 2(3): 207–219.

ABEL, W.
1934 "Finger- und Handlinienmuster," in *Wissenschaftliche Ergebnisse der Deutschen Grönland-Expedition Alfred Wegener 1929 und 1930/1931*, volume six, *Anthropologie und Zoologie*. Edited by K. Wegener, 43–65. Leipzig: Brockhaus.

BOURLIÈRE, F., S. PAROT
1962 Le vieillissement de deux populations blanches vivant dans des conditions écologiques très différentes. Étude comparative. *Revue Française d'Études Cliniques et Biologiques* 7: 629–635.

BROZEK, J., W. KINZEY
1960 Age changes in skinfold compressibility. *Journal of Gerontology* 15: 45–51.

CRENIER, E.
1966 La prédiction du poids corporel "normal." *Biométrie Humaine* 1: 10–24.

CROGNIER, E.
1969 Données biométriques sur l'état de nutrition d'une population africaine tropicale: Sara de Tchad. *Biométrie Humaine* 4:37–55.
CUMMINS, H., C. MIDLO
1943 *Finger prints, palms and soles: an introduction to dermatoglyphics.* Philadelphia: Blakiston.
DAMON, A., R. F. GOLDMAN
1964 Predicting fat from body measurements: densitometric validation of ten anthropometric equations. *Human Biology* 36:32–44.
DUCROS, A.
1971 Adiposité et densité corporelle d'une population arctique (Eskimo Ammassalimiut). *L'Anthropologie* 75:605–620.
DUCROS, J., A. DUCROS
1972 Nombre de crêtes et dessins des doigts d'Ammassalimiut (Scoresby-Sund, Groënland oriental) et des populations eskimo et asiatiques. *L'Anthropologie* (Paris) 76:711–726.
GESSAIN, R.
1959 Dermatoglyphes digitaux et palmaires des Eskimo d'Angmassalik. *Bulletins et Mémoires de la Société d'Anthropologie de Paris* 10: 233–250.
1960 Contribution à l'anthropologie des Eskimo d'Angmassalik. *Meddelelser om Gronland* 161:1–167.
HUIZINGA, J.
1965 Reflectometry of the skin in Dogons. *Proceedings Koninklijk Nederlands Akademie van Wetenschappen* 68:289–296.
HULSE, F.
1967 Selection for skin colour among Japanese. *American Journal of Physical Anthropology* 27:143–156.
LAUGHLIN, W. S.
1966 "Genetical and anthropological characteristics of Arctic populations," in *The biology of human adaptability*. Edited by P. T. Baker and J. S. Weiner, 468–495. Oxford: Clarendon.
LE BIDEAU, G.
1959 La méthode anthropométrique d'évaluation de la graisse corporelle. Mesure du pli cutané. *Bulletins et Mémoires de la Société d'Anthropologie de Paris* 10:302–325.
LEGUÈBE, A.
1961 Contribution à l'étude de la pigmentation chez l'homme. *Bulletin de l'Institut Royal des Sciences Naturelles de Belgique* 37:1–29.
NAGAMINE, S., S. SUZUKI
1964 Body composition of Japanese. *Human Biology* 36:8–15.
OSBORNE, R. H., F. V. DE GEORGE
1959 *Genetic basis of morphological variations.* Cambridge, Mass.: Harvard University Press.
PASCALE, L. R., M. I. GROSSMAN, H. S. SLOANE, T. FRANKEL
1956 Correlation between thickness of skinfolds and body density in 88 soldiers. *Human Biology* 28:165–176.
ROBERT, J.
1971 Les Ammassalimiut émigrés au Scoresbysund. Étude demographique et socio-économique de leur adaptation (Côte orientale du Groenland, 1968). *Cahiers du C.R.A. n° 11–12, Bulletins et Mémoires de la Société d'Anthropologie de Paris* 8:5–135.

RODAHL, K.
1952 Basal metabolism of the Eskimo. *Journal of Nutrition* 48:359–368.
1957 Human acclimatization to cold. *Transcript of the Fifth Conference on Cold Injury*. Josiah Macy Foundation.

SLOME, C., B. GAMPEL, J. H. ABRAMSON, N. SCOTCH
1960 Weight, height and skinfold thickness of Zulu adults in Durban. *South African Medical Journal* 34/24:505–509.

WEINER, J. S., G. HARRISON, R. SINGER, R. HARRIS, W. JOPP
1964 Skin colour in southern Africa. *Human Biology* 36:294–307.

Natural Selection and Physical Adaptation to a Biotype of Tropical Africa

EMILE CROGNIER

An anthropological study of the Sara populations of the southwestern part of the Republic of Chad revealed that in this region of Africa the natural population and ethnic unity should coincide (Hiernaux 1969). A detailed investigation in one of these settlements, Sara Madjingay (8° 50′ N. and 17° 40′ E.) resulted in the collection of anthropometric and demographic data on the populations of four villages (Crognier 1973, 1977a, 1977b).

The extensive research which has focused on human morphological adaptation to climatic variables shows that the most crucial requirement for existence in hot climates is the ability to eliminate by improved thermolysis the excess calories produced by the metabolic processes (Weiner 1964). Human morphological adaptation seems to follow an ecological pattern, outlined by Bergmann (1847) and Allen (1877): in hot climates the surface area of the body is increased in proportion to the volume (Schreider 1963).

The morphology of the Sara Madjingay, like that of other groups of tropical Africans, evidences extensive adaptation (see Table 1). The Sara Madjingay are typified by an elongated skeletal frame. Out of a sample of 312 African groups, they rank tenth in stature, fourth out of fifty-eight groups in the length of the lower extremities, and first out of eighty-eight for the length of the trunk.

Our purpose here is to assess the degree to which the Sara have succeeded in adapting their morphology to their environment. In order to do this we have used as a guide the relative position of the population within its total possible adaptive range on the assumption that natural selection has favored individuals in a community which for a long time has remained in a particular ecological setting.

Most anthropometric characteristics are polygenic. Therefore, they are

Table 1. Main anthropometric characteristics of the young adult population (between 20 and 45 years old) in the Sara Madjingay ethnic group

	Males			Females		
	n	$\bar{x}$	s.d.	*n*	$\bar{x}$	s.d.
Stature	346	1739	68.3	405	1640	66.8
Iliocristal height	346	999	52.4	403	939	52.3
Tibial height	283	500	28.9	318	466	31.2
Arm length	284	804	38.7	319	735	40.3
Humeral length	285	330	19.1	319	304	19.4
Forearm length	285	278	17.8	319	254	16.6
Sitting-height	348	890	30.4	406	846	31.9
Biacromial breadth	348	384	18.2	406	344	17.2
Chest breadth	285	270	13.5	319	245	15.0
Chest depth	285	193	12.4	319	176	11.3
Biiliac diameter	348	264	13.1	406	258	14.3
Weight (kilograms)	348	66.1	7.3	403	58.0	7.0
Head length	348	190	6.6	406	183	6.2
Head breadth	348	154	5.5	406	147	5.2
Bizygomatic breadth	348	144	5.5	406	134	4.9
Total face height	348	121	6.9	405	111	6.8
Nose height	348	51.7	3.8	406	48	3.8
Nose breadth	285	43.3	3.1	319	39	2.7

influenced by natural selection even when the average magnitude of a trait has reached a point of adaptive equilibrium. Selection has in fact two functions: (1) the evolution of the average of a trait by penalizing those individuals farthest from the optimal state (directional selection); and (2) the reduction of the range of phenotypic variability when the average of a trait is close to the equilibrium point (stabilizing selection). Both types of selection usually occur simultaneously.

Karn and Penrose (1952) studied birth weights of newborn infants in London and found that the death-rate during the first four weeks of life increases as the weight deviates from the average value. Highest survival rates occurred among infants whose weight was slightly above average. This indicates that two factors of selection are at work, resulting in an essentially stabilizing function with a slight directional component.

Unfortunately some difficulties arose when we used this method to study the Sara Madjingay:

1. Selective patterns are usually subtle and can only be perceived if the group is observed over an extended period of time, or if one is studying a very large population. Due to time limitations, our survey was limited to several dozen family groups.

2. Our analysis cannot be based directly on the subjects we examined. They were adults in the reproductive period of life and the selection operating on them would have little effect at the genetic level. A survey of the adult differential survival rate could at most indicate what phenotypes

are most influenced by selection, but this would require a longer period of research and has not been done yet. It is also impossible to evaluate the size of phratries due to a lack of sufficient genealogical data.

We are therefore forced to study the effect of selection by focusing on parental couples. Here we have information concerning fecundity as well as offspring survival rates, although the small size of the sample reduces the precision of these two parameters.

TRAITS OF PARENTAL COUPLES

Number of Informants

Our survey was based on 450 couples, but we excluded endogamous couples, second marriages, and couples in which the age of the woman was less than twenty because data concerning fecundity are too variable in the fifteen to nineteen age group.

The final investigation was limited to 240 couples.

Appraisal of Morphology

Each couple was assigned a morphological rating computed in proportion to the total sample. Only one method seemed applicable in this case, although it tended to reduce the amount of information we were able to obtain from our data: each individual was compared to the other subjects of his sex. A distribution scale extending from one to six was set up to cover the overall physical range. Since a minimal number of subjects deviated by more than two units from the norm, we put those who deviated more than one unit from the norm into the same class. The results we obtained for stature are shown in Table 2.

Table 2. Distribution scale for stature

		1669 mm.		1739 mm.		1804 mm.	
Males	(1)	$-1\,s$	(2)	$\bar{x}$	(3)	$+1\,s$	(4)
		1570 mm.		1639 mm.		1701 mm.	
Females	(1)	$-1\,s$	(2)	$\bar{x}$	(3)	$+1\,s$	(4)

There are sixteen class combinations of couples: M1–F1, M1–F2, M2–F2, M2–F3, M3–F3, M2–F1, M3–F2, M3–F4, M4–F4, M2–F4, M1–F3, M4–F1, M4–F3, M4–F2, M3–F1, M1–F4.

Reproduction Indices

A couple's fecundity rate cannot be strictly calculated until the woman has reached the end of the reproductive period of her life. Unfortunately the limited number of people in the small communities we studied did not allow us to comply with this rule. Moreover, because of the high death rate, a large proportion of individuals die before they have reached the post-reproductive stage.

For this reason we assigned to each woman, and therefore to each couple, a fecundity rating which is the ratio of the total number of children she has conceived to the average number of children conceived by other women in her category (married only once, excluding cases of intermarriage) and in her age class (an age class is defined as a five-year period). Since all the Sara girls get married at the same age (seventeen years), it is unnecessary to modify the fecundity index by any references to the length of married life.

The survival index for children is obtained for each couple by the following percentage:

$$\frac{\text{overall number of children conceived } \textit{minus} \text{ overall number of deceased children}}{\text{overall number of children conceived}}$$

The coefficient obtained is then compared to a similar coefficient computed for each female age-class yielding the survival index.

THE INFLUENCE OF SELECTIVE FORCES

Differential Fecundity

Our evaluation of the distribution of the fecundity index is based on the observation of three types of parental couples: (1) couples morphologically situated at the lower limit of the distributional scale, those of the 1–1 and 1–2 classes; (2) couples close to the average, classes 2–3; and (3) couples situated at the upper limit, those of the 3–4 and 4–4 classes. The symmetrical distribution of each of the traits observed was tested for both sexes by Snedecor's symmetry test (1966). Weight was the only feature that showed clear asymmetry; in this case we converted individual values into decimal logarithms, which yielded a Gaussian distribution.

We evaluated the relative fecundity of each category of couples by counting the couples whose fecundity index was superior or equal to unity, and those for whom it was lower.

Next, we determined for each group the probable chances of deviation from a neutral distribution of fecundity. This is defined as the point where the number of couples more prolific than their age/class average equals the number of those which are below average. We used the chi-square test for this purpose.

The results are shown in Table 3. They are based on values for morphological traits we believe indicate morphological adaptation to climatic variations. The probable agreement of the observed with the expected distribution is only mentioned when it is uncertain (≤20 percent).

It appears that few features clearly indicate significant variations in the fecundity rate. Note, however, that couples of average height have a slightly higher fecundity rate. This tendency toward stabilization is reflected in measurements of sitting height, and is clearly indicated by the total arm length ($P<0.02$), and the total leg length ($P<0.001$) (not shown), and the intermembral length ($P<0.001$). The same stabilizing selection is evident for bi-acromial breadth ($P<0.05$) and bi-iliac breadth ($P<0.05$).

The overall impression is therefore that in this population, selection is operating in a manner which will stabilize morphological traits near existing values. However, since there are differences in the fecundity rate of couples situated farthest from the average, it is possible that physical evolution is not yet finished. Thus, a clear decline in fecundity has been observed in couples with overdeveloped arms. While this tendency does not appear with respect to the length of the leg, it reappears when the total length of arms and legs is calculated. Here a selective disadvantage appears to exist for individuals with larger total values; however, the significance of this result remains unclear.

Although we noted that the selective effect tends to act in favor of average couples for some traits, this is not true for thoracic depth, where both higher and lower values appear to be favored. If we compare the combined fecundity rate for couples situated in the upper and the lower ranges, the probability of conformity is found to be about 2 percent with a χ^2 of 5.66.

Weight is one of the traits showing no clear tendency. This is not surprising since weight varies widely within a life-span; its value at the time of the examination is not necessarily permanent. Moreover, the heavier subjects are the socially privileged individuals who work less than the others and get better medical care.

We calculated the selective value of the proportional relationship:

$$\frac{\text{upper extremity length } \textit{plus} \text{ lower extremity length} \times 2}{\text{weight}}$$

Although this relationship was established by Schreider (1963) to compare African groups, it does not provide us with any significant information, possibly due to variability of the weight factor.

Table 3. Conformity of the distribution of fecundity to the theoretical neutral distribution: for couples situated at the lower limit of the distribution (S=short), for average couples (M=mean), and for couples at the upper limit (H=high)

		$F \geqslant 1$	$F<1$	χ^2	Probability of conformity (1 degree of freedom)
Stature	S	23	19	0.38	–
	M	28	17	2.69	$0.20<P<0.10$
	H	22	20	0.09	–
Sitting height	S	18	21	0.23	–
	M	37	22	3.81	$P \neq 0.05$
	H	15	18	0.27	–
Arm length	S	11	14	0.36	–
	M	34	17	5.67	$0.02<P<0.01$
	H	8	19	4.48	$0.05<P<0.02$
Forearm length	S	21	16	0.67	–
	M	25	24	0.02	–
	H	11	15	0.61	–
Leg length	S	16	17	0.03	–
	M	38	12	13.52	$P<0.001$
	H	22	15	1.32	–
Biacromial diameter	S	22	23	0.02	–
	M	39	22	4.74	$0.05<P<0.02$
	H	20	19	0.02	–
Thoracic breadth	S	19	16	0.26	–
	M	33	25	1.10	–
	H	16	19	0.26	–
Thoracic depth	S	13	24	3.27	$0.10<P<0.05$
	M	22	20	0.10	–
	H	18	10	2.39	$0.10<P<0.05$
Biiliac diameter	S	20	28	1.33	–
	M	39	22	4.74	$0.05<P<0.02$
	H	17	20	0.24	–
Weight log.	S	20	18	0.10	–
	M	29	22	0.96	–
	H	21	18	0.23	–
Nose height	S	26	19	1.09	–
	M	36	31	0.37	–
	H	18	16	0.12	–
Intermembral length	S	15	12	0.33	–
	M	39	12	14.29	$P<0.001$
	H	12	20	2.00	$0.20<P<0.10$
Length m./log. weight	S	10	12	0.18	–
	M	28	24	0.31	–
	H	14	13	0.04	–

Differential Survival

The method here is the same as the one used in the study of the fecundity rate with the fecundity index replaced by the survival index. The results are shown in Table 4. There are slightly fewer subjects than in the preceding survey, since sterile couples, whose survival index would have been zero, were not included.

Offspring survival yielded few significant relationships. A weak advantage is sometimes found for offspring of couples with average lower arm length, thoracic depth, biiliac breadth, and nasal length, and greater length of the extremities seems to constitute a negative factor.

The situation in couples situated at the upper and lower limits is analogous. Only two characteristics show notable variation: (1) the sum of total leg and total arm length reveals that couples with short limbs experience greater offspring survival than couples with long limbs; and (2) the ratio of the intermembral index to the weight index indicates greater viability for offspring of short parents.

In spite of a lack of precision in some of the demographic data, certain variables do appear to be related to physical characteristics. Tables 3 and 4 indicate that most anthropometric traits tend to stabilize in their present standard form. This may be interpreted as evidence that the Sara Madjingay ethnic group has achieved an effective morphological adaption to its environment.

Although the main thrust of the selection process is toward stabilization of physical features, certain variables appear subject to directional selection. This is seen mostly in length measurements (specifically arm length, intermembral length, and the latter's ratio to weight), and to a lesser extent, in transversal measurements (a high value for the thoracic depth corresponding to increased fecundity).

We suspect some selective advantage for couples with shorter limbs in an ethnic group whose main feature is linearity. A more extensive survey would be required in order to test the likelihood of such selection.

We divided the population into several age groups and found that for each sex, older individuals (forty-five years or more) have longer limbs than young adults. We noted that the most important demographic change between the two generations was a gradual decline of endogamy in village communities (Crognier 1973). The frequent consequence of this type of event is heterosis which becomes apparent in two ways: (1) modification of the phenotypes of the exogamous individual's offspring, and (2) increase in the fecundity of exogamous couples and in the viability of their offspring.

We may now look at the same population with another question in mind: can the phenomenon of heterosis account for the selective advantage of short individuals?

Table 4. Conformity of the distribution of the survival index to the theoretical neutral distribution: for couples situated at the lower limit of the distribution (S=short), for average couples (M=mean), and for couples at the upper limit (H=high)

		$S \geqslant 1$	$S<1$	χ^2	Probability of conformity (1 degree of freedom)
Stature	S	23	19	0.38	–
	M	22	23	0.02	–
	H	21	21	0	–
Sitting height	S	17	17	0	–
	M	22	31	1.53	–
	H	12	18	1.20	–
Arm length	S	15	10	1.00	–
	M	25	24	0.02	–
	H	10	15	1.00	–
Forearm length	S	17	17	0	–
	M	28	19	1.72	$0.20<P<0.10$
	H	11	13	0.17	–
Leg length	S	22	17	0.64	–
	M	21	27	0.75	–
	H	22	15	1.32	–
Biacromial length	S	24	15	2.08	$0.20<P<0.10$
	M	31	23	1.18	–
	H	15	20	0.71	–
Thoracic breadth	S	17	15	0.12	–
	M	27	21	0.75	–
	H	16	14	0.13	–
Thoracic depth	S	16	10	1.38	–
	M	21	13	1.88	$0.20<P<0.10$
	H	12	14	0.15	–
Biiliac diameter	S	17	16	0.02	–
	M	31	21	1.88	$0.20<P<0.10$
	H	14	19	0.76	–
Weight log.	S	18	20	0.10	–
	M	28	20	1.33	–
	H	22	13	2.31	$0.20<P<0.10$
Nose height	S	16	16	0	–
	M	34	22	2.57	$0.20<P<0.10$
	H	14	16	0.13	–
Intermembral length	S	16	10	1.38	–
	M	17	31	4.08	$0.05<P<0.02$
	H	11	19	2.13	$0.20<P<0.10$
Length m./weight log.	S	15	5	5.00	$0.05<P<0.02$
	M	19	27	1.39	–
	H	12	15	0.33	–

MORPHOLOGY AND EXOGAMY

The adults were classified into two categories: (1) those whose parents were both born in the same village, and (2) those whose parents were born in different villages. Again, related couples were excluded. In order to avoid error due to changes in anthropometric traits associated with age the sample was limited to young adults. The two categories were compared by computing their averages and the standard deviation for the major anthropometric traits, especially those concerning climatic adaptation. The *t* deviations from the norm were then estimated on the basis of Student's test.

Sara males exhibit morphological changes which may be ascribed to exogamy (Table 5), but which occur in traits which also play an important part in climatic adaptation. Individuals born of exogamous parents are shorter than those born of endogamous parents. This change is seen in decreases in the average values for height and for arm and leg length. When measured compositely, the probability that these traits are heterogeneous from one group to another is low. This is supported by the differences between measurements of the distal segments of the members.

One of the length characteristics, sitting height, is changing in the opposite direction. It appears that trunk height is significantly greater in offspring of exogamous parents. Thus, these individuals have a longer trunk and shorter members than the progeny of endogamous matings.

Exogamous marriage has had the same result among females (Table 6). The trend in females of exogamous parents also tends toward a shortening of linear dimensions and is marked by a diminution of lower arm length. Tibial height also is reduced but the difference is not significant. The length of the trunk is greater as is hip breadth. This change in the hips does not occur in males. Even weight increases perceptibly but the difference between the two groups is so small that it is difficult to attach any significance to it.

The measurements of the head are not modified by exogamy, in either sex, but the nose appears to decrease in height and increase in breadth. This change is especially noticeable among women where the nasal index becomes significantly higher.

It thus appears that a breakdown of endogamy is associated with physical changes. These are limited to traits which affect the thermolytic processes of the body. This is the case for the distal segments of the arms and legs. The essential role played by these portions of the extremities in calorie loss has already been shown (Weiner 1964). Trunk length is also important since it is the part of the body where calorie loss is most difficult.

Endogamy may have been associated with less biological vigor through

Table 5. Average measurements of the main characteristics of males according to the exogamous or endogamous status of their parents

	Endogamous			Exogamous			t	P homogeneity
	n	$\bar{x}$	s.d.	n	$\bar{x}$	s.d.		
Stature	176	1740	68.70	98	1731	68.67	1.00	0.32
Sitting height	176	884	31.71	98	894	30.48	2.51*	0.02
Arm length	175	806	37.17	97	800	38.82	1.16	0.25
Humeral length	176	331	18.65	98	329	20.13	0.85	0.40
Forearm length	176	280	19.15	98	275	15.45	2.34*	0.02
Iliocristal height	175	997	52.00	97	993	52.53	0.60	0.55
Tibial height	175	503	27.94	97	492	28.54	3.05*	$0.01<P<0.001$
Tibial length	176	414	23.94	97	409	25.39	1.74	0.08
Biacromial diameter	177	383	18.15	98	382	19.51	0.42	0.68
Thoracic depth	177	193	12.68	98	190	12.01	1.89*	0.06
Biiliac diameter	177	262	14.48	98	263	13.25	0.54	0.59
Weight	177	66.0	11.20	98	66.1	11.10	0.05	0.96
Head length	177	189.4	6.36	98	189.7	6.61	0.40	0.69
Head breadth	177	153.7	5.04	98	153.6	5.44	0.25	0.80
Cephalic index	177	80.9	3.05	98	80.6	2.94	0.74	0.46
Nose height	177	51.1	3.97	98	50.4	3.46	1.44	0.15
Nose width	177	42.6	3.09	98	43.1	2.82	1.35	0.18
Nasal index	177	83.6	9.19	98	85.3	8.18	1.62	0.11

Table 6. Average measurements of the main metric characteristics of females according to the exogamous or endogamous status of their parents

	Endogamous			Exogamous			t	P homo-geneity
	n	$\bar{x}$	s.d.	n	$\bar{x}$	s.d.		
Stature	189	1637	70.89	112	1641	63.32	0.46	0.65
Sitting height	190	842	35.38	115	850	26.70	2.07*	0.04
Arm length	190	737	40.31	114	734	38.20	0.82	0.41
Humeral length	190	305	19.42	115	303	18.57	0.67	0.51
Forearm length	190	256	16.52	115	252	15.52	2.18*	0.03
Iliocristal height	188	936	54.57	113	935	50.70	0.14	0.90
Tibial height	188	468	31.01	115	463	30.69	1.37	0.17
Tibial length	189	389	26.02	115	387	23.77	0.60	0.55
Biacromial diameter	190	345	19.12	115	343	16.05	0.64	0.53
Thoracic depth	190	175	11.74	115	176	10.34	0.31	0.76
Biiliac diameter	190	255	14.64	115	259	12.79	2.20*	0.03
Weight	188	56.9	11.40	115	58.6	11.10	1.28	0.20
Head length	190	181.9	6.19	115	182.1	6.02	0.22	0.83
Head breadth	190	147.4	5.03	115	146.7	5.35	0.12	0.90
Cephalic index	190	80.7	3.16	115	80.2	3.08	1.35	0.18
Nose height	190	47.9	3.60	115	47.1	4.00	1.97*	0.05
Nose breadth	190	38.7	2.85	115	39.0	2.67	1.17	0.25
Nasal index	190	80.7	7.98	115	83.1	8.70	2.32*	0.02

a reduction in the recombination of genes thereby allowing selective forces to exert continued pressure on morphological traits. The increment of exogamy on the contrary would have produced increased vigor, which would explain the results shown in Tables 5 and 6.

The investigation of variations in fecundity and survival among exogamous and endogamous couples also raised the possibility that exogamous individuals may have an advantage in fecundity. However, this advantage would be weak and there is a 70 percent chance that the differences we observed are mere fluctuations in the different samplings.

CONCLUSION

The trend toward stabilization of the main anthropometrical features of the Sara Madjingay around their current average values could be interpreted as a sign that through their initial genetical possibilities, they should have attained a morphological structure adaptive in their environment.

Some biological advantages seem to have been gained from increased interchanges of genes. There is a trend for metric features to change, producing a morphology which is less characteristic of the group. This perhaps indicates that the selective pressure exerted by the environment on anatomical traits has been reduced to a level insufficient to offset the genetic effect of the new "vigor" generated by exogamy.

REFERENCES

ALLEN, J. A.
1877 The influence of physical conditions in the genesis of species. *Radical Review* 1: 108–140.

BERGMANN, C.
1847 Uber die Verhältnisse der Wärmeökonomie des Thiere zu ihrer Grösse. *Göttinger Studien* 3:595–708.

CROGNIER, EMILE
1973 Adaptation morphologique d'une population africaine au biotope tropical: les Sara du Tchad [Morphological adaptation of an African group to the Tropical biotope: the Sara of Chad]. *Bulletins et Mémoires de la Société d'Anthropologie de Paris*: 1–150.
1977a Marriages, migrations and the biological population in a Sara tribe from Chad. *Journal of Human Evolution* 6:159–168.
1977b Assortative mating for physical features in an African population from Chad. *Journal of Human Evolution* 6:105–114.

HIERNAUX, JEAN
1969 Investigations anthropobiologiques au Moyen-Chari (République du Tchad). Préliminaire à des recherches multi-disciplinaires [Anthropological and biological research in Middle-Chari (Republic of Chad). An introduction to multidisciplinary studies]. *Homo* 20:1–11.

KARN, M. N., L. S. PENROSE
1952 Birth weight and gestation time in relation to maternal age, parity and infant survival. *Annals of Eugenics* 16:147–164.

SCHREIDER, EUGENE
1963 "Anthropologie, physiologie et variations climatiques" ["Anthropology, physiology and climatic variations"], in *Physiologie et psychologie en milieu aride* [Physiology and psychology in the arid zone]. Paris: UNESCO.

SNEDECOR, GEORGE
1966 *Statistical methods applied to experiments in agriculture and biology*, sixth edition. Ames: Iowa State University Press.

WEINER, JOHN
1964 "Human ecology," in *Human biology, an introduction to human evolution, variation and growth*. Edited by G. A. Harrison, J. S. Weiner, J. M. Tanner, N. A. Barnicot, 401–508. Oxford: Clarendon.

Territorial Morphofunctional Studies in the Soviet Union: the Problem of Adaptive Types

T. I. ALEKSEEVA

The last decade has witnessed the onset of more or less systematic studies of human adaptability. Several published collections are largely devoted to the problems of genetic and psychological adaptation under Arctic, equatorial, and high altitude conditions (Malhotra 1966; Baker and Weiner 1966; Cohen 1969; Agadzhanian and Mirrakhimov 1970). As far as the problem of biological adaptation is concerned, researchers mostly focus on the effects of cold and high altitude conditions on the human organism.

It has been noticed that the Arctic aboriginal groups (Eskimos, Alaskan Indians, Saams) have a higher basal metabolism and a greater resistance to cold than Europeans who find themselves in low temperature conditions. The high altitude groups, however, regardless of racial origin, display a reduced gas metabolism, a higher level of lactic acid in the blood, hypotonia, and increased red cell counts. Examples are the Pamir mountaineers, Indians, aboriginal Tibetans, and South American Indians.

The above-mentioned investigations deserve credit above all for having established that the organism's response to environmental influences is genetically determined.

Until some twelve years ago, Soviet anthropologists had all but ignored problems of biological adaptability. What little research had been done was far from systematic; for instance, Ginzburg's work on the Pamir Tajiks (1937) and Grigorova's work on the Nenetz (1933). The prerogative in the study of environmental effects on man belonged to physicians and medical geographers. Yet most of these scholars actually interpreted adaptability as a process of acclimatization, the adaptation of individuals to the natural-social environment, i.e., phenotypic variability. At the same time, the biological concept of adaptability implies a feature arising at the population level and fixed in the process of selection.

The generally accepted biological concept of adaptability as one of the forms of phylogenetic development (Severtsov 1949) seems to be applicable to man with one reservation: at all stages of evolution, man has invariably adapted to his environment along two channels — biological and social — with the latter always predominant. As I see it, adaptability as applied to man should imply a state of equilibrium of a given population with the environment, such as is characterized by a high survival rate, stable morphofunctional status, and optimal harmony of the cultural and economic elements with the geographic environment.

Soviet anthropology traces its first systematic program of adaptability research to the year 1961. Extensive work has since been done on the Saams, forest Nenetz, Chukchi, Eskimos, and Buryats, as well as on aged Russians of the Transbaikal area, Tajiks of the Fergana Valley and Pamir foothills, and Russians living in various geochemical zones of the east European plains. Researchers have concentrated on the demographic characteristics of populations, on the nature of morphological and functional variability and the quantitative characteristics of morphofunctional indices, social and geographic environmental conditions, and prerequisites for genetic and social adaptability.

Mankind is a panecumenic species; hence the enormous variability of the indices that characterize the human species.

Individual, intragroup, and intergroup variability of anthropological indices seems to be the principal prerequisite for adaptability and its most potent mechanism. Geographic variability of anthropological characteristics provides numerous examples illustrating their connection with the environment. Note for example, the concentration of kinky hair, dark skin, large values of the nasal index, and expansion of the mucous surface of the lips under equatorial conditions; the reduced body weight and increased evaporation surface area in hot climates; or the reduced pigmentation level in northern areas.

If one adds to the foregoing the known relationship between the degree of skeletal mineralization and size on the one hand, and the geochemical conditions on the other (Pavlovskiĭ 1970), the relationship between basal metabolism and the hemoglobin levels with landscape specificity (Roberts 1960; Shchekochikhina 1970), it will become clear that the known facts amply prove the dependence of anthropological characteristics of the environment.

It is also known that these characteristics are genetically determined as well. Apparently, the environmental and genetic aspects of an index do not conflict. And once a particular characteristic is no longer sensitive to environmental influences, in any population there may appear individuals for whom such influences are the most favorable. Such an advantage is subject to stabilization by selection, for its carriers have the best chances for the perpetuation of their strain.

Physiological characteristics are found to be the most responsive to environmental pressures. In spite of the fact that the intragroup coefficients of correlation between the morphological and functional indices are relatively small, proving the relative independence of the functional and somatic features of the organism (Alekseeva 1970; Volkov-Dubrovin 1969) at the intergroup level, these relationships fall into a definite pattern.

Thus, for instance, the negative relationship between the degree of skeletal mineralization and skeleton size, as well as the relationship of this characteristic to the geochemical situation, lead to the assumption that massive skeletal forms are evolved in zones with a reduced level of bone forming minerals and that gracile types are formed in zones with a more favorable geochemical status. Diminution of the somatic body components is taken to be associated with a lower level of gas metabolism and higher immunity properties. Thus, not only the functional status, but also the morphological status display adaptive shifts.

One is led to assume that as the globe was increasingly populated, the variability in human populations progressively grew and consequently so did the adaptability potential. Since different races and peoples inhabit different geographic areas, we can assume the existence of an adaptability differential, the so-called "adaptability reserve" (Alekseeva 1971). From the biological standpoint, the history of mankind as a species is a process of accumulation of adaptability potential.

Let us now consider the specific morphofunctional characteristics of the Soviet population comprising different ethnic groups, living in different climatic and geochemical conditions, and representing two races — Europoid and Mongoloid.

Along with the estimation of the absolute values of morphological and functional characteristics for individual populations, a method was employed whereby the morphofunctional characteristics of various groups could be more or less objectively evaluated and compared. Most physiological characteristics, particularly those of a metabolic nature, represent integral values of the metabolic processes occurring in the organism as a whole. Since the groups examined differ considerably in terms of weight, which can also be evaluated as an integral quantity, one is well advised to correlate functional characteristics with weight. Some morphological characteristics, too, have been correlated with body weight and length in order to equalize the body size differential which in our case is generally associated with the different racial origins of the groups. In addition, the genetically very homogeneous population of Russians from the village of Rozhdestvenka, Oboyansk district, Kursk region, was used as a kind of scale. In this reference group, the values of the principal morphological and physiological characteristics are assumed to be equal to 100, with the rest of the groups considered relative to it. The relative morphofunctional characteristics are given in Table 1.

The somewhat definitive geographic localization of morphofunctional characteristics suggests the existence of certain adaptive types typical of certain climatic zones.

Thus, for instance, there exists an *Arctic* type featuring an increased evaporation surface area, a more intensive gas metabolism, a higher respiratory function, relatively high blood serum-immune protein and cholesterol levels, and a higher than normal level of minerals in the organism. Such a set of features readily suggests a response to a humid and cold climate and oxygen deficiency. Characteristically, these features are inherent in different Arctic populations, whatever their racial origin may be. Table 1 presents the data on the forest Nenetz and the Saams.

A *foothills* type exhibits a reduced gas metabolism and respiratory function, a relative reduction in the fat-free body mass, proteins and cholesterol in the blood serum; but like the Arctic type, the foothills type features a higher level of immune proteins and intensified skeletal mineralization. This type is represented by the Pamir foothills' Tajiks.

A *continental* type is characterized by a marked reduction of the fat-free body mass through increased fat deposits, a sharp rise in the skelia index (ratio of torso to leg length) not found in any of the previous groups, a higher relative blood serum protein level, and a lower level of minerals in the skeleton. The higher skelia index (relative leg shortness) and the increased fat component in the body may be explained as adaptations to very low winter temperatures. The lack of intensified gas metabolism in this connection is a problem still to be solved.

As for the relatively reduced skeletal mineralization, whereby the continental type can be easily distinguished from all others, it is likely that here we are dealing with a manifestation of an intragroup law: skeletal mineralization and the quantity of the body's fat mass are negatively correlated. The Transbaikal Buryats are typical representatives of the continental type.

Each adaptive type is conditioned by a specific environment and hence can be encountered only in specific environmental conditions. Therefore, the greater the number of territories having specific environmental conditions covered in an investigation, the more reliable is the assumption that such types exist. Unfortunately, there are still very few factual data on the morphofunctional characteristics of the world population.

It is very unfortunate that foreign literature contains practically no data on groups examined by different systems, whereby their morphofunctional status could be characterized. What little material is available was obtained by integrating the results of several authors who worked with different samples of the same ethnic groups. Thus brief morphofunctional characteristics were obtained for the subequatorial and equatorial zones,

Table 1. Brief morphofunctional characteristics of the adaptive types (in percent of the control values)

Characteristic	Arctic	Continental	Foothills	Equatorial		
Typical groups	Saams; Nenetz	Buryats	Tajiks	Bushmen;	Bantu;	Indians
Body surface area	103.7	96.7	100.0	114.9	103.7	107.4
Body circumference to length	101.8	103.8	96.4	–	–	–
Skelia index (torso to leg length ratio)	100.5	108.5	103.4	–	–	–
Fat-free body mass	102.1	85.1	97.1	108.1	105.5	105.3
AB phase (gas metabolism to body weight ratio)	99.0	100.0	96.9	–	–	–
Circulation flow rate to body length ratio	95.4	93.2	95.5	–	–	–
Total proteins to body weight ratio	109.7	100.8	97.6	135.5	108.8	111.2
Albumins to body weight ratio	100.0	103.6	75.9	79.5	107.2	92.8
Gamma-globulins to body weight ratio	150.0	112.5	137.5	325.0	150.0	190.2
Cholesterol to body weight ratio	101.7	105.9	81.2	97.2	96.8	111.1
Skeleton mineralization to body weight ratio	101.7	75.0	108.3	–	–	–

respectively, Indians (Henrotta 1966; Gopalan and Ramnathan 1957), and Bantu and Bushmen (Tobias 1966; Wyndham 1966) (see Table 1). All these groups are characterized by a relative increase in the evaporation surface area, as well as increased total proteins and gamma-globulins of the blood serum. The Bushmen exhibit these characteristics particularly strongly; those of the Bantu are the least pronounced; the Indians occupy an intermediate position in this respect. In addition to the above characteristics, it should be noted that the equatorial and subequatorial groups have a lower level of basal metabolism (Roberts 1960).

I am of the opinion that an *equatorial* adaptive type, existing in several versions, can also be defined.

Quite possibly, as data is further accumulated, the number of adaptive types will increase. In all probability, adaptive types form a real category. The history of their evolution, however, is still obscure. We can only suggest that they might have evolved throughout the entire history of mankind, possibly both predating and paralleling the processes of race formation. That they could accompany, in some cases, the processes of ethnic history is also a possibility that must not be lightly discounted. The correlation of adaptive processes with race — and ethnogenesis — seems to be a highly intriguing subject for future anthropological investigations.

REFERENCES

AGADZHANIAN, N. A., M. M. MIRRAKHIMOV

1970 *Gory i rezistentnost' organizma* [Mountains and resistance of the organism]. Moscow: Nauka.

ALEKSEEVA, T. I.

1970 "Izmenchivost' osnovnykh komponentov tela cheloveka v zavisimosti ot urovnia lipidov i belkov v syvorotke krovi" ["Variability of the basic components of the human body dependent on the level of lipides and proteins in the blood serum"], in *Morfo-fiziologicheskie issledovaniia v antropologii*, 53–63. Moscow: MGU.

1971 Geograficheskaia izmenchivost' soderzhaniia kholesterina v syvorotke krovi cheloveka (k vliianiiu sredy i rasy) [Geographic variability of the cholesterol content of human blood serum (toward the influence of environment and race)]. *Voprosy antropologii* 38.

ALEXEEV, V. P.

1971 Chelovek: biologiia i sotŝiologicheskie problemy [Man: biology and sociological problems]. *Priroda* 8:37–53.

BAKER, P. T., J. S. WEINER, *editors*

1966 *The biology of human adaptability.* Oxford: Clarendon.

COHEN, I. A., *editor*

1969 *Man in adaptation: the biosocial background.* Chicago: Aldine.

GINZBURG, V. V.

1937 Gornye tadzhiki [Mountain Tajiks]. *Trudy Instituta antropologii, ètnografii i arkheologii* 16, anthropological series 2. Moscow and Leningrad.

GOPALAN, C., K. S. RAMNATHAN
1957 Studies on blood cholesterol levels of healthy men of different socioeconomic groups in India. *Indian Medical Research Memoirs* 45(4).

GRIGOROVA, O. P.
1933 Kartina krovi v sviazi s usloviiami zhizni i avitaminozom u russkogo i nenetŝkogo naseleniia Novoĭ Zemli [Blood picture in connection with living conditions and avitaminosis among the Russian and Nenetz population of Novaya Zemlya]. *Antropologicheskiĭ zhurnal* 1–2.

HENROTTA, I. G.
1966 "A biological study of Indian populations at different levels of economic growth," in *Human adaptability to environments and physical fitness.* Edited by M. S. Malhotra, 395–406. Madras: Defense Institute.

MALHOTRA, M. S., *editor*
1966 *Human adaptability to environments and physical fitness.* Proceedings of a symposium held at New Delhi, September, 1965. Madras: Defense Institute.

PAVLOVSKIĬ, O. M.
1970 Osobennosti dinamiki kostnoĭ plotnosti v polo-vozrastnom i ėtnoterritorial'nom aspektakh [Peculiarities of the dynamics of bone density in its sex-age and ethnoterritorial aspects]. *Morfo-fiziologicheskie issledovaniia v antropologii.* Moscow.

ROBERTS, D. F.
1960 "Basal metabolism, race and climate," in *Readings on race.* Edited by S. M. Garn. Springfield: Charles C. Thomas.

SEVERTSOV, A. N.
1949 Morfologicheskie zakonomernosti evoliutŝii [Morphological characteristics of evolution]. *Sochineniia* 5. Moscow, Leningrad.

SHCHEKOCHIKHINA, L. K.
1970 Materialy k izucheniiu polo-vozrastnykh i territorial'nykh variatŝii urovnia gemoglobina v krovi zdorovogo cheloveka [Materials toward the study of sex-age and territorial variations in the hemoglobin level of the blood of healthy man]. *Voprosy antropologii* 36.

TOBIAS, P. V.
1966 "The peoples of Africa south of the Sahara," in *The biology of human adaptability.* Edited by P. T. Baker and J. S. Weiner, 111–200. Oxford: Clarendon.

VOLKOV-DUBROVIN, V. P., N. S. SMIRNOVA
1969 Opyt izucheniia vzaimozavisimosti priznakov, otnosiashchikhsia k razlichnym sistemam chelovecheskogo organizma [An attempted study of the interdependence of traits relating to various systems of the human organism]. *Voprosy antropologii* 32.

WYNDHAM, C. H.
1966 "Southern African ethnic adaptation to temperature and exercise," in *The biology of human adaptability.* Edited by P. T. Baker and J. S. Weiner, 201–244. Oxford: Clarendon.

Adaptation of Morphology to Climate in Central Africa

JEAN HIERNAUX

A previous publication (Hiernaux 1956), which was based on the analysis of a matrix of Mahalanobis's D^2 distances between fifteen populations of Rwanda, Burundi, and eastern Zaire, has shown the influence of biotope on the pattern of anthropometric differentiation in an area located in Zaire, west of the Great Lakes rift. The western part of this area is covered with equatorial rain forest, whereas its eastern part, higher in altitude, is covered with savanna and grassland. Temperature, air moisture, and solar irradiation at ground level differentiate the climate of the two biotopes. Five populations inhabit the forest section of the area, and four populations its savanna section (see Figure 1). Historical data assign a relatively recent origin in the savanna east of the Great Lakes rift to all nine populations. As an average, the D^2 interpopulational distances are greater between biotopes than within them. An especially high original affinity of populations living in the same biotope does not seem to be the cause of this pattern. Compared with the savanna populations on a number of metrical attributes analyzed individually, the forest populations are nearer to the Mbuti Pygmies, longtime inhabitants of the same forest. They are farther from the Tutsi pastoralists, who settled a few centuries ago in Rwanda and Burundi, east of the Zairan area, among a majority of Hutu who are related to the populations of this area. The presence of the Mbuti in the forest and of the Tutsi at the eastern border of the savanna section suggest an explanation. The anthropometric differentiation between the biotopes may be explained either by the impact of a gene flow from the Tutsi on the savanna populations or the impact of a gene flow from the Mbuti on the forest populations, or both. However, the distribution of ABO allele frequencies, which greatly differ in Tutsi and Mbuti, contradicts such an explanation. In terms of the ABO system, the forest populations and the savanna populations are equidistant from

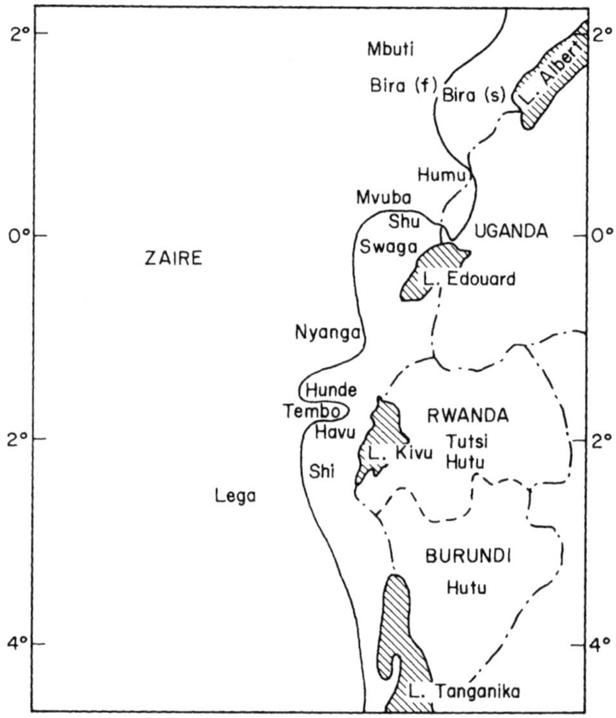

Figure 1. Geographical position of the sixteen populations (the heavy black line represents the forest–savanna boundary)

both the Tutsi and the Mbuti. The savanna populations diverge from the Tutsi on the basis of the MN system as well. Apparently, only the influence of the environment can explain the anthropometric differentiation between the two biotopes.

This paper extends the former analysis to three populations of Zaire whose territory broadens the previously circumscribed area: (1) the Humu, who live in the forested Semliki plain near the Uganda border and north of the Mvuba, and form the Zairan section of a bi-national ethnic entity which, in Uganda, is called Amba; (2) the Bira, who inhabit the Ituri forest in symbiosis with the Mbuti; and (3) the Bira, who live in the savanna east of the former. Tradition assigns to the Humu an earlier habitat located in the savanna of modern Uganda. A similar earlier habitat is assigned to the Bira: a subgroup of them would have remained in the savanna, while the others would have penetrated into the equatorial forest (Sporcq 1972).

The present analysis also considers the Tutsi of Rwanda and the Hutu of Rwanda and of Burundi, who were already included in the matrix of D^2 distances published in 1956. The analysis of this matrix had then suggested the existence of a gene flow from the Tutsi to the Hutu of Rwanda and, to a lesser degree, to the Hutu of Burundi; in Zaire this gene flow could be recognized in the Shi only, in whom it was still more modest.

Lastly, the present study includes the Mbuti Pygmies in the matrix of multidimensional distances, owing to their importance to the conclusions of the previous work. It was not possible, however, to include them in the matrix of D^2 distances, in the absence of the individual data collected by Gusinde (1948) on the only large sample of male Mbuti (510 adults) to have been measured. Gusinde published anthropometric means only, ruling out the possibility of computing the C_H^2 distance which requires knowledge of standard deviations. Consequently, a measure of distance, which requires the knowledge of means only, had to be used here. The one used is that of Womble (1951), modified into the Δg distance (Hiernaux 1965). For a set of populations and a set of variables which largely intersect the ones considered here, it is highly correlated with C_H^2 (r = 0.99) and with D^2 (r = 0.96) (Hiernaux 1965).

All samples considered here have been measured by myself except for the two Bira samples, measured by Sporcq, who was trained in the same techniques in my laboratory; and the Mbuti sample, measured by Gusinde, who used explicitly similar techniques. Interobserver variance can therefore account for only a very small part of the diversity which is studied here.

METHOD

The 1956 study analyzed the matrix of distances by logical means directly, without passing through a graphical representation of the pattern of distances which would constitute a "biometrical map" of the populations. The present study, however, is based on such a representation (Figure 2), extracted from a map bearing thirty-two populations of sub-Saharan Africa, which J. C. Gower kindly computed according to principal coordinates analysis (Gower 1966) from a matrix of Δg distances. These were based on a set of ten variables, which represent various aspects of morphology: stature, biacromial diameter, biiliocristal diameter, head length and breadth, bizygomatic diameter, total face height, nose height and width, and lip thickness. This map and the justification of the set of variables are published elsewhere (Hiernaux 1973).

To be sure, a map directly computed from the matrix of distances between the sixteen populations of Figure 2 would not be identical to the one extracted here from a map locating twice this number of populations. However, the qualitative nature of the question to which we seek an answer by surveying the map (the clearness of the influence of the biotope, the relative position of the populations in relation to Mbuti and

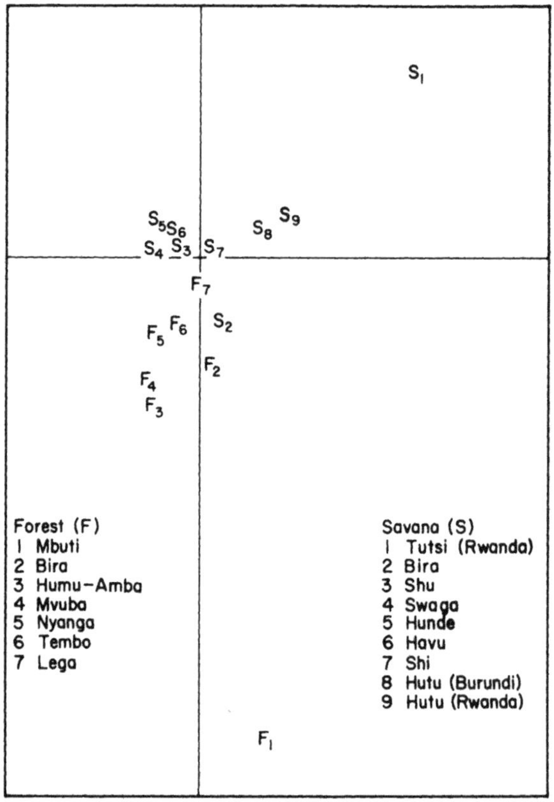

Figure 2. Biometric map of the sixteen populations produced by J. C. Gower from a matrix of Δg distances. The axes are the first two latent vectors of principal coordinates analysis

Tutsi, the trace of a gene flow from the Tutsi) allows for a slight loss in the accuracy of a representation. In part, the representation depends on the set of variables, and it is unavoidably distorted by being reduced from a ten-dimensional space of the computed distances to the two-dimensional space imposed by the figure.

RESULTS

Influence of the Biotope

In Figure 2, the set of Zairan savanna populations (S2 to S7) and the set of forest populations (F2 to F7) are nearly totally disjointed. Their intersection is restricted to the savanna Bira (S2) and the Lega (F7). And even the savanna Bira diverge from the forest Bira (F2) in the direction of the other savanna populations.

Position of the Twelve Zairan Populations in Relation to the Tutsi and the Mbuti

The set of six savanna populations is nearer to the Tutsi (S1) than the set of forest populations. The axis of their dispersion is approximately perpendicular to a line joining the Tutsi and the center of this dispersion. The six savanna populations do not differ much in distance to the Tutsi. Of them all, the Shi are the nearest to the Tutsi.

The set of six forest populations is nearer to the Mbuti Pygmies (F1) than the set of savanna populations. The Mbuti, who are the most distant from the Tutsi, are more remote from the nearest population than the latter. The other populations of the Zairan forest are not equidistant from the Mbuti: the Humu (F3), Mvuba (F4), and forest Bira (F2) are nearer to the Mbuti than the Nyanga (F5) and Tembo (F6) and still more so than the Lega (F7).

Position of the Hutu of Rwanda and Hutu of Burundi in Relation to the Tutsi

The Hutu of Rwanda (S9) and those of Burundi (S8) lie approximately on a line joining the Tutsi (S1) to the Shi (S7), much nearer to the latter.

DISCUSSION

Only one interpretation of the relative positions of the Hutu and the Tutsi forces itself upon the mind: a gene flow taking its source in the Tutsi and presenting decreasing values in the Hutu of Rwanda, the Hutu of Burundi, and the Shi successively (all these populations live in a similar biotope). The 1956 study already explained the pattern of D^2 distances between the four populations by such a gene flow. But this aspect of a much larger matrix of distances would probably have escaped analysis had not the data collected on the history of the populations and on the structure of their mate exchanges drawn attention to it. The quasi-linear S1–S9–S8–S7 disposition is striking when surveying Figure 2.

The proximity of the six Zairan savanna populations to a perpendicular to the Tutsi–Shi line indicates that, beyond the Shi, the gene flow from the Tutsi is nonexistent or, at least, has no differentiating effect. This was already the conclusion of one of the analyses of the matrix of D^2 distances. It is attained here through reading the biometric map and is strengthened by considering one more Zairan population, the savanna Bira.

The Tutsi–Zairan savanna/Zairan forest populations cline, which the analysis of the D^2 matrix had revealed, and the extension of this cline toward the Mbuti which univariate analysis indicated, show themselves very clearly on the biometric map. The addition of the Humu and the forest Bira increases the separation of the two ecological sets, whereas the addition of the savanna Bira decreases it.

No more than the D^2 matrix can the biometric map disclose the nature of this cline, either through a unipolar or bipolar gene flow, through the influence of the biotope on morphology, or through the synergy of both mechanisms. However, the argument put forth in 1956 for rejecting gene flow as a major factor of the observed cline is strengthened by the inclusion of the Bira in the present analysis. By their ABO allele frequencies, the forest Bira are more distant from the Mbuti and closer to the Tutsi than the savanna Bira (contrary to the morphological cline); the two Bira do not, however, differ significantly from each other (Sporcq 1972).

The present analysis therefore, more firmly than the 1956 one, leads one to look upon the influence of biotope on morphology as the main factor of the noted cline. This was observed at the level of the phenotype. Does it also concern the genotype? There are good reasons to believe it does. It shows itself not only from the multidimensional viewpoint that is considered here, but also from the viewpoint of the variables, whether the original measurements or transformations for making them independent are used (Hiernaux 1956). The probability that the forest biotope could modify, by directly influencing growth, such diverse characteristics as stature, face, and nose proportions, relative lower limb length, and lip thickness, in a way to make all of them converge toward the Pygmy mean appears to be very low. It is clear that the morphological distinction of the Pygmies is largely genetic. The Mbuti seem to represent an extreme genetic adaptation of human morphology to the climate of the equatorial forest, whereas the Tutsi seemingly show a marked genetic adaptation of their morphology to a hot and dry climate (Hiernaux 1968).

In this context, the partial convergence of the morphology of the Zairan populations, who migrated from the eastern savanna into the equatorial forest, toward the morphology of the Pygmies, who have inhabited the forest for many more generations, seems to result, in large part, from a process of pygmyization through genetic adaptation. The inequality of the distances from the other forest populations to the Mbuti suggests that this process has not gone as far in all populations. An inequality of gene flow from the Mbuti cannot, however, be excluded as a minor factor of the differentiation of the forest populations.

Methodologically, the present study emphasizes the interest of graphic representations of the pattern of multidimensional distances for analyzing the morphological diversity of human populations. To be sure, different maps will be produced for the same set of populations according to the distance statistics, the set of variables, and the method used for reducing the space to two dimensions. It is, however, striking that Jardine's map

(1971) based on an algorithm of Kruskal to a matrix of K-dissimilarity coefficients for thirty-two characters between the fifteen populations of my 1956 study, differs little from the one Gower (1966) drew up through principal coordinates analysis from D^2 distances for nine variables between the same populations (Hiernaux 1972). Jardine's map also is similar to the one that could be extracted for the same populations from the map showing thirty-two populations which Gower drew up from Δg distances for ten characters (Hiernaux 1973), of which Figure 2 is another extract.

SUMMARY

Analysis of the anthropometric diversity of the populations occupying an area which includes Rwanda, Burundi, and, in Zaire, a savanna section and a forest section was carried out in 1956 from a matrix of multidimensional distances. The present study is an extension of that work using a bidimensional map which J. C. Gower drew up from a matrix of distances which concerns three more populations.

The map shows a nearly complete separation of the set of six Zairan savanna populations from the set of six forest populations. The first set is nearer to the Rwanda Tutsi, whereas the second set is nearer to the Mbuti Pygmies of the equatorial forest of Zaire. When this pattern is compared to the distribution of ABO allele frequencies, genetic adaptation to climate appears to be the main factor of the morphological differentiation according to biotope.

Gene flow from the Tutsi, which the analysis of the matrix of distances had suggested, shows itself on the map by a quasi-linear disposition of the four concerned populations.

These results stress the interest of anthropometric maps for the analysis of human diversity.

REFERENCES

GOWER, J. C.

1966 Multivariate analysis and multidimensional geometry. *Statistician* 17:13–28.

GUSINDE, M.

1948 *Urwaldmenschen am Ituri*. Vienna: Springer.

HIERNAUX, J.

1956 *Analyse de la variation des caractères physiques humains en une région de l'Afrique centrale: Rwanda-Urundi et Kivu.* Tervuren: Musée Royal du Congo Belge.

1965 Une nouvelle mesure de distance anthropologique entre populations, utilisant simultanément des fréquence géniques, des pourcentages de

traits descriptifs et des moyennes métriques. *Comptes-Rendus de l'Académie des Sciences de Paris* 260:1748–1750.
1968 *La diversité humaine en Afrique subsaharienne.* Bruxelles: Institut de Sociologie de l'Université Libre de Bruxelles.
1972 "The analysis of multivariate biological distances between human populations: principles, and applications to sub-Saharan Africa," in *The assessment of population affinities in man.* Edited by J. S. Weiner and J. Huizinga, 96–114. Oxford: Clarendon.
1973 "Numerical taxonomy of man: an application to a set of 32 African populations," in *S. S. Sarkar memorial volume*. Edited by A. Basu, A. K. Ghosh, S. K. Biswas, and R. Ghosh, 151–161. Calcutta: Orient Longman.

JARDINE, N.
1971 Patterns of differentiation between human local populations. *Philosophical Transactions of the Royal Society of London* 263:1–33.

SPORCQ, J.
1972 Les Bira de la savane et les Bira de la forêt. *Bulletins et Mémoires de la Société d'Anthropologie de Paris* 9:97–120.

WOMBLE, W. H.
1951 Differential systematics. *Science* 114:315–322.

The Differential Geography of Races

V. P. ALEXEEV

The mapping of racial characteristics has long concerned physical anthropologists. As early as the turn of the century, reports by Ripley (1900) and Deniker (1900) included maps of the distribution of stature and cranial index in European populations. However, it was not until the work of Chepurkovskii (1913) that the mapping of properties was recognized as an effective method of racial analysis, rather than a mere illustration for works in physical anthropology.

Chepurkovskii's suggestion of the geographical method of racial analysis was an evaluation of geographical clines in relation to variations in individual trait gradations, their underlying dynamics, and occasional discontinuities. On this basis, local combinations of traits were identified, enabling the analysis of racial variation in historical perspective.

In the mapping of continuous variables, there has been little, if any, progress over the past several decades. In compiling anthropological maps various cartographic techniques are employed, such as the use of symbols (Lundman 1968) or a continuous background (Biasutti 1959). Whatever the method, the data analyzed represent diverse samples, obtained at various times without reference to either their representativeness or their distribution. This accounts for the lively debate still going on as to the intervals into which a series should be divided, whether or not a quadratic deflection may be usable, and whether it is worthwhile to use equal intervals or ones which decrease as distance from the median value increases, etc.

When mapping simple hereditary characteristics for which gene frequencies can easily be calculated, it has long been customary to use the method of isogenes, or, more precisely, isogenic lines. The most complete maps of this kind, dealing with the D locus in the Rh system, were published by Mourant (1954, 1958).

The isogenic method seems worthwhile whenever the samples are numerous and continuously distributed throughout the territory being mapped. However, in some cases, the method gives an impression of continuous distribution when, in actual fact, discontinuities exist; in other cases, it gives an impression of discontinuity where none exists. Few samples and uneven coverage of the mapped territory lead to maps which are, at best, schematic in character and, at worst, largely inaccurate. An example of this can be found on the map of *q* genes in Siberia (Rychkov 1965). In the instance of mapping the gene frequencies of individual characteristics, the samples vary in size and demographic structure. The selection of the sample is probably a more important factor in these studies than in ones of metrical variation.

Emphasis on the population approach to the study of racial variation, though widely favored, still lacks sufficient data to establish beyond a doubt its effectiveness as a methodological approach to the field of racial analysis. The problem, to a great extent, derives from the fact that the material available was collected without regard to populational principles and, not infrequently, in direct violation of them. The populational structure of mankind is barely emerging as an object of research and has greatly altered in many areas of the world. For this reason, the populational approach has had little impact on the study of racial geography, although its application promises to be singularly effective, particularly where the findings of research on a small sample can be extended to a larger group. The situation is much the same in paleoanthropology and craniology, where the fragmentary nature of available data makes it necessary to calculate summary means for populations of large territories or archeological cultures (Figure 1).

The first deduction to be made from the persistent application of the population concept of race is that any large race is the sum of populations, large and small, not, as typological thinking would imply, a sum of individuals. Therefore, maps derived from samples of nations or large ethnic groups give a poor picture of reality. When an administrative district is used as a unit of cartography, as with the distribution maps of somatological and craniological features in Ireland (Hooton and Dupertuis 1955), in the territory of the east European plain (*Proiskhozhdenie* 1965; Alexeev 1969), the Caucasus (Abdushelishvili 1964) and Japan (Levin 1971), the result is an approximation of the population approach, provided there is some knowledge of population structure.

Through employment of the population approach, it becomes clear that race is not so much a sum, as an aggregation, of population. This implies that population characteristics follow the distribution pattern affected by the operation of genetic mechanisms within each racial entity. The level of differentiation of traits, the degree of morphological expression, and the hierarchic relationship of populations, all are fundamental

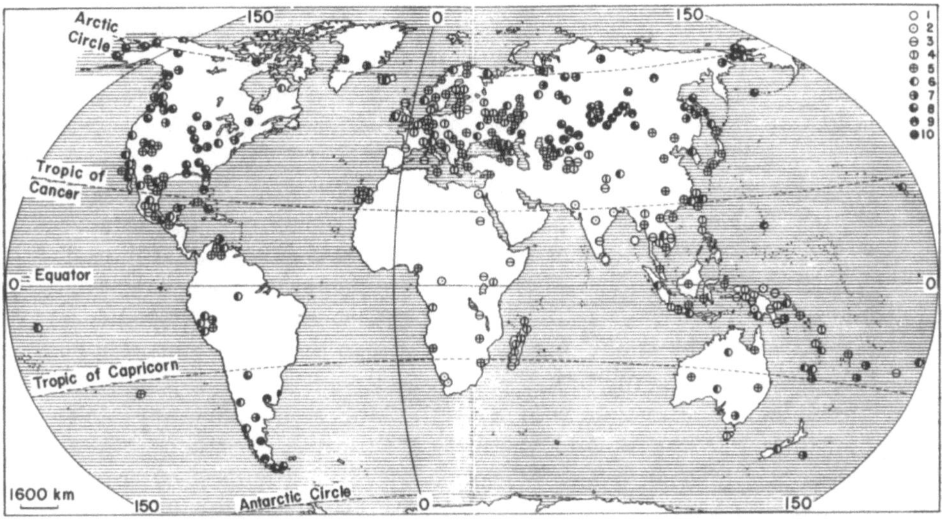

Figure 1. Geographical distribution of bizygomatic breadth (craniological data) *Key:* Male skulls (in millimeters): (1) 121.0–123.7; (2) 123.8–126.5; (3) 126.6–129.3; (4) 129.4–132.1; (5) 132.2–134.9; (6) 135.0–137.7; (7) 137.8–140.5; (8) 140.6–143.3; (9) 143.4–146.1; (10) 146.2–148.9

characteristics of racial entities, making it possible to perceive their structures. Without such information, mean values for certain characteristics are of little value.

By making certain assumptions it is possible to hypothesize the means by which race formation occurs, i.e., the differentiation level and degree of morphological specificity (Alexeev 1967). Typological variability occurs when reproductive isolation prevails between breeding populations and racial entities of larger or smaller size come into being. When this happens, intergroup correlations for many characteristics are much lower than intragroup correlations. Larger racial entities are, more often than not, clearly distinguishable from each other.

Another mechanism of race formation involves intensification of local variability. Under this mode, the only genetic barriers may be those generated by the inertia of overcoming distances and, consequently, the breeding populations overlap one another. Intensification of local variability does not usually have the effect of giving rise to well-defined superpopulational racial entities. Finally, racial variation can arise through continued directional changes in characteristics, leading to the emergence of a complex which is sharply contrasted to that of adjacent populations, in the absence of other raciogenetic processes. In terms of geographic distribution the latter mode appears much more limited than the former two. To illustrate our necessarily general statements, an example will be useful. The east European plain has been described as a zone characterized by local variability and the Caucasus as an area where typological variability occurs. Among thirteen Russian craniological series, collected during the late eighteenth and nineteenth centuries, biological distances between pairs were calculated using the Penrose formula as revised by Knussmann (1962).

Similar calculations were performed using thirteen Caucasian series which were roughly contemporary with the Russian series. The distances between Caucasian populations proved, on the average, to be substantially larger in size, shape, and summary expression. Thus, when typological variability occurs, it is possible to discern greater morphological specificity of populations than occurs through local variability. This is fully consistent with the initial assumptions.

From the above, one can infer that differential racial geography is but a populational geography and, therefore, will reflect variations of mean values for individual populations and aggregations thereof, rather than extensive contingents of racial or ethnic character. It is in this sense that it deserves to be termed differential geography, as it gives a differentiated rather than a generalized description of the physical anthropology of an area. Thus, a repeat anthropological survey of the globe's population, paralleling studies of population structure in the countries concerned, represents a priority objective of contemporary racial study. Such studies have been started in many countries, including the Soviet Union.

I shall next consider how the study of individual racial entities at the population level can advance our understanding of their geography. The Turkmen represent a large and homogenous racial group, whose linguistic individuality is associated with characteristic craniological traits. In racial classifications, this group has been variously called an "East Mediterranean," "Caspian," or "Transcaspian" race. However, all researchers in the field emphasize the morphological distinctiveness of the Turkmen. Information regarding the distribution of the ABO phenotypes among Turkmen, which made it possible to calculate gene frequencies for them appeared many years ago (Ginzberg 1937). But it is doubtful that these data were applicable to the Turkmen nation as a whole.

Until recently, the Turkmen fell into more than twenty clan-tribal groups. The memory of the clan-tribal identity is still alive to some extent today. These tribal subdivisions are not endogamous and do not constitute, therefore, populations in the precise sense of the word. Rather they are exogamous, being connected with one another through a system of cross-marriages. Yet the very fact of their existence suggests that they could have played, at one time or another, the role of populations, and that they are probably relics of the populational breakdown of a previous epoch; otherwise their origin and functional social significance will remain obscure. A postgraduate student of mine, O. B. Babakov, carried out an anthropological survey of the Turkmen nation at the level of clan-tribal divisions, which included collection of blood-group data. As an example, I supply the map of MN system gene distribution (Figure 2). Although on the whole the picture seems fairly stable, individual clan-tribal groups differ substantially from the peripheral groups. The ABO distribution follows the same pattern (Figure 3). Thus, it is safe to say that the extrapolation of summary data over the entire territory populated by the Turkmen results in a drastic schematization of the nation's actual racial composition.

A still greater schematization occurs when the extrapolation of summary data is attempted, using paleoanthropological material, especially since we are unaware yet of its potential. The concept of the "Cro-magnon" type, which includes the populations of the Neolithic and Bronze ages in the steppe regions of the Soviet Union has been broadly applied in Soviet paleoanthropological studies (Debets 1948). A universally large malar diameter of nearly 138–140 millimeters on the cranium (N45) has been regarded as a distinguishing characteristic of the type. However, if this trait is mapped not on the basis of summary means for cultures which may extend for hundreds of miles, but on the basis of means for graves, or at least from smaller areas, we can see considerable

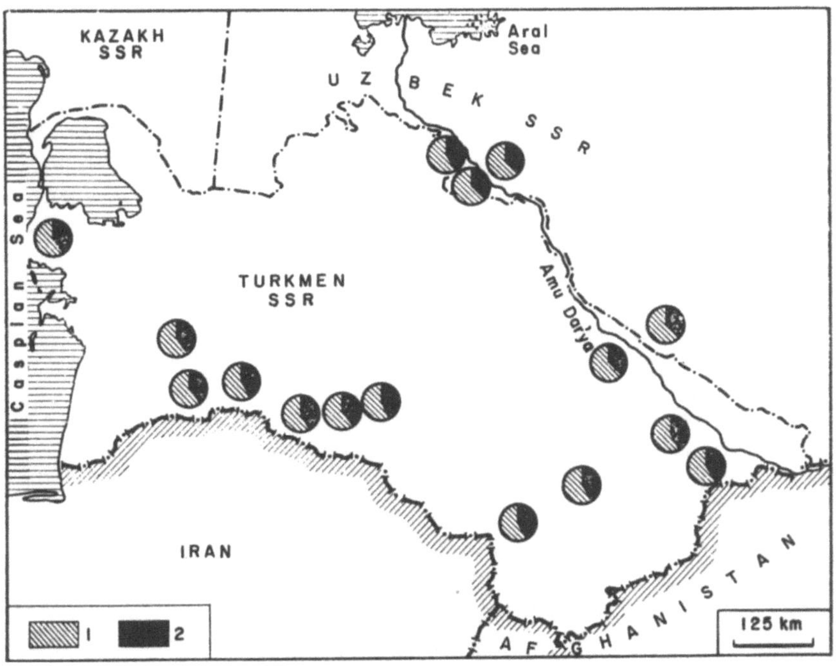

Figure 2. Geographical distribution of MN genes among Turkmen
Key: (1) concentration of m gene; (2) concentration of n gene

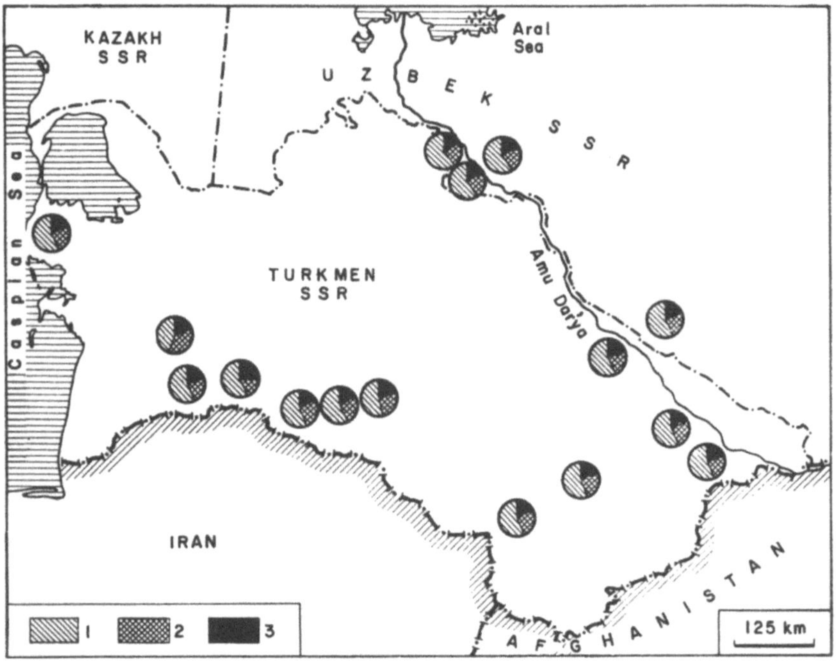

Figure 3. Geographical distribution of ABO genes among Turkmen
Key: (1) concentration of r gene; (2) concentration of p gene; (3) concentration of q gene

geographical distribution of the trait. This dispersion is augmented to some extent by gaps in the chronological sequence, which ranges from Neolithic to middle Bronze, but we are compelled to make do because of the scarcity of material. Together with very broad-faced groups within the Europoid population of the steppe regions in the Soviet Union, groups with a modest facial breadth were also fairly widely represented (Figure 4). Similar differences are observed for other properties. It is not altogether impossible that the "Cro-magnon" type, if approached from the populational standpoint, will give way to more differentiated estimates.

The differential or populational geography of races provides other new information as well. By following persistently the principles of differential geography, a substantial new element is introduced into the reconstructions of the genealogies of human races while many taxonomic delusions characteristic of the typological method are avoided. For example, the South-Siberian race is usually said to comprise the Kazakhs, Kirghizs, and the nationalities of the Altai-Sayan range. They have no integral habitation now, nor are they likely to have had any in ancient times, and they differ from one another by the ratio of the Europoid to Mongoloid components. Therefore, it is possible to identify within the

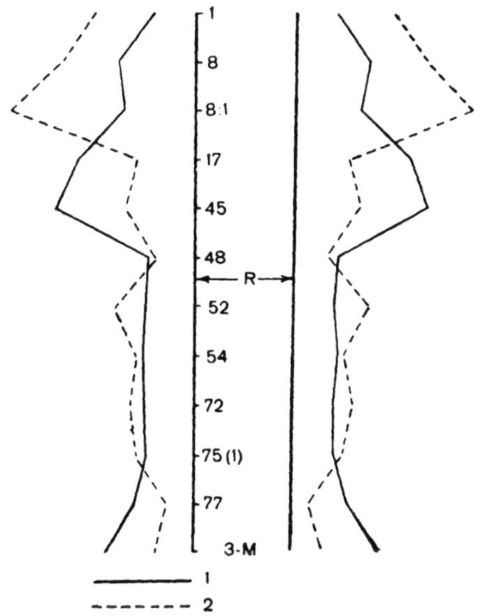

Figure 4. Craniological differentiation of Mesolithic, Neolithic, and Aeneolithic populations of Europe and southern Siberia in comparison with modern European peoples.
Key: Male skulls: (1) variability of ancient populations; (2) variability of modern populations; (3) standard deviation for each trait, adopted for 100 percent

local South-Siberian race three population groups: Kazakhstan, Near-Tien-Shan, and Altai-Sayan.

As another example: some classifications combine, on the strength of morphological similarity, the Tibetan and Arctic races which are then considered as relic forms of a proto-Mongoloid complex. However, even long-standing genetic isolation was bound to have generated differences in numerous traits with simple hereditary bases. (Support for this belief exists in data on blood-group variations exhibited by Tibetans, which precludes their classification with Arctic peoples.)

Dozens of examples of this kind can be cited. As this is not possible here, I supply instead variations of racial classification (Figure 5) and its associated system of centers should represent original racial or "breeding grounds" of race formation (Figure 6) which, to some extent, reflect the principles presented herein.

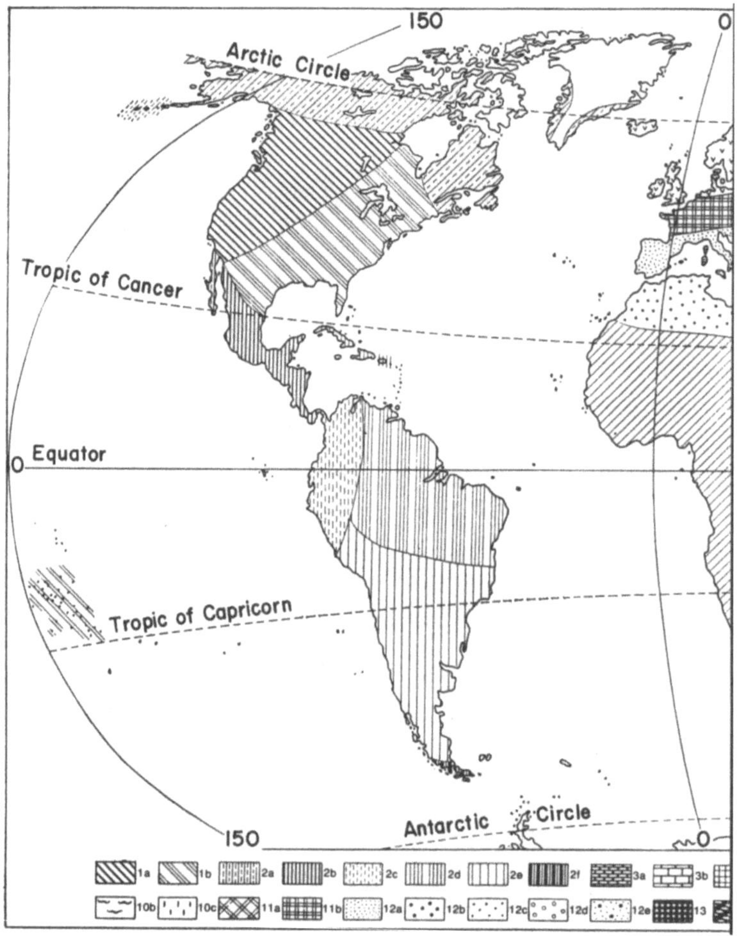

Figure 5. Probable distribution of races before European colonization
Key: EASTERN AMERO-ASIATIC DIVISION
I. *Amerindian branch*, local races and groups:
1. North American local race; 1a. Pacific group; 1b. Atlantic group; 2. Central and South American local race; 2a. Californian group; 2b. Central American group; 2c. Andean group; 2d. Amazonian group; 2e. Patagonian group; 2f. Semi-Marginal group
II. *Asiatic branch*, local races and groups:
3. Southern Mongoloid local race; 3a. Island group; 3b. Continental group; 4. Eastern Mongoloid local race; 4a. Far Eastern group; 4b. Amur and Sakhainian group; 5. Arctic local race; 5a. Continental group; 5b. Island group; 6. Tibetan local race; 7. Northern Asiatic local race; 7a. Central Asiatic group; 7b. Baikal or forest group; 8. Southern Siberian group; 8a. Altaj and Sajanian group; 8b. Tien Shanian group; 8c. Kazakhstanian group; 9. Uralian local race; 9a. Western Siberian group; 9b. Sub-Uralian group

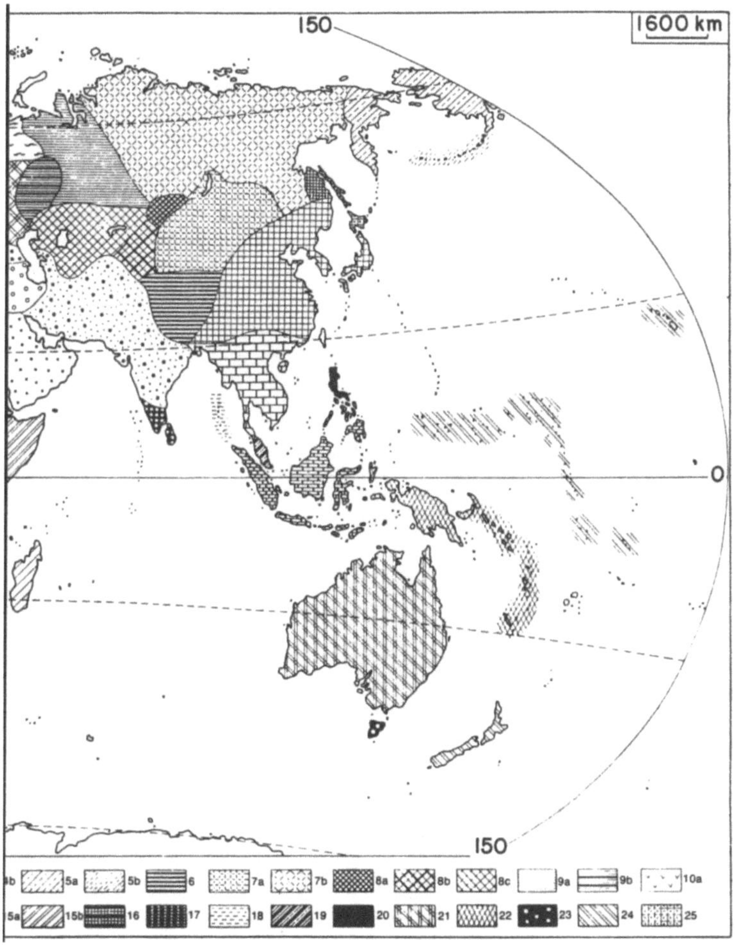

WESTERN EURO-AFRICAN DIVISION

III. *European branch*, local races and groups:

10. Baltic or northern European local race; 10a. Western Baltic; 10b. Eastern Baltic group; 10c. Lapponoid group; 11. Central European local race; 11a. Central and Eastern European group; 11b. Western European group; 12. Mediterranean or Southern European local race; 12a. Western Mediterranean group; 12b. Balkan and Caucasian group; 12c. Arabian and African group; 12d. Near Eastern group; 12e. Afghanistan and Indian

IV. *African branch*, local races and groups:

13. Southern Indian local race; 14. Ethiopian local race; 15. Negroid local race; 15a. Sudanian group; 15b. Eastern African group; 16. Central African local race; 17. Southern African local race

V. *Pacific branch*, local races:

18. Andaman local race; 19. Negrito Continental local race; 20. Negrito Philippine local race; 21. Australian local race; 22. Melanesian local race; 23. Tasmanian local race; 24. Polynesian local race; 25. Ainu or Kuril local race

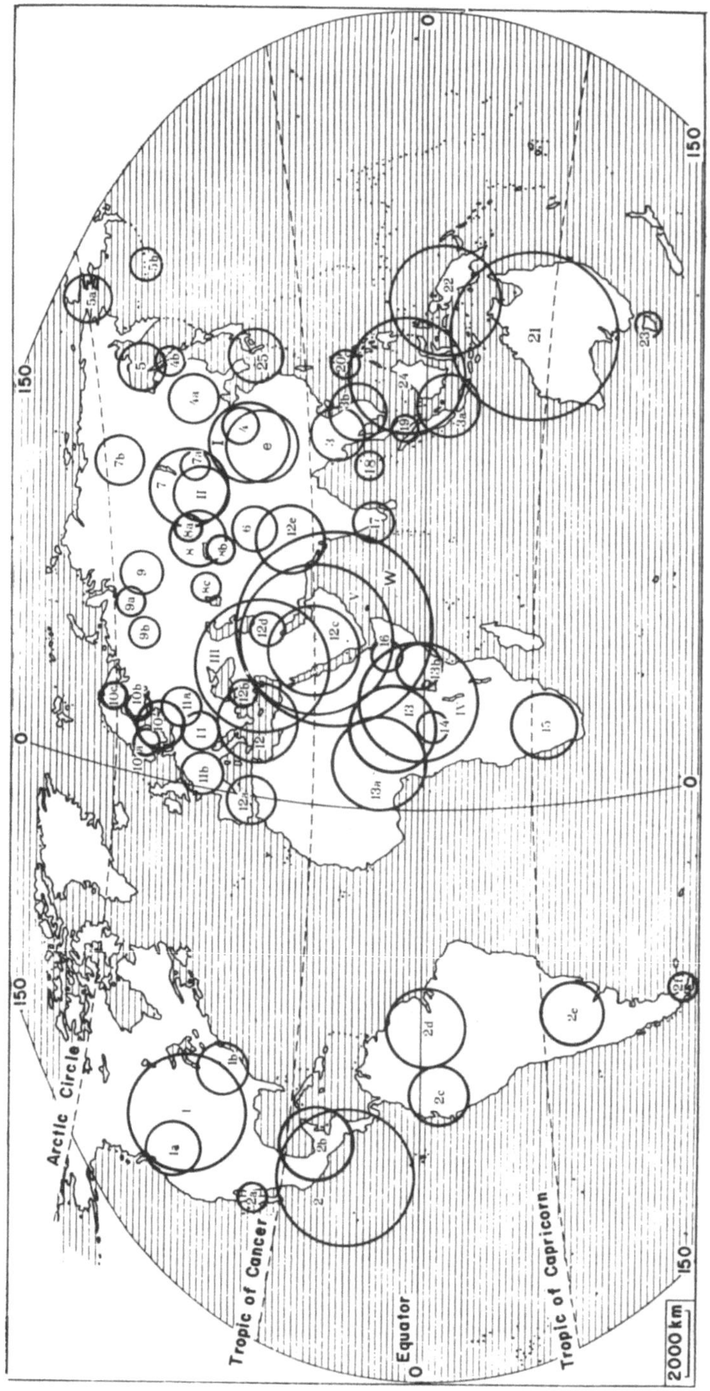
Arctic Circle
Tropic of Cancer
Equator
Tropic of Capricorn
2000 km
150
0
150
0
150
0
150

Figure 6. Probable distribution of race formative centers
Key: (E) EASTERN PRIMARY CENTER
I. *Asiatic coastal secondary center*
1. North American tertiary center: 1a. Pacific quaternary center; 1b. Atlantic quatenary center; 2. Central and South American tertiary center; 2a. Californian quaternary center; 2b. Central American quaternary center; 2c. Andean quaternary center; 2d. Amazonian quaternary center; 2e. Patagonian quaternary center; 2f. Semi-marginal quaternary center
II. *Asiatic continental secondary center*
3. Southeastern Asiatic tertiary center; 3a. Island quaternary center; 3b. Continental quaternary center; 4. Eastern Asiatic tertiary center; 4a. Far Eastern quaternary center; 4b. Amur and Sakhalinian quaternary center; 5. Kamchatka and Okhotsk tertiary center; 5a. Beringian continental quaternary center; 5b. Aleutian island quaternary center; 6. Tibetan tertiary center; 7. Central Asiatic tertiary center; 7a. Transbaikal and Mongolian quaternary center; 7b. Northern Asiatic quaternary center; 8. Southern Siberian tertiary center; 8a. Altai and Sajanian quaternary center; 8b. Tien Shanian quaternary center; 8c. Kazakhstanian quaternary center; 9. Transuralian tertiary center; 9a. Western Siberian quaternary center; 9b. Suburalian quaternary center
(W) WESTERN PRIMARY CENTER
III. *European secondary center*
10. Baltic tertiary center; 10a. Western Baltic quaternary center; 10b. Eastern Baltic quaternary center; 10c. Kola quaternary center; 11. Central European tertiary center; 11a. Central and Eastern European quaternary center; 11b. Western European quaternary center; 12. Mediterranean tertiary center; 12a. Western Mediterranean quaternary center; 12b. Balkan and Caucasian quaternary center; 12c. Arabian and African quaternary center; 12d. Near Eastern quaternary center; 12e. Afghanistan and Indian quaternary center
IV. *African secondary center*
13. African tropical tertiary center; 13a. Sudanian quaternary center; 13b. Eastern African quaternary center; 14. Central African tertiary center; 15. Southern African tertiary center; 16. Ethiopian tertiary center; 17. Southern Indian tertiary center
V. *Asiatic and African secondary center*
18. Andaman tertiary center; 19. Malacca tertiary center; 20. Philippine tertiary center; 21. Asiatic and Australian tertiary center; 22. Melanesian tertiary center; 23. Tasmanian tertiary center; 24. Asiatic and Pacific tertiary center; 25. Korean and Japanese tertiary center

REFERENCES

ABDUSHELISHVILI, M. G.
1964 *Antropologiia drevnego i sovremennogo naseleniia Gruzii* [Anthropology of the ancient and modern populations of Georgia]. Tbilisi.

ALEXEEV, V. P.
1967 Modusy rasoobrazovaniia i geograficheskoe rasprostranenie genov rasovykh priznakov [Modes of race development and the geographic distribution of genes of racial traits]. *Sovetskaia ėtnografiia* 1.
1969 *Proiskhozhdenie narodov Vostochnoĭ Evropy (kraniologicheskoe issledovanie)* [Origin of the peoples of Eastern Europe (a craniological investigation)]. Moscow.

BIASUTTI, R.
1959 *Razze i popoli della terra.* Turin.

CHEPURKOVSKII, E. M.
1913 Geograficheskoe raspredelenie formy golovy i t͡svetnosti krest'ianskogo naseleniia preimushchestvenno Velikorossii v sviazi s kolonizat͡sieĭ eë slavianami [Geographic distribution of head-shape and pigmentation in the peasant population predominantly of Great Russia in connection with the latter's colonization by the Slavs]. *Izvestiia Obshchestva liubitelei estestvoznaniia, antropologii i ėtnografii* 124 (2).

DEBETS, G. E.
1948 Paleoantropologiia SSSR [Paleoanthropology of the USSR]. *Trudy Instituta ėtnografii AN SSSR (novaia seriia)* 4.

DENIKER, J.
1900 *Les races et les peuples de la terre.* Paris.

GINZBURG, V. V.
1937 Izogemoaggliutinat͡siia u turkmen [Isohemoagglutination among the Turkmen]. *Antropologicheskiĭ zhurnal* 2.

HOOTON, E., C. DUPERTUIS
1955 The physical anthropology of Ireland.

KNUSSMANN, R.
1962 "Moderne statistische Verfahren in der Rassenkunde" in *Die neue Rassenkunde.* Stuttgart: Gustav Fischer.

LEVIN, M. G.
1971 *Ėtnicheskaia antropologiia Iaponii* [Ethnic anthropology of Japan]. Moscow.

LUNDMAN, B.
1968 *Geographische Anthropologie.* Stuttgart.

MOURANT, A.
1954 *The distribution of the human blood groups.* Oxford.

MOURANT, A. *et al.*
1958 *The ABO blood groups; comprehensive tables and maps of world distribution.* Oxford.

Proiskhozhdenie
1965 Proiskhozhdenie i ėtnicheskaia istoriia russkogo naroda po antropologicheskim dannym [Origin and ethnic history of the Russian people according to anthropological data]. *Trudy Instituta ėtnografii AN SSSR (nov. seriia)* 88.

RIPLEY, W.
1900 *The races of Europe.* London: Kegan, Trench, Trübner.

RYCHKOV, JU. G.
1965 Nereotorje aspekti serologicheskih issledoranii v antropologii [Some aspects of serological investigations in anthropology]. *Voprosi antropologii* 19:95–105.

Blood Levels of Some Biochemical Components among Negroid Individuals with Hemoglobin A and AS

V. CORRENTI, G. SPEDINI, M. CRESTA, F. VECCHI, and E. CAPUCCI

This article presents some of the results of a bioanthropological survey carried out in Dahomey by the Anthropology Institute of the University of Rome.

A total of 417 adult males, aged eighteen to sixty years, was examined from a biochemical and hematological point of view: blood levels of Na^+, K^+, Ca^{++}, Mg^{++}, cholesterol, urea, and nitrogen; blood groups ABO, MN, Rh; Hp, Gc systems, and red cell acid phosphatase were tested. Hemoglobin type determinations were also carried out on 390 adult males and 381 children aged two to fourteen. The following methods were used: Na^+ and K^+, by flame photometry; Ca^{++} and Mg^{++} by a titrimetric method, using EDTA; urea nitrogen by urease; cholesterol by the Liebermann-Bouchard method; proteins with biuret and paper electrophoresis; Hb-types by cellulose polyacetate electrophoresis.

The frequency of the various Hb-types is shown in Table 1. From this table we see that the frequency of the HbS gene is about 11 percent, similar to what is well known for other populations of that African area.

As for HbAS frequences, Table 2 shows that there are no significant

Table 1. Frequencies of Hb-types

Hb-types	*n*	%	Gene frequencies
A	554	71.86	
AS	159	20.62	HbA = 0.8579
AC	49	6.35	HbS = 0.1066
S	2	0.26	HbC = 0.0355
SC	6	0.78	
C	1	0.13	Total 1.0000
Total	771	100.00	

$\chi^2_{(3)} = 7.25$ $P > 0.05$

Table 2. Hb-types in various age groups

		Hb-type frequencies (%)					
Age	*n*	A	AS	AC	S	C	SC
2–7	159	75.49	19.49	3.78	0.62	–	0.62
8–17	222	71.17	18.92	7.65	0.45	0.45	1.36
18–27	138	71.74	19.56	7.25	–	–	1.45
28–37	145	67.59	24.13	8.28	–	–	–
38–47	73	72.70	23.20	4.10	–	–	–
48–56	34	76.47	20.59	2.94	–	–	–

differences among various age groups. Dealing only with the blood biochemical characters, the comparison between HbA and HbAS groups is shown in Tables 3 and 4. Table 3 shows that the levels of Na^+ and K^+ are comparable in the two groups and that they are similar to those usually related to the Caucasoid populations. The Mg^{++} levels of HbAS people are slightly lower than those of the HbA group, with a statistical probability close to the limits of significance ($P<0.1$); the Ca^{++} levels are 20 percent lower, in both groups, than those detected in Caucasoid populations. This last result agrees with that already shown by Walker (1967). As for urea nitrogen, the HbAS group shows an average value which is slightly higher than the HbA one. This difference is not statistically significant, but this does not rule out the possibility that the higher urea values in individuals with HbAS may be attributed to the presence, in some of these subjects, of a renal subpathology.

The urea levels of both groups are considerably lower than those found in Caucasoid populations (24 mg. percent) and the distribution of these values in frequency groups shows that about 10 percent of the subjects present levels under 11 mg. percent, a value close to the minimum compatible with a positive nitrogen balance. The low urea levels may be attributed to the low level of protein intake. A food consumption survey

Table 3. Levels of sodium, potassium, calcium, magnesium, urea nitrogen, and cholesterol in the blood of Dahomeian people with HbA and HbAS

	HbA			HbAS		
	n	Mean	s.d.	*n*	Mean	s.d.
Sodium mEq/l	264	142.92	18.03	81	142.47	17.59
Potassium mEq/l	264	4.236	0.640	81	4.346	0.870
Calcium mEq/l	243	4.024	0.680	77	4.035	0.600
Magnesium mEq/l	239	1.694	0.284	77	1.609	0.275
Urea mg percent	273	18.23	5.79	82	19.29	5.58
Cholesterol mg percent	273	144.19	31.69	82	140.31	31.94

Table 4. Levels of proteins in the blood of Dahomeian people with HbA and HbAS

	HbA	(*n* = 249)	HbAS	(*n* = 75)
	Mean	s.d.	Mean	s.d.
Total proteins (g/100 ml)	6.88	1.00	6.91	1.01
Albumin (g/100 ml)	3.00	0.48	3.10	0.50
Alpha 1 globulin (g/100 ml	0.33	0.10	0.34	0.12
Alpha 2 globulin (g/100 ml)	0.53	0.14	0.51	0.15
Beta globulin (g/100 ml	0.82	0.20	0.80	0.18
Gamma globulin (g/100 ml)	2.20	0.53	2.16	0.50
Albumin/globulin	0.79	0.16	0.83	0.17

on fifty-six families showed that the total protein intake for the adult male is about 55 grams per day (practically all from maize and only 10.5 grams of animal origin).

No difference was found between the HbA and HbAS groups with regard to cholesterol, and the average levels are considerably lower than those of Caucasoid populations. Also, in this case the cause is very probably of nutritional origin, as in the Dahomeian diet the level of lipids is rather low (about 68 grams daily per person, 24 percent of total calories).

The comparison of the serum proteins pattern between individuals with HbA and HbAS in Table 4 indicates that it is not possible to show any significant difference between the two groups. Further, and in line with what is already known, the albumin values are rather low if compared with those of Caucasoid populations, while the gamma-globulin values appear to be quite high. From what has been stated, one can conclude that in these highly malaric areas the selective process which has favored and continues to favor the high frequency of HbAS people (Allison 1954) operates independent of the serum-biochemical characters of people carrying this mutation.

REFERENCES

ALLISON, A. C.

1954 Protection afforded by the sickle cell trait against subtertian malaria infection. *British Medical Journal* 1:290.

WALKER, A. R. P.

1967 Interpretation of biological data on one ethnic or regional group may not be equally applicable to other groups. *American Journal of Clinical Nutrition.* 20:1025.

Genogeography of Blood Factors in the Transcaucasus

A. A. VORONOV

In the Soviet Union the natural environment of the subtropical lowlands of the Transcaucasus corresponds highly with natural zones of abnormal hemoglobin distribution. In view of this, the author set up a number of expeditions (1964–1967) to study the regularities in geographic distribution of abnormal hemoglobins and blood groups.

The Talyshes, a people of Iranian origin, inhabiting the Lenkoran lowland (the Lenkoran and Astara districts) near the southern coast of the Caspian sea were examined as well as a group of Azerbaijanians in the Kura lowland (the Barda district). Other Azerbaijanians were studied in the Alpine Shemakha district (1,200 meters above sea level). We also studied Georgians (Kakhetians from the Gurjaani district) and Azerbaijanians (the Nukha district) in the Alazen-Egrichai valley, in northern Transcaucasus. It adjoins the Main Caucasian range, and the border between Georgia and Azerbaijan divides it in the middle. Two groups of Georgians (Gurians in the Lanchkhuti district and Megrelians in the Abashi district) as well as a group of Abkhazians (the Ochamchiri district) were examined in the Kolkhida lowland. We also examined a group of Armenians in Armenia (the Ashtarak district).

Three subraces of the Indian-Mediterranean race are represented anthropologically in the Caucasus; the Anterior–Asiatic, the Mediterranean–Balkan, and the Indian–Pamirian ones. The Mediterranean–Balkan subrace is represented here by a small Pontian race (Bunak 1932), the Indian–Pamirian by a Caspian group (Bunak 1947), the Anterior–Asiatic by an Anterior–Asiatic or Armenoid one (Luschan 1911). According to Abdushelishvili's scheme (1963, 1964) these small races are represented in the Caucasus by their local variants: the Caspian group by its West Caucasian variant, the Pontiian by an Adygi group, the Anterior–Asiatic by Kolkhian, Iberian, Armenoid, and Assiroid variants. In addition to all these, there also exists an autochtonic anthropological type in the Alpine Great Caucasus, i.e., the Caucasian group (Natishvili and Abdushelishvili 1954).

In this way the Kolkhian race was under study in Kolkhida, the Iberian race in the Alazan-Egrichai valley, and the West Caspian race in the Kura Lowland.

DISTRIBUTION OF ABNORMAL HEMOGLOBINS IN THE TRANSCAUCASUS

The study of genogeography, even on widely spread abnormal hemoglobins and thalassemia, is always a very difficult task because of the criteria upon which the research must be based. We agreed upon the following minimum of initial laboratory tests for our investigations in the Transcaucasus: (1) microscopic analysis of blood smear; (2) definition of erythrocytic osmotic resistance; (3) hemoglobin electrophoresis with a HbA_2 qualitative definition; and (4) a HbF quantitative definition. Additional tests were made following the method of Lehmann and Ager (1961).

An increased HbA_2 was regarded as diagnostic of the heterozygous form of a B-thalassemia (the F-form of thalassemia is not found in the Transcaucasus). Morphologically changed erythrocytes (target shape, poikilocytosis, anisocytosis) and enlarged osmotic resistance in erythrocytes were considered as additional evidence of the diagnosis. As a result of these vigorous requirements, some cases of heterozygous thalassemia might have passed unobserved (most likely, 10 to 15 percent of all the cases studied). But in view of the fact that ours was the first research of this kind in the Soviet Union, we preferred to risk understatement of the actual frequency of genes rather than lower our standard for detection of thalassemia. We carried out mass field tests for HbA_2 qualitative content by means of paper electrophoresis. Voronov's 1969 data thus obtained were similar to those of starch-block electrophoresis.

Nine regions along the entire Transcaucasus from the Black Sea to the Caspian Sea were the objects of our study (see Table 1). A great number of B-thalassemia cases was discovered in the lowland, 9.4 percent in the Alazan-Egrichai valley Azerbaijanians (see Table 1 and Figure 1). The most peculiar finding of our research was the ethnic dependence of thalassemia distribution. Only 1 percent of cases showed heterozygous thalassemia in Georgia and Armenia, even in the extreme lowlands, among the three ethnic groups in Kolkhida (Gurians, Megrelians, Abkhazians) and the Armenians in Armenia (see Table 1 and Figure 1).

This ethnic dependence was most vivid in the Alazan-Egrichai valley.

Table 1. Abnormal hemoglobin distribution in the Transcaucasus

Nationality	Administrative district	Geographic region	n	Thalassemia frequency (percentage)	Thalassemia genic frequency	HbD frequency (percentage)	HbD genic frequency
1. Talyshes	Lenkoran Astara	Lenkoran lowland	340	3.5	0.018	2.5	0.013
2. Azerbaijanians	Barda	Kura lowland	102	6.9	0.036	0.0	0.000
3. Azerbaijanians	Shemakha	Alpine Caucasus	171	1.3	0.007	0.0	0.000
4. Azerbaijanians	Nukha	Alazan–Egrichai valley	265	9.4	0.049	0.3	0.000
5. Kakhetians	Gurdjaani		200	1.0	0.005	0.0	0.000
6. Gurians	Lanchkhuti	Kolkhida lowland	81	1.2	0.006	0.0	0.000
7. Megrelians	Abashi		72	0.0	0.000	0.0	0.000
8. Abkhazians	Ochamchiri		87	1.2	0.006	0.0	0.000
9. Armenians	Ashtarak	Armenian upland	180	1.7	0.009	0.5	0.002

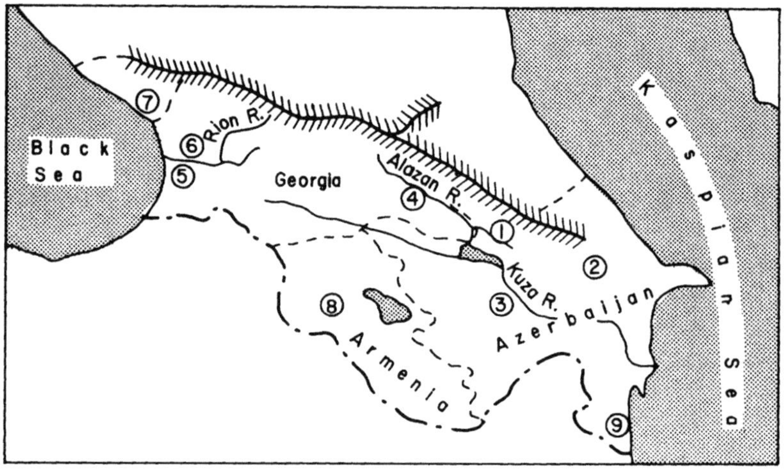

Figure 1. Map of the studied population groups in the Transcaucasus
1. Azerbaijanians of the Nukha district; 2. Azerbaijanians of the Shemakha district; 3. Azerbaijanians of the Barda district; 4. Kakhetians of the Gurdjaani district; 5. Gurians of the Lanchkhuti district; 6. Megrelians of the Abashi district; 7. Abkhazians of the Ocham-chiri district; 8. Armenians of the Ashtarak district; 9. Talyshes of the Lenkoran district

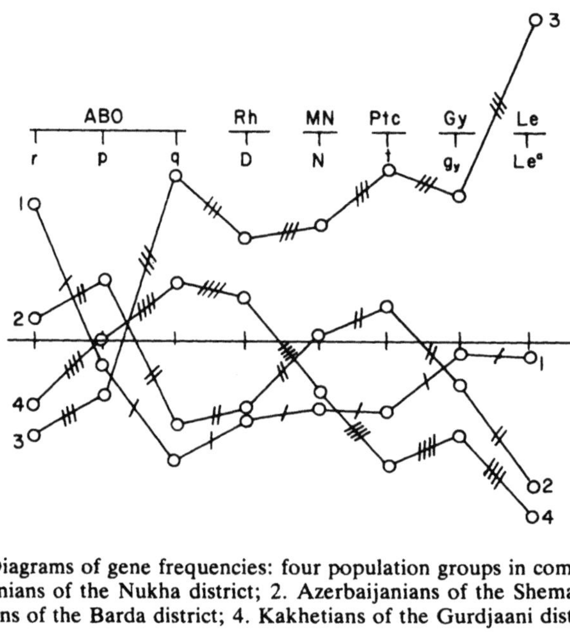

Figure 2. Diagrams of gene frequencies: four population groups in comparison
1. Azerbaijanians of the Nukha district; 2. Azerbaijanians of the Shemakha district; 3. Azerbaijanians of the Barda district; 4. Kakhetians of the Gurdjaani district

The valley is narrow, some 20–30 kilometers wide and about 200 kilometers long. Georgia lies to the west, and Azerbaijan to the east. Within this valley, under the same ecological conditions, Georgians showed 1.8 percent cases of thalassemia, and Azerbaijanians 9.4 percent. The Georgians under study live in the Gurdjaani district, 40 kilometers west of the border; Azerbaijanians come from the Nukha district, 40 kilometers east of the border. The statistical treatment of this difference shows it to be significant: $t = 4.36$ ($n = 436$ and $P<0.1$ percent).

We found a large amount of thalassemia in Azerbaijanians, the incidence being limited by ethnic and genetic barriers. It should be pointed out that in the past Georgia was widely Christian and Azerbaijan mainly Moslem. Marriages between Georgians and Azerbaijanians were extremely rare. Anthropologically, the population of the valley is homogeneous and belongs to the Iberian race (the eastern variant). Their homogeneous gene pool is indicated by the blood group distributions. Such ethnic difference in gene frequencies in adjacent territories, where the population has common anthropological features and a similar gene pool, would indicate that thalassemia is not an inherent and ancient feature of these regions, but that it spread after the appearance of an ethnic barrier between Georgia and Azerbaijan. This phenomenon could be attributed to a mutation recently developed in the territory of Azerbaijan (in which case it would probably not have reached such high gene frequencies). Another possibility is that the thalassemia gene was brought into Azerbaijan by gene flow.

The Azerbaijanian population has experienced the infusion of many ethnic elements, including Greeks, Romans, Arabs, Turks, and Mongolians. It appears that the B-thalassemia gene spread in the Transcaucasus when ethnic groups in the territory of Azerbaijan were consolidating; and the old ethnic barriers, which hindered the progress of the gene, were destroyed while some new ones arose. Most likely this process took place during the Moslem invasions when the population of Azerbaijan became Turkicized and a new state was formed. There are two possible explanations for the appearance of the B-thalassemia gene during that period. It could have entered Azerbaijan from the southeast (perhaps from the southern coast of the Caspian sea), or it could have come through gene flow from Turkic tribes which, in huge numbers, settled down in the territory of Azerbaijan, especially in the Kura-Araks lowland.

To establish the origin of the thalassemia gene is as difficult as reconstructing any genogeographic processes of the past. Our results seem to support the ideas presented by L. Brumpt at the VI Congress of Anthropological and Ethnological Sciences held in Paris in 1962. He argued that the large distribution of the thalassemia gene in peoples of Asian origin ("horse-breeders") could not be attributed to the migration of Mediterranean races (Brumpt 1962). It seems to us also that the hypothesis of global diffusion of thalassemia from the west does not have sufficient support (Gatto 1960; Zaino 1964; Romer and Dorken 1956). However, the data presently available are not sufficient to deny the possibility of its initial entry into Central Asia from the Mediterranean area. By the way, the great infrequency of thalassemia in Kolkhida is puzzling if the Mediterranean area is regarded as its point of origin. Kolkhida has had migratory connections with Adriatic peoples from time immemorial.

We also examined an Alpine region (1,200 meters above sea level) closely adjacent to the Kura lowland (the Shemakha district). The frequency of heterozygous thalassemia there proved to be low — 1.3 percent. The degree of difference with lowland regions is statistically significant. Our observation supports the hypothesis of the influence of altitude on thalassemia distribution (Siniskalko et al. 1966; Choremis et al. 1963; Curtain et al. 1962). Without detailed discussion of the selective reasons for thalassemia distribution, we should like to emphasize the fact that in the past all the Transcaucasian lowlands under study, i.e., those of Lenkoran, Kura, Alazan-Egrichai, and Kolkhida were characterized by severe endemic malaria.

Besides thalassemia the sample from the Lenkoran lowland produced HbD, also discovered in Talyshes, in 2.5 percent of the 340 subjects under examination. It should be remembered that Talyshes are of Iranian origin. Only a single case of HbD was found in over 500 Azerbaijanians from the Kura lowland and the Alazan-Egrichai valley. We may therefore believe that HbD is limited to Talyshes. More field research is necessary for more exact results.

BLOOD GROUP DISTRIBUTION IN THE TRANSCAUCASUS

Blood groups of the ABO, MN, Rhesus, Lewis, and Gy systems were studied. We also tested for taste sensitivity to PTC. Gene frequencies of traits of low selection intensity were studied in the districts where cases of hemoglobin abnormalities were examined to learn something about the gene pool of the populations.

The distribution of hemoglobin abnormalities and blood groups was studied on samples taken in five to six dispersed villages of the region in question.

The results thus obtained are presented in Tables 2 through 7.

The ABO System

When analyzed, ABO system frequencies show that the Kolkhian group

Table 2. Phenotypes and ABO system gene frequency distribution in the Transcaucasus

Nationality	Administrative district	*n*	A	B	AB	O	r	p	q	χ^2	*P*
1. Azerbaijanians	Nukha	287	38.3%	16.7%	4.9%	40.1%	0.6375	0.2471	0.1153	0.399	0.529
2. Azerbaijanians	Shemakha	180	40.0%	19.5%	7.8%	32.8%	0.5749	0.2780	0.1470	0.055	0.830
3. Azerbaijanians	Barda	147	51.7%	11.6%	7.5%	29.2%	0.5393	0.3607	0.1002	0.020	0.891
4. Kakhetians	Gurdjaani	196	43.9%	13.8%	9.7%	32.7%	0.5518	0.3245	0.1236	3.673	0.056
5. Megrelians	Abashi Mikha-Tsakai	195	33.8%	5.6%	1.5%	59.0%	0.7672	0.1962	0.0367	1.820	0.179
6. Gurians	Lanchkhuti	98					0.7599	0.1432	0.0968	0.231	0.633
7. Abkhazians	Ochamchiri	91					0.7744	0.1829	0.0427	1.589	0.208

(Megrelians, Gurians, Abkhazians) is characterized by its low frequencies of gene q and considerable presence of gene r (see Table 2). Similar data were also obtained by other scientists. Georgians and Azerbaijanians from the Alazan-Egrichai valley, as well as Azerbaijanians from the Shemakha district, show similar gene frequencies.

Azerbaijanians of the Barda district (the West Caspian anthropological group) are characterized by a higher frequency of gene p and low level of gene r.

The MN System

This system, as usual, showed smaller differences in the gene frequencies of the populations (see Table 3). The West Caspian group stands somewhat apart with its lower frequency of gene m (0.5543).

The Rhesus System

Only the gene D frequency was determined in this system. The obtained data correspond to the frequencies found in the Mediterranean race (see Table 4). In the West Caspian group (Azerbaijanians from the Barda district) the number of Rh negative individuals falls to 4.6 percent.

The Lewis System

We used both the anti-Le(a) and anti-Le(b) serums (see Table 5). To make the results of phenotypic analysis suitable for further calculations we determined frequencies of a hypothetical gene Le^a which was supposed to exhibit a proportional dependence.

The West Caspian group (Azerbaijanians of the Barda district) is distinguished by this system also.

The Gy System

The world distribution of this system has not been widely investigated. The system was discovered by W. Boyd (Boyd et al. 1959). We used the conglutination reaction developed by M. Potapov (1967). Gene frequencies were calculated in accordance with a supposed two allele system of Gy, gy (Potapov 1967). Again, Azerbaijanians from the Barda district are found to be somewhat different from the rest, showing a larger frequency of the Gy allele (see Table 6).

Table 3. Phenotypes and MN system gene frequency distribution in the Transcaucasus

Nationality	Administrative district	Number in study	M	N	MN	m	n	χ^2	*P*
1. Azerbaijanians	Nukha	248	41.9%	15.7%	42.3%	0.6315	0.3685	0.891	0.3468
2. Azerbaijanians	Shemakha	180	38.9%	20.0%	41.1%	0.5944	0.4056	3.930	0.0458
3. Azerbaijanians	Barda	147	25.8%	15.0%	59.2%	0.5543	0.4457	4.000	0.0455
4. Kakhetians	Gurdjaani	196	37.2%	12.2%	50.5%	0.6250	0.3750	1.129	0.2879
5. Megrelians	Abashi	192	44.3%	16.7%	39.1%	0.6385	0.3615	2.026	0.1547
	Mikha-Tsakai								
6. Gurians	Lanchkhuti	98				0.6123	0.3877	0.029	0.878
7. Abkhazians	Ochamchiri	91				0.7088	0.2912	22.289	0.00+

Table 4. Phenotypes and Rhesus system gene frequency distribution in the Transcaucasus

Nationality	Administrative district	Numbers in sample	Rh(D+)	Rh(D−)	D	d
1. Azerbaijanians	Nukha	191	91.6%	8.4%	0.7106	0.2894
2. Azerbaijanians	Shemakha	176	92.0%	8.0%	0.7179	0.2821
3. Azerbaijanians	Barda	136	95.4%	4.6%	0.7851	0.2149
4. Kakhetians	Gurdjaani	195	94.4%	5.6%	0.7625	0.2375
5. Megrelians	Mikha-Tsakai	121	89.3%	10.7%	0.6723	0.3277

Table 5. Phenotypes and Lewis system gene frequency distribution in the Transcaucasus

Nationality	Administrative district	Number in study	Le(a+)	Le(B+)	Le(a−B)	Le•
1. Azerbaijanians	Nukha	203	27.1%	43.8%	29.1%	0.5105
2. Azerbaijanians	Shemakha	181	23.2%	38.6%	38.2%	0.4717
3. Azerbaijanians	Barda	141	42.6%	37.6%	19.9%	0.6523
4. Kakhetians	Gurdjaani	184	19.6%	55.4%	25.0%	0.4424
5. Megrelians	Mikha-Tsakai	112	22.3%	55.4%	22.3%	0.4624

Table 6. Phenotypes and Gy system frequency distribution in the Transcaucasus

Nationality	Administrative district	Number in samples	Gy(+)	Gy(−)	Gy	gy
1. Azerbaijanians	Nukha	187	75.9%	24.1%	0.5095	0.4905
2. Azerbaijanians	Shemakha	181	68.5%	31.5%	0.4388	0.5612
3. Azerbaijanians	Barda	142	80.9%	19.1%	0.5640	0.4340
4. Kakhetians	Gurdjaani	176	71.0%	29.0%	0.4617	0.5383

PTC Taste

The capacity to taste PTC was tested using two dilutions, 1:5,000 and 1:8,000. Only the 1:5,000 dilution data were used for genetic calculations. Azerbaijanians of the Barda district show higher frequencies of gene t (nontaster) (see Table 7).

An attempt was made to evaluate the degree of "biological distances" in all of the systems taken together. Figure 2 contains a diagram in which the ABO system is characterized by all three alleles: r, p, q; the Rhesus system is presented by gene D; the MN system by gene N; the PTC by gene t (ptc); the Gy system by gene Gy; the Lewis system by a hypotheti-

Table 7. Phenotypes and PTC taste sensitivity gene frequency of distribution in the Transcaucasus

Nationality	Administrative district	Number in samples	PTC (+) 1:8000	PTC (+) 1:5000	T	*t*
1. Azerbaijanians	Nukha	196	69.9%	81.1%	0.5655	0.4345
2. Azerbaijanians	Shemakha	180	67.2%	77.9%	0.5286	0.4714
3. Azerbaijanians	Barda	158	50.0%	70.9%	0.4605	0.5395
4. Kakhetians	Gurdjaani	205	66.8%	82.9%	0.5868	0.4132

cal gene Le^a. The ordinate axis shows the mean arithmetical value of gene frequencies in all four regions under examination. Positive differences for each region were marked upwards; negative ones downwards. For comparison we have taken Georgians and Azerbaijanians of the Iberian race [Georgians from Kakheti (4), Azerbaijanians from the Nukha district (1), Azerbaijanians from the Shemakha district (2)] as well as Azerbaijanians from the Barda district (3) belonging to the Caspian anthropological group. This diagram vividly shows that the curve representing Azerbaijanians from the Barda district (the Caspian group) considerably differs from the others.

All groups were correlated by pairs with the help of f^2 technique, each system taken separately. Despite their different ethnic origins, Georgians from Kakhetia and Azerbaijanians from the Nukha and Shemakha districts demonstrate homogeneity at statistically significant levels in all these systems. Out of the eighteen correlations made, only one failed to show significance at the 5 percent level. According to these correlations, Azerbaijanians of the Caspian anthropological group proved to be different in most of their genetic systems from the other peoples.

To evaluate this distance we used R. Knussmann's suggested formula (1962). The results are presented in Table 8 and Figure 3. The groups from the foothills of the Central Caucasus, all belonging to the Iberian race, are quite close to each other (x_1 – 0.52; x_2 – 0.56; x_3 – 0.72). The Caspian race is distinguished from each of the others by doubled or trebled values (x_6 – 1.36; x_7 – 1.45; x_8 – 1.52). The largest distance is observed between Azerbaijanians of the Caspian race and Megrelian Georgians of the Kolkhian anthropological group (x_{10} – 2.16).

As we see, despite the existence of the ethnic barrier between Georgia and Azerbaijan for a long time (at least ten centuries), the gene pool has remained essentially homogeneous. On the other hand, the mixing process within ethnic groups was not intensive enough to eliminate the differences in gene pools of various races which composed an ethnic community. Such conclusions are possible only when sufficient numbers of genetic traits are analyzed together, so as to avoid the influence of

Table 8. Evaluation of the distance between studied population groups according to their blood groups, on the Knussmann formula

	Azerbaijanians of the Nukha district	Azerbaijanians of the Shemakha district	Azerbaijanians of the Barda district	Kakhetians of the Gurdjaani district	Megrelians of the Abashi district
Azerbaijanians of the Nukha district	–				
Azerbaijanians of the Shemakha district	0.52	–			
Azerbaijanians of the Barda district	1.45	1.52	–		
Kakhetians of the Gurdjaani district	0.72	0.56	1.36	–	
Megrelians of the Abashi district	1.13	1.02	2.16	1.64	–

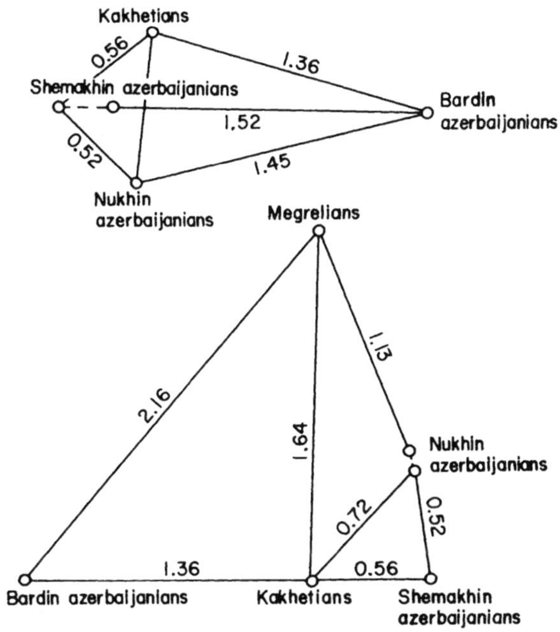

Figure 3. "Biological distances" between the populations in the Transcaucasus

selection causing deviations of gene frequencies. If in evaluating the differences in the gene pool we examine only one genetic system, it may lead to erroneous conclusions. Besides, even with a dispersed sampling we must always consider population size and structure in order to avoid the influence of local fluctuations of the gene frequencies arising from the nature and size of breeding populations on mean values for biological tracts.

The example of thalassemia distribution in the Transcaucasus shows how fast the spread of a trait with a selective advantage is. Because such traits spread so fast, their distribution can reflect ethnic barriers and, at the same time, not coincide with the distribution of traits characterizing more ancient populational distributions. Contrary to this, the gene frequencies of blood groups proved to be so stable that their mean figures could reflect different anthropological compositions.

REFERENCES

ABDUSHELISHVILI, M. G.

1963 Antropologiia drevnego i sovremennogo naseleniia Gruzii [The anthropology of the ancient and current population of Georgia]. *Avtoref, dokt. diss.* Tbilisi.

1964 *Antropologiia drevnego i sovremennogo naseleniia Kavkaza* [The anthropology of the ancient and current population of the Caucasus], Moscow.

BOYD, W., D. GREEN, D. FUJINAGA, J. DRABIK, E. WASZCZENKO-ZACHARCHENKO

1959 A blood factor, possibly new, detected by extracts of arachis hypogaea. *Vox Sang* 4:6.

BRUMPT, L.

1962 Signification anthropologique de la Thalassemia. *Proceedings of VI International Congress of Anthropological and Ethnological Sciences.* Paris.

BUNAK, V.

1932 Neues Material zur Aussonderung anthropologischer Typen under der Bevölkerung Osteuropas. *Zeitschrift der Morphologische Anthropologie* 30(3).

1947 Antropologicheskiĭ sostav naseleniia Kavkaza [The anthropological composition of the population of the Caucasus]. *Vestnik Gosudarstvennogo muzeia Gruzii* 8-A.

CHOREMIS, C., PH. FESSAS, C. KATTAMIS *et al.*

1963 Three inherited red-cell abnormalities in a district of Greece. *Lancet* 1:907.

CURTAIN, C. C., CHEV KIDSON, D. C. GAJDUSEK, J. G. GORMAH

1962 Distribution pattern, population genetics and anthropological significance of thalassemia and abnormal hemoglobins in Melanesia. *American Journal of Physical Anthropology* 20(4):457.

GATTO, J.

1960 Origine della Thalassemia. *Proceedings of the Seventh International Society of Hematology.* Rome.

KNUSSMANN, R.

1962 Contribution to *Die neue Rassenkunde.* Edited by I. Schwidetsky Stuttgart.

LEHMANN, H., J. A. M. AGER

1961 The laboratory detection of abnormal haemoglobins. *Association of Clinical Pathologists Broadsheet* (n.s.) 33(6).

LUSCHAN, F.

1911 "The early inhabitants of Western Asia," Huxley memorial lecture for 1911. London.

NATISHVILI, A. I., M. G. ABDUSHELISHVILI

1954 *Osnovnye itogi antropologicheskikh ėkspeditsiĭ 1950, 1951 i 1952 gg. Instituta ėksperimental'noĭ morfologii AN Gruz. SSR, Vyezdnaia nauchnaia sessiia IEM v g. Batumi v 1954 g.* [Basic results of the anthropological expeditions of 1950, 1951 and 1952 of the Institute of experimental morphology of the AN of the Georgian SSR, visiting scientific session of the IEM, held in Batumi in 1954]. Tbilisi.

POTAPOV, M. I.

1967 Novoe metod reaktsii kongiutinatsii dlia vyiavleniia gruppovogo faktora Gy krovi cheloveka [A new method of conglutination reaction for discovering group factor Gy in human blood]. *Laboratornoe delo* 1.

ROMER, M. A., U. DORKEN

1956 Die neuen Erkenntnisse über die geographische Verbreitung der Talassemia. *Folia haematologica* 1:96.

SINISKALKO, M., L. BERNINI, G. FILIPPI, B. LATTE, P. KHAN, S. POMELLI, M. PATTATSI

1966 Populiatsionnaia genetika variantov gemoglobina, talessemii i nedostatochnosti gliukoza-6-fosfat-degidrogenizy v sviazi s obsuzhdeniem maliariĭnoĭ gipotezy [The population genetics of variants of hemoglobin, thalassemia and deficiencies of glucose-6-phosphate-dehydrogenesis in connection with the discussion of a Malaria hypothesis]. *Biulletin voz*, 34(3).

ZAINO, E. C.
1964 "Paleontologic Thalassemia," in *Problems of Cooley's Anemia*. New York.

Morphological Peculiarities of the Body Frame in Different Ethnoterritorial Groups

N. S. SMIRNOVA

An important objective of research in one area of anthropology is the identification of constellations of traits characteristic of certain groups. Such research is aimed at substantiating and illustrating the concept of group adaptation, long an important one in anthropology. Major emphasis is placed on the study of laws governing adaptation of human populations to the natural and social environment (Mair 1968; Roginskii 1969; Shmal'gauzen 1969; Ginsburg and Laughlin 1966).

As a part of studies in the biological adaptation of man, the Institute of Anthropology of Moscow University has been carrying out a systematic investigation of different ethnoterritorial groups in the Soviet Union (Alekseeva 1972; Alekseeva et al. 1971). We shall dwell here on only one aspect of that work, the morphological characterization of peculiarities of body-frame. It must be noted that morphological status of each group is of a complex and differentiated nature. One is immediately struck by a high degree of intragroup polymorphism of morphological properties (see Table 1). The values for the coefficients of variation in Table 1 demonstrate the extent of polymorphism which exists. In each population it is possible to find virtually the entire range of values for most traits (see Table 2) from brachymorphy to dolichomorphy in body proportions and from gracility and asthenia to athleticism and a high degree of adiposity in body frame constitution (see Table 3). It is well known in anthropology that polymorphisms in morphological structure are manifest even in groups homogeneous for sex, age, ethnic type, diet, and labor activities (Bunak 1923; Elsner 1963; Eysenk 1953; Smirnova 1971; Zenkevich 1934). This polymorphism is twofold in origin: the natural and the social. Prominent among the natural factors are autochtonous genetic processes which account for the birth of homozygous children to heterozygous parents.

Table 1. Morphological characteristics of the population under study

Feature	Ethnic group	Men n	Men M±m(M)	Men S±m(S)	Men C±m(C)	Women n	Women M±m(M)	Women S±m(S)	Women C±m(C)
Stature (in cm)	Tajiks	120	167.35±0.53	5.83±0.37	3.48±0.22	101	155.16±0.50	5.00±0.35	3.22±0.22
	Buryats	136	165.42±0.48	5.65±0.34	3.41±0.21	201	154.85±0.36	5.12±0.25	3.30±0.16
	Chukchi	70	165.08±0.60	5.12±0.42	3.10±0.26	82	152.64±0.52	4.68±0.37	3.06±0.23
	Wood Nenetz	50	160.08±0.94	4.16±0.66	2.60±0.26	42	147.62±0.94	6.24±0.66	4.23±0.45
Weight (in kg)	Tajiks	108	62.48±0.87	9.05±0.62	14.48±0.98	100	53.81±0.82	8.18±0.58	15.20±1.07
	Buryats	136	63.82±0.71	8.27±0.50	12.96±0.78	190	56.20±0.67	9.29±0.47	16.53±0.85
	Chukchi	66	63.30±0.93	7.60±0.66	12.00±1.12	75	54.52±0.84	7.24±0.59	13.27±1.08
	Wood Nenetz	53	57.38±0.71	5.18±0.50	9.02±0.87	39	47.30±1.17	7.34±0.83	15.51±1.75
Circumference of chest (in cm)	Tajiks	110	89.44±0.55	5.72±0.39	6.39±0.43	101	83.10±0.58	5.91±0.41	7.11±0.50
	Buryats	135	89.12±0.46	5.33±0.32	6.07±0.37	191	84.88±0.48	6.63±0.34	7.91±0.40
	Chukchi	69	92.66±0.66	5.50±0.47	5.93±0.50	80	87.96±0.58	5.24±0.41	5.96±0.47
	Wood Nenetz	53	88.64±0.62	4.52±0.44	5.09±0.49	41	83.18±0.68	4.38±0.48	6.26±0.58
Surface of the body (in m^2)	Tajiks	110	1.72±0.00	0.15±0.00	8.72±0.58	100	1.53±0.00	0.13±0.00	8.49±0.85
	Buryats	136	1.72±0.01	0.12±0.00	7.10±0.43	187	1.57±0.01	0.15±0.00	9.55±0.70
	Chukchi	66	1.71±0.02	0.13±0.00	7.60±0.67	75	1.55±0.01	0.13±0.00	8.38±0.68
	Wood Nenetz	51	1.64±0.01	0.09±0.00	5.48±0.54	39	1.41±0.02	0.13±0.00	9.21±1.04
Length of torso (in cm)	Tajiks	110	78.34±0.36	3.78±0.25	4.82±0.32	102	72.58±0.31	3.07±0.22	4.22±0.30
	Buryats	133	78.92±0.28	3.20±0.26	4.11±0.25	184	74.10±0.21	2.90±0.15	3.97±0.21
	Chukchi	64	77.28±0.38	2.46±0.27	3.18±0.28	79	72.60±0.26	2.36±0.18	3.25±0.25
	Wood Nenetz	43	75.22±0.44	2.90±0.31	3.85±0.41	15	72.88±0.63	2.36±0.44	3.23±0.62
Length of leg (in cm)	Tajiks	110	89.52±0.36	3.80±0.25	4.24±0.28	101	82.26±0.36	3.59±0.25	4.36±0.29
	Buryats	133	85.29±0.31	3.54±0.22	4.15±0.25	184	79.24±0.27	3.62±0.19	4.57±0.24
	Chukchi	64	88.30±0.44	3.54±0.31	4.00±0.35	79	80.04±0.37	3.32±0.26	4.14±0.33
	Wood Nenetz	43	84.40±0.51	3.30±0.36	3.90±0.43	15	77.40±0.60	3.98±0.75	5.14±0.99
Length of arm (in cm)	Tajiks	110	73.30±0.33	3.48±0.23	4.74±0.32	101	66.18±0.27	2.74±0.19	4.14±0.29
	Buryats	133	71.96±0.27	3.09±0.19	4.29±0.26	182	66.66±0.24	3.30±0.17	4.95±0.26
	Chukchi	64	72.06±0.36	2.92±0.25	4.05±0.35	79	65.44±0.31	2.76±0.22	4.22±0.33
	Wood Nenetz	43	70.82±0.47	3.10±0.33	4.37±0.47	15	65.12±0.92	3.46±0.65	5.31±1.02
Width of shoulder (in cm)	Tajiks	113	37.84±0.18	1.95±0.12	5.15±0.34	102	33.63±0.15	1.51±0.10	4.49±0.31
	Buryats	135	37.43±0.15	1.74±0.11	4.64±0.28	191	34.30±0.12	1.64±0.08	4.78±0.24
	Chukchi	70	38.39±0.21	1.88±0.15	4.89±0.42	82	34.89±0.15	1.40±0.10	4.01±0.31

Width of pelvis (in cm)	Tajiks	113	28.51±0.13	1.41±0.09	4.94±0.33	102	28.38±0.15	1.48±0.10	5.21±0.36
	Buryats	137	27.50±0.12	1.49±0.08	5.42±0.33	189	27.98±0.15	2.05±0.10	7.32±0.37
	Chukchi	70	28.90±0.17	1.44±0.12	4.98±0.42	82	29.06±0.15	1.43±0.10	5.08±0.39
	Wood Nenetz	50	27.90±0.22	1.60±0.16	5.73±0.57	40	27.90±0.21	1.34±0.15	4.80±0.53
Transversal diameter of chest (in cm)	Tajiks	113	26.80±0.16	1.72±0.11	6.41±0.43	102	24.47±0.16	1.59±0.11	6.49±0.45
	Buryats	135	27.96±0.14	1.65±0.09	5.90±0.35	189	24.92±0.12	1.69±0.08	6.78±0.35
	Chukchi	70	27.96±0.21	1.77±0.15	6.33±0.53	82	26.05±0.15	1.40±0.10	5.37±0.42
	Wood Nenetz	52	27.23±0.21	1.50±0.15	5.50±0.55	40	25.57±0.26	1.66±0.18	6.49±0.73
Longitudinal diameter of chest (in cm)	Tajiks	113	19.96±0.16	1.69±0.11	8.46±0.56	102	17.81±0.15	1.53±0.11	8.59±0.60
	Buryats	135	20.25±0.12	1.37±0.08	6.76±0.41	188	18.06±0.12	1.69±0.08	9.35±0.48
	Chukchi	70	21.86±0.21	1.79±0.15	8.18±0.69	82	19.68±0.19	1.80±0.14	9.14±0.71
	Wood Nenetz	51	21.46±0.19	1.40±0.14	6.52±0.65	40	18.98±0.26	1.67±0.18	8.79±0.98
Width of shoulder epiphysis (in mm)	Tajiks	113	69.40±0.42	4.52±0.30	6.51±0.43	101	59.28±0.32	3.24±0.23	5.46±0.38
	Buryats	137	68.54±0.30	3.51±0.21	5.12±0.31	189	60.71±0.26	3.62±0.18	5.96±0.31
	Chukchi	68	69.08±0.41	3.38±0.30	4.89±0.42	80	60.34±0.34	3.06±0.24	5.07±0.40
	Wood Nenetz	48	65.74±0.44	3.10±0.31	4.72±0.48	21	58.80±0.71	3.16±0.50	5.37±0.83
Width of forearm epiphysis (in mm)	Tajiks	113	56.66±0.28	3.00±0.20	5.29±0.35	102	49.66±0.26	2.60±0.18	5.23±0.36
	Buryats	136	58.18±0.28	3.24±0.20	5.56±0.34	190	51.38±0.22	3.12±0.15	6.07±0.31
	Chukchi	68	58.99±0.35	2.88±0.25	4.88±0.42	80	52.29±0.33	3.00±0.23	5.73±0.45
	Wood Nenetz	48	58.46±0.41	2.84±0.29	4.85±0.41	21	52.02±0.81	3.64±0.57	7.00±1.09
Width of femoral epiphysis (in mm)	Tajiks	113	95.72±0.37	4.00±0.26	4.17±0.27	102	88.18±0.49	4.94±0.34	5.60±0.39
	Buryats	137	95.28±0.42	2.87±0.29	5.10±0.31	188	89.28±0.42	5.83±0.29	6.53±0.33
	Chukchi	68	97.88±0.46	3.84±0.35	3.92±0.33	80	89.26±0.47	4.24±0.33	4.75±0.37
	Wood Nenetz	48	93.86±0.49	3.42±0.34	3.64±0.37	21	86.02±1.07	4.82±0.75	5.60±0.87
Width of shin epiphysis (in mm)	Tajiks	113	69.76±0.50	5.30±0.36	7.59±0.50	102	62.08±0.34	3.42±0.24	5.50±0.38
	Buryats	136	69.62±0.30	3.49±0.21	4.99±0.30	190	63.10±0.21	3.00±0.14	4.75±0.24
	Chukchi	68	71.26±0.40	3.30±0.29	4.63±0.39	80	63.92±0.31	2.82±0.22	4.41±0.34
	Wood Nenetz	48	68.00±0.56	3.87±0.40	5.69±0.58	21	62.12±1.02	4.56±0.71	7.34±1.14
Circumference of waist (in cm)	Tajiks	111	80.28±0.72	7.56±0.50	9.41±0.63	98	78.44±0.88	8.72±0.62	11.10±0.79
	Buryats	137	79.44±0.63	7.72±0.43	9.80±0.59	187	75.48±0.70	9.56±0.50	12.66±0.65
	Chukchi	72	81.28±0.80	6.84±0.56	8.41±0.70	76	78.88±0.94	8.24±0.65	10.44±0.84
	Wood Nenetz	54	78.44±0.57	4.20±0.40	5.35±0.51	33	71.40±1.13	6.52±0.80	9.13±1.12

Table 1. *(continued)*

Feature	Ethnic group	Men				Women			
		n	M±m(M)	S±m(S)	C±m(C)	*n*	M±m(M)	S±m(S)	C±m(C)
Circumference of buttocks (in cm)	Tajiks	105	91.64±0.49	5.08±0.34	5.54±0.38	102	91.28±0.71	7.08±0.49	7.75±0.54
	Buryats	137	93.16±0.45	5.28±0.31	5.76±0.34	187	94.12±0.53	7.36±0.37	7.82±0.40
	Chukchi	67	93.96±0.51	4.20±0.35	4.47±0.38	75	96.92±0.60	5.19±0.42	5.35±0.43
	Wood Nenetz	51	89.56±0.55	3.96±0.39	4.42±0.44	33	88.60±0.78	4.52±0.55	5.10±0.62
Circumference of shoulder (in cm)	Tajiks	113	27.54±0.23	2.40±0.16	8.71±0.58	102	26.11±0.27	2.78±0.19	10.64±0.74
	Buryats	134	28.62±0.18	2.14±0.12	7.47±0.45	187	27.31±0.23	3.16±0.16	11.57±0.60
	Chukchi	67	28.51±0.09	1.77±0.06	6.20±0.53	79	27.78±0.26	2.37±0.18	8.53±0.67
	Wood Nenetz	51	16.29±0.20	1.47±0.14	5.59±0.55	35	24.13±0.42	2.52±0.30	10.44±1.24
Circumference of hip (in cm)	Tajiks	110	50.87±0.52	5.42±0.37	10.65±0.72	102	51.45±0.47	4.76±0.33	9.25±0.65
	Buryats	135	52.52±0.33	3.94±0.23	7.50±0.46	185	54.00±0.35	4.82±0.24	8.92±0.46
	Chukchi	66	52.52±0.43	3.52±0.30	6.70±0.58	77	53.26±0.44	3.92±0.31	7.36±0.58
	Wood Nenetz	44	48.68±0.49	3.27±0.34	6.72±0.71	35	48.30±0.67	3.96±0.48	8.19±0.97
Circumference of shin (in cm)	Tajiks	111	34.43±0.23	2.49±0.16	7.23±0.48	102	32.85±0.28	2.83±0.19	8.61±0.60
	Buryats	135	34.00±0.20	2.12±0.14	6.24±0.38	186	33.20±0.20	2.02±0.14	6.09±0.31
	Chukchi	67	34.20±0.23	1.90±0.16	5.55±0.48	79	32.65±0.23	2.09±0.16	6.40±0.50
	Wood Nenetz	51	31.50±0.28	2.05±0.20	6.51±0.65	36	30.22±0.36	2.19±0.25	7.24±0.85
Average fat column (in mm)	Tajiks	100	9.27±0.33	3.39±0.23	36.56±2.58	100	13.44±0.37	3.76±0.26	27.97±0.98
	Buryats	137	9.84±0.32	4.40±0.22	44.72±2.70	192	15.18±0.36	5.04±0.25	33.20±1.69
	Chukchi	71	8.24±0.29	2.50±0.20	30.33±2.56	81	31.28±0.39	3.52±0.27	26.50±2.07
	Wood Nenetz	48	6.98±0.20	1.41±0.14	20.20±2.06	20	10.80±0.69	3.02±0.49	27.96±4.48
Amount of hypodermic fat (in kg)	Tajiks	100	5.14±0.34	3.36±0.24	65.37±4.62	100	8.36±0.34	3.36±0.24	40.19±2.80
	Buryats	137	5.54±0.26	3.04±0.18	54.87±3.31	189	9.46±0.31	4.33±0.21	45.77±2.35
	Chukchi	66	4.40±0.26	2.16±0.18	49.09±4.28	74	7.78±0.36	3.16±0.25	40.61±3.33
	Wood Nenetz	46	3.13±0.14	0.98±0.09	31.31±3.26	18	5.83±0.58	2.38±0.41	40.82±6.20
Overall amount of fat (in kg) by Mateiki	Tajiks	107	9.14±0.35	3.66±0.25	40.04±2.70	100	13.28±0.50	5.04±0.35	37.95±2.60
	Buryats	136	10.40±0.48	5.60±0.34	53.84±3.26	188	15.68±0.52	7.12±0.36	45.40±2.33
	Chukchi	66	9.04±0.44	3.56±0.31	39.38±3.42	74	13.54±0.50	4.30±0.36	31.75±2.61
	Wood Nenetz	46	6.90±0.21	1.43±0.15	29.48±2.13	18	10.12±0.88	3.66±0.62	36.16±6.11
Percentage of fat to weight of body	Tajiks	111	14.92±0.40	4.30±0.28	28.82±1.93	100	24.24±0.47	4.72±0.33	19.47±1.37
	Buryats	137	17.16±0.49	5.80±0.34	33.80±2.04	187	27.24±0.56	7.72±0.39	28.34±1.46
	Chukchi	66	14.06±0.46	3.78±0.33	26.88±2.33	74	24.84±0.61	5.26±0.43	21.17±1.74
	Wood Nenetz	46	12.18±0.34	2.34±0.24	19.21±2.00	18	19.56±1.42	5.88±1.00	30.06±5.08

Table 2. Range of variability of morphological status with reference to some features. Extreme values observed in groups under review

Feature	Ethnic group	Men		Women	
		Min.	Max.	Min.	Max.
Stature in cm	Tajiks	149.0	185.0	141.2	168.0
	Buryats	151.3	180.4	143.4	168.2
	Chukchi	153.4	176.6	142.2	166.8
	Wood Nenetz	152.9	168.8	131.6	158.8
Weight in kg	Tajiks	51.0	100.0	40.0	87.0
	Buryats	47.8	91.0	40.0	89.6
	Chukchi	46.2	96.2	42.0	79.8
	Wood Nenetz	43.0	69.3	29.9	61.0
Width of shoulders in cm	Tajiks	31.5	42.0	28.9	36.8
	Buryats	34.2	42.2	30.8	38.6
	Chukchi	32.0	42.0	32.0	38.2
	Wood Nenetz	32.0	40.8	30.0	36.2
Circumference of chest in cm	Tajiks	78.0	110.0	72.0	98.2
	Buryats	76.6	107.2	72.2	105.0
	Chukchi	80.0	113.9	76.0	106.0
	Wood Nenetz	76.9	98.2	74.9	91.9
Average fat column in mm	Tajiks	3.0	19.0	7.3	24.6
	Buryats	5.0	25.0	5.4	31.3
	Chukchi	3.5	16.6	7.7	23.0
	Wood Nenetz	2.8	10.1	7.4	19.5

Table 3. Percentage of relative incidence of basic types of body-frame among men in groups under review

Type of body-frame	Ethnic group			
	Tajiks	Buryats	Chukchi	Wood Nenetz
Asthenic	28.4	23.6	11.4	17.0
Athletic	20.6	15.0	31.5	32.1
Brachymorphic	9.8	24.3	4.2	–

Yet the high degree of variability within groups (Table 2) cannot conceal the qualitative differences separating groups from each other. This is because morphological characteristics within each group are in optimum conformity with local natural conditions, thus producing an entity with a number of specific features of its own. This entity and the uniqueness of each group's morphological features can be viewed as a result of the impact of the main variability vector which acts at the population level. Let us consider data on morphology from the viewpoint of adaptability to different conditions of the biological and geographical environment for the following groups: the Tajiks (the Tajik Soviet Socialist Republic, Isfaghin District, village of Chorm); the Buryats (the Buryat Autonomous SSR, the eastern region of the Trans-Baikal area, villages of the Barguzin Depression); the Chukchi (the Chukotsk Coast, settlements of Uelen and Lorino); and the Nenetz (Western Siberia, Pur basin, settlement of Tarno-Saleh). It should be kept in mind that the Chukchi and Nenetz inhabit arctic areas; the Buryats, the continental belt; and the Tajiks, the foothills of the south. The morphological data were collected in 1967, 1968, 1969, and 1970.

The groups surveyed were made up of natives related through marriage. Subjects were able-bodied adults, twenty to fifty years of age (the average age ranging from thirty-two to thirty-six). The groups were homogeneous ethnically and in socioeconomic status and diet. Table 1 shows the data for a number of morphological features in the form of basic statistical parameters.

The Nenetz and Chukchi, despite all the differences between them (Table 1), have a number of common features, as do all aboriginal populations of the northern and northeastern parts of Siberia. These include: a relatively small stature; high values for specific surface weight (28–29 centimeters per kilogram) of the body; large absolute and relative thorax size, with characteristically large anterior-posterior diameter; relatively broad shoulders; massive epiphysis of the bones of the extremities; and weak (in the Nenetz) or medium (in the Chukchi) growth of adipose tissue. Large circumferences of the chest, waist, and buttocks relative to the length of the body produce a body frame typical of these ethnic groups. In other words, the variability vector of the morphological status of the arctic belt groups is slanted (in the case of the Nenetz and Chukchi) toward a mesomorphic body frame. This peculiarity is only observed as a trend at the group level. It should be noted that the morphology of the Nenetz and Chukchi groups is generally characterized by the narrowing of the variability range compared to other groups — the Buryats, the Tajiks (see Table 1). This suggests the existence of mechanisms limiting the degree of variability in a number of systems in populations under extreme environmental conditions.

The Buryats exhibit decreased lean body mass associated with intensified growth of adipose tissue (Tables 1 and 3). Among the surveyed groups, the Buryats show the greatest average values for body weight and a relatively small surface per kilogram of weight (26.4 centimeters in men; 26.0 centimeters in women). They are relatively short-legged (leg length amounts to 51.6 percent of stature in men; 51.2 percent in women). The brachymorphic type is commonly observed in both men and women. These peculiarities of morphology may be viewed as a result of adaptation to low temperatures and the dry climate deep in the hinterland of Asia. Apparently, the continental environment produced a variability factor slanted toward the endomorphic features which conventionally include an intensified growth of adipose tissues, the conic shape of the thorax, and the brachymorphic type of body frame.

An analysis of intragroup variability of morphological features, such as the skeletal, muscle, and adipose tissue, reveals its relationship to the distribution of types of body frame (and constitutional types) in different populations (Heath and Carter 1968, 1969).

Thus, a higher degree of variability of the adipose component in the Buryats (see Table 1) is due to the fact that instead of asthenic types, the group includes a substantial number of brachymorphic types (24 percent of the persons surveyed). However, the uneven distribution of features related to fat accumulation (the average fat column, or fat complement including subcutaneous fat) complicates analysis of both intra- and intergroup variability for this feature. The environment of the Tajiks, who inhabit the foothills of the Pamirs, differs in biological and geographical conditions from those of the arctic and continental belts. The Tajiks are generally tall, with a high incidence of flattened chests, and relatively weak development of the muscle and fat components combined with relatively narrow shoulders and long legs. Thus, the Tajiks are characterized by a predominance in their morphological status of ectomorphic features (i.e., relative elongation of the body and its segments with a weak development in width and depth), which is typical of inhabitants of mountainous or hot climate regions. As in the other groups, these features are revealed through statistical averages for characteristics or by way of the distribution of features, i.e., degree of variability. One could cite more examples of this nature. In our opinion, an analysis of the complex mechanisms producing variability in morphological properties must take into account the interaction of two tendencies. These are the centrifugal tendency, which results in polymorphism, and the centripetal vector, which reflects a set of conditions all acting in one direction and thus gradually transforming an anthropological type, imparting specific features to it which promote adaptation (Alekseev 1970).

REFERENCES

ALEKSEEV, V. P.
1970 O dvukh protivopolozhnykh tendentsiiakh v rasoobrazovanii [On two opposing tendencies in race formation]. *Voprosy antropologii* 35.
ALEKSEEVA, T. I.
1972 "Biologicheskie aspecty izucheniia adapiatsii u cheloveka" ["Biologi-

cal aspects of the study of adaptation in man"], in Symposium *Antropologiia 70*, 103–129. Moscow: Moscow State University.

ALEKSEEVA, T. I., V. P. VOLKOV-DUBROVIN, O. M. PAVLOSKII *et al.*

1971 Antropologicheskie issledovaniia v Zabaikal'e v sviazi s problemoi adaptatsii u cheloveka (morfologiia, fiziologiia i populiatsionnaia genetika) [Anthropological research in the Trans-Baikal region in connection with the problem of adaptation in man (morphology, physiology and population genetics)]. Parts I and II. *Voprosy antropologii* 36/37.

BUNAK, V. V.

1923 Neskol'ko dannykh po voprosu o tipichnykh constitutsiiakh cheloveka [Some data on the question of human constitutional types]. *Russkii antropologicheskii zhurnal* 1.

ELSNER, P. W.

1963 Skinfold thickness in primitive peoples native to cold climate. *Annals of the New York Academy of Sciences* 110:11.

EYSENK, H. J.

1953 *The structure of human personality.* London: Methuen.

GINSBURG, B. B., W. S. LAUGHLIN

1966 The multiple bases of human adaptability and achievement: a species point of view. *Eugenics Quarterly* 13:3.

HEATH, B. H., J. E. CARTER

1968 Sovremennye metody somatotipirovaniia [Contemporary methods of somatic typology]. Part I. *Voprosy antropologii* 29.

1969 Sovremennye metody somatotipirovaniia [Contemporary methods of somatic typology]. Part II. *Voprosy antropologii* 33.

MAIR, E.

1968 *Zoologicheskii vid i evoliutsiia* [*Zoological species and evolution*]. Moscow: Paese.

1970 Morfo-fiziologicheskie issledovaniia v antropologii [Morphophysiological research in anthropology]. Moscow: Moscow State University.

ROGINSKII, IA. IA.

1969 "O tipakh kharaktcra i ikh znachenii v teorii antropogeneza" ["On character types and their meaning in the theory of anthropogenesis"], in *Problemy antropogeneza*, chapter 6. Moscow.

SHMAL'GAUZEN, I. I.

1969 *Problemy darvinizma* [*Problems of Darwinism*]. Leningrad: Academy of Sciences.

SMIRNOVA, N. S.

1971 Opyt morfo-funktsional'noi kharakteristiki osnovnykh tipov teloslozheniia muzhchin [Attempt at a morpho-functional characterization of basic male body types]. *Voprosy antropologii* 38.

ZENKEVICH, P. I.

1934 Fizicheskii tip gornykh i lugovykh mari [The Physical type of the mountain and meadowland Mari]. *Mariiskaia avtonomnaia oblast'* 8–12.

Effects of Isolation on Genetic Variability: Macaque Populations as Model Systems

LAURA NEWELL MORRIS and PETER E. NUTE

INTRODUCTION

In studying the genetics of one or a group of organisms, it is all too easy to adopt an atomistic point of view, focusing, for example, on a few loci whose products are readily detected by biochemical means. This approach is understandable, especially when applied to human populations, since the investigator is often hampered by lack of time and personnel and cannot obtain data relevant to more than a few genetic parameters. In addition, some physical anthropologists and population geneticists share the attitude, traceable to the early use and overuse of blood groups as surrogates for anthropometric measurements, that as much information can be gained from the study of a few genetic loci as might be obtained from a more broadly based investigation.

One outgrowth of this restricted approach is the tendency to regard organisms as simple sums of the discrete effects of genes at unrelated loci. This view completely disregards the fact that organisms and/or populations comprise highly complex, well-integrated biological systems. Although several theorists have stressed the evolutionary importance of gene interaction and coadapted gene complexes with respect to both individuals and populations (Birdsell 1972; Mayr 1963), the results of genetic analyses are seldom couched in holistic terms.

Of great importance to the present analysis is the concept that the survival value of any particular gene depends not only upon those environmental conditions external to the organisms in question, but also upon the compositions of the genomes and gene pools in which the gene exists. This unifying view implies that all traits are, in a broad sense,

Supported in part by the United States Public Health Service National Institutes of Health grants RR 00166 and GM 15253.

polygenic, and underlies our assertion that genetically complex and simple Mendelian traits should share many similarities in their responses to conditions that alter gene frequencies and generate or restrict variability in natural populations. Our hypothesis is supported by the consistent results of those few studies in which serological and/or biochemical analyses were combined with observations of gross morphology (Birdsell 1950; Pollitzer 1958).

We have chosen several species of the genus *Macaca* as the subjects of our study (see Figure 1). Three reasons underlie this choice: (1) a large

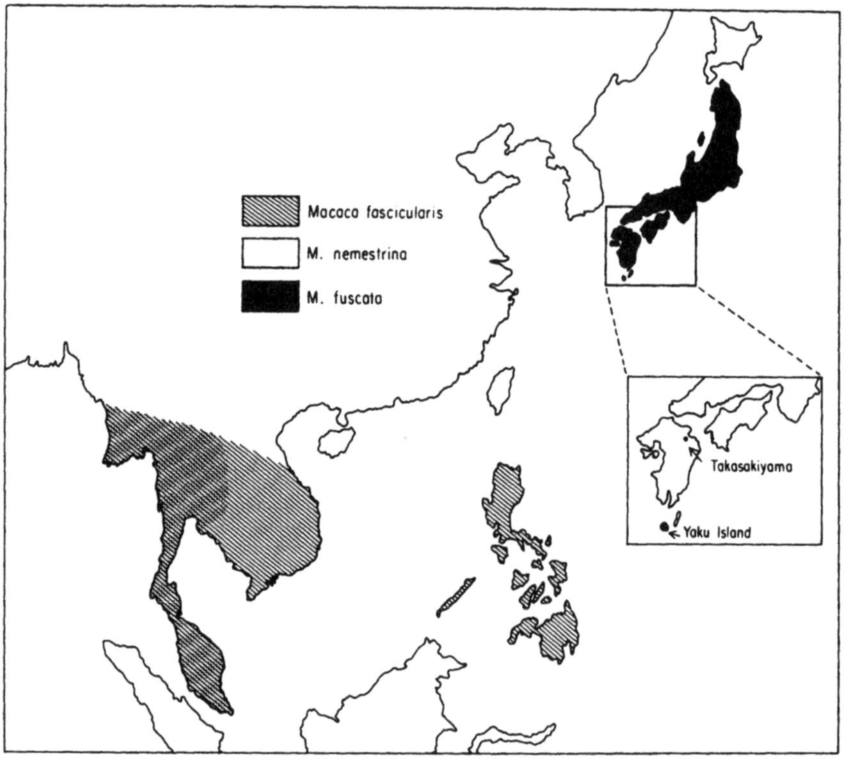

Figure 1. The geographical distributions of the mainland and insular macaques discussed in this paper. Only the two groups of *M. fuscata* can be specifically located within the range depicted (inset)

body of data derived from comparative genetic studies of different species, as well as of different populations of the same species, is available for analysis; (2) macaque populations are well suited to the study of the effects of isolation on the compositions of gene pools and provide a model for the analysis of similar effects in human populations; (3) the results of these studies provide information of importance to the study of human evolutionary history.

We shall discuss the extent to which variability generated in part by gene duplication, variability produced by alleles at single loci, and variability manifested in a system of morphological traits are differentially distributed among mainland and insular populations of macaques.

ANALYSIS OF DATA

Macaque Hemoglobins: Correlation of Phenotype with Genotype

The existence of duplicate loci provides a source of genetic variability of a different nature from that ascribed to multiple alleles at a single locus. While one can readily infer genotype from phenotype in most cases of multiple allelism, the introduction of duplicate loci often renders such correlations difficult. This problem is encountered in the case of the hemoglobin α chains of several species of *Macaca*. By first considering the hemoglobin phenotypes of animals in which duplicate α-chain loci are known to exist, we shall construct a model whose application permits assessment of α-chain genotypes according to the hemoglobin phenotypes found in several species of macaques.

The basic elements of our model are derived from consideration of the frequencies and kinds of hemoglobin phenotypes in *Macaca arctoides* (= *speciosa*). Kouba and Kitchen (1970) found that each of 150 macaques (*M. arctoides*) had electrophoretically fast (HbF) and slow (HbS) hemoglobins which differed in the structures of their α chains. In addition, Ishimoto et al. (1970) found 189 FS phenotypes among 191 animals of this species; the remaining two individuals had HbS alone.

The large proportion of apparent heterozygotes (99.4 percent) among the total of 341 animals is most readily accounted for by postulating that the α^{S} gene may exist singly or in close linkage with the α^{F} gene, and that nearly all members of this species are homozygous for the linked genes. The two animals with HbS alone are presumed to be homozygous for a single α^{S} allele. The latter allele must be present at a low frequency among *M. arctoides*, but one might expect to find animals heterozygous for the single α^{S} allele and the linked α^{S}–α^{F} genes among 341 individuals.

How might these heterozygotes be identified when present in a group of animals, most of which produce both hemoglobins? Presumably, hemoglobin phenotypes of animals with one α^{F} and two α^{S} genes would differ from animals with two of each type of gene (that is, homozygotes for the linked genes) according to the relative amounts of HbF and HbS present in their red cells. Of the 150 animals examined by Kouba and

Kitchen (1970), six (4 percent) had HbF and HbS in an approximate ratio of 30 F: 70 S; the remaining 144 had the hemoglobins in a 60 F: 40 S ratio. Ishimoto et al. (1970), upon examination of 52 animals of FS phenotypes, found that all had levels of HbF between 55 and 75 percent, with a majority (39) falling between 60 and 70 percent Hb F.

In the absence of any evidence that unstable hemoglobins or β-chain variants are present, we can assume that the relative amounts of HbF and HbS are proportional to the amounts of α^S and α^F chains present in the erythrocytes. On this basis, we conclude that the least-common phenotype, 30 F: 70 S, is produced by heterozygosity for the linked loci and a single α^S allele and that the 60 F: 40 S phenotype is produced by homozygosity for the linked loci.

It is tempting to assume that both partners in a gene duplication would govern the synthesis of equal amounts of chains. Were this the case, HbF and HbS would each comprise about 50 percent of the total hemoglobin in animals homozygous for the duplication. However, quantitative disparities in the amounts of hemoglobin chains produced by duplicate loci are common among primates (Nute, Buettner-Janusch, and Buettner-Janusch 1969; Nute and Stamatoyannopoulos 1971a, 1971b; Boyer et al. 1969; Wade and Barnicot 1971; Huisman et al. 1972; de Jong 1971). In short, there appear to be pronounced "position effects" bearing on the levels of activity of closely linked loci. This observation leads to further refinement of our model by the introduction of two assumptions, both of which were initially suggested by Barnicot, Wade, and Cohen (1970). First, one of a pair of linked α-chain genes governs the synthesis of twice as much chain as does its partner; second, two α-chain genes, when linked, govern the synthesis of the same amount of chain as does a single α allele. Adding these assumptions to the model, we can correlate genotypes with phenotypes as illustrated in Table 1.

Table 1. Correlation of hemoglobin α-chain phenotype with genotype in *Macaca arctoides*

Phenotypes observed (percent of each Hb)	60 F:40 S		30 F:70 S		100 S
Genotypes[a]	$\frac{\alpha^F}{33}$	$\frac{\alpha^S}{17}$	$\frac{\alpha^F}{33}$	$\frac{\alpha^S}{17}$	$\frac{\alpha^S}{50}$
	$\frac{\alpha^F}{33}$	$\frac{\alpha^S}{17}$		$\frac{\alpha^S}{50}$	$\frac{\alpha^S}{50}$
Percent α^F predicted	66		33		
Percent α^S predicted	34		67		100

[a] Numbers listed beneath the line representing each gene refer to the percentage of total α chain produced by each gene. It should be noted that the contributions of the loci residing in each of the two homologous chromosomes add up to 50 percent of the total.

The Major Hemoglobins of Macaca fascicularis (= irus): *Application of the Model*

The hemoglobins of *M. fascicularis* have been more intensively studied than have those of any other cercopithecoid. Several major hemoglobins have been described, all of which differ in the structures of their α chains. Barnicot, Huehns, and Jolly (1966) and Barnicot, Wade, and Cohen (1970) designated the various chains α^A, α^Q, and α^P. Wade, Barnicot, and Huehns (1970) and Boyer et al. (1972) found that the α^A chains from some animals were heterogeneous; as there are no data available on their distributions among populations of *M. fascicularis*, we shall refer to them jointly as the α^A chain.

A summary of the data on the major hemoglobin phenotypes of *M. fascicularis* appears in Table 2. The absence of animals homozygous

Table 2. Distribution of hemoglobin phenotypes among insular and mainland *Macaca fascicularis*: number of animals of each phenotype[a]

Phenotype	Q	AQ	A	QP	AP	P	AQP	References
Insular								
Philippines	0	118	0	0	0	0	0	Ishimoto, Toyomasu, and Uemura 1968; Ishimoto et al. 1970
Mainland								
Cambodia	0	2	111	0	0	0	0	Ishimoto, Toyomasu, and Uemura 1968
Vietnam	2	34	77	2	8	1	0	Barnicot, Huehns, and Jolly 1966
Malaysia	0	371	83	21	24	1	3	Barnicot, Huehns, and Jolly 1966; Barnicot, Wade, and Cohen 1970; Ishimoto et al. 1970
Thailand	1	75	132	11	32	3	0	Barnicot, Huehns, and Jolly 1966; Ishimoto et al. 1970
Total Mainland	3	482	403	34	64	5	3	

[a] Only results of studies carried out after Barnicot, Huehns, and Jolly (1966) differentiated HbP from HbQ are included in this analysis. The Hb's F and S described by Ishimoto, Toyomasu, and Uemura (1968) and Ishimoto et al. (1970) are equated with HbQ and Hb A, respectively.

for HbA or HbQ among the Philippine *M. fascicularis* indicates that, whenever the α^Q gene is present, it is linked to the α^A gene. Ishimoto et al. (1970) measured the relative proportions of HbA and HbQ in fifty-seven Philippine *M. fascicularis*; fifty-one animals had HbQ in excess of 50 percent, with a mode between 65 and 70 percent. In the remaining six macaques, HbQ amounted to only 30 to 40 percent of the total hemoglobin. According to our model, the two classes of phenotypes found among Philippine *M. fascicularis* are produced by homozygosity for the linked α^Q and α^A genes (65–70 percent Q: 30–35 percent A) and heterozygosity for the linked genes and a single a^A allele (30–40 percent Q: 60–70 percent A).

Although duplicate genes on a single chromosome are not allelic to a single gene on the homologous chromosome in the strictest sense, the duplicate is expected to act as such in gametogensis. Hence, it is possible to calculate frequencies of duplicate and single genes in the same fashion as one calculates allelic frequencies. Basing our calculations on the data of Ishimoto et al. (1970), we find that the frequency of the duplication is 0.95. Philippine *M. fascicularis* thus appear to be polymorphic for the linked and single genes, but the frequency of the single a^A gene is sufficiently low (0.05) to identify it as the atypical allele.

A third hemoglobin, HbP, is found among mainland *M. fascicularis*. The distribution of HbP among the various phenotypes indicates that the single a^A and a^P genes are allelic (Wade, Barnicot, and Huehns 1967; Barnicot, Huehns, and Jolly 1966; Barnicot, Wade, and Cohen 1970). The highly complex collection of phenotypes found among mainland populations (see Table 2) might have arisen through a series of events similar to those depicted in the following scheme:

1. Alleles in ancestral gene pool; $\underline{a^Q} \leftarrow \underline{a^A}$
   Duplication
2. Later gene pool contains: $\underline{a^Q\,a^A}$ $\underline{a^Q}$ $\underline{a^A}$
   Mutation
3. Contemporary gene pool contains: $\underline{a^Q\,a^A}$ $\underline{a^P}$ $\underline{a^Q}$ $\underline{a^A}$

The linked genes and single alleles segregate, according to the proposed model, to produce the genotypes and phenotypes listed in Table 3. If all of the genotypes listed in Table 3 were present in a group of animals, one would expect the proportion of HbQ in AQ individuals to vary over a wide range, with values for HbQ at or near 33, 50, 66, 83, and 100 percent. Because of possible overlapping of amounts of a^Q produced by different genotypes (see Figure 2), calculations of gene frequencies are subject to error.

Barnicot, Huehns, and Jolly (1966) and Ishimoto et al. (1970) measured the proportions of HbQ and HbA in 157 mainland *M. fascicularis* of AQ phenotype. The resultant distribution of HbQ is essentially bimodal, with clustering about 35–40 percent and 65–70 percent (see Figure 2). Phenotypes containing 45–55 percent HbQ may represent relatively small numbers of animals heterozygous for the single a^A and a^Q alleles, while the few animals with HbQ in excess of 75 percent may indicate heterozygosity for the linked a^A and a^Q genes and the single a^Q allele (see Table 3).

Table 3. Correlations of phenotypes with proposed genotypes in *Macaca fascicularis*[a]

Genotype	$\frac{a^Q}{33}\ \frac{a^A}{17}$	$\frac{a^Q}{33}\ \frac{a^A}{17}$	$\frac{a^Q}{33}\ \frac{a^A}{17}$	$\frac{a^Q}{33}\ \frac{a^A}{17}$	$\frac{a^A}{50}$
	$\frac{a^Q}{33}\ \frac{a^A}{17}$	$\frac{a^A}{50}$	$\frac{a^Q}{50}$	$\frac{a^P}{50}$	$\frac{a^A}{50}$
Phenotype	66 Q/34 A	33 Q/67 A	83 Q/17 A	33 Q/17 A/50	P 100 A
Genotype	$\frac{a^Q}{50}$	$\frac{a^P}{50}$	$\frac{a^A}{50}$	$\frac{a^A}{50}$	$\frac{a^P}{50}$
	$\frac{a^Q}{50}$	$\frac{a^P}{50}$	$\frac{a^Q}{50}$	$\frac{a^P}{50}$	$\frac{a^Q}{50}$
Phenotype Q	100 Q	100 P	50 Q/50 A	50 A/50 P	50 P/50 Q

[a] See footnote to Table 1.

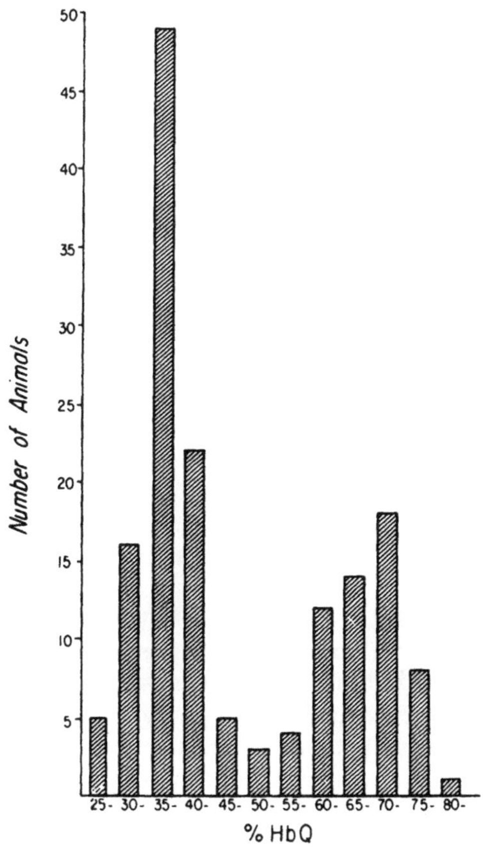

Figure 2. Percentage of HbQ in 157 mainland *M. fascicularis* of AQ phenotype

The large number of AQ phenotypes and the paucity of a^Q homozygotes among mainland populations indicate that the a^Q gene is most frequently found in linkage with the a^A gene. Linkage of the a^Q and a^A genes is confirmed by the existence of animals with a^A, a^Q, *and* a^P chains, given the allelic nature of the a^A and a^P genes. The proportions of HbQ (38 percent) HbA (22 percent) and HbP (40 percent) were measured in one of these animals (Barnicot, Wade, and Cohen 1970) and are in close agreement with those predicted by the proposed model (33 percent Q, 17 percent A, 50 percent P; see Table 3). It is likely that some HbP was lost during isolation of this polymerized component by gel filtration.

There are obvious discrepancies between the frequencies of QP phenotypes reported among Malaysian *M. fascicularis* that cannot be explained by our model, Ishimoto et al. (1970) described 21 QP phenotypes among 262 animals examined; Barnicot, Huehns, and Jolly (1966), and Barnicot, Wade, and Cohen (1970) found no QP phenotypes among 241 animals from the Malay Peninsula. It is possible that the difficulties inherent in distinguishing the QP from the P phenotype might contribute to misidentification, but this seems unlikely in view of the absence of animals with HbQ alone and the near absence of animals with only HbP from both Malaysian groups. Until further data derived from more extensive analyses of the QP phenotypes are available, these puzzling observations can only produce speculation involving additional hemoglobin variants which mimic HbQ or HbP in electrophoretic properties, further duplication producing linked a^Q and a^P genes, or gene introgression through hybridization between *M. fascicularis* and neighboring species of macaques.

Hemoglobin Variation: Insular Versus Mainland M. fascicularis

It is difficult to account for the distributions of the phenotypes found among various populations of mainland *M. fascicularis*. First, there are obvious discrepancies between the gene pools of different mainland populations. For example, 111 of 113 Cambodian animals were homozygous for the single a^A allele, while most (74 percent) Malaysian *M. fascicularis* had both HbQ and HbA (see Table 2). The frequencies of A and AQ phenotypes in Thailand and Vietnam appear to bridge the gap between those found in Malaysia and Cambodia.

However, when the hemoglobin phenotypes of mainland *M. fascicularis* are compared with those of insular populations, differences in degree of variability are observed which parallel those described below for alleles at single loci. High levels of variability characterize that portion of a species occupying a mainland range; insular populations show a lesser degree of variability. Among mainland *M. fascicularis*, a great degree of phenotypic variation is produced by various combinations of the linked a^A and a^Q genes and the single a^A, a^Q, and a^P alleles, all of which exist at polymorphic frequencies (>0.01). It is clear that the insular (Philippine) *M. fascicularis* are more uniform than mainland groups with respect to their hemoglobin genotypes. In some respects, their phenotypic identity is complete, since all possess both hemoglobins A and Q.

In this context, it is noteworthy that the hemoglobins of 217 *M. fuscata* (from Japan) and 101 *M. cyclopis* (from Taiwan) showed no signs of variation; all yielded a single electrophoretic species of hemoglobin identical to that of HbA in *M. fascicularis* (Ishimoto et al. 1970). However, it may be premature to describe *M. fuscata* and *M. cyclopis* as monomorphic at hemoglobin loci, as more elaborate screening procedures may demonstrate variation. For example, *M. nemestrina* from the Malay Peninsula and *M. mulatta*, once thought to be monomorphic at hemoglobin loci, show a-chain variation similar to that involving the linked and nonlinked a^A and a^Q genes of *M. fascicularis* when screened by sufficiently rigorous techniques (Nute and Pataryas 1974; Basch 1972).

Allelic Variability: Insular Versus Mainland Macaque Populations

Of the many allelic systems that have been demonstrated within populations of macaques, we shall consider only those listed in Table 4. Other systems were eliminated from consideration because (1) the number of animals examined was so small that the absence of allelic variation did not justify the diagnosis of monomorphism, or (2) there were difficulties in distinguishing between the presence of multiple alleles and/or multiple loci or in correlating phenotypes (in these cases, patterns produced by electrophoresis of specific proteins) with genotypes.

Most variants of human proteins differ from their normal counterparts by single amino acid substitutions whose origins can best be explained by postulating the past occurrence of point mutations at their respective loci (Harris 1970). Since the variation described for each type of protein listed in Table 4 is presumed to reflect corresponding variation at a single genetic locus, we assume that most, if not all, of the variation arose in this way.

In general, one views alleles at a single locus as comprising a polymorphic system if each of their frequencies in a gene pool exceeds 0.01. Electrophoretic variants present at lower frequencies are not included in Table 4 as they may be isolated, familial abnormalities that are in no way characteristic of the population in question. Inspection of Table 4 indicates that mainland populations harbor more genetic variability than do insular populations of macaques.

This difference is particularly impressive with respect to the transferrins of *M. fascicularis*. The apparent monomorphism in Philippine members of this species is striking when contrasted with the high degree of variability apparent in mainland populations, among which there are at least nine different transferrin alleles present at polymorphic frequencies. To a lesser extent, the distributions of variant forms of 6-phosphogluconate dehydrogenase, carbonic anhydrase I, carbonic anhydrase II, and NADH diaphorase fit the pattern exhibited by the transferrins.

Table 4. Number of alleles present at polymorphic frequencies (>0.01)[a]

Genetic systems	Insular populations *M. fuscata* (Japan)	*M. fascicularis* (Philippines)	Mainland populations *M. fascicularis*	*M. nemestrina*
Transferrin				
Ishimoto, Toyomasu, and Uemura 1967	1 (260)	—	—	—
Prychodko et al. 1969	2 (394)	1 (232)	9 (484)	5 (274)
Lai 1972	—	—	7 (358)	—
6-phosphogluconate dehydrogenase (6-PGD)				
Prychodko et al. 1971	1 (294)	1 (156)	3 (798)	2 (450)
Carbonic anhydrase I (CAI)				
Tashian et al. 1971	2 (240)	1 (148)	2 (794)	3 (402)[b]
Carbonic anhydrase II (CAII)				
Tashian et al. 1971	1 (186)	2 (32)	2 (132)	2 (58)
Methemoglobin reductase (NADH diaphorase)				
Ishimoto 1971	1 (538)	1 (54)	2 (188)	3 (50)

[a] Numbers of alleles tested (twice the number of animals) appear in parentheses.
[b] All animals showing the CAI (0) phenotype were excluded from consideration, since it is not yet known if this phenotype is the result of allelic variation at the CAI locus or of repression mediated by action at another locus.

Data from other macaque groups not considered in detail here provide additional evidence for reduced variability in insular populations. Prychodko et al. (1969) found only two transferrins in sixty-five Taiwanese *M. cyclopis*; in contrast, they found eleven molecular forms of transferrin at polymorphic frequencies among 309 mainland *M. mulatta*.

Dermatoglyphic Analyses: Comparative Uses

Dermatoglyphic analyses have long served as means for comparing human populations phenotypically (see Biswas 1961). Some investigators have, in fact, suggested that dermatoglyphic comparisons provide as much information about population affinities as do the more simply inherited Mendelian traits (Rife 1954; Roberts et al. 1968). Moreover, this use of dermatoglyphic systems is illustrated by the complementary results of studies involving combinations of dermatoglyphic analyses with analyses of other morphological or Mendelian traits from the same population(s) (Newman 1960; Chai 1971).

Nonhuman primates possess volar dermatoglyphic patterns that can be described by standard human dermatoglyphic nomenclature. Midlo and Cummins (1942) have analyzed in detail the dermatoglyphic similarities and differences among several primate taxonomic groups down through the generic level. However, they held that interspecific differences, if they exist, would be of little use in differentiating species. Later studies of species within the genera *Presbytis*, *Colobus*, *Cercopithecus*, and *Macaca* (Brehme 1967a, 1967b, 1968; Morris and Kerr 1974) have shown that dermatoglyphic differences do exist between species of the same genus, and that dermatoglyphic comparisons at this taxonomic level can yield valuable information about the adaptation and divergence of species. Apart from our present comparisons of the dermatoglyphic patterns of *M. fuscata fuscata* and *M. f. yakui*, there are no data on which to base comparisons of conspecific populations of nonhuman primates.

Dermatoglyphic Pattern Frequencies and MSD

The dermatoglyphic data used here were obtained from three distinct groups (see Figure 1): 42 *M. f. yakui* from Yaku Island, Japan; 56 *M. f. fuscata* from the Takasakiyama group of Kyushu Island, Japan (Iwamoto 1964); and 79 *M. nemestrina* from mainland Malaysia (Morris and Kerr 1974). The frequencies of the various pattern types were calculated for each of the seven configurational areas of the palm. Patterns from the right and left hands of both sexes were combined, giving a sample number double the number of animals used. Each of the several pattern types has been assigned to one of four general categories: open field, loop, double loop, and whorl. The frequencies at which the four categories of pattern types appear in each of the seven configurational areas of *M. f. yakui*, *M. f. fuscata*, and *M. nemestrina* are listed in Table 5. It is obvious that there are differences among the three groups of animals, although these variations do not appear vastly different from those that have been reported for various human populations. However, it is clear that particular configurational areas of *M. f. yakui* are as different from those of *M. f. fuscata* as they are from the corresponding areas of *M. nemestrina*.

The pattern frequencies offer rough indications of the dermatoglyphic similarities and dissimilarities among the three macaque groups. Although we would agree with Birdsell (1972) that no single figure can

Table 5. Dermatoglyphic pattern frequencies of seven configurational areas

	Thenar				Hypothenar proximal				Hypothenar distal			
	O	L	D	W	O	L	D	W	O	L	D	W
M. nemestrina	0.94	0.06	—	—	—	0.82	0.17	—	0.94	0.66	—	—
M. fuscata yakui	0.17	0.83	—	—	—	+	—	—	+	—	—	—
M. fuscata fuscata	0.67	0.33	—	—	—	0.48	0.30	0.23	0.69	0.29	—	0.03

Interdigitals	I				II				III				IV			
	O	L	D	W	O	L	D	W	O	L	D	W	O	L	D	W
M. nemestrina	—	0.45	0.16	0.38	—	0.29	0.08	0.64	—	0.03	—	0.96	—	0.19	0.11	0.69
M. fuscata yakui	—	0.84	0.05	0.12	—	—	—	+	—	0.08	—	0.90	—	—	—	+
M. fuscata fuscata	—	0.65	0.26	0.08	—	—	—	+	0.05	0.23	—	0.72	—	—	—	+

Key: (+) indicates pattern fixation; (—) indicates absence of pattern. Pattern types: O = open field; L = loop; D = double loop; W = whorl. Any pattern not found in more than one hand is not included, as such single occurrences are not considered important in any assessment of phenotypic variability.

adequately summarize and measure the extent of biological variability between populations, calculation of the mean square dermatoglyphic distance (MSD) as described by Hughes (1967) provides an overall statement of the differences in pattern frequencies between any two populations. The mean square distances (MSD) for all configurational areas are averaged to obtain a total MSD (see Table 6). Again, *M. f. yakui* is dermatoglyphically distinct, showing a significantly greater MSD from *M. nemestrina* (MSD = 1125) than does *M. fuscata* (MSD = 485) (see Figure 3).

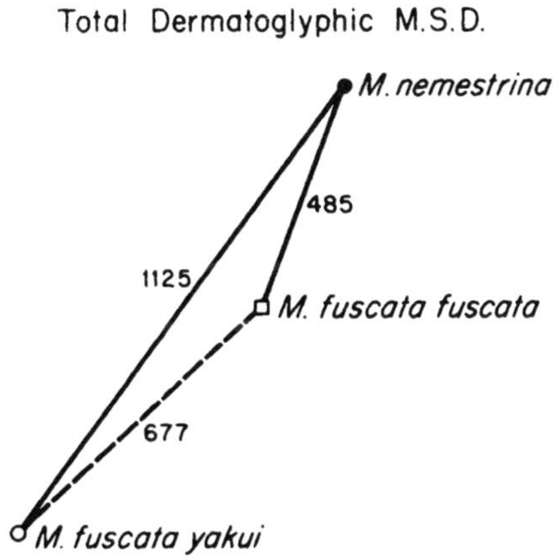

Figure 3. The degree of dermatoglyphic divergence between three groups of macaques, as measured by the mean square distance (MSD)

A second observation emerges when the dermatoglyphic data are analyzed in terms of range of phenotypic variability (variability is here defined as the number of general categories of pattern types exhibited within a population). The mainland Malaysian *M. nemestrina* show more variation of pattern type than do the groups of *M. fuscata* (see Table 5). Patterns are fixed in four of the seven configurational areas of *M. f. yakui* and in two of the areas in *M. f. fuscata*; no pattern fixation was observed in *M. nemestrina*.

Variability of Growth and Development

In addition to simple frequencies of whorls, loops, and open fields, dermatoglyphic pattern types reflect the developmental stages through which volar pads have passed during prenatal life. According to Midlo and Cummins (1942):

> The volar pads are very properly studied in conjunction with the dermatoglyphs which figure their surfaces. . . . It is probable that the directions of epidermal

Table 6. Dermatoglyphic mean square distance (MSD) based on seven configurational areas [a]

Group	Th	Hy^p	Hy^d	I	II	III	IV	Total MSD
MSD from a (*M. nemestrina*)								
a *M. nemestrina*								
b *M. fuscata yakui*	5990.8	185.5	22.2	585.8	732.0	15.4	345.8	1125
c *M. fuscata fuscata*	761.8	583.2	392.3	348.3	694.0	251.3	364.2	485
MSD from b (*M. fuscata yakui*)								
b *M. fuscata yakui*								
c *M. fuscata fuscata*	2480.0	1309.9	601.6	204.9	0.54	142.7	0.71	677

[a] The MSD's were calculated from the total number of pattern types and subtypes present, rather than from the four general categories shown in Table 5.

ridges, hence their fashioning of patterns, are conditioned by mechanical processes in growth. Irregular reliefs of the volar surfaces owe their existence to differential growth, and there is every indication that epidermal ridge directions are but another expression of the same growth processes. Thus, a local pattern of ridges, such as an apical pattern of a digit, possesses a distinctive ridge configuration owing to the growth of the pad in which it is seated.

If the developmental association between volar pad morphology and dermal ridge patterning is accepted, then we have in the range and distribution of dermatoglyphic pattern types within a population a rough means of assessing the range of variability expressed throughout a part of prenatal development. For example, *M. nemestrina* shows developmental polymorphism at all configurational areas, as expressed by range of pattern types, whereas *M. f. yakui* is developmentally more uniform at the hypothenar proximal, hypothenar distal, and interdigitals II and IV. Conceivably, all animals in the latter population undergo uniform development during prenatal life with respect to these particular areas.

DISCUSSION

Several studies have shown that insular populations tend to harbor less genotypic and phenotypic variability than do related mainland populations (Dobzhansky 1957; Goodman et al. 1965; Prychodko et al. 1969). In addition, these and other studies have demonstrated inverse relationships between island size and amount of genetic diversity within resident populations and the degree of genetic differentiation between their populations (Kramer and Mertens 1938; Lowe 1955). For example, the investigation of Dobzhansky (1957) of chromosomal variability in populations of *Drosophila willistoni* from Central and South America and the West Indies showed a decreased average number of heterozygous inversions per female in island populations (1.78) as compared with mainland populations (5.25), with the lowest number from the island of St. Kitts (0.20) and the highest from Santa Marta, Colombia (7.21). In addition, Dobzhansky correlated insular size and amount of genetic diversity; for example, females from Cuban populations contain the greatest number of inversions (2.67) while those from St. Kitts, the smallest island, approach monomorphism.

Insular and mainland populations of macaques differ in extent of variability in much the same way as do insular and mainland populations of *D. willistoni*. Where multiple alleles exist at single loci, mainland *M. fascicularis* far outstrip their Philippine counterparts in the numbers of different alleles they possess and, hence, the numbers of phenotypes that can be produced through segregation. For example, the nine transferrin alleles found by Prychodko et al. (1969) among mainland

M. fascicularis are potentially capable of combining to produce forty-five different electrophoretic phenotypes.

By contrast, Philippine members of this species are phenotypically and genotypically identical. A similar situation is apparent in the case of the hemoglobin α-chain genes. The three single α-chain alleles, α^A, α^Q, and α^P, among mainland *M. fascicularis*, can produce six different α-chain phenotypes. However, the addition of the tandemly linked pair of genes, α^Q-α^A, introduces an element of quantitative variability without altering the structural nature of any chain. In this way, the three different α chains participate in the production of ten different hemoglobin phenotypes (see Table 3).

The only phenotypic variation described for the hemoglobins of Philippine *M. fascicularis* is of a quantitative nature. More than 90 percent of the animals showed the 65–70 Q: 35–40 A phenotype, indicative of homozygosity for the tandemly linked α^Q-α^A genes. All other Philippine *M. fascicularis* were of the 30–40 Q: 60–70 A phenotype. Thus, although two structurally different α chains exist among the Philippine animals, variability is limited, and these insular macaques are close to achieving monomorphism for the duplicate genes.

With respect to the palmar dermatoglyphics, insular *M. fuscata* show less total pattern variability than do mainland *M. nemestrina* (see Table 5). Furthermore, *M. f. yakui* from Yaku Island show less variability than do *M. f. fuscata* (Takasakiyama group) from the larger island of Kyushu. If the developmental hypothesis of dermatoglyphic pattern formation is accepted, fixation of pattern type as exhibited by the groups of *M. fuscata* reflects uniformity of the prenatal growth and development of the palmar pad.

It is difficult to discuss growth and development in genetic and evolutionary terms. Developmental polymorphisms have been described for human populations, but the effects of genetic isolation on growth and development have not been conclusively documented (Hunt 1966). Yet as Van Valen (1973) has stated, "A plausible argument could be made that evolution is the control of development by ecology." Certainly, considerations of growth and development as reflected in dermatoglyphic patterns, single loci, and linked loci lead to the same conclusions: insular populations tend to exhibit less variability than do mainland populations, and the manner in which the former have responded to isolation is reflected at several levels of biological organization.

The question of why related mainland and insular populations differ has been discussed in detail by Mayr (1963). He rejects genetic drift and natural selection as insufficient explanations for the evolutionary divergence of insular populations. Mayr's explanation for the differences between insular and mainland populations stresses the concept of an integrated gene pool and its reaction to gene flow. In a polytypic species, gene flow has two important consequences. First, opportunities for new heterozygous genotypes are maximized in populations that continually receive new alleles from other populations. Second, gene flow acts to maintain the specific identity of populations by preventing the local establishment of gene complexes in response to local selective pressures.

When a population becomes isolated from the mainstream of gene flow the evolutionary pressures on this population change. Generally these changes are discussed in terms of the founder principle (see Mayr 1963; Dobzhansky 1957). One can justifiably invoke the founder principle as an explanation of genetic discontinuities between ancestral and descendant populations when the latter derive from a small number of individuals bearing a nonrandom sample of the alleles present in the former. Many insular populations may owe their uniqueness to an initial founder effect. However, isolated populations may also be created by the fragmentation of a previously continuous range. The direction of divergence and the rapidity with which it will occur in newly isolated populations are unpredictable. Fission of the range of a species, as opposed to the operation of the founder effect, may provide for different rates of evolutionary change. However, the ultimate fate of isolated populations, whatever their origin, is the same: heterozygosity, whether measured by complete fixation of a single gene at a locus or by a reduction in the total number of alleles at a given locus, will decrease.

The genetic divergence and reduced variability characteristic of insular macaque populations is probably not a consequence of the founder effect, but rather the result of subdivision of the range of a polytypic species and the concurrent removal of these populations from the mainstream of gene flow. The twelve extant species of *Macaca* recognized by Napier and Napier (1967) represent, in Simons' view (1963), the descendants of a single species that spanned the Old World as recently as one to three million years ago. It is noteworthy that *M. fascicularis*, *M. nemestrina*, *M. mulatta*, and *M. arctoides*, whether from insular or mainland populations, all possess duplicate hemoglobin α-chain loci. Similarly, the transferrins, 6-phosphogluconate dehydrogenases, and carbonic anhydrases I and II of insular macaques are electrophoretically identical to certain forms of these proteins found among mainland populations. These observations strongly suggest a relatively recent genetic communality of insular and mainland populations of macaques.

An Old World-wide species range would have supported numerous populations inhabiting diverse environments and differing in the extent of their isolation from each other. Clearly, such a situation would tend to favor the establishment of genetic differences between various populations while permitting the flow of genes from one end of the range to the other. Under these conditions, one suspects that genotypic (and hence phenotypic) differences between populations accumulated to the extent that the single ancestral species became highly polytypic. However, gene flow between contiguous populations would have maintained conspecific identity of these populations throughout the range.

Several macaque populations were isolated as postglacial Pleistocene elevations of sea level submerged the continental shelves connecting the present islands of Japan, the Philippines, Indonesia, and Malaysia to the mainland, thus interrupting gene flow between these areas. Just as gene flow previously tended to neutralize the effects of selection on various gene pools, thereby inhibiting evolutionary divergence, so the absence of gene flow between mainland and newly isolated populations permitted more rapid divergence of the latter groups from each other and from mainland populations. In this context, the reduced variability within insular populations as compared with mainland populations is viewed as a consequence of the operation of natural selection within a more restricted geographic range, while the variability characteristic of mainland populations reflects the continual flow of genes between populations dispersed over a wider area.

The conclusions drawn from consideration of the data presented above may be applied to questions concerning the course of human evolution. Unlike *Macaca*, *Homo* has remained a widespread monospecific genus. In spite of the existence of geographic barriers between human populations, continual species-wide gene flow has been maintained by cultural developments that facilitated human mobility and communication. However, one important question remains unanswered: have any human populations experienced loss of variability (as distinct from the loss of heterozygosity attributable to consanguinity) to the extent described for insular populations of macaques? A satisfactory answer requires that numerous genetic systems be investigated in several human populations that have experienced little gene flow for long periods of time.

REFERENCES

BARNICOT, N. A., E. R. HUEHNS, C. J. JOLLY

1966 Biochemical studies on hæmoglobin variants of the irus macaque. *Proceedings of the Royal Society*, B, 165:224–244.

BARNICOT, N. A., P. T. WADE, P. COHEN

1970 Evidence for a second hæmoglobin α-locus duplication in *Macaca irus*. *Nature* 228:379–381.

BASCH, R. S.

1972 Hæmoglobin polymorphism in the rhesus macaque. *Nature New Biology* 238:238–240.

BIRDSELL, J. B.

1950 Some implications of the genetical concept of race in terms of spatial analysis. *Cold Spring Harbor Symposia on Quantitative Biology* 15:259–314.

1972 The problem of the evolution of human races: classification or clines? *Social Biology* 19:136–162.

BISWAS, P. C.
1961 Ethnogeographic variations in dermatoglyphics. *Proceedings of the Scandinavian International Congress on Human Genetics* 196: 1430–1449.

BOYER, S. H., E. F. CROSBY, T. F. THURMON, *et al.*
1969 Hemoglobins A and A_2 in New World primates: comparative variation and its evolutionary implications. *Science* 166:1428–1431.

BOYER, S. H., A. N. NOYES, C. F. TIMMONS, R. A. YOUNG
1972 Primate hemoglobins: polymorphisms and patterns. *Journal of Human Evolution* 1:515–543.

BREHME, H.
1967a Hautleisten System der Palma und Planta von sechs Arten der asiatischen Schlankaffen (*Presbytis*). *Zeitschrift für Morphologische Anthropologie* 59:109–145.
1967b Untersuchungen am Hautleisten System der Palma und Planta von *Colobus polykomos*, *Colobus badius*, *Colobus verus* und *Nasalis larvatus*. *Folia Primatologica* 6:243–283.
1968 Zur Variabilität des Hautleisten Systems der Meerkatzen (*Cercopithecus*) und des Husarenaffen (*Erythrocebus patas*). *Folia Primatologica* 9:41–67.

CHAI, C. K.
1971 Analysis of palm dermatoglyphics in Taiwan indigenous populations. *American Journal of Physical Anthropology* 34:369–376.

DE JONG, W. W. W.
1971 Chimpanzee foetal haemoglobin: structure and heterogeneity of the γ chain. *Biochimica et Biophysica Acta* 251:217–226.

DOBZHANSKY, T.
1957 Genetics of natural populations XXVI. Chromosomal variability in island and continental populations of *Drosophila willistoni* from Central America and the West Indies. *Evolution* 11:280–293.

GOODMAN, M., A. KULKARNI, E. POULIK, E. REKLYS
1965 Species and geographic differences in the transferrin polymorphism of macaques. *Science* 147:884–886.

HARRIS, H.
1970 *The principles of human biochemical genetics.* New York: American Elsevier.

HUGHES, D. R.
1967 Finger dermatoglyphics from Nuristan, Afghanistan. *Man* 2:121–125.

HUISMAN, T. H. J., W. A. SCHROEDER, W. H. BANNISTER, J. L. GRECH
1972 Evidence for four nonallelic structural genes for the γ chain of human fetal hemoglobin. *Biochemical Genetics* 7:131–139.

HUNT, E. E.
1966 "The developmental genetics of man," in *Human development*. Edited by Frank Falkner. Philadelphia: W. B. Saunders.

ISHIMOTO, G.
1971 Electrophoretic patterns of red cell NADH diaphorase in macaque monkeys. *Journal of the Anthropological Society of Nippon*. 79:382–387.

ISHIMOTO, G., T. TANAKA, H. NIGI, W. PRYCHODKO
1970 Hemoglobin variation in macaques. *Primates* 11:229–241.

ISHIMOTO, G., T. TOYOMASU, K.-I. UEMURA
1967 Serum transferrins of Japanese macaques: comparisons with other species of monkeys. *Primates* 8:29–34.
1968 Intraspecies variations of red cell enzymes and hemoglobin in *Macaca irus*. *Primates* 9:395–408.

IWAMOTO, M.
1964 Morphological studies of *Macaca fuscata*. Dermatoglyphics of the hand. *Primates* 5:53–73.

KOUBA, E. O., H. KITCHEN
1970 A comparative structural study on the hemoglobins in *Macaca speciosa*. *Archives of Biochemistry and Biophysics* 140:415–424.

KRAMER, G., R. MERTENS
1938 Rassenbildung bei west-istrianischen Inseléidechsen in Abhängigkeit von Isolierungsalter und Arealgrösse. *Arch. Naturgeschichte* 7:189–234.

LAI, L. Y. C.
1972 Transferrins in *Macaca irus*. *Folia Primatologica* 17:193–201.

LOWE, C. H., JR.
1955 An evolutionary study of island faunas in the Gulf of California, Mexico, with a method for comparative analysis. *Evolution* 9:339–344.

MAYR, E.
1963 "Change of genetic environment and evolution," in *Evolution as a process*. Edited by J. Huxley, A. C. Hardy, and E. B. Ford. New York: Collier.

MIDLO, C., H. CUMMINS
1942 *Palmar and plantar dermatoglyphics in primates*. American Anatomical Memoirs, number twenty. Philadelphia: Wistar Institute of Anatomy and Biology.

MORRIS, L. N., B. A. KERR
1974 Genetic variability in macaques: correlations between biochemical and dermatoglyphic systems. *Journal of Human Evolution* 3:223–235.

NAPIER, J. R., P. H. NAPIER
1967 *A handbook of living primates*. New York: Academic Press.

NEWMAN, M. T.
1960 Population analysis of finger and palm prints in highland and lowland Maya Indians. *American Journal of Physical Anthropology* 18:45–58.

NUTE, P. E., V. BUETTNER-JANUSCH, J. BUETTNER-JANUSCH
1969 Genetic and biochemical studies of transferrins and hemoglobins of *Galago*. *Folia Primatologica* 10:276–287.

NUTE, P. E., H. A. PATARYAS
1974 Multiple fetal and adult hemoglobins in Malaysian *Macaca nemestrina*: consequences of α-chain variation. *American Journal of Physical Anthropology* 40:17–25.

NUTE, P. E., G. STAMATOYANNOPOULOS
1971a Evidence for duplication of the haemoglobin γ chain locus in *Macaca nemestrina*. *Nature New Biology* 229:145–147.
1971b The ontogenesis of hemoglobins in *Macaca nemestrina*. *Blood* 38:108–115.

POLLITZER, W. S.
1958 The Negroes of Charleston; a study of hemoglobin types, serology and morphology. *American Journal of Physical Anthropology* 16:241–263.

PRYCHODKO, W., M. GOODMAN, E. POULIK, *et al.*
1969 Geographic variations of transferrin allelic frequencies in continental

and insular macaque populations. *Proceedings of the Second International Congress of Primatology* 2:103–108.

PRYCHODKO, W., M. GOODMAN, B. M. SINGAL, *et al.*

1971 Starch-gel electrophoretic variants of erythrocyte 6-phosphogluconate dehydrogenase in Asian macaques. *Primates* 12:175–182.

RIFE, D. C.

1954 Dermatoglyphics as ethnic criteria. *American Journal of Human Genetics* 6:319–327.

ROBERTS, D. F., E. M. COOPE, F. S. JACKSON, *et al.*

1968 Digital dermatoglyphics of a Lunana sample from North Bhutan. *Man* 3:5–19.

SIMONS, E. L.

1963 Some fallacies in the study of hominid phylogeny. *Science* 141: 879–889.

TASHIAN, R. E., M. GOODMAN, V. E. HEADINGS, *et al.*

1971 Genetic variation and evolution in the red cell carbonic anhydrase isozymes of macaque monkeys. *Biochemical Genetics* 5:183–200.

VAN VALEN, L.

1973 Review of *Evolutionary biology*, volume 6, edited by Theodosius Dobzhansky, Max K. Hecht, and William C. Steere. *Science* 180:488.

WADE, P. T., N. A. BARNICOT

1971 Studies on the minor hæmoglobin components of anthropoid apes and New World monkeys. *Comparative Biochemistry and Physiology* 39B:659–671.

WADE, P. T., N. A. BARNICOT, E. R. HUEHNS

1967 Possible duplication of hæmoglobin *a*-chain locus in the irus macaque. *Nature* 215:1485–1487.

1970 Structural studies on the major and minor hæmoglobin of the monkey *Macaca irus*. *Biochimica et Biophysica Acta* 221:450–466.

SECTION TWO

Growth and Development

Introduction

WILLIAM A. STINI

It is clear from the papers comprising Section One that human populations possess characteristics that stamp them as residents of particular areas of the world. It is less certain, however, that populations known for certain characteristics will inevitably exhibit them. Variability within populations is great enough that there are very few traits that are not expressed in some individuals in a manner more characteristic of another population. This is true for traits under the control of a number of genes, such as body size, proportion, and skin color. It is also true for single-gene traits, such as the blood group antigens, where frequencies shift from place to place but total absence of one or another allele is the exception rather than the rule. In the case of blood group antigens, the selective pressures underlying gene frequency changes are poorly understood, but we do know that an individual blood type A/B or M/N retains that blood type for a lifetime. Environmental stress will not alter the expression of such traits within the lifetime of the individual, and phenotype accurately reflects genotype. This is not so for many human traits. Great deviation from the genotypic pattern is possible in traits that depend on extended periods of growth and development to attain full phenotypic expression. Human plasticity is itself an adaptive strategy of great importance in allowing our species to inhabit most of the world's habitat zones without producing reproductively isolated populations.

Much of the plasticity humans exhibit is a product of altered growth and development under stress, deviations from maximum values being greatest in cases of early and prolonged stress. Because of this capacity for alterations in the growth process, resulting in varying adult values for many characteristics, we cannot be sure that the phenotype most often observed in a specific climatic zone is truly a genetic adaptation. Many contemporary studies of human variation focus on the interaction of genotype and environment in producing characteristic phenotypes. Therefore, the study of human growth and development has much to offer students of human biology, including those concerned with evolutionary processes.

The following papers provide a sampling of the range of applications to which the study of human growth and development may lend itself.

Variations in Female Physique

O. G. EIBEN

One of the main objectives of physical anthropology and human biology is to examine individual variability and differentiation. One area of variability is physique. Up to the present, research has been concerned primarily with males. Thus a detailed study of the female physique seems justified.

MATERIAL AND METHODS

The anthropometric investigations reported in this paper were carried out on the most outstanding women athletes of Europe, as well as on female students of the Physical Education College of Budapest and of the Teachers' College of Szombathely, West Hungary. The first group attained the developmental stage through special physical training for athletic activity on all-European level. The second group included individuals selected on the basis of physique, representing an average population. Within the first group several subgroups were identified according to athletic events. The women athletes, as groups of individuals pursuing various well-differentiated athletic events, were used as models.

The women athletes were examined at the Eighth European Athletic Championship in Budapest in 1966. Altogether 125 women athletes from Albania, Belgium, Denmark, France, the German Democratic Republic, the German Federal Republic, Great Britain, Greece, Hungary, Iceland, Italy, the Soviet Union, and Yugoslavia were examined. Their ages varied between fifteen and thirty-six years (mean = 23.8 years). All belonged to the Europoid races.

For comparative data, 139 girls who had passed their entrance examination in 1966 at the Physical Education College of Budapest were examined. Their ages varied between eighteen and twenty-five years (mean = 18.7 years). An examination of the same kind was performed in 1964 on 179 female students of the Szombathely Teachers' College. They ranged in age from eighteen to twenty-two years (mean = 19.9 years). With special considerations in mind, twenty-six women fencers, members of the selected and second-line teams of Hungary, were included in the examination. Their age varied between eighteen and thirty-four years (mean = 23.0 years).

In the course of the study an extensive anthropometric program was carried out, according to Martin's measuring technique (Martin and Saller 1957). The data were processed by a computer. The usual parameters of fifty-three measured and calculated anthropometric characteristics were computed. Further, total and partial correlation analysis, as well as analysis by means of generalized coordinates, were performed.

RESULTS AND DISCUSSION

The detailed analysis of the body measurements, the valuation of the calculated relative body measurements and indices, as well as the correlation examinations are dealt with in other papers by this author (Eiben 1972). Further analysis of the data yielded by the examination is performed with generalized coordinates.

Working Hypothesis

Morphological differences can also be demonstrated by comparing several body measurements. The working hypothesis of the author can be set up as follows:

the subgroups made up of individuals who successfully participate in some sport (athletic event) are distinguishable on the basis of their physique, i.e., there is a connection between physique and successful participation in sports, as well as in the event pursued.

Examination of these groups should constitute a test of this hypothesis.

Initiation of Generalized Coordinates

The introduction of certain generalized coordinates dependent upon body measurements is justified in that, strictly speaking, there are no "pure types" in Kretschmer's system (1967). Anyone is an "alloy of physique." If we know k body measurements of the examined persons and wish to compare them in their totality, then a point of the k-dimensional space can be assigned to each of them. Because body measurements are not independent of one another, the points representing the examined persons are arranged on a so-called hypersurface. This also means that (discounting smaller statistical fluctuations) less than k body measurements also determine the physique of the examined person. The points representing the persons are not arranged at random in a k-dimensional parallelepipedon determined by extreme body measurements, since the body measurements are correlated.

According to the original system of coordinates, a k coordinate belongs to each of the body measurements, and the coordinates are perpendicular to one another. I intend to present a method in which a new system of coordinates, the "principal axes of inertia," can be obtained and used for solving the outlined problems of morphology.

The Method of Dyad Separation

The first main direction is determined in an iterative way. Let us consider matrix M, of which the rows represent the subjects, and the columns, the body measurements. This is what one intends to represent with the least possible error as one dyad. The next step involves an M' (residual) matrix, the elements of which are, by that stage, generally small. The dyad is, in fact, a matrix determined by means of a column vector and a row vector: the factors of the dyad. The dyad is the so-called dyadic product of these two components, i.e., a matrix, of which the rows are the scalar multiples of the second component. The scalar multipliers are the elements of just the first component. (Csébfalvi (1969) gives a more detailed discussion of dyads; for a discussion of the first application of dyad separation in anthropology, see Eiben 1972). Since the vectors intersect orthogonally, the various dyads do not contain unidirectional components. Consequently the examinees can be represented in an orthogonal coordinate system.

The advantage of the method outlined here is that it is simple to program for computation technique, it permits economic storage of a great number of body measurement data for many subjects, and recovery of the data within the limits of the errors of measurements.

The Anthropological Content of the Dyads

In the vertical column of the first dyad, the figures belonging to the examined persons are the so-called coordinates of the individuals referring to the main direction: the properties of greatest importance and discriminating power. These coordinates are, of course, not identical with body measurements; they are instead certain linear functions of all body measurements. The nature of these functions is decided by the row vector of the dyad which contains, in a sense, the average proportions of the body measurements in the first dyad.

The anthropological importance of the first dyad is easy to understand if one considers the row vector as representing average body measurements and the column vector as the size of the subject (general size and shape factor). The second and third dyads bear all of the residual matrices (the remaining mass of information) that they can express.

Differentiation of the Physique by Means of Normal Components

The body measurements of the examined persons can be resolved to "normal components".

The first normal component (I) reflects the measurements of the subject; its coefficient referring to the subject can be considered an expression of "size and shape." The second normal component (II) indicates the deviation from the average body measurements, its coefficient, the morphological form and robustness of the body, while the third normal component (III) shows the secondarily differentiating measurements and their coefficient, muscularity.

For a further analysis of the morphological characters, the data of the second and third normal components are used. In this respect, the information conveyed by these components is greater than that of the first normal component. The constructed ellipsoidic regions of dispersion show the results of the study: in the planar coordinate system of the second and third normal components, the subgroups made up of successful athletes (who therefore possess various physical abilities) disperse to different fields (Figures 1 through 6).

If special attention is given to the data collected from fencers and the magnitude of error overall, it is possible to consider normal components as fixed for the group.

Under such circumstances, if new subjects are added, recalculation of the components from the enlarged data set is unnecessary. As can be seen from the control examination, this would not cause substantial change in values of the normal components. When adding a subject, generalized coordinates of his/her body measurements can be calculated by simple scalar multiplication.

The author does not intend to go into further detail of the mathematical aspects of the method here. By analysis of the intrinsic relationships of body measurements, a plane can be obtained, within which, upon per-

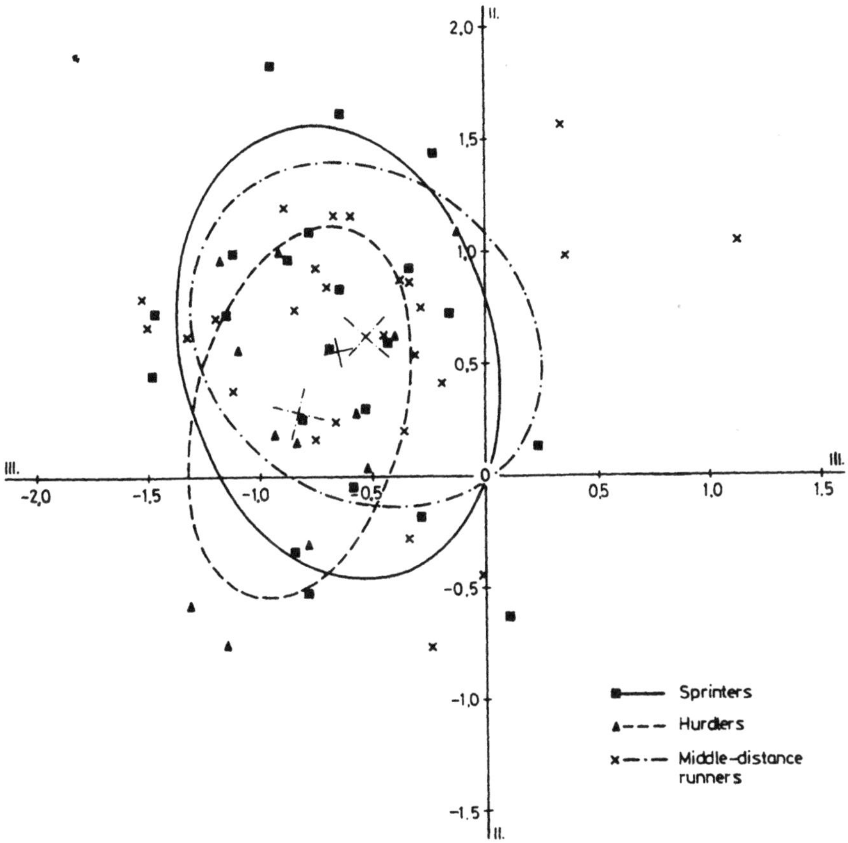

Figure 1. Distribution of female runners on normal components II and III

pendicular projection, subjects involved in athletic events will be found in different fields.

Connection Between Body Measurements and Physique

The figures presented above are as significant for human biologists and sports anthropologists as for researchers in physical education or sports trainers. The question is, which morphological marks and what body measurements characterize women athletes and female college students found in one or another field of the planar coordinate system. This is determined by the body measurement vectors (Figure 7). The length measurements: stature (4), suprasternal height (5), shoulder height (6), elbow height (7), wrist height (8), finger height (9), height of the anterior

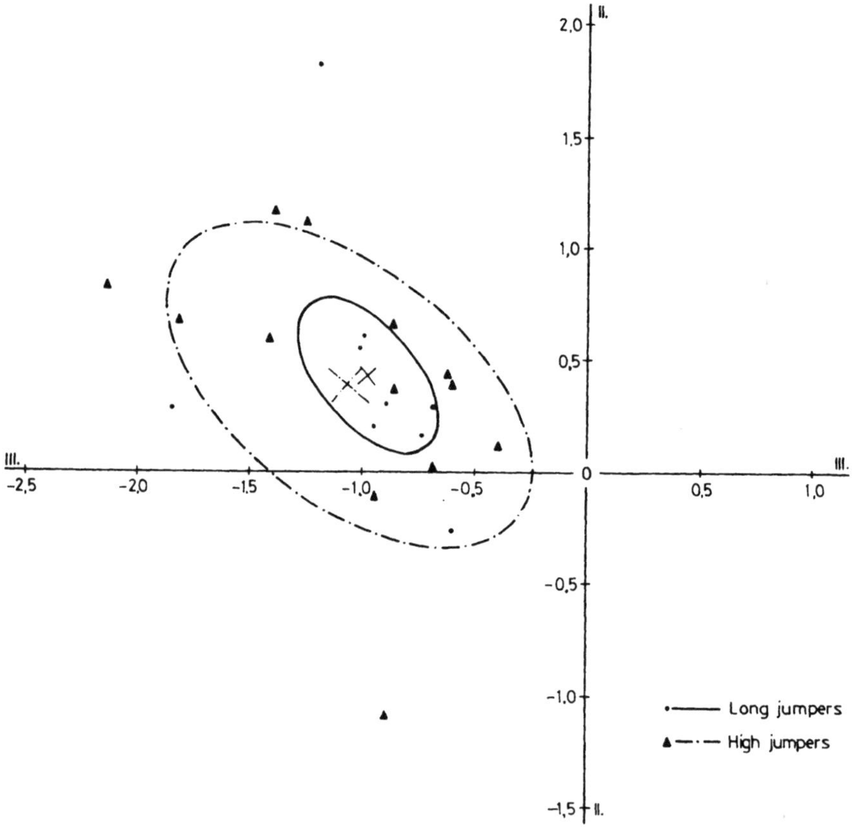

Figure 2. Distribution of female jumpers on normal components II and III

superior iliac spine (10), knee height (11), and the span (3) affect the direction of the upper left field.

The width measurements and circumferences of the upper part of the trunk: shoulder width (13), bideltoid width (14), back width (19), chest circumference (20), the circumferences of the extremities: upper-arm circumference (23), forearm circumference (24), wrist circumference (25), thigh circumference (26), calf circumference (27), as well as body weight (1) affect the lower left field.

The measurements of width and circumference of (mainly the upper part of) the trunk: bispinal width (15), bitrochanter width (16), chest breadth (17), chest depth (18), abdomen circumference (21), and ankle circumference (28) affect the lower right field.

Only sitting height (2) and ankle height (12) come into the upper right field; these will be disregarded in the remainder of this study.

Relying on the body measurement vectors, one could provisionally

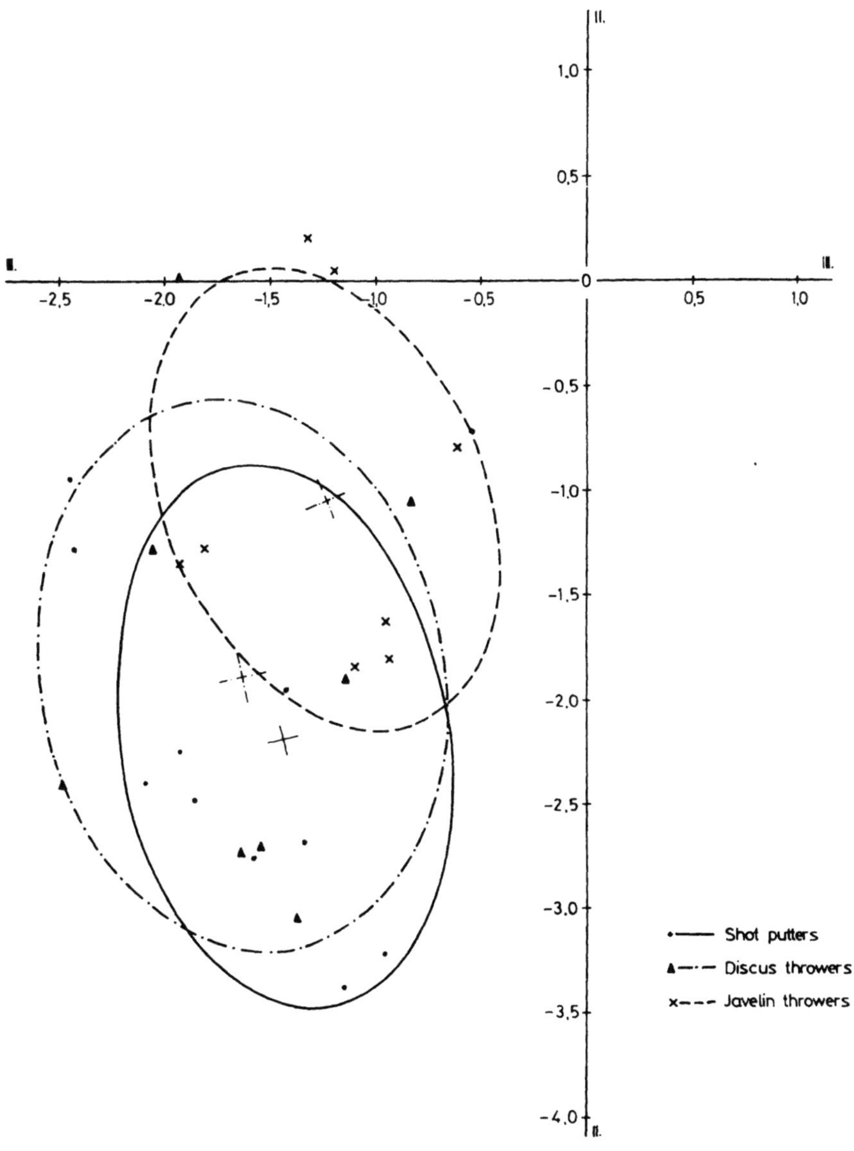

Figure 3. Distribution of female throwers on normal components II and III

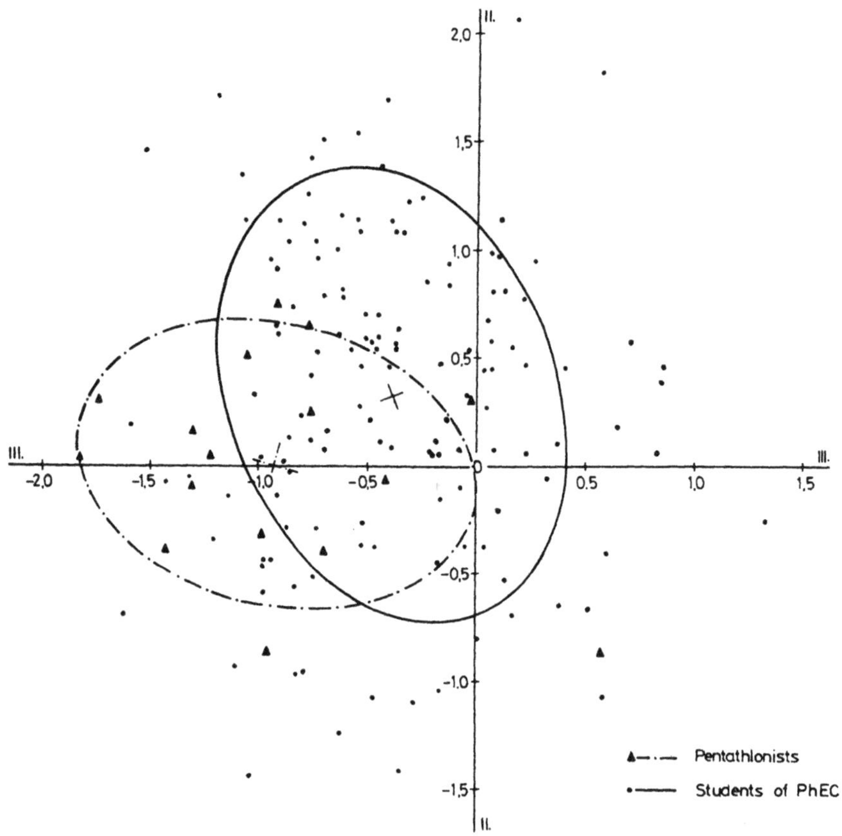

Figure 4. Distribution of female pentathlonists and physical education majors on normal components II and III

call the upper left field the *linearity field*; the lower left field, the *muscle field*; and the lower right field the *viscera field*. The main directions of the vectors of the three fields divide the planar coordinate system in such a way that if one turns it clockwise about 120 degrees, the picture resembles the triangle applied in Sheldon's somatotyping method (Sheldon, Stevens, and Tucker 1940). In the author's system the viscera direction corresponds to endomorphy, the muscle direction of mesomorphy, and the linearity direction to ectomorphy (Figure 8).

Accordingly, a remarkable connection could be found between Sheldon's somatotyping method and the author's method, which analyzes physique by means of normal components, objective components, i.e., founded on mathematical principles.

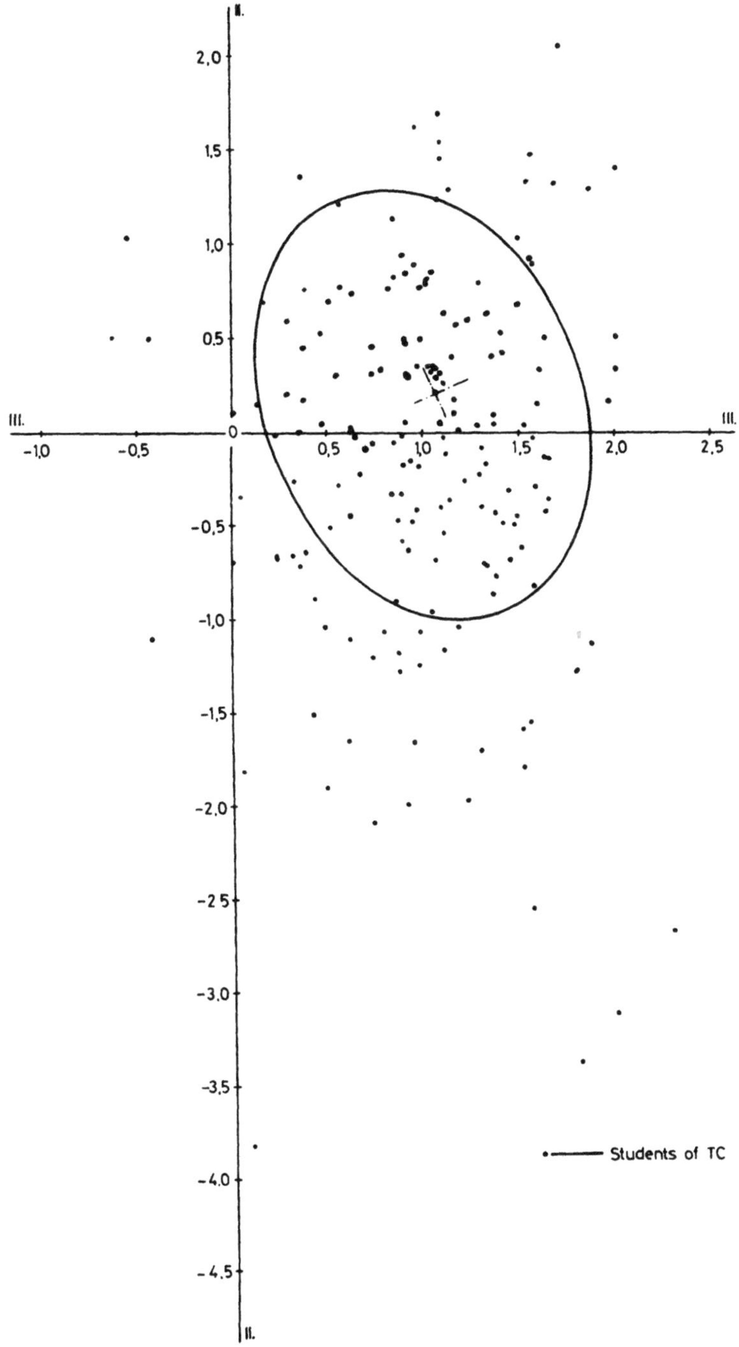

Figure 5. Distribution of female students of teachers' college on normal components II and III

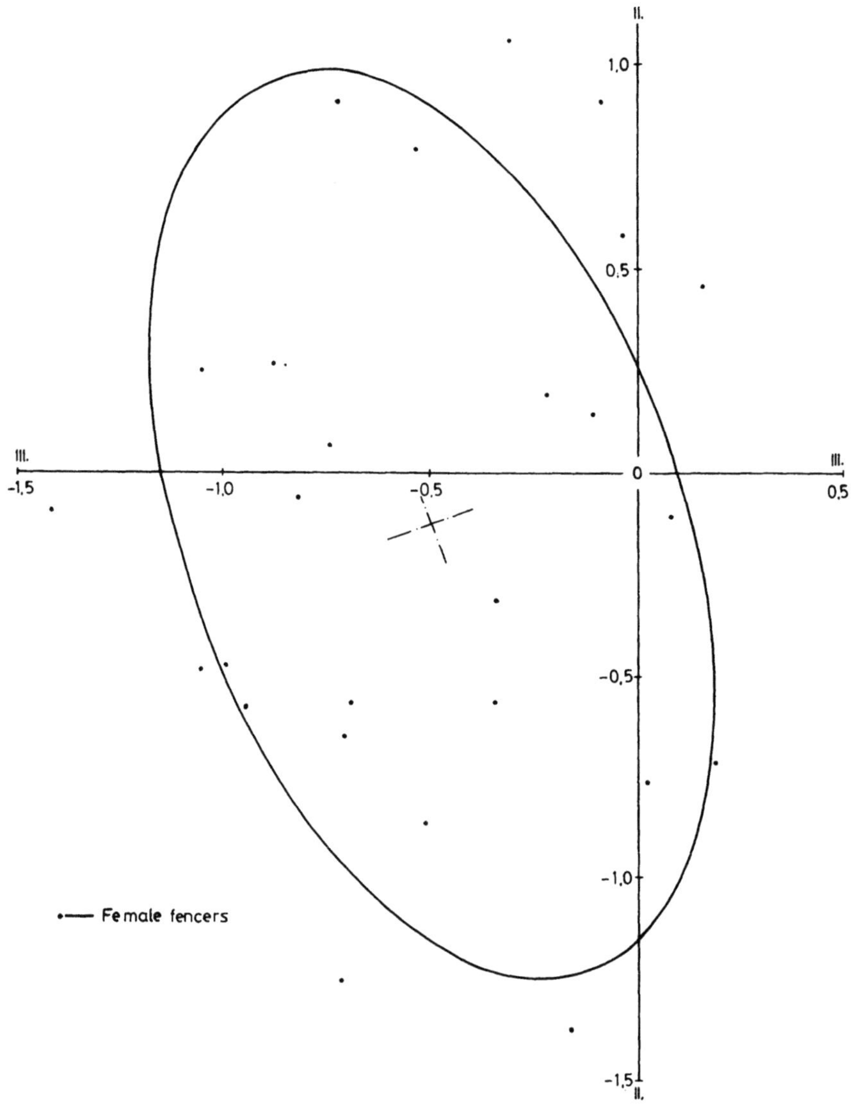

Figure 6. Distribution of female fencers on normal components II and III

Figure 7. Body-measurement vectors: (1) weight; (2) sitting height; (3) span; (4) stature; (5) suprasternal height; (6) shoulder height; (7) elbow height; (8) wrist height; (9) finger height: (10) height of ant. sup. iliac spine; (11) knee height; (12) ankle height; (13) shoulder width; (14) bideltoid width; (15) bispinal width; (16) bitrochanter width; (17) chest breadth; (18) chest depth; (19) back width; (20) chest circumference; (21) abdomen circumference; (22) trochanter circumference; (23) upper arm circumference; (21) abdomen circumference; (22) trochanter circumference; (23) upper arm circumference; (24) forearm circumference; (25) wrist circumference; (26) thigh circumference; (27) calf circumference; (28) ankle circumference

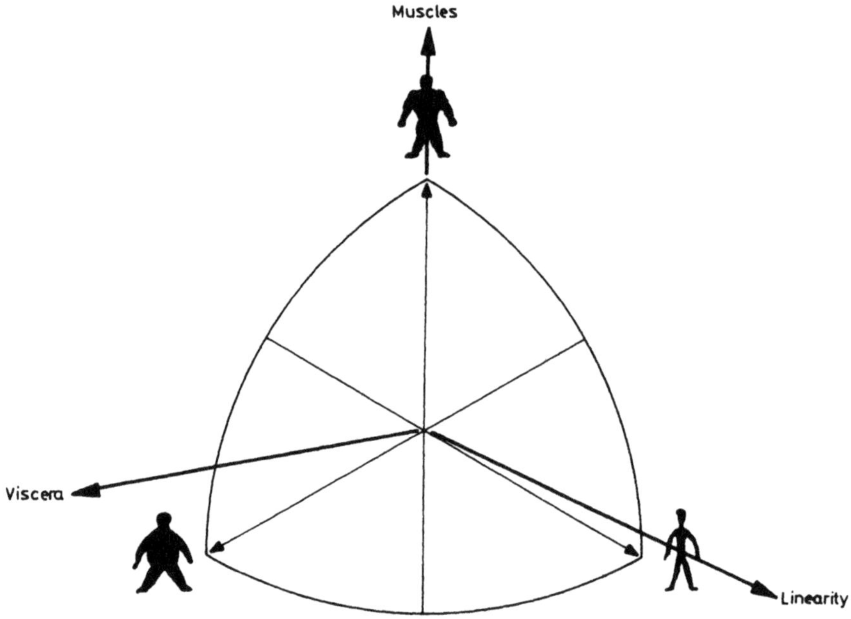

Figure 8. A comparison of the main directions of body-measurement vectors and Sheldon's three components

REFERENCES

CSÉBFALVI, K.
1969 Über numerische Probleme der Behandlung von Matrixen hoher Ordnung. Mathematical Symposium, Braunschweig.

EIBEN, O. G.
1972 *The physique of women athletes.* Budapest: The Hungarian Scientific Council for Physical Education.

KRETSCHMER, E.
1967 *Körperbau und Charakter* (twenty-fifth edition). Berlin: Springer.

MARTIN, R., K. SALLER
1957 *Lehrbuch der Anthropologie* (third edition). Stuttgart: Fischer.

SHELDON, W. H., S. S. STEVENS, W. B. TUCKER
1940 *The varieties of human physique.* New York: Harper.

Body Composition of Japanese Children: Factor Analysis of Development

S. KANDA

Studies of constitutional and nutritional status on human bodies have led to the concept of body composition. Scientific measurement of body composition requires such complicated methods that estimations of composition are limited to well equipped institutes. On the other hand, somatometry has been conducted by many investigators in the field, often with the hope of utilizing data to estimate body composition. The author has applied factor analysis of somatometrical measurements to the study of body composition. Specifically, an attempt has been made to pursue developmental changes of composition on a somatometrical basis.

MATERIAL AND METHODS

Japanese children of five to fifteen years of age — 604 boys and 611 girls — living in and near Osaka City, were measured in 1969 to obtain the following: stature; leg length (standing height of right anterior superior iliac spine); biacromial breadth; biiliac breadth; thickness of skinfold 1, over triceps; thickness of skinfold 2, subscapular; upper arm circumference; chest circumference; and body weight. The measurements were made following Martin's method. The thickness of skinfolds was measured by Harpenden's caliper on two points which indicated skinfold 1 and 2, respectively.

For a normal distribution of measurements of skinfolds and calculation of arm diameter, the following transformations were applied:

Skinfold transform = 100 × log 10 (reading in 0.1 mm. − 18);
Arm diameter = (arm circumference) / 3.1416 − (thickness of skinfold 1).

The factor analysis and varimax rotation were applied on the previous nine items, arm diameter, and age. Extracted factor loadings over 0.40 in absolute value were used in interpreting factors. The original group was divided into several groups depending upon age, stature, and skinfold 1. After analysis, the extracted factor patterns were compared.

RESULTS AND COMMENTS

Analysis on Boy and Girl Groups

The total sample is divided into male and female groups. Factor analysis of the two groups reveals three factors. The distribution of high factor loadings of males is similar to that of females (Table 1).

Table 1. Factor loadings from five- to fifteen-year-olds

	Boys			Girls		
	I	II	III	I	II	III
Stature	98	05	−15	95	25	−12
Leg length	98	07	−10	95	23	−11
Biacromial breadth	94	05	−25	91	32	−18
Biiliac breadth	94	09	−16	91	35	−11
Arm diameter	79	16	−58	67	46	−58
Skinfold 1	−12	96	10	22	95	02
Skinfold 2	37	80	−30	39	88	−08
Chest circumference	89	19	−37	82	48	−24
Arm circumference	72	40	−55	57	70	−42
Weight	91	20	−32	83	49	−24
Age	96	04	−09	92	29	−08

The first factor has significant loadings on the items of body framework and age, except for the thickness of skinfold. The biological meaning of the loading distribution indicates different natures between the thickness of skinfold and body framework. Thickness of skinfold 2 (subscapular) makes a moderate contribution to the first factor. The first factor may be interpreted as physique and age.

The second factor reveals marked loadings on the two skinfolds in boys and girls. The loadings on circumference of girls are more significant than those of boys. This difference in distribution of loadings in the second factors between boys and girls seems to be associated with high adipose deposit in girls. The factor may be interpreted as adiposity.

The third factor has significant loadings in arm diameter and arm circumference. A muscular development of the upper arm may be influenced by this factor. This factor will be interpreted as circumference of the upper arm.

Contribution of the second factor (adiposity) to a total variance proves to be more significant in girls (30.7 percent) than in boys (17.5 percent). This factor shows sexual difference of adipose deposition. Age has significant correlations with the items of physique but shows few relations with adipose or upper arm circumference. Aging influences are not revealed in this analysis, and another analysis is intended to separate factors of physique from those of age.

Analysis on Age Groups

The original measurements are classified according to chronological ages. After the analyses, three main factors are extracted from all groups. They are named physique, subcutaneous adipose and circumference, and age factors (Table 2). Sometimes the second factor may be divided into subcutaneous adipose and circumference factors. Such division in boys is possible in six groups out of ten and in girls in four groups, three of which are older (twelve, thirteen, and fourteen year age groups) (Table 3). Until eleven years of age, female arm circumference is more highly correlated with other items, such as chest circumference, than male.

Table 2. Factor loadings from eight-year-old boys

	I	II	III
Stature	91	−15	26
Leg length	79	−01	31
Biacromial breadth	87	−14	06
Biiliac breadth	82	−17	09
Arm diameter	57	−65	−22
Skinfold 1	−02	−90	02
Skinfold 2	13	−87	−05
Chest circumference	69	−52	−13
Arm circumference	39	−86	−18
Weight	79	−58	00
Age	18	03	92

Table 3. Factor loadings from twelve-year-old girls

I	II	III	IV
12	94	13	12
16	93	14	−08
21	73	−05	42
32	50	20	50
14	19	−10	93
94	02	15	18
88	06	07	38
55	28	03	65
57	12	−01	78
49	52	12	65
−06	13	98	−00

The age factor is expressed differently in boys than in girls. Sexual differences are found in loadings of the various circumferences. It is probable that the biological nature of circumference consists of an adipose component and a framework component.

Analysis on Stature Groups

Boys' and girls' stature was divided at 10 centimeter intervals between 100.0 centimeters and 169.9 centimeters, and seven groups were analyzed for each sex.

In boys, four factors (framework, adipose, circumference and breadth, and age) are extracted in each group. But distributions of significant loadings reveal specificity of each group. If the correlation coefficients between stature and age exceed 0.40, significant loadings for stature appear in two factors, i.e., in the pure height factor and the framework factor (Table 4). In this case, loadings on age are associated with the pure height factor. The nature of stature consists of two components from the factorial viewpoint, i.e., the development of pure height which is markedly influenced by age, on the one hand, and development which is closely related to the breadth and circumference of the body, on the other hand.

In girls, separation of stature loadings into two factors does not take place in any but the 140 centimeter group. Significant loadings of age are involved in the body framework factor, where the correlation coefficient between stature and age exceeds 0.40. In the 150 centimeter group, a significant loading of age appears in the adipose factor (Table 5). This distribution may indicate a significant adiposity in this group. This trend is not seen in the 160 centimeter group. Girls with stature over 160 centimeters exceed the mean stature for age fifteen by one standard deviation (mean = 153.96 centimeters; s.d. = 4.99 centimeters [Kanda 1971]). Therefore, it is conceivable that these girls do not reveal a normal physique.

These groupings reveal the factorial natures of stature and the relations between height and age.

Table 4. Factor loadings from 120 centimeter height, boys

	I	II	III	IV
Stature	17	−77	13	49
Leg length	08	−85	16	28
Biacromial breadth	47	−31	−08	59
Biiliac breadth	32	−08	−03	84
Arm diameter	95	−15	−09	09
Skinfold 1	11	06	90	14
Skinfold 2	30	−16	78	−04
Chest circumference	80	−17	18	26
Arm circumference	91	−11	29	13
Weight	68	−43	32	25
Age	27	−82	−09	−19

Table 5. Factor loadings from 150 centimeter height, girls

	I	II	III	IV
Stature	03	89	21	−15
Leg length	−00	93	01	17
Biacromial breadth	55	43	−14	−45
Biiliac breadth	28	22	27	−83
Arm diameter	90	02	−12	−08
Skinfold 1	72	−07	55	−08
Skinfold 2	74	−04	55	−07
Chest circumference	85	14	17	−26
Arm circumference	95	−04	22	−09
Weight	88	23	21	−22
Age	10	−01	84	−12

Analysis of Skinfold 1 Groups

A marking using 2.0 millimeter intervals is made for skinfold 1 with values from 3.0 millimeters to 28.9 millimeters. Measurements of skinfolds reveal a skewed distribution with less boys than girls having skinfolds of the greater thicknesses. Intervals occupied by small numbers were excluded from analysis. The analysis divided boys and girls into five and seven groups, respectively. As a result, three main factors are obtained, i.e., general physique and age, skinfold 1, and circumference factors. Correlation coefficients between skinfold 1 and 2 reveal a marked decrease in comparison with values indicated in the previous analysis. The highest loading of skinfold 1 appear almost always in the second factor, and that of skinfold 2 will be traced as follows. In thin skinfold 1 groups of boys and girls, i.e., less than 7.0 millimeter groups, marked loadings of skinfold 2 appear in the general physique factor (Table 6). Following addition of skinfold 1, significant loadings on skinfold 2 frequently appear in the third factor with significant loadings of circumferences (Table 7). Male groups with over 11.0 millimeter skinfold 1 and females over 17.0 millimeter reveal significant loadings on skinfold 1 and 2 in the second factor and significant loadings on circumferences (Tables 8 and 9). Girls with skinfold 1 from 11.0 millimeters to 12.9 millimeters differ with respect to the tendency for the second factor to involve loadings on skinfolds 1 and 2. Skinfolds show irregular factorial relations with age. Therefore, a detailed discussion of skinfolds on the basis of age and skinfold 1 is warranted.

Table 6. Factor loadings from 5.0 to 6.9 millimeter skinfold 1, girls

	I	II	III
Stature	90	−12	−38
Leg length	90	−13	−38
Biacromial breadth	86	−04	−45
Biiliac breadth	88	−04	−37
Arm diameter	51	02	−85
Skinfold 1	13	95	−03
Skinfold 2	69	33	−12
Chest circumference	81	01	−53
Arm circumference	51	11	−85
Weight	83	−02	−52
Age	88	−14	−34

Table 7. Factor loadings from 9.0 to 10.9 millimeter skinfold 1, girls

	I	II	III
Stature	94	−02	−31
Leg length	94	−02	−28
Biacromial breadth	90	−06	−38
Biiliac breadth	91	−00	−36
Arm diameter	70	−05	−64
Skinfold 1	17	98	02
Skinfold 2	39	33	−78
Chest circumference	88	00	−43
Arm circumference	70	04	−63
Weight	89	01	−43
Age	92	01	−32

Analysis on Age and Skinfold Groups

The boys and girls are classified by age and thickness of skinfold 1. For the classification care is taken to insure an adequate number of each group reflecting increases in skinfold 1. As shown in Table 10, skinfold 1 is analyzed for four age categories. (Age trends are seen in Figure 1 also.)

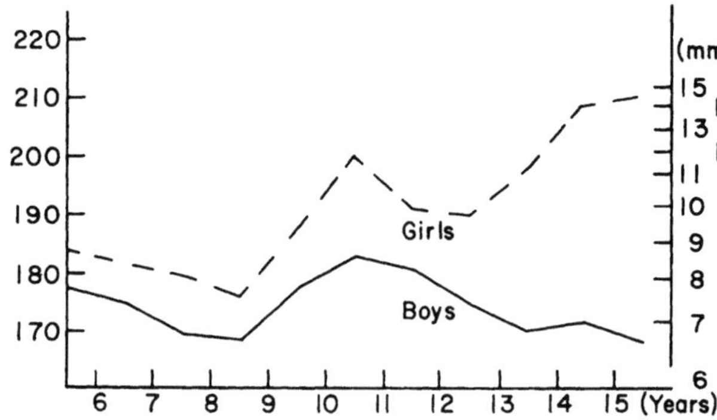

Figure 1. Developmental changes of thickness of skinfold 1 for Osaka sample, 1969

Table 8. Factor loadings from over 11.0 millimeter skinfold 1, boys

	I	II	III
Stature	98	11	−14
Leg length	98	09	−05
Biacromial breadth	94	06	−31
Biiliac breadth	92	19	−22
Arm diameter	67	35	−65
Skinfold 1	01	95	10
Skinfold 2	29	85	−13
Chest circumference	84	39	−35
Arm circumference	60	57	−55
Weight	87	35	−31
Age	91	17	−22

Table 9. Factor loadings from over 17.0 millimeter skinfold 1, girls

	I	II	III
Stature	95	−04	−23
Leg length	91	−08	−19
Biacromial breadth	78	03	−49
Biiliac breadth	89	23	−17
Arm diameter	24	06	−95
Skinfold 1	01	92	−13
Skinfold 2	19	80	−25
Chest circumference	54	38	−67
Arm circumference	20	46	−84
Weight	63	28	−67
Age	79	26	−17

Table 10. List of classified groups on the basis of age and thickness of skinfold 1 (in millimeters)

	5 to 7 years	8 to 9 years	10 to 11 years	12 to 15 years
Boys	3.0–5.9	–6.9	–6.9	3.0–4.9
	6.0–6.9	7.0–	7.0–	5.0–5.9
	7.0–7.9			6.0–6.9
	8.0–8.9			7.0–8.9
	9.0–			9.0–
Girls	–6.9	–7.9	–9.9	5.0–9.9
	7.0–7.9	8.0–	10.0–	10.0–10.9
	8.0–8.9			11.0–12.9
	9.0–9.9			13.0–14.9
	10.0–			15.0–

ANALYSIS OF FIVE- TO SEVEN-YEAR AGE GROUPS. In boys, four factors are extracted, i.e., body framework, circumferences, skinfold 1, and skinfold 2 (Table 11). Significant loadings of skinfold 2 reveal different distributions from that of skinfold 1. The loadings of skinfold 1 and 2 of the 3.0 to

5.9 millimeter group are significant in the third factor (Table 12). The correlation coefficient between the two skinfolds is moderately significant (0.555) and the individuals of this group would usually be considered lean. In over 9.0 millimeter groups, significant loadings of skinfold 2 appear in the second factor as is the case with circumference (Table 13), indicating adipose deposition in the trunk region.

Girls reveal four factors as do boys. In the over 10.0 millimeter groups, significant loadings of skinfold 1 and 2 and circumference are found in the second factor. These loadings again suggest adipose deposition.

ANALYSIS OF EIGHT AND NINE YEAR AGE GROUPS. In boys, the groups of over 7.0 millimeters reveal significant loadings of skinfold 1 and 2, and circumference in the second factor. These loadings suggest adipose deposition in the trunk.

Two groups of girls reveal similar distributions of loadings. Significant loadings of two skinfolds and moderate circumference loadings are found in the second factor. It seems that an adipose deposition is occurring in these groups.

Table 11. Factor loadings from 8.0 to 8.9 millimeter skinfold 1 in five- to seven-year-old boys

	I	II	III	IV
Stature	95	23	−11	04
Leg length	91	15	−15	06
Biacromial breadth	65	34	−43	−07
Biiliac breadth	79	34	22	12
Arm diameter	26	95	−13	−04
Skinfold 1	−21	−04	−01	97
Skinfold 2	−14	06	−95	02
Chest circumference	59	70	−24	09
Arm circumference	24	95	−13	04
Weight	77	59	−16	01

Table 12. Factor loadings from 3.0 to 5.9 millimeter skinfold 1 in five- to seven-year-old boys

	I	II	III	IV
Stature	96	−00	05	−16
Leg length	96	−16	−01	−03
Biacromial breadth	82	24	−06	−11
Biiliac breadth	32	13	09	−89
Arm diameter	28	94	−07	−11
Skinfold 1	09	−14	86	−26
Skinfold 2	09	27	87	04
Chest circumference	43	38	28	−61
Arm circumference	30	92	11	−16
Weight	80	37	08	−38

ANALYSIS OF TEN AND ELEVEN YEAR AGE GROUPS. In boys, significant loadings of skinfold 1 and 2 appear in the second factor of the two groups. Adiposity of physique is more pronounced in the group with over 7.0 millimeters of skinfold 1 than a lean skinfold group.

In girls, significant loadings of skinfold 1 and 2 are shared in the second and third factors of the less than 9.9 millimeter group, but they are included in the second factor only of the 10.0 millimeter group. No trend of adipose deposition is evident in the group.

The number of individuals eight to nine and ten to eleven years of age is insufficient for classification into several skinfold classes. Therefore, no detailed analysis of adiposity is attempted.

ANALYSIS OF THE TWELVE TO FIFTEEN YEAR AGE GROUP. In boys, the trend of distribution of significant loadings of skinfold is similar among those five to nine years of age. A group with less than 4.9 millimeter skinfold has significant loadings of two skinfolds in the third factor without marked loadings on other items. This group may be considered lean. In groups of 5.0 millimeters to 8.9 millimeters, significant loadings of skinfold 1 appear as a unique factor and that of skinfold 2 appears in other factors with marked loadings on circumferences of the trunk and arm. Adipose deposition is observed in groups of 5.0 millimeters to 8.9 millimeters of skinfold 1. The group with thickest skinfolds (over 9.0 millimeters) shows prominent adipose deposition and marked loadings on skinfold 1 and 2, and on circumferences appear as a unique factor (Table 14). In girls, distributions of marked loadings of skinfold 1 and 2 do not resemble those of boys. No factorial traits are extracted from loadings of skinfold 1 in the less than 14.9 millimeter groups. The groups of over 15.0 millimeter skinfold 1 are characterized by adiposity as are boys with over 9.0 millimeter (Table 15.)

Table 13. Factor loadings from over 9.0 millimeter skinfold 1 in five- to seven-year-old boys

	I	II	III
Stature	97	−01	04
Leg length	94	00	09
Biacromial breadth	97	01	−04
Biiliac breadth	91	01	10
Arm diameter	69	67	03
Skinfold 1	−22	28	93
Skinfold 2	−24	83	27
Chest circumference	83	47	01
Arm circumference	55	72	34
Weight	86	29	11

Table 14. Factor loadings from over 9.0 millimeter skinfold 1 in twelve- to fifteen-year-old boys

	I	II	III
Stature	93	16	−29
Leg length	96	14	−13
Biacromial breadth	77	11	−53
Biiliac breadth	56	28	−40
Arm diameter	36	33	−86
Skinfold 1	07	94	−18
Skinfold 2	14	90	−22
Chest circumference	56	42	−66
Arm circumference	30	58	−74
Weight	69	44	−55

Table 15. Factor loadings from over 15.0 millimeter skinfold 1 in twelve- to fifteen-year-old girls

	I	II	III	IV
Stature	12	91	−15	−28
Leg length	10	95	−08	−05
Biacromial breadth	44	26	−11	−77
Biiliac breadth	05	38	32	−79
Arm diameter	97	07	04	−01
Skinfold 1	15	−03	92	−01
Skinfold 2	34	−01	75	−02
Chest circumference	77	−00	35	−36
Arm circumference	83	04	50	−00
Weight	76	40	34	−22

DISCUSSION

Anthropometric data matrices used in factor analysis involve several configurations of items. The age item is part of the data matrix. An assessment of biological age was carried out by Jalavisto (1963) on Finnish women, aged twenty to eighty-three years. From the twenty-six items, six factors were extracted. Rotation of raw factors did not produce a clear-cut, pure age factor. The significant loadings on age may be associated with alteration of central nervous system function, loss of connective tissue elasticity, and over-weight. The above-mentioned relations between factors and age suggest intimate relations among them. Makkonen (1963) extracted an aging factor from the analysis of data on Finnish women aged between forty-four and eighty-three. He noticed that a given measurement may have a different meaning in the young than in the old, and it is advisable to restrict analysis to narrow age ranges. This author agrees. Therefore, in the present study, all members are subdivided into groups according to age, stature, thickness of skinfold 1, and age and skinfold 1. Special factors associated with the items used for subgrouping have emerged from the factor analysis. And relations between special factors and other items are simplified for interpretation.

Cross-sectional data without the age item were studied factorially by McCloy (1940) and Mullen (1940). McCloy indicated the constancy of factor patterns among age groups. Mullen studied growth of girls grouped into age categories of seven, nine, eleven, thirteen, fifteen, and seventeen years. He concluded that the general factor derived from various age groups is fairly consistent, although at the earlier ages, linear measurements correlate more highly with the general factor, and at later ages the cross-sectional measures have higher loadings. This conclusion indicates changes of factor pattern with age. In the present study, loadings of thickness of skinfolds reveal more prominent developmental changes than other items, and adiposity merits special consideration.

The results of application of radiographic methods to the analysis of human physique are shown by Tanner (1965). He postulates independent factors as follows: skeletal frame size, limb bone width, limb bone length, muscle width, and fat thickness. Tanner's studies showed the values of factor analysis in studies of physical composition.

The factor analysis of body composition was carried out previously by the present author (Kanda 1968), and differences in factor pattern were identified between spinners and students.

The present study indicates the possibility of connecting the analysis of growth with that of body composition.

CONCLUSION

Nine anthropological measurements were made on Japanese boys and girls aged from five to fifteen years. Eleven items were analyzed by factor analysis. All members were classified by age, stature, skinfold 1, and skinfold 1 and age, and the following results were obtained. Physical development of the individuals may be assessed on the basis of physical framework, skinfold, and age. The age factor was excluded from analysis. Circumferences consist of an adipose component and framework component. The adipose component reveals sex differences. Stature is associated with development of pure height and that of breadth and circumferences of the body. The development of pure height is influenced by age. The analysis of single skinfold groups indicates that the possession of over 11.0 millimeter skinfold 1 in boys and over 17.00 millimeters in girls characterizes adiposity.

Adiposity of boys is diagnosed by the thickness of skinfold 1 over 9.0 millimeters. Adiposity of girls is diagnosed by thickness of skinfold 1 over 10.0 millimeters in a five to seven year age group, and over 15.0 millimeters in a twelves to fifteen year age group. Leanness of boys is diagnosed by thickness of skinfold 1 less than 6.0 millimeters in a five to seven year group, and less than 4.9 millimeters in a twelve to fifteen year group. Therefore, adiposity is diagnosed differently depending upon age and sex.

REFERENCES

JALAVISTO, EEVA

1963 On the assessment of biological age I. A factor analysis of physiological measurements in old and young women. *Annales Academiae Scientiarium Fennicae* 100:5–34.

KANDA, S.

1968 Body composition of Japanese women. *Proceedings of the VIIIth International Congress of Anthropological and Ethnological Sciences* 1:55–58.

1971 Measurements of physical growth of the Japanese. *Medical Journal of Osaka University* 21:197–213.

MAKKONEN, T.

1963 On the assessment of biological age II. A factorial study of aging in postmenopausal women. *Annales Academiae Scientiarium Fennicae* 101:3–20.

McCLOY, C. H.

1940 An analysis for multiple factors of physical growth at different age levels. *Child Development* 11:249–277.

Genetics of Human Morphological Characters

C. SUSANNE

The study of heredity of polygenic, morphological characters consists of estimating the influence of the genotype as well as the environment on these phenotypes. Comparison of morphological characters by the method of analysis of variance permits the total variation to be partitioned into different factors. Regression of the subject's measurements on those of his father and mother facilitates isolation of a dominance factor.

MATERIAL

The present study is concerned with inheritance of morphological and anthropological characters in a sample of the Belgian population. Our sample is socially homogeneous: 65.7 percent are university educated men and 23.2 percent are university educated women. The sample consists of 132 families with parents and 287 adult children; 551 individuals altogether. For each individual we collected thirty-six measurements, taken on the left side as recommended by Martin and Saller (1957) and Twiesselmann (1952).

Sex and age differences are well known (Susanne 1968, 1971). A number of cross-sectional studies have shown that practically no anthropological measurement remains stable throughout adulthood (Marquer and Chamla 1961; Parot 1961; Hooten and Dupertuis 1951; Howells 1970; Susanne 1971).

In view of these phenotypical variations with age, an effort was made to eliminate the effects due to age differences of the individuals compared. The most adequate method seemed to be the standardization of the measurements for age and sex: these standardized measurements

$\left(y_i = \frac{x_i - \bar{x}}{s_x}\right)$,

where x_i is the absolute measurement, $\bar{x}$ and s_x, respectively, the average value and the standard deviation for the age class have an average equal to zero

$$(\bar{y} = \frac{\Sigma\ (x_i - \bar{x})}{ns_x} = 0$$

because

$\Sigma\ (x_i - \bar{x}) = 0)$ and a standard deviation equal to

$$1\left(s^2 y = \frac{\Sigma y_i^2}{n} - \bar{y}^2 = \frac{1}{s_x^2} \cdot \frac{\Sigma (x_i - \bar{x})^2}{n} = 1\right).$$

We have shown (Susanne 1971) that correlation coefficients are generally lower when they are calculated with absolute values instead of standardized measurements. The variance is indeed higher with absolute values than with standardized measurements: the superposition of two clusters of points representing two correlations between two different pairs of related individuals is indeed incomplete with absolute values, but is complete with standardized measurements around the average 0 and with a variance of 1. Table 1 shows the difference of correlation between absolute values and standardized measurements for some of the studied measurements. In a comparison of variation between sibs and parents, the utilization of standardized measurements is also necessary.

Table 1. Comparison of correlation coefficients for absolute values of some anthropological measurements and standardized measurements

	Parent–child	Sibs	Mid parent–child
Weight (absolute)	0.25	0.38	0.41
(standard)	0.34	0.54	0.46
Stature (absolute)	0.26	0.30	0.45
(standard)	0.51	0.59	0.63
Physio-face height			
(absolute)	0.27	0.29	0.47
(standard)	0.40	0.35	0.50

ANALYSIS OF VARIANCE

A two-way analysis of variance can be used to test the significance of the difference between two factors of variation, variation between and within sibships.

Considering l families and n_i children in the family i, the total number of children is then equal to

$$\sum_{i=1}^{l} n_i = N.$$

For a measurement Y the total sum of squares, equal to

$$\sum^{l}\sum^{n_i} Y^2 - N\bar{Y}^2$$

with $N - 1$ degrees of freedom, can be partitioned into the variation between sibships and the variation within sibships.

The variation between sibships, let us say between the means of the measurement in each sibship, can be calculated by the sum of squares between sibships

$$\sum^{l} n_i \bar{Y}_i^2 - N\bar{Y}^2$$

with $l-1$ degrees of freedom and the variation within sibships by the sum of squares within sibships

$$\sum^{l}\sum^{n_i} Y^2 - \sum^{l} n_i \bar{Y}_i^2$$

with $N - l$ degrees of freedom. If the variation between sibships is higher than the variation within sibships, as we expected for measurements determined by genetical factors, this difference can be tested by F test with $l - 1$ and $N - l$ degrees of freedom.

$$F = \frac{\text{sum squares between}}{l-1} \Big/ \frac{\text{sum squares within}}{N - l}$$

RESULTS

We used this method for the thirty-six measurements on the whole sibship; the different measurements of brothers and sisters were standardized for age and sex. Table 2 gives the sum of squares between and within sibships as well as the results of the test F in our sample. Table 2 also compares these results with those of Howells (1966), the only author who applied the F method on several measurements. We see that the value F is significant in our sample for all studied measurements.

This indicates that genetic factors, but perhaps also some environmental factors, are responsible for a greater similarity of anthropometrical measurements of brothers and sisters than those of children from different sibships. Thus, the grouping of children in families cannot be equivalent to random samples of children from a homogeneous population.

Table 2. Analysis of variance, mean square and *F* test

	Mean square between sib-ships d. f. 116	Mean square within sib-ships d. f. 156	*F* d. f. 116, 156	*F* (Howells 1966) d. f. 171, 511
Weight	1.71	0.53	3.25	2.47
Stature	1.78	0.41	4.30	4.47
Sternal height	1.77	0.42	4.26	
Sitting height	1.56	0.60	2.59	3.13
Arm length	1.70	0.48	3.54	
Upper arm length	1.43	0.59	2.43	3.23
Forearm length	1.56	0.54	2.88	3.89
Middle finger length	1.81	0.53	3.41	
Biacromial diameter	1.42	0.60	2.36	
Biiliac diameter	1.58	0.48	3.33	
Bitrochanteric diameter	1.64	0.42	3.95	
Upper arm circumference (relaxed)	1.50	0.57	2.63	2.38
Upper arm circumference (flexed)	1.52	0.60	2.56	
Wrist circumference	1.66	0.52	3.18	3.31
Ankle circumference	1.52	0.61	2.51	3.11
Calf circumference	1.26	0.78	1.61	
Neck circumference	1.39	0.54	2.59	
Head length	1.50	0.69	2.19	3.64
Head breadth	1.81	0.84	2.15	3.83
Frontal breadth	1.37	0.60	2.27	3.04
Bizygomatic breadth	1.50	0.60	2.49	2.81
Bigonial breadth	1.41	0.57	2.49	3.39
Physio-face height	1.43	0.58	2.48	
Nasion-gnathion height	1.54	0.57	2.70	
Nasion-stomion height	1.52	0.58	2.64	
Nose height	1.38	0.70	1.97	4.66
Nose breadth	1.46	0.63	2.30	2.13
Internal biocular breadth	1.56	0.62	2.53	
External biocular breadth	1.55	0.65	2.37	
Interpupillary breadth	1.52	0.77	1.99	
Lips height	1.36	0.51	2.68	
Mouth breadth	1.41	0.60	2.33	
Ear height	1.35	0.58	2.33	1.80
Ear breadth	1.40	0.57	2.46	1.88
Nose depth	1.64	0.74	2.22	
Head height	1.60	0.58	2.77	

However, *F* values can be very different depending upon the measurements, which form a hierarchy of influence of genetic factors. Figure 1 represents the heritability of the seventeen body measurements illustrating a decreasing order of the *F* values. The most correlated measurements are stature and suprasternal height, two longitudinal measurements. The *F* values are a little lower for two other longitudinal measurements, arm length and middle finger length, as well as two diameters of the pelvis, biiliac and bitrochanteric, also weight and wrist circumference. For three other longitudinal measurements, the *F* values are still lower: forearm length, upper arm length, and sitting height. For these last characters, the

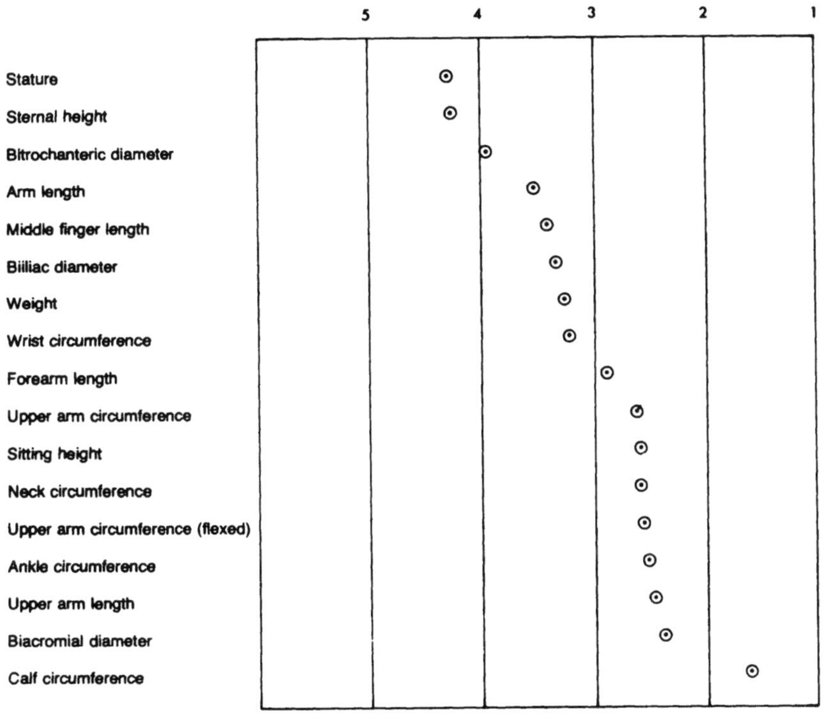

Figure 1. Heritability of body measurements: *F* values

measurement errors are larger, leading to an increase of the variances within sibships (the middle square within sibships is equal to 0.59 for the upper arm length, 0.54 for the forearm length, and 0.60 for the sitting height, but is equal to only 0.48 for the arm length) and thus a diminution of the *F* values.

For the biacromial diameter and for several circumferences the *F* values are also very low, as with upper arm circumference (relaxed and flexed), neck, ankle, and calf. In such a hierarchy, the more a character is influenced by genetic factors and the less it is influenced by environmental factors, the higher the *F* values.

We see, for instance, that the genetic component of variation is very small for all circumferences, except the wrist diameter. The genetic determination of phenotype of osseous circumferences is higher than of circumferences where muscles and fat predominate, as shown by comparison of wrist and upper arm circumferences, or ankle and calf circumferences.

Figure 2 uses *F* values to illustrate the hierarchy of genetic components of nineteen face and head measurements. However the variation of these *F* values is less than the variation of those for body characters, and *F* values of face and head measurements are smaller than those for body

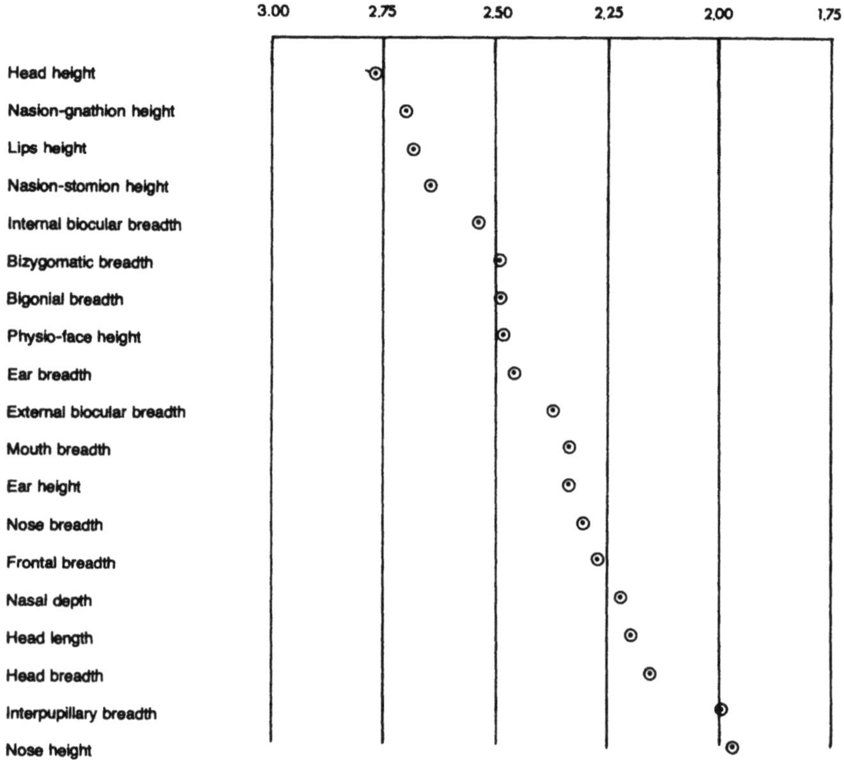

Figure 2. Heritability of face and head measurements: *F* values

measurements. *F* values for face and head measurements range between values 2 and 3, equivalent to the heritability level of limb circumferences. The heritability level of some longitudinal measurements is higher than for facial breadth, as is the case for head height, nasion-gnathion height, lip height, and nasion-stomion height. Values for physiognomical and nose height are lower, the measurement errors of these characters being relatively large, increasing the within sibship variance and reducing the *F* values. It is evident that the position in this hierarchy of the six longitudinal measurements is higher (mean: 2.54) than that of breadth (nose, minimum frontal, bizygomatic, bigonial, internal biocular, external biocular, interpupillary, mouth: mean 2.35), and still higher than the *F* level of head breadth and head length (mean: 2.17).

ANALYSIS OF VARIANCE

This method was first applied by Fisher and Gray (1937) on stature, head length, head breadth, and bizygomatic breadth. Grab (1956) applied the same analysis of variance to the same four measurements, and Tanner and Healy (1956) used it for stature. The method is based on a separation of the differences between sibships into three portions. In the same way that the total sum of squares may be divided into deviations within sibships and differences between sibship means, the latter sum of squares likewise may be divided again into a sum of squares relative to the linear regression, the deviations from regression, and a product term. This method is not based on linear regression on the two parents independently, but on regression on father, mother, and also on possible deviations due to dominance. The formula of multiple regression is:

$Y = b_0 + b_1x_1 + b_2x_2 + b_3x_1x_2$

where Y stands for the child's measurement, x_1 and x_2, respectively, for the father's and mother's measurements.

The term in x_1x_2 represents the non-additive component of parental influences. Dominance or interaction between genes may cause this nonadditivity. A positive b_3 term corresponds to a negative bias in dominance, where the heterozygote more nearly resembles the smaller than the larger homozygote. A negative b_3 term corresponds to a positive bias in dominance. In Table 3, we find the mean square of these three terms: deviation from regression, linear regression, and term x_1x_2.

ANALYSIS OF THE TERM OF THE LINEAR REGRESSION

As for the F test, there is no doubt of the significance of this term. The quotient of the mean square of linear regression and of the mean square within sibships is always very high and highly significant. Figure 3 shows that this quotient lets us determine a hierarchy of genetic components. This hierarchy slightly modifies the one established with the classical F values which did not take into account the measurements of both parents.

In Table 4, we find the equations of multiple regression. The efficacy of regression is evident for all measurements. The paternal and maternal regression coefficients, respectively b_1 and b_2, are all significant, except for the maternal regression coefficient of nose height. The efficacy of regression is illustrated by the sum of the regression coefficients $b_1 + b_2 + (\bar{x}_1 + \bar{x}_2)b_3$ which are all significant. This factor is strongly correlated with the factor of linear regression, as Figure 3 shows.

ANALYSIS OF SEX DIFFERENCE BETWEEN THE REGRESSION COEFFICIENTS b_1 AND b_2

By the comparison of the difference $b_2 - b_1 - (\bar{x}_2 - \bar{x}_1)b_3$ with the standard error, we test the eventual superiority of one of both regres-

Table 3. Analysis of variance: separation of intersibship variation

	Mean square deviation from regression d. l. 113	Mean square linear regression d. l. 2	Term x_1x_2 d. l. 1
Weight	1.17	31.48	2.84
Stature	0.85	54.47	0.60
Sternal height	0.80	57.41	0.21
Sitting height	1.04	31.22	0.08
Arm length	0.85	51.38	0.13
Upper arm length	0.85	35.78	0.01
Forearm length	0.91	39.06	2.00
Middle finger length	0.82	58.73	0.70
Biacromial diameter	0.89	31.59	0.93
Biiliac diameter	0.82	45.57	1.01
Bitrochanteric diameter	0.95	40.30	0.26
Upper arm circumference (relaxed)	1.18	20.22	1.71
Upper arm circumference (flexed)	1.23	18.33	2.67
Wrist circumference	0.92	42.50	4.00*
Ankle circumference	1.01	27.96	0.02
Calf circumference	1.05	13.12	0.01
Neck circumference	1.03	21.32	2.32
Head length	1.11	24.89	0.03
Head breadth	1.21	37.86	0.01
Frontal breadth	0.83	33.38	0.03
Bizygomatic breadth	0.90	33.66	5.47*
Bigonial breadth	0.77	39.08	0.01
Physio-face height	0.86	34.77	0.49
Nasion-gnathion height	1.07	29.26	0.002
Nasion-stomion height	1.17	23.19	0.03
Nose height	1.10	18.71	0.02
Nose breadth	1.02	27.19	0.09
Internal biocular breadth	1.04	31.88	1.49
Eternal biocular breadth	0.95	35.37	3.83*
Interpupillary breadth	0.89	38.61	0.28
Lips height	0.92	26.34	2.39
Mouth breadth	1.12	18.24	2.13
Ear height	0.95	24.92	0.49
Ear breadth	1.02	21.76	4.18*
Nose depth	1.16	22.71	3.63
Head height	0.88	42.34	0.22

Key: * These terms are significant ($P<0.05$)

sions. The difference in regression (see Table 4) in favor of the mother (for fourteen measurements) or of the father (for twenty-two measurements), though sometimes large, is often not significant. Exceptions are: middle finger length, biacromial diameter, and length of head where b_2 is significantly higher than b_1, and for ankle diameter and nose height where b_1 is higher than b_2.

With respect to most of the measurements, this study is the first application of this method, and caution is advisable in interpreting the results. Indeed, comparison with other author's results does not confirm the significance of the difference for the head length (see Table 5).

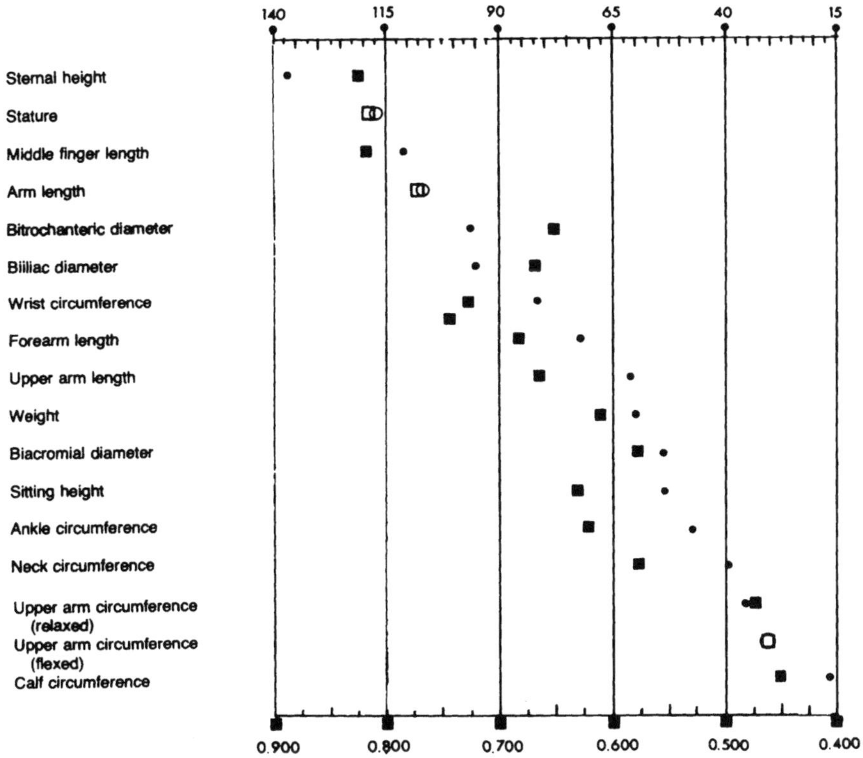

Figure 3. Heritability of body measurements: quotient mean square of linear regression and of mean square within sibships (●) and sum of regression coefficients (■)

ANALYSIS OF THE DOMINANCE FACTOR b_3

The significance of this term is tested by the ratio of the mean square for the dominance term to that for deviation from regression. As Table 4 shows, seventeen terms are positive, indicating a negative bias in dominance; that is, a situation in which the heterozygotes more nearly resemble the smaller than the larger homozygotes. Nineteen terms are negative, indicating that the heterozygotes more frequently resemble the larger homozygotes. Consequently, the term b_3 is not systematically positive or negative. It was found significant only for wrist diameter, bizygomatic breadth, external biocular breadth, and ear breadth. Before we conclude that dominance factors influence these measurements, more comparative material is needed. We must interpret the results cautiously since comparison with the results of Fisher and Gray (1937), Grab (1956), and Tanner and Healy (1956) does not confirm the significance of the b_3 term for bizygomatic breadth. However, for the other measure-

Table 4. Equations of multiple regression. Efficacy of regression and analysis of sex difference

	Equation : $y = b_0 + b_1x_1 + b_2x_2 + b_3x_1x_2$	Efficacy $b_1+b_2+(x_1+x_2)b_3$	Sex difference $b_2-b_1-(x_2-x_1)b_3$
Weight	$y = -0.17 + 0.32x_1 + 0.29x_2 + 0.10x_1x_2$	0.61	−0.03
Stature	$y = -0.04 + 0.43x_1 + 0.39x_2 - 0.04x_1x_2$	0.81	−0.04
Sternal height	$y = -0.05 + 0.41x_1 + 0.41x_2 - 0.03x_1x_2$	0.82	−0.01
Sitting height	$y = +0.02 + 0.31x_1 + 0.33x_2 - 0.02x_1x_2$	0.63	0.02
Arm length	$y = -0.08 + 0.41x_1 + 0.36x_2 + 0.02x_1x_2$	0.77	−0.05
Upper arm length	$y = -0.08 + 0.35x_1 + 0.32x_2 - 0.004x_1x_2$	0.67	−0.04
Forearm length	$y = +0.09 + 0.42x_1 + 0.27x_2 - 0.11x_1x_2$	0.68	−0.15
Middle finger length	$y = -0.001 + 0.30x_1 + 0.52x_2 + 0.05x_1x_2$	0.82	0.21*
Biacromial diameter	$y = -0.10 + 0.13x_1 + 0.45x_2 + 0.06x_1x_2$	0.58	0.33*
Biiliac diameter	$y = -0.15 + 0.30x_1 + 0.37x_2 + 0.07x_1x_2$	0.67	0.06
Bitrochanteric diameter	$y = -0.11 + 0.28x_1 + 0.37x_2 + 0.03x_1x_2$	0.65	0.09
Upper arm circumference (relaxed)	$y = -0.11 + 0.27x_1 + 0.21x_2 + 0.10x_1x_2$	0.48	−0.06
Upper arm circumference (flexed)	$y = -0.14 + 0.27x_1 + 0.20x_2 + 0.12x_1x_2$	0.46	−0.07
Wrist circumference	$y = +0.04 + 0.40x_1 + 0.32x_2 - 0.14x_1x_2$	0.73	−0.07
Ankle circumference	$y = +0.04 + 0.42x_1 + 0.20x_2 + 0.01x_1x_2$	0.62	−0.23*
Calf circumference	$y = -0.04 + 0.27x_1 + 0.18x_2 - 0.01x_1x_2$	0.45	−0.09
Neck circumference	$y = -0.09 + 0.23x_1 + 0.35x_2 + 0.11x_1x_2$	0.58	0.11
Head length	$y = -0.0004 + 0.19x_1 + 0.38x_2 - 0.01x_1x_2$	0.56	0.19*
Head breadth	$y = -0.08 + 0.40x_1 + 0.28x_2 - 0.01x_1x_2$	0.68	−0.12
Frontal breadth	$y = +0.04 + 0.37x_1 + 0.26x_2 - 0.01x_1x_2$	0.63	−0.12

Table 4. (*continued*)

Bizygomatic breadth	$y = -0.07 + 0.37x_1 + 0.25x_2 - 0.14x_1x_2$	0.60	−0.12
Bigonial breadth	$y = -0.04 + 0.34x_1 + 0.30x_2 + 0.01x_1x_2$	0.65	−0.04
Physio-face height	$y = -0.06 + 0.36x_1 + 0.29x_2 + 0.04x_1x_2$	0.65	−0.07
Nasion-gnathion height	$y = -0.09 + 0.37x_1 + 0.23x_2 - 0.003x_1x_2$	0.60	−0.13
Nasion-stomion height	$y = -0.09 + 0.31x_1 + 0.20x_2 - 0.01x_1x_2$	0.51	−0.11
Nose height	$y = -0.04 + 0.36x_1 + 0.03x_2 - 0.01x_1x_2$	0.39	−0.33*
Nose breadth	$y = -0.05 + 0.34x_1 + 0.31x_2 + 0.02x_1x_2$	0.65	−0.03
Internal biocular breadth	$y = +0.01 + 0.24x_1 + 0.39x_2 - 0.07x_1x_2$	0.63	0.16
External biocular breadth	$y = +0.08 + 0.30x_1 + 0.40x_2 - 0.12x_1x_2$	0.71	0.10
Interpupillary breadth	$y = +0.06 + 0.34x_1 + 0.35x_2 + 0.03x_1x_2$	0.68	0.01
Lips height	$y = +0.005 + 0.23x_1 + 0.34x_2 + 0.10x_1x_2$	0.57	0.11
Mouth breadth	$y = -0.06 + 0.18x_1 + 0.34x_2 - 0.09x_1x_2$	0.52	0.16
Ear height	$y = -0.04 + 0.28x_1 + 0.33x_2 + 0.05x_1x_2$	0.61	0.05
Ear breadth	$y = -0.07 + 0.27x_1 + 0.25x_2 + 0.12x_1x_2$	0.52	−0.01
Nose depth	$y = +0.04 + 0.35x_1 + 0.24x_2 - 0.13x_1x_2$	0.58	−0.10
Head height	$y = -0.06 + 0.38x_1 + 0.33x_2 - 0.03x_1x_2$	0.70	−0.05

Key: * These differences are significant ($P<0.05$)

ments there is a good agreement with the values calculated by others (see Table 5).

Table 5. Comparison of sex differences ($b_2 - b_1 - (\bar{x}_2 - \bar{x}_1)b_3$) and b_3 values (term in x_1x_2) with results of other authors

	Sex differences			Term b_3		
	Fisher	Grab	Susanne	Fisher	Grab	Susanne
Stature	0.08	0.18	−0.04	0.00	0.00	−0.04
Head length	0.04	0.02	0.19*	0.00	0.00	−0.01
Head breadth	0.03	0.20	−0.12	0.01	0.00	−0.01
Bizygomatic breadth	0.01	0.10	−0.12	0.01	0.03	−0.14*

* These differences or terms are significant ($P<0.05$)

CONCLUSION

For all studied measurements, the F ratio of the mean square between sibships and that within sibships is highly significant. Undoubtedly, genetic factors are responsible for the greater similarity of sibs than of random children from the same population, with respect to a number of anthropometric characters. However, we cannot deny that environmental factors contribute also to the similarity between sibs: the prenatal and postnatal environment is more similar in a sibship than between relatives of different generations. Thus, the F values are influenced by environmental factors for a number of measurements. The contribution of the environment is smaller in the comparison between parent and child than between two sibs. Thus, it is possible that an analysis of variance, where the variation is expressed in terms of the measurements of the parents, partitions the variance in a manner related to genetic factors. We examined familial data by the method of analysis of variance (Fisher and Gray 1937), based on a multiple regression. The partial regression coefficients obtained by this method are significant except for the maternal regression of nose height. We were able to determine a hierarchy of genetic components (see Figure 3). The efficacy of regression is illustrated by the sum of the regression coefficients. These regression coefficients are also correlated with linear regression factors, calculated by partition of variation factors. The same partition of the differences between sibships permits evaluation of a term derived from both parents' measurements which indicates a possible dominance factor.

REFERENCES

FISHER, R. A., H. GRAY
1937 Inheritance in man: Boas' data studied by the method of analysis of variance. *Annals of Eugenics* 8:74-93.

GRAB, B.
1956 Etude de la transmission héréditaire de certains caractères anthropométriques à l'aide des méthodes statistiques. *Journal de Génétiques Humaine* 5:120–166.

HOOTEN, E. A., C. W. DUPERTUIS
1951 Age changes and selective survival in Irish males. *Study of Physical Anthropology* 2:1–129.

HOWELLS, W. W.
1966 Variability in family lines vs. population variability. *Annals of the New York Academy of Sciences* 134:624–631.
1970 Hutterite age differences in body measurements. *Papers of the Peabody Museum of Archaeology and Ethnology* 57:1–123.

MARQUER, P., M. C. CHAMLA
1961 L'évolution des caractères morphologiques en fonction de l'âge chez 2089 français de 20 à 91 ans. *Bulletins et Mémoires de la Société d'Anthropologie de Paris* 2:1–78.

MARTIN, R., K. SALLER
1957 *Lehrbuch der Anthropologie*, part one. Stuttgart: G. Fisher.

PAROT, S.
1961 Recherches sur la biométrie du vieillissement humain. *Bulletins et Mémoires de la Société d'Anthropologie de Paris* 2:299–341.

SUSANNE, C.
1968 Les changements de la taille liés à l'âge (Phénomènes normaux de sénescence). *L'Anthropologie* 72:297–316.
1971 Hérédité des caractères anthropologiques polyfactoriels. Méthodologie des études familiales. *Bulletins et Mémoires de la Société d'Anthroplogie de Paris* 7:169–190.

TANNER, J. M., J. R. HEALY
1956 The genetics of human morphological characters. *Advances of Science* 13:192–194.

TWIESSELMANN, F.
1952 Aide-mémoire d'anthropométrie. *Mémoires de l'Institut Royal Colonial Belge* 25:1–61.

Lean Body Mass in Young, Adult, and Advanced Age Groups

JANA PAŘÍZKOVÁ

Extensive cross-sectional studies in normal, healthy subjects, ages seven to eighty years, with comparable relative weight revealed previously characteristic ontogenetical changes in absolute and relative amounts of lean body mass (LBM) and depot fat during ontogeny (Pařízková 1959, 1961, 1963, 1968, 1972a; Pařízková and Eiselt 1966). A densitometric method was used for our measurements. In males, the proportion of LBM increased steadily during childhood and adolescence, with only a short and temporary decrease just before the prepubertal growth acceleration. The peak of LBM proportion was achieved approximately at the age of twenty years and was followed by a steady decline (Pařízková and Eiselt 1966; Pařízková 1968, 1972a, 1977). In females the peak of LBM proportion was observed at the beginning of puberty and was followed by a decrease of LBM proportion during the remainder of life. Sexual differences in body composition, i.e., lower proportion of LBM in females, were already significant during childhood, appearing most markedly at the end of puberty and remaining significant in advanced age as well (Pařízková 1972a). The peak in LBM development in males corresponded to the peak in aerobic capacity of the organism as described by Åstrand (1967). Changes in body composition during development and aging seem to correspond approximately, *inter alia,* to the changes in the functional capacity to supply the tissues of the organism with sufficient amounts of oxygen (Pařízková 1972a, 1977).

In another cross-sectional study, changes in body composition and aerobic capacity of the organism were examined in different age groups of males (n = 156) from 10.8 to 73.7 years (Pařízková et al. 1972). Highest values of body weights were found in a group with a mean age of 35.4 years. Relative weight increased continuously from the youngest to the oldest age category. The proportion of LBM was again highest at the age of eighteen to twenty-five (Table 1). The absolute amount of LBM increased up to the age of eighteen, remained unchanged until 44.7 years, but later decreased significantly. Absolute values of aerobic capacity, ascertained as maximal oxygen uptake during graded work load on a horizontal treadmill (Šprynarová 1968), was also highest at the age of eighteen. In the third and fourth decade, this did not change significantly.

Table 1.

Age (years) $\bar{x}$	10.8	14.8	18.0	25.3	35.4	44.7	66.0	73.7
s.d.	0.4	0.4	0.5	3.3	3.1	1.8	2.0	3.0
Height (cm) $\bar{x}$	144.4	169.1	179.0	176.2	175.3	176.1	171.5	168.1
s.d.	4.0	5.7	6.7	3.8	7.6	5.7	9.5	4.9
Weight (kg) $\bar{x}$	36.7	58.4	70.8	68.3	76.0	75.6	71.5	70.7
s.d.	3.4	8.6	5.8	9.0	8.4	8.2	9.3	7.1
LBM (%) $\bar{x}$	84.0	84.6	89.3	89.2	83.7	82.5	78.7	78.0
s.d.	5.8	4.5	6.2	5.1	4.0	7.0	6.1	5.0
LBM (kg) $\bar{x}$	30.6	49.3	63.0	60.6	63.4	62.0	55.9	54.8
s.d.	1.5	5.5	3.5	4.7	5.0	4.7	5.3	4.0
Max O_2/kg LBM (ml) $\bar{x}$	53.8	63.7	52.1	51.6	51.1	41.8	24.3	20.8
s.d.	4.8	4.3	6.8	8.0	5.9	6.3	5.3	8.2

A more marked decrease was seen in the group with the mean age of 44.7 years and apparently thereafter (Macková 1968). Maximal oxidative capacity of the organism was evaluated according to the data on maximal oxygen uptake per kilogram LBM. These values were highest at the age of 14.8 years. At the age of eighteen the maximal oxygen uptake related to LBM was already significantly lower and did not change thereafter until the age of 44.7. The increase of the absolute amount of LBM of 14.8 to 18.0 years was therefore not paralleled by a corresponding increase in maximal oxygen uptake (Table 1). Marked decrease of the aerobic capacity appeared in the fifth decade of life, with further decline in the aerobic capacity, and in absolute and relative amount of LBM. Depot fat increased absolutely and relatively. The deterioration of the relative aerobic capacity, however, preceded that of LBM (Pařízková et al. 1972). Maximal oxygen uptake relative to LBM changed as did basal oxygen uptake relative to LBM, declining steadily with age(Behnke 1953). Other studies have shown a decrease of oxygen saturation of the blood, or lower P_{O_2} in the tissues (Sirotinin 1972). These changes parallel an increase of depot fat in the organism and decrease of LBM. The lack of oxygen in the tissues causes a shift in DPN/DPNH which favors the deposition of fat in the organism (Pařízková 1972a).

The decline in the values of maximal oxygen uptake relative to LBM shows a decrease in the intensity of metabolic activity. In our previous studies a significant relationship between maximal oxygen uptake relative to LBM and maximal oxygen pulse was found (Fischer, Pařízková, and Dýcka 1971). It was therefore possible to suppose a significant relationship between the transport capacity of the circulation and the aerobic capacity of the LBM. The decrease of the transport capacity of the circulation could be one of the most important causes of the decline in the level of the metabolic processes in the tissues. Therefore, in another study the density of capillaries in a skeletal muscle (the major component of LBM) and their relationship to the number of muscle fibers (by means of PAS reaction) was determined together with LBM, circumferential measures of the extremities, and the aerobic capacity (Pařízková et al. 1971). For the evaluation of age changes ($n = 16$) groups of men with markedly differing ages were compared and biopsies from the quadriceps muscle were taken. There was a significant difference in the number and diameter of the muscle fibers according to age (Table 2). In young men,

Table 2.

	$\bar{x}$	s.d.	$\bar{x}$	s.d.
Age (years)	20.7	0.3	72.5	2.9
Height (cm)	181.0	2.2	169.2	3.9
Weight (kg)	75.5	6.4	69.3	9.6
LBM (%)	88.8	4.6	81.7	3.3
LBM (kg)	67.3	8.0	56.4	5.5
Arm circumference (cm)	31.5	1.2	30.1	3.1
Thigh circumference (cm)	55.1	3.3	51.9	4.4
Max O_2/kg LBM (ml)	47.1	7.1	21.7	3.4
Capillaries/mm^2	270.1	99.2	314.7	99.2
Muscle fibers/mm^2	329.2	45.4	532.8	154.1
Capillary: fiber ratio	0.81	0.22	0.59	0.08

the capillary-to-fiber ratio was higher than in old men. This means a better capacity for the supply of oxygen and nutrients to muscle fibers. The relative and absolute amount of LBM, aerobic capacity, and circumferential measures of the extremities were all higher in the group of young men. In this group there existed also a significant relationship between the aerobic capacity of the organism and the capillary-to-fiber ratio in the quadriceps muscle ($r = 0.850$) (Pařízková et al. 1971). It is apparent that together with changes in body composition, i.e., absolute and relative amount of LBM, there occur significant changes in functional characteristics of the organism (which mostly precede the morphological changes) and the microstructure of tissues, i.e., skeletal muscle.

In advanced age, a constant body weight and LBM seem to characterize a good state of health and performance capacity. In another longitudinal study we found no changes in body composition in a selected group of men in the seventh and eighth decade of life ($n = 55$) which could be followed longitudinally during eight to ten years (Pařízková and Eiselt 1971). These men, with constant weight and body composition, also did not change in aerobic capacity and other parameters of functional capacity (Eiselt and Pařízková 1971). Obviously a delay in LBM involution and a slowing down of its decrease in advanced age run parallel with satisfactory health prognosis and relatively good level of physical fitness. This further indicates an important relationship between body composition and other functional characteristics in this age period.

REFERENCES

ÅSTRAND, P. O.
1967 Physical fitness in adolescence. *Clinical Pediatrics* 6:234.
BEHNKE, A. R., JR.
1953 The relation of lean body weight to metabolism and subsequent systematisations. *Annals of the New York Academy of Sciences* 56:1095.
EISELT, E., J. PAŘÍZKOVÁ
1971 Depotfett und aktive Körpermasse unter Berichtung des Einflusses der Physiotherapie. *Zeitschrift für Altersforschung* 24:259.
FISCHER, A., J. PAŘÍZKOVÁ, Z. DÝCKA
1971 *Zeitschrift Internazionale für angewandte Physiologie und Arbeitsphysiologie* 21:269.
MACKOVÁ, E.
1968 Changes in functional and efficiency indicators between the 18th and 50th year of man. *Physiologia bohemoslovaca* 17:279.
PAŘÍZKOVÁ, J.
1959 Lean body mass. *Československá fyziologie* 8:1.
1961 Age trends in fatness in normal and obese children. *Journal of Applied Physiology* 16:173.
1963 The impact of age, diet and exercise on man's body composition. *Annals of the New York Academy of Sciences* 16:661.
1968 Compositional growth in relation to metabolic activity. *Proceedings of the XIIth International Congress of Pediatrics*, Mexico City, volume 1:32.
1972a Interrelationships of physical activity and nutrition to body composition and fitness. *Proceedings of the First International Symposium "Alimentation et travail."* Edited by Vitell G. Débry and R. Blayer, 85. Paris: Librarie Masson.
1972b Senile changes of lean body mass and depot fat in men in 7th and 8th decade of life. Proceedings of the IXth International Congress of Gerontology, Kiev. *Abstracts*, volume 3:243.
1977 *Body fat and physical fitness.* The Hague: Martinus Nijhoff, Medical Division.
PAŘÍZKOVÁ, J., E. EISELT
1966 Body composition and anthropometrical indicators in old age and the influence of physical exercise. *Human Biology* 38:351.
1971 A further study of changes in somatic characteristics and body composition of old men after 8–10 years (a longitudinal study). *Human Biology* 43:318.
PAŘÍZKOVÁ, J., E. EISELT, Š. ŠPRYNAROVÁ, M. WACHTLOVÁ
1971 Body composition, aerobic capacity and density of muscle capillaries in young and old men. *Journal of Applied Physiology* 31:323.

PAŘÍZKOVÁ, J., Š. ŠPRYNAROVÁ, E. MACKOVÁ, E. EISELT
1972 Changes in the aerobic capacity as related to lean body mass in the ontogeny of man. *Physiologia bohemoslovaca* 22:425.
SIROTININ, N. N.
1972 Hypoxia at elderly and senile age. Proceedings of the IXth International Congress of Gerontology, Kiev. *Proceedings*, volume 1:254.
ŠPRYNAROVÁ, Š.
1968 Maximum oxygen consumption during work-load on a treadmill. *Physiologia bohemoslovaca* 17:383.

Skeletal Maturation of Japanese–American Hybrids

KUNIHIKO KIMURA

This study forms a part of the comprehensive anthropological investigation on Japanese–American hybrids which has been carried out since 1949 (Suda 1968). The hybrids of this study have been cared for at the Elizabeth Sanders Home at Oiso, Kanagawa in Japan. Most of them were born to the servicemen of the occupation forces and Japanese women after the end of World War II. According to their father's racial characteristics, they were divided into two groups: the white and Negro hybrids. However, it has been rather difficult to obtain accurate information about the father's racial background, particularly in the case of Negro inheritance. Therefore, the racial determination of their father was based on their mother's statement, and, if available, on the child's outstanding physical traits. Since 1958, I have taken charge of the study on skeletal maturation based on the radiographs of hand and wrist. The present paper deals with the findings on the skeletal maturation in male Japanese–American white hybrids.

MATERIALS AND METHOD

The sample consists of fifty-seven boys who were longitudinally investigated every spring and autumn from 1958 to 1965. The series for some of the children was occasionally interrupted, and some subjects left the home halfway through the study. Therefore, the samples were dealt with semi-longitudinally, rather than longitudinally. The number of samples in each age group from four to eighteen years is shown in Table 1. Age groups were formed as follows: for example, the seven-year-old group consists of children aged six years seven months to seven years six months. The control sample consists of 492 Japanese boys aged from zero

Table 1. Means and variances of the maturity scores by the Tanner-Whitehouse and the Kimura methods in the male Japanese–American white hybrids

		Tanner–Whitehouse		Kimura	
Age	n	$\bar{x}$	U^2	$\bar{x}$	U^2
4	1	87		31.1	
4.5	2	85.5	924.50	29.4	19.22
5	4	85.3	1086.25	29.3	22.91
5.5	4	148.0	6990.00	35.3	80.63
6	5	160.6	5026.30	37.6	58.98
6.5	7	195.1	3592.82	41.0	49.79
7	8	202.6	4460.55	42.0	54.26
7.5	15	235.1	3065.99	45.4	49.86
8	20	265.0	3753.10	48.3	50.21
8.5	22	298.5	4068.07	52.0	54.80
9	24	321.8	4660.02	55.2	52.68
9.5	28	361.1	5561.34	58.9	42.12
10	34	381.0	5972.48	60.6	35.43
10.5	37	429.2	8164.68	63.3	32.87
11	43	467.9	9871.84	66.3	22.19
11.5	43	514.4	12675.56	69.3	22.32
12	46	574.6	16905.46	71.8	20.66
12.5	47	649.6	17539.04	75.0	25.91
13	43	714.9	19056.60	77.2	25.44
13.5	45	789.8	16775.31	80.9	27.27
14	40	844.8	13272.57	83.1	31.98
14.5	36	892.6	11007.57	87.0	28.09
15	28	927.0	6383.01	89.6	25.67
15.5	23	944.3	3159.90	90.7	22.46
16	20	954.5	2778.16	92.1	19.56
16.5	17	981.9	520.76	94.3	15.44
17	5	980.0	1144.00	94.9	21.96
17.5	2	999.0	2.00	98.5	0.00
18	2	999.0	2.00	99.8	0.13

to eighteen years who were part of cross-sectional studies in Tokyo in 1963 and 1970 and in Sapporo in 1959 (Kimura 1972a). Comparison was made with Australian data reported by Roche, Davila, and Eyman (1971).

Currently, several methods of assessment of skeletal maturation are in use. One of the most widely used qualitative procedures is the "skeletal age" method of Greulich and Pyle (1950, 1959). Another is the "maturity score" method of Acheson (1954) and Turner, Whitehouse, and Healy (1962). (In the Tanner–Whitehouse method skeletal age equivalents for point scores are available.) In the present paper, the Tanner–Whitehouse and the Kimura method (1972b) have been used. In the Kimura method, equal emphasis is given to three areas of skeletal growth (the phalanges and metacarpals, the carpals, and the radius and ulna) in calculating a total maturity score, adopting the indicators by Greulich and Pyle (1959) for the point scores of each bone and epiphysis. The self-weighting maturity scores have been used for the analysis of skeletal maturity in the Tanner–Whitehouse method. Therefore, skeletal maturity is expressed by maturity scores valued from 0 to 1,000 in the Tanner-Whitehouse method and those from 0 to 100 in the Kimura method, respectively. The hybrid and control Japanese samples were all assessed by the author alone.

RESULTS AND DISCUSSION

Table 1 shows the means and variances of the maturity scores by the Tanner–Whitehouse and Kimura methods in the male hybrids aged from four to eighteen years. Figure 1 compares the distance curves of the means of maturity score by the Tanner–Whitehouse method plotting against age among the hybrids, Japanese, and Australians. Only the Australians attained the fiftieth percentile of the scores for the age groups from two to fourteen years. Further, Figure 2 shows the annual increment curves of the maturity scores in these groups.

The maturity graphs generally show the S-figured pattern, but their courses are not necessarily similar to each other. For example, the matu-

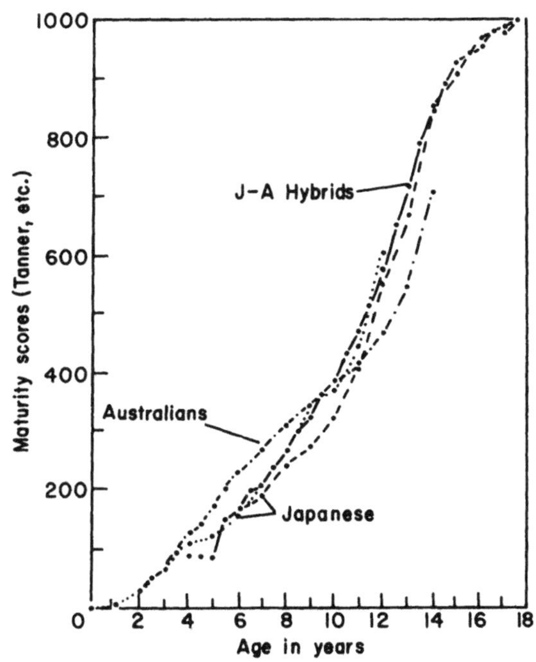

Figure 1. Distance curves of the maturity score by the Tanner–Whitehouse method in the hybrids and others

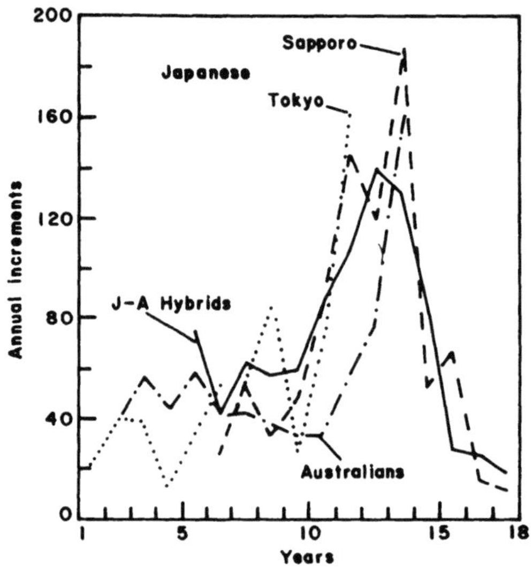

Figure 2. Annual increment curves of the maturity score by the Tanner–Whitehouse method in the hybrids and others

rity score of 200 is attained at the age of about 5.5 years in the Australians and at the age of six to seven years in the Japanese and the hybrids. However, the maturity score of 500 is attained at the age of about 12.4 years in the former and at the age of eleven to twelve years in the latter. The adolescent spurt may begin at or after the age of about eleven years in the Japanese and the hybrids, while it occurs later after the age of thirteen years in the Australians.

Figure 3 shows the maturity graphs, plotting the mean maturity score obtained by the Kimura method against chronological age of the Japanese in Tokyo and Sapporo, and of the hybrids. This maturity graph yields straighter lines than the maturity curve of the point score using the Tanner–Whitehouse method in Figure 1. Consequently, a significant linear relation is obtained between the age in months (x) and the maturity score (y). The following regression equations have been estimated from the samples in each group: in the hybrids (n = 640), aged from 69 to 218 months,

$y = 0.46\,x + 5.28,\ Sy{\cdot}x^2 = 29.0979;$

in the Japanese in Tokyo (n = 274), aged from 79 to 150 months,

$y = 0.48\,x + 3.32,\ Sy{\cdot}x^2 = 35.3372;$

and in the Japanese in Sapporo (n = 135), aged from 76 to 218 months,

$y = 0.47\,x + 2.35,\ Sy{\cdot}x^2 = 23.9311.$

No statistically significant differences are recognized for the regressions of maturity scores on age in these three groups. Figure 4 shows the regression curves for the groups. It may be estimated from each regression equation that the maturity score at the age of seventy months is 35.76

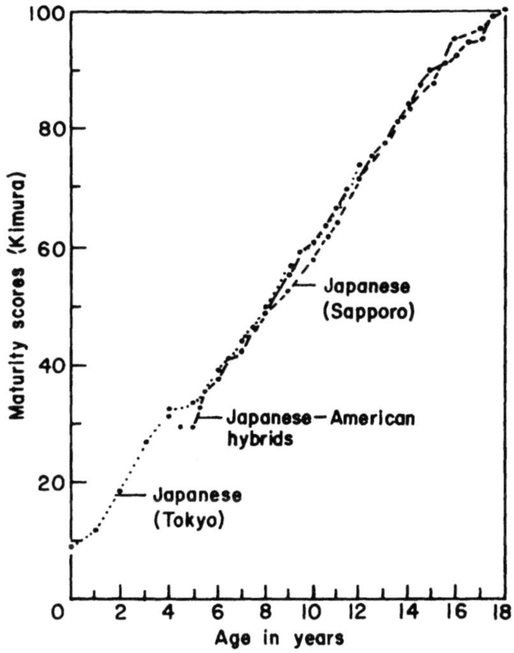

Figure 3. Distance curves of the maturity score by the Kimura method in the hybrids and the Japanese

in the Japanese in Tokyo, 35.25 in those in Sapporo, and 37.66 in the hybrids. The score at the age of 220 months may be calculated as 108.92, 105.75, and 107.05 in each group, respectively.

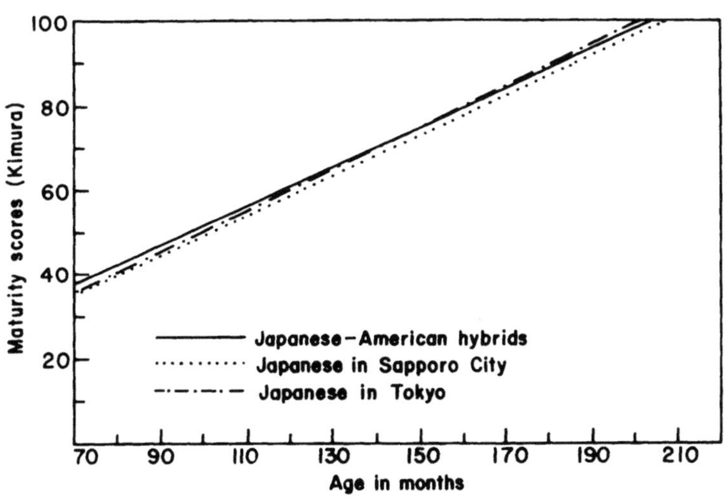

Figure 4. Regression curves between age (x) and maturity score by the Kimura method (y) in the hybrids and the Japanese

CONCLUSION

These findings concerning skeletal maturation in Japanese and Japanese–American hybrids have been confirmed: Skeletal maturation in the Japanese shows some logical differences, boys in Tokyo maturing more rapidly by about one year than those in Sapporo; Japanese mature more slowly than Australians during infancy and childhood, but show a more rapid adolescent spurt; hybrids may mature more or less rapidly at childhood compared to the Japanese, and their adolescent spurt occurs more rapidly than the Australians'. From these findings it may be pointed out that the Japanese–American hybrids show more similarity to the Japanese than to the whites in skeletal maturation.

REFERENCES

ACHESON, R. M.

1954 A method of assessing skeletal maturity from radiographs. *Journal of Anatomy* 88:498–508.

GRUELICH, W. W., S. I. PYLE

1950 *Radiographic atlas of skeletal development of the hand and wrist.* Stanford: Stanford University Press.

1959 *Radiographic atlas of skeletal development of the hand and wrist.* (Second edition). Stanford: Stanford University Press.

KIMURA, K.

1971 On the skeletal maturation of the Japanese–American hybrids — a preliminary report. *Journal of Anthropological Society of Nippon* 79:21–29.

1972a Skeletal maturation in Japanese according to the Oxford and Tanner–Whitehouse methods. *Acta Anatomica Nipponica* 47.

1972b Skeletal maturation in Japanese — a new analytical method. *Journal of Anthropological Society of Nippon* 80.

ROCHE, A. F., G. H. DAVILA, S. L. EYMAN

1971 A comparison between Greulich–Pyle and Tanner–Whitehouse assessments of skeletal maturity. *Radiology* 98:273–280.

SUDA, A.

1968 Introduction to our anthropological studies on the Japanese–American hybrids. *Journal of Anthropological Society of Nippon* 76:89–94.

TANNER, J. M., R. H. WHITEHOUSE, M. J. R. HEALY

1962 *A new system for estimating skeletal maturity from the hand and wrist, with standards derived from a study of 2600 healthy British children. Part II: The scoring system.* Paris: International Children's Center.

Growth and Environment Studies in Czechoslovakia

MIROSLAV PROKOPEC

From a number of countries comes good evidence that at present the growth rate of children and height of adults is increasing (Tanner 1955). Some countries have even been able to follow, on the basis of systematic measurements, growth changes since the middle or the end of the last century and thus to document this almost general phenomenon. A number of acceptable theories have been advanced to explain this trend on a genetic and environmental basis.

The purpose of this paper is to outline some findings of this sort in Czechoslovakia, consisting of the Czech and Slovak Socialist Republics. Its territory — some 128,000 square kilometers — extends from west to east. The population is 15 million, one-quarter of which is children under fourteen years. The proportion of children is lower in the west in the Czech Socialist Republic (CSR) which includes Bohemia and Moravia than in the Slovak Socialist Republic (SSR) in the east.

RESULTS

Data on the height of the male population, the trend toward a higher stature, and the geographical distribution of body height comes from measurements taken of conscriptees by the Ministry of National Defense (see Figure 1). The mean height of eighteen-year-old males has risen during the last fifty years by 7.9 centimeters. The increment was approximately 1.3 centimeters per decade from 1921 to 1960. Since 1960 this trend has increased to 2.9 centimeters per decade. If, on the basis of these trends, we were to forecast the mean height increment in young adult males for the next twenty years, we should probably choose a half-way figure, i.e., about 2.1 centimeters per decade. This judgment is

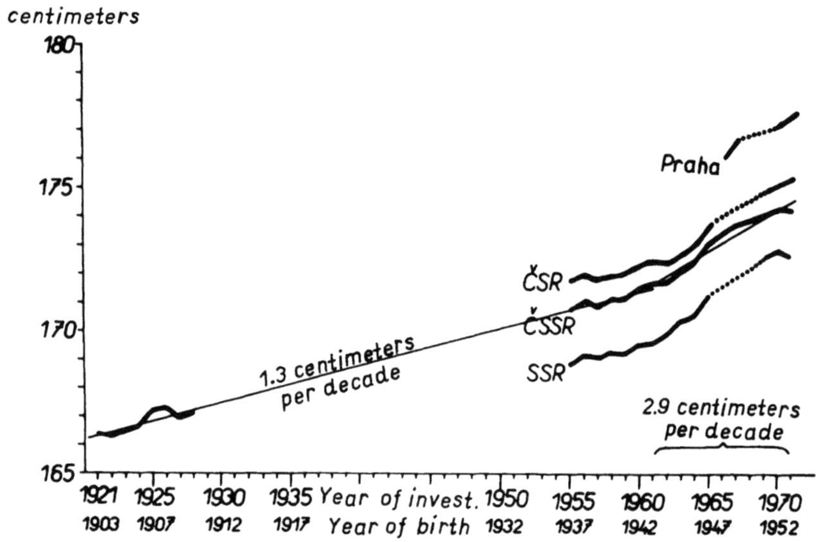

Figure 1. Trend in height in eighteen-year-old men

supported by the fact that eighteen-year-old males from Prague already attain a mean height of about 178 centimeters which, on the basis of this forecast, can be expected in young men in other parts of the country after the twenty-year period. It cannot be claimed, however, that the difference between the mean height of eighteen-year-old males in 1921 and 1971 (7.9 centimeters) simply signifies a difference in the height reached in adulthood. Growth acceleration is only one of the factors involved. We do not exactly know at what point growth ceased fifty years ago, and we do not know exactly when it does today. But the point at which mean final height at adulthood is reached is shifting toward a younger age. It was reached at about age twenty-five several decades ago and is reached at twenty years or earlier today.

The geographic distribution of tall and short stature in Czechoslovakia is presented in Figures 2 and 3. The percentages of persons shorter than 168 centimeters (short stature) and taller than 180 centimeters (tall stature) have been calculated for the various administrative districts. These limits have been selected so that they roughly correspond with the limits of plus and minus one standard deviation from the mean of 174 centimeters, taking six centimeters as the standard deviation. Theoretically, we would expect about 16 percent of the cases to be below and above these limits. In actuality, however, these districts have less than 5 percent or more than 30 percent of individuals with short or tall stature. The proportions of short statures in these districts have been grouped in 5 percent intervals from 10 to 30 percent, and in one group above 30 percent. There are six groups altogether. For tall stature, we have again

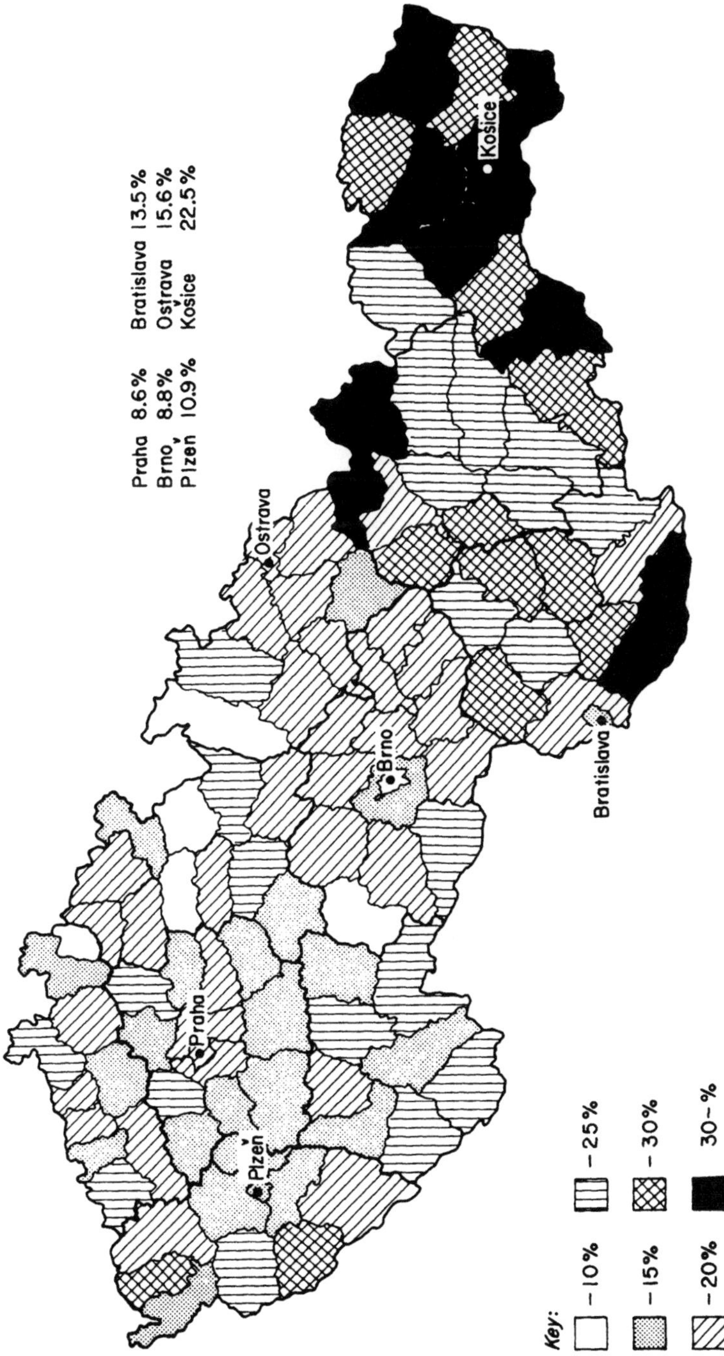

Figure 2. Stature below 168 centimeters in Czech and Slovak eighteen-year-old men in various districts and in large cities

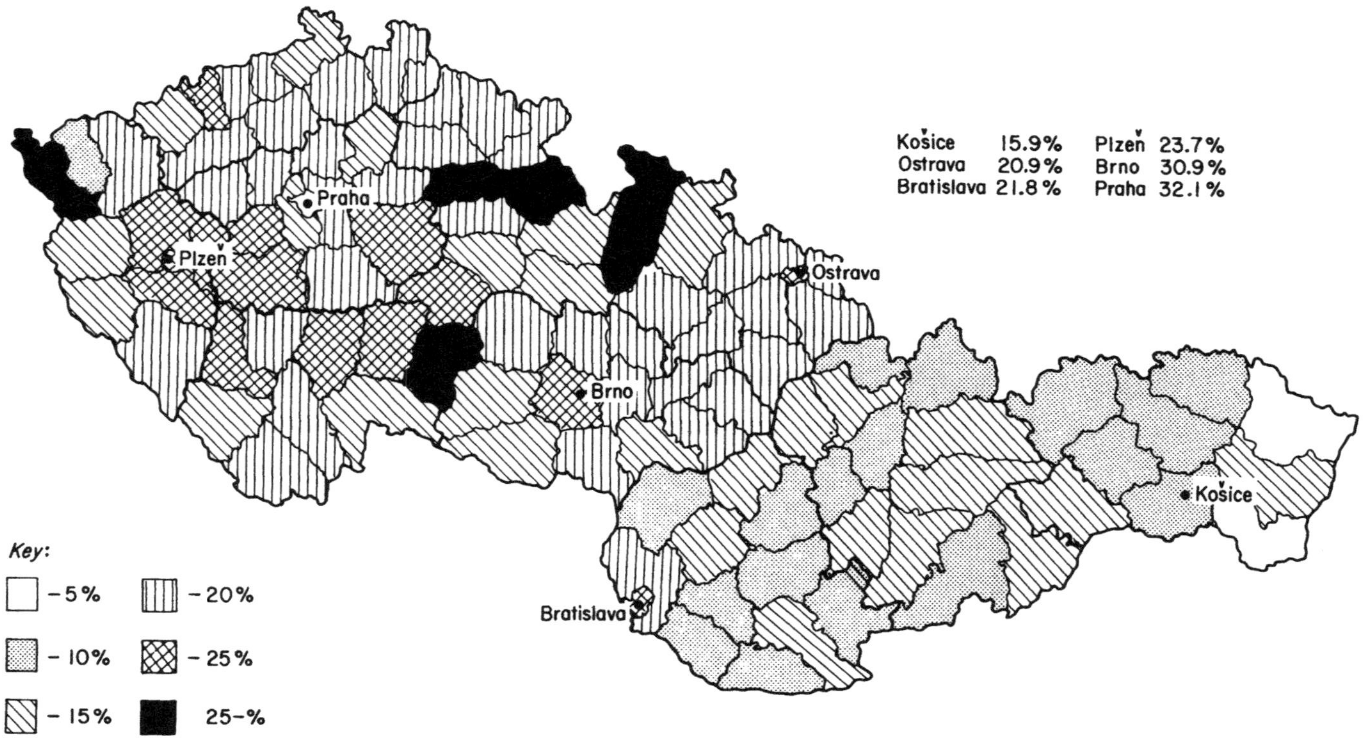

Figure 3. Stature over 180 centimeters in Czech and Slovak eighteen-year-old men in various districts and in large cities

formed six groups according to the percentage of occurrence: one group to 5 percent followed by groups at 5 percent intervals to 25 percent, and the last group includes cases above 25 percent. Each group was given its own designation which was used in drawing the graph. Large cities are presented independently with exact calculated percentages of persons of tall or short stature.

Tall stature is more frequent in the western part of the country and short stature in the eastern part. Height in the CSR exceeds the countrywide mean by 0.5 to 1.0 centimeters whereas height in the SSR falls 1.0 to 2.0 centimeters below the countryside mean. This applies to both districts and cities. Only one district in the Czech provinces, Sokolov, has fewer than 10 percent of its heights above 180 centimeters; whereas in Slovakia, there is only one district, Bratislava (outskirts), with more than 15 percent of its heights above this figure. Tall stature in Prague is twice as frequent as in Košice, about one-third more frequent than in Ostrava or Bratislava, and one-quarter more frequent than in Plzeň. Prague and Brno do not differ much in the occurrence of tall stature (1.2 percent). There are about twice as many short men in Košice as in Plzeň and a little less than twice as many in Ostrava as in Prague. The outstanding position of the capital, Prague, and probably also of some other large cities, as regards the body height of young males, is shown in Table 1, which gives mean heights for eighteen-year-old males in 1971. The mean height is slightly under 178 centimeters in Prague, about 175 centimeters in the Czech districts as a whole, 173 centimeters in the West Slovakia and Central Slovakia regions, and 172 centimeters in East Slovakia. The mean height of eighteen-year-old males in all of Czechoslovakia is 174

Table 1. Regional distribution of body height in eighteen-year-old males in Czechoslovakia (1971)

Region	Mean height (cm)	Standard deviation (cm)	Rounded up mean height (cm)
Prague–City	177.57	6.67	178
Middle Bohemia	174.85	6.53	175
Western Bohemia	174.94	6.73	175
Northern Bohemia	174.69	6.61	175
Southern Bohemia	174.91	6.97	175
Eastern Bohemia	174.87	6.61	175
Southern Moravia	175.21	6.53	175
Northern Moravia	174.76	6.59	175
Western Slovakia	173.08	6.48	173
Middle Slovakia	172.65	6.49	173
Eastern Slovakia	171.84	6.51	172
CSSR	174.25	6.72	174
CSR	175.33	6.66	175
SSR	172.60	6.51	173

centimeters, 175 centimeters in the Czech provinces as a whole, and 173 centimeters in Slovakia.

The mean height of 625 adult males (aged twenty to forty-nine) from different districts of Bohemia, Moravia, and Slovakia, as ascertained from 1967 to 1971, was 172.1 centimeters (Prokopec 1976).

Persons above twenty years of age are, on the average, shorter with every increasing year of age, and the overall trend of stature decrease with age in adults corresponds to the trend of mean stature increase in the past decades, but in the opposite direction. For every ten years after the age of twenty, the mean height among the contemporary male population decreases by 1.25 centimeters (Figure 4).

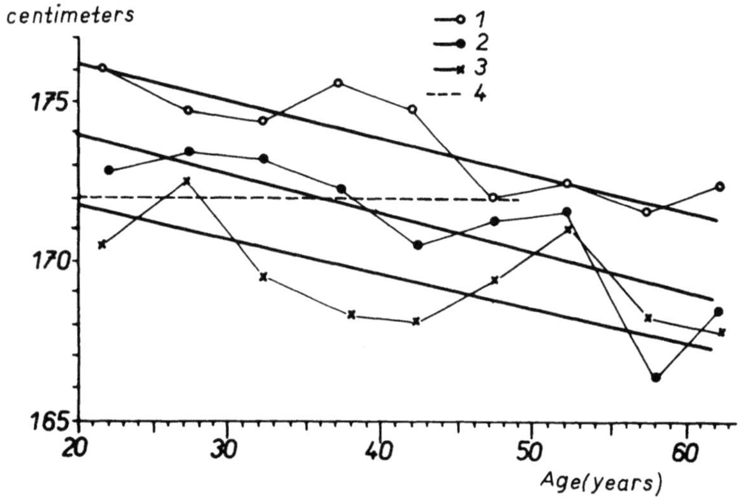

Figure 4. Height for adult Czech and Slovak males: 1, clerks; 2, workers; 3, agriculturists; 4, mean height in the age group from twenty to forty-nine years

Mean Height Varies between Professions

Mean height is greatest among persons with a sedentary lifestyle (e.g., clerks) while the mean height of manual workers in industry coincides with the overall mean. Agricultural workers, with some exceptions, display the lowest mean body height. The older classes with a lesser body height show good correspondence with the mean height which applied at the time their growth stopped. Hence, individual body height does not seem to change so much with age as has been presumed; marked changes occur only after fifty years of life.

The bodily growth of children and youths has been systematically measured in Czechoslovakia since 1951 within countrywide anthropometric studies of representative samples taken at ten-year intervals. These studies were initiated and organized by the late Professor V. Fetter (Fetter, Prokopec, and Suchý 1965; Fetter et al. 1967). They have been continued by his co-workers and pupils associated with four institutes: the Institute of Hygiene and Epidemiology, Prague; Faculty of Pedagogics, Charles University, Prague; Faculty of Natural Sciences, Charles University, Prague; and the Research Institute of Hygiene, Bratislava. Each study included more than 200,000 children who were randomly selected from medical centers and schools and measured by pediatricians and child nurses or teachers of biology and physical education at schools according to precise procedures and training. The data from each republic were treated separately. The curve expressing the changes in height and weight with age are presented in Figures 5 through 12. In 1971, the study of Czech provinces included 120,000 children (60,000 up to six years and 60,000 from six to eighteen; half boys and half girls) who were divided into the following groups: twelve (monthly) groups to one year of age; four (three-month) groups up to the age of two years; four (six-month) groups from two to four years; and fifteen (annual) groups from four to nineteen years. There was a total of thirty-five age groups (Prokopec, Suchý and Titlbachová 1973).

In contrast to previous studies in which the limits of these groups fell in the middle of the respective intervals (e.g., the group of six-year-old children included children from 5.5 to 6.5 years of age), the 1971 study relied on the method of decimal division of the years as recommended by the World Health Organization (e.g., age group of 5.0 to 5.99 years, etc.).

Height of Children Aged Zero to Six Years

From the age of six months, the mean heights of boys measured in 1971 begins to diverge from the 1961 curve, and thereafter the higher mean values are maintained. We were able also to compare the results on three-year-olds with those from the 1951 study. The shift in height values from 1961 to 1971 from three to six years of age is higher than that recorded between 1951 and 1961. A similar picture is presented by curves for girls which lie below the curve of the boys by about 1.7 centimeters at the age of one year, and 0.4 centimeters at the age of six. As compared with 1951, the increment of the mean heights between three and six years was 1.5 to 2.3 centimeters in boys and 1.6 to 3.0 centimeters in girls in 1971. The respective increments from 1961 to 1971 were 1.0 to 1.5 centimeters in boys and 1.0 to 2.0 in girls (Figures 5 and 6).

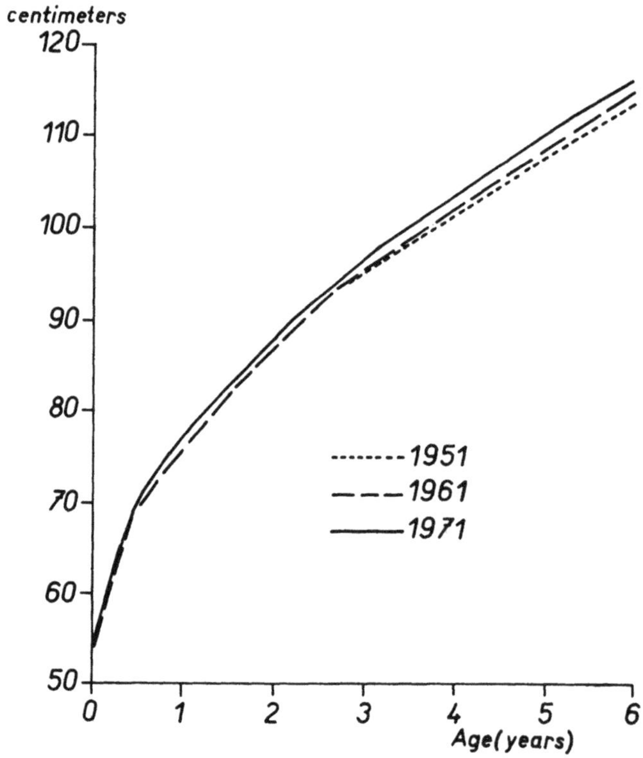

Figure 5. Height for Czech boys 1951–1971

Weight of Children Aged Zero to Six Years

The 1971 curve for boys up to three years does not differ much from that of 1961, whereupon the two curves diverge. The difference increases with age and amounts to 0.6 kilograms at six years. As compared with the 1951 curve which coincides with the 1961 curve during the three to five-year interval, the difference at six is about one kilogram. The weight curve for girls is roughly similar to that for boys. The 1961 and 1971 curves diverge at the end of the first year. The difference increases progressively to 0.7 kilograms in relation to 1961 and 1.0 kilogram in relation to 1951 at age six. On the average, boys at the age of one year are 0.7 kilograms heavier than girls and 0.4 kilograms heavier at the age of six (Figures 7 and 8).

Height of Boys and Girls Aged Six to Eighteen Years

In addition to curves from 1951, 1961, and 1971 studies, the figures contain curves for school-age children (seven to fourteen) from a study

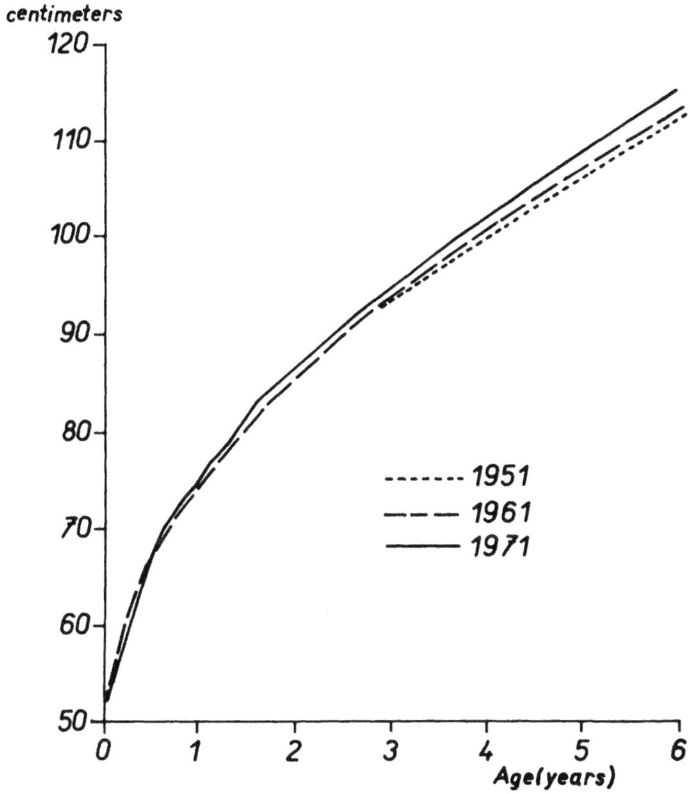

Figure 6. Height for Czech girls 1951–1971

performed on more than 100,000 children in 1895 by Professor Matiegka (1933). The figures also include some results obtained in Prague in 1971 with a limited number of children (about 10,000, ages six to eighteen), especially the higher grades. The male height curves for 1951, 1961, 1971, and Prague (1971) lie above each other at more or less equal intervals with the exception of the fifteen-to-eighteen-year segment where the Prague data display considerable variation because of the small number of children in the respective age groups. As compared with 1961, the mean values for 1971 have increased by about 1.0 centimeter up to fifteen years and 2.0 centimeters up to eighteen years. As compared with 1951, the 1971 values have risen by about 2.0 centimeters up to eleven years and 3.0 to 5.0 centimeters up to eighteen years. Growth had thus accelerated by about two to three months of age since 1961 and by about eight months since 1951. As compared with 1895, the growth acceleration at the age of seven to fourteen years amounts to two to two and one-half years, or in other words, the difference between mean heights in children of equal age amounts to 10 to 15 centimeters.

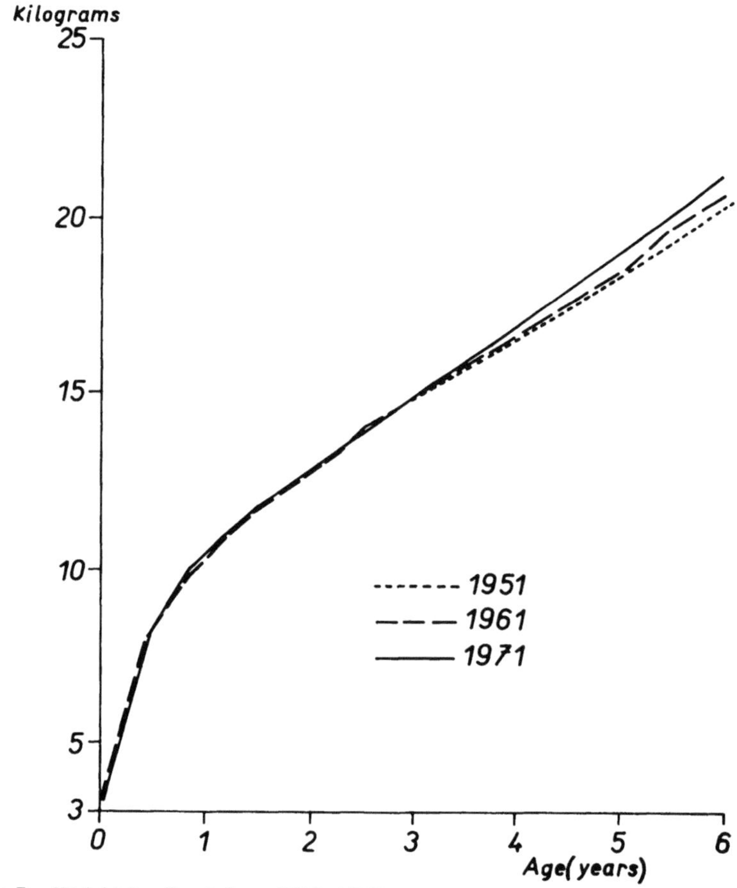

Figure 7. Weight for Czech boys 1951–1971

The female curves obtained in these studies also lie above each other at regular, 1.0- to 2.0-centimeter intervals, except for the curve for Prague girls representing ages above 15.5 years (a small number of cases) and the 1961 curve for eighteen-year-olds. In 1961 girls did not display an increase in height after the age of sixteen, whereas in 1971 the height curve continues up to eighteen.[1] At present (1971) equal height values are attained by girls a quarter of a year earlier than in 1961; one-third to one year earlier than in 1951, and two to two and one-half years earlier than in 1895.

According to the 1971 investigation, the girls from 11.5 to 13.5 years of age were taller on the average than boys, the maximal difference between

[1] In the 1961 study, the fifteen- to eighteen-year segment displayed a plateau that was not confirmed in 1971. This apparently was a consequence of the war and postwar years, during which the groups studied in 1961 had spent their early childhood. (The fifteen- to eighteen-year-olds were born from 1943 to 1946.)

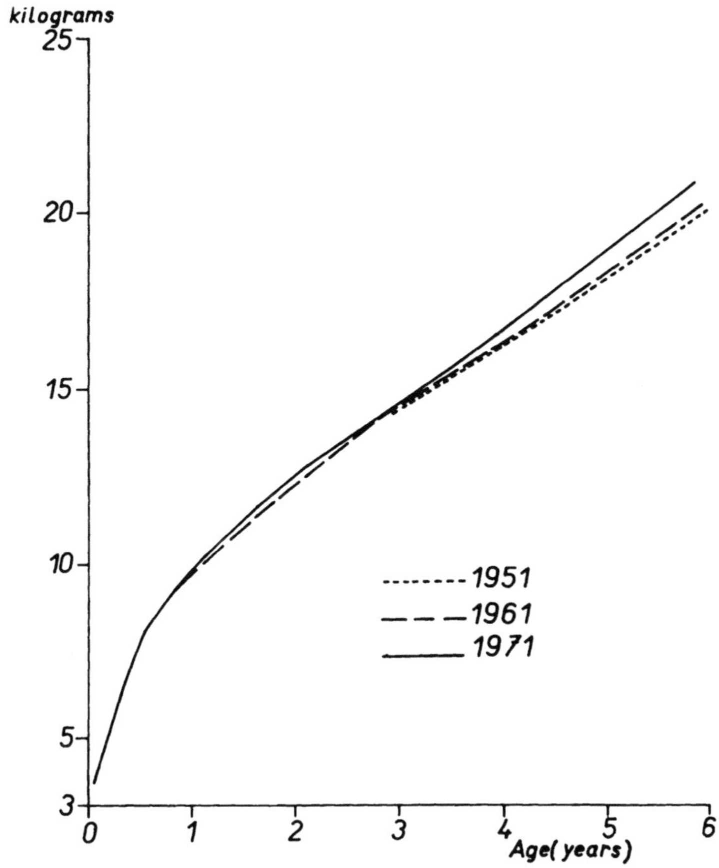

Figure 8. Weight for Czech girls 1951–1971

the means being at the age of 12.5 (2.5 centimeters). At the age of 18.5 years, the difference between the girls and the boys means amounts to 12.7 centimeters in favor of the boys.

The results demonstrate that the mean height at every age has increased as compared with ten years ago. The considerably higher mean values for Prague boys and girls are evidence that the possibilities of mean height acceleration and increase have not yet been exhausted (Figures 9 and 10).

Weight of Boys and Girls Aged Six to Eighteen Years

Up to 15.5 years the 1971 weight curve for boys passes at equal distances between the 1961 curve and the curve for Prague boys. The difference amounts to approximately 0.5 kilograms and 2.0 kilograms at the younger and older ages, respectively. The 1951 curve runs lowest at

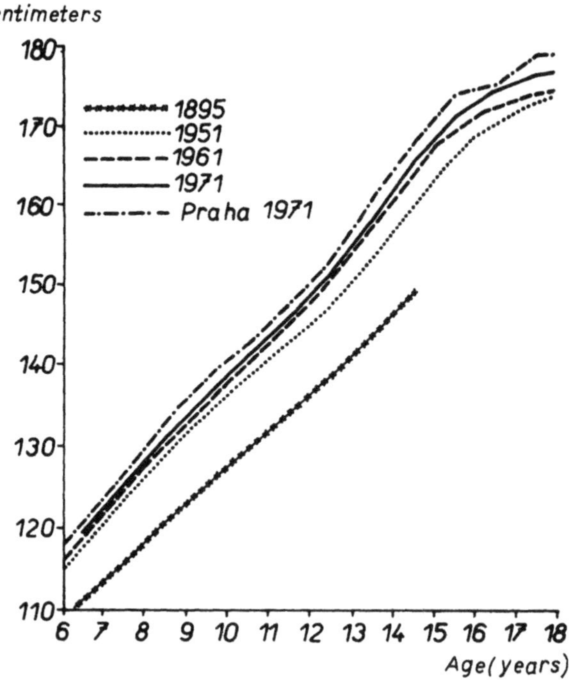

Figure 9. Height for Czech boys 1895–1971

approximately the same distance up to nine years and then declines toward still lower values. From the age of sixteen, the 1961 weight curve for boys forms a plateau and the curve for the Prague sample has an irregular course owing to the small number of children measured. As compared with 1951, the mean weight of boys aged six to eighteen years has increased by 1.0 to 4.0 kilograms in 1971. As compared with 1895, the difference at the ages of seven and fourteen amounts to about 4.0 and 13.0 kilograms, respectively, which means that an equal mean weight is reached by Czech boys today 1.6 to 2.3 years earlier than in 1895.

The position of the weight curves for girls is similar to that for boys with the sole difference being that at age eighteen the curves for 1951, 1961, and 1971 converge at a value of 59.0 kilograms. The weight for girls at eighteen has practically not changed since 1951, probably because of deliberate weight reduction. The only change is that the same weight is attained at a lower age. Prague girls in the six to sixteen and one-half year age group are the heaviest. As compared with 1895, Czech girls today (1971) reach equal weight values 2.0 to 2.3 years earlier. At an equal age this gives a mean difference of 4.0 to 11.0 kilograms.

According to the 1971 survey, the girls are heavier on the average than boys between 9.5 and 14.5 years of age, the maximal difference between

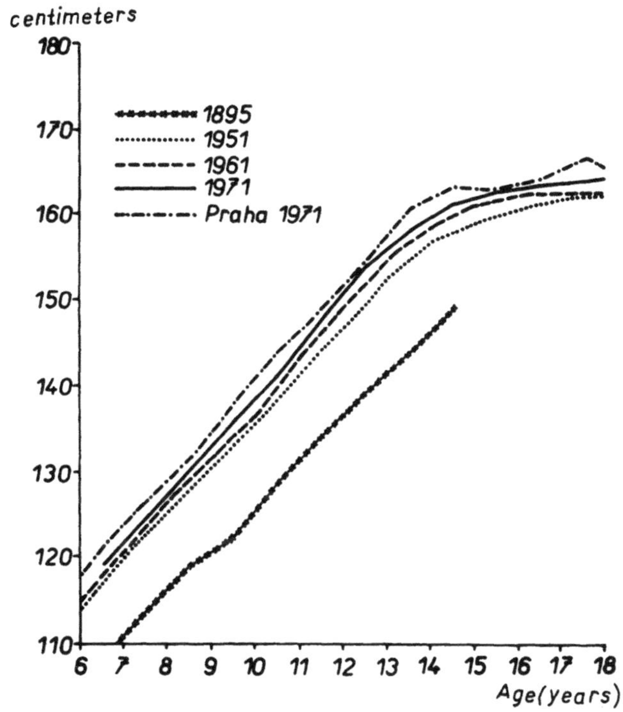

Figure 10. Height for Czech girls 1895–1971

the means being 2.9 kilograms at the age of 12.5. Boys exceed girls in weights on the average by 9.4 kilograms at the age of 18.5 (Figures 11 and 12).

On the whole, the height and weight curves for boys and girls, with the exception of Prague children, are smooth, which testifies to the reliability of data collection and processing procedures. These curves also confirm the strict laws of growth (Tables 2, 3, and 4).

DISCUSSION

The eighteen-year-old males whose heights were dealt with in the first part of this report were born in 1953. They grew up under the conditions existing in the CSR, after the widespread introduction of antibiotics, and they were not affected by the war. A sober estimate is that by 1991, the average height of eighteen-year-old males will reach approximately 178 centimeters. Also, it can be judged from the stable difference between the sexes (about 12 centimeters in height) that the mean height of the eighteen-year-old females in 1991 will be 166 centimeters. By this time

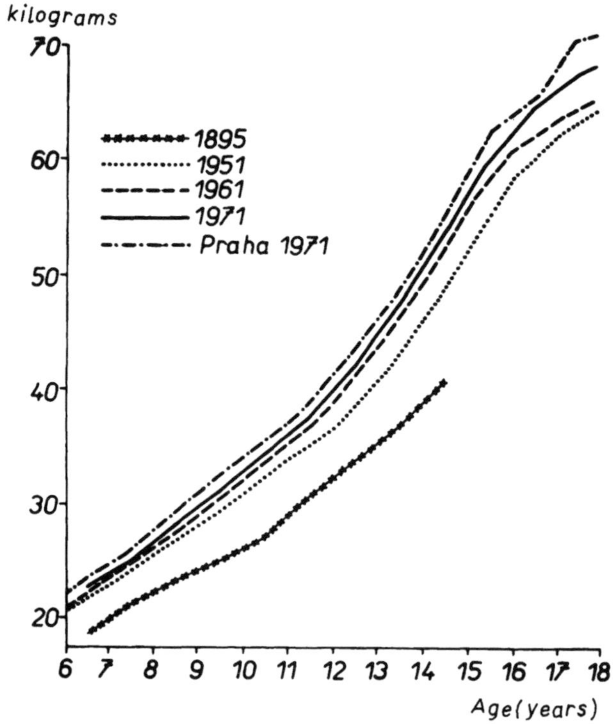

Figure 11. Weight for Czech boys 1895–1971

all children and youths in the country will probably reach the mean height of the Prague children of today. This trend is explained by the opening up of genetic isolates and an increase in the number of heterozygotes in the population (according to some authors they display a tendency toward greater height). All these trends are a consequence of technological progress and the concentration of people in cities and industrial centers lowering the mean number of children in the family etc.

We can see that the distribution of short and tall stature varies considerably among the districts of Czechoslovakia with an overall tendency toward height increase. In separate studies (Prokopec 1959, 1970) we pointed out that children's heights also depended on the order of birth in the family and the profession, income, and height of the parents. Thus, the height of the present young generation is not merely a reflection of the living conditions but also of an inborn disposition. It can be said that potential growth is always higher than the actual height attained. However, the latter approximates most closely the potential growth where inhibitory influences (deficiency of proteins and vitamins, diseases affecting the growth of long bones, etc.) have been abolished by improved medical care and favorable living conditions. The limit of potential

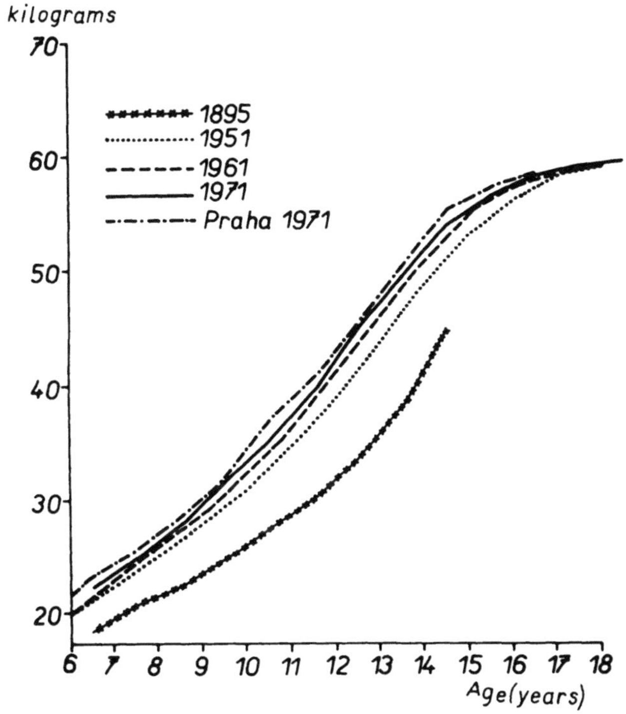

Figure 12. Weight for Czech girls 1895–1971

growth is apparently given by heredity, but a child growing up under favorable conditions has a chance of approximating it. Of course, the limit is lower in children of short parents and higher in those of tall or taller parents; and individual exceptions do occur.

The geographical distribution of tall and short stature also reflects both social and ethnic factors. The present state of height distribution is a result, to a certain extent, of such factors as isolation, migration (colonization of certain regions by people with different bodily properties), hybridization (especially present in cities and industrial centers), different incomes, and the influences of cultural traditions, education, medical care, modern technology, the use of electricity, reduction of heavy physical labor, fast transportation, prolongation of the working day, better incomes, hygienic housing, etc. This set of influences operates unevenly and, moreover, on a heterogeneous sub-stratum, for presumably the settled agricultural population in the past differed from the populations bound to the crafts, trade, administrative apparatus, etc. The varied history of colonizations, migrations, and wars probably also influences the physical characteristics of even some of the present-day inhabitants of certain regions.

Table 2. Countrywide growth study, 1971, Czechoslavakia (Czech districts)

Age (months)	Boys					Girls				
	n	Height		Weight		*n*	Height		Weight	
		Mean	s.d.	Mean	s.d.		Mean	s.d.	Mean	s.d.
0– 0.9	423	53.21	2.78	3.88	0.54	393	52.34	2.43	3.68	0.78
1– 1.9	565	56.15	3.05	4.66	0.69	580	55.31	2.95	4.38	0.59
2– 2.9	607	59.76	3.45	5.64	0.82	603	58.81	3.16	5.27	0.75
3– 3.9	583	63.32	3.65	6.58	0.85	548	61.70	3.48	6.07	0.82
4– 4.9	620	66.25	3.42	7.42	0.88	615	64.21	3.07	6.76	0.82
5– 5.9	606	68.28	3.41	8.01	0.97	598	66.56	3.46	7.44	0.89
6– 6.9	661	70.19	3.15	8.50	0.97	649	68.26	3.31	7.95	0.90
7– 7.9	634	71.46	3.36	8.92	1.05	625	69.75	3.23	8.30	0.95
8– 8.9	583	72.49	3.01	9.25	1.06	597	70.89	3.00	8.67	1.01
9– 9.9	597	73.67	3.30	9.69	1.13	611	72.21	3.48	9.12	1.13
10–10.9	569	75.01	3.33	10.00	1.12	540	73.43	3.55	9.33	1.11
11–11.9	616	76.01	3.22	10.23	1.13	635	74.29	3.33	9.54	1.33

Table 3. Countrywide growth study, 1971, Czechoslavakia (Czech districts)

Age in years	Boys					Girls				
	n	Height		Weight		*n*	Height		Weight	
		Mean	s.d.	Mean	s.d.		Mean	s.d.	Mean	s.d.
1–1.24	1739	78.39	3.73	10.80	1.23	1769	76.71	3.67	10.10	1.23
1.25–1.49	1764	81.25	3.80	11.32	1.27	1737	80.04	3.71	10.79	1.28
1.50–1.74	1812	83.95	3.95	12.04	1.42	1772	82.78	3.78	11.33	1.28
1.75–1.99	1612	86.56	3.88	12.55	1.40	1641	85.34	3.89	11.94	1.37
2.00–2.49	2402	90.35	4.22	13.38	1.55	2442	89.26	4.25	12.86	1.58
2.50–2.99	2336	94.63	4.29	14.44	1.66	2282	93.63	4.42	14.00	1.73
3.00–3.49	2397	98.50	4.49	15.50	1.83	2365	97.71	4.35	15.10	1.91
3.50–3.99	2285	102.06	4.46	16.48	1.97	2288	101.32	4.55	16.02	2.04
4.00–4.99	3172	107.34	5.18	18.13	2.45	3167	106.67	5.65	17.69	2.46
5.00–5.99	3150	113.95	5.18	20.30	2.80	3145	113.39	5.31	19.96	2.95
6.00–6.99	2390	119.64	5.41	22.68	3.33	2422	119.21	5.41	22.29	3.52

Table 4. Countrywide growth study, 1971, Czechoslovakia (Czech districts)

Age in years	Boys					Girls				
	n	Height		Weight		*n*	Height		Weight	
		Mean	s.d.	Mean	s.d.		Mean	s.d.	Mean	s.d.
6.00– 6.99	2390	119.64	5.41	22.68	3.33	2422	119.21	5.41	22.29	3.52
7.00– 7.99	2433	125.24	5.62	25.16	4.08	2582	124.85	5.93	25.02	4.28
8.00– 8.99	2347	130.91	5.98	28.19	4.70	2470	130.12	6.09	27.79	5.06
9.00– 9.99	2552	136.13	6.20	31.19	5.44	2530	135.78	6.29	31.27	5.90
10.00–10.99	2614	141.10	6.48	34.55	6.34	2662	141.31	7.05	35.10	6.99
11.00–11.99	2989	145.77	6.77	37.64	6.90	3004	147.64	7.23	39.73	8.41
12.00–12.99	2357	151.15	7.57	42.03	8.00	2561	153.62	7.45	44.93	8.86
13.00–13.99	2461	158.10	8.56	47.41	9.29	2648	158.13	6.65	49.70	8.82
14.00–14.99	2508	164.99	9.02	53.59	10.23	2643	161.02	6.28	53.86	8.34
15.00–15.99	2338	170.53	7.73	59.89	9.51	2343	162.53	5.90	56.51	7.91
16.00–16.99	2405	173.82	7.02	64.32	8.97	2136	163.14	5.87	58.08	7.91
17.00–17.99	1702	175.89	6.71	67.29	8.69	1769	163.69	5.61	59.08	7.44
18.00–18.99	586	176.91	6.29	68.86	8.37	694	164.22	5.61	59.51	7.45

The differences in bodily height between the individual professional groups can also be explained in terms of hybridization. For example, among clerks and workmen (compared to agricultural workers), there is a greater chance that each parent comes from a different place. Other factors include the selection of mates from particular professions, isolation, influence of traditions in the different professions (family size, dietary habits, housing style). Differences similar to those found among adults were also discovered in the children (Prokopec 1955, 1970).

The postwar studies of representative samples demonstrated that children and youths of a given age had greater height and weight in each consecutive decade, but that the increment in school children during the decade of 1961 to 1971 was lower than during 1951 to 1961. It seems that higher growth limits correspond with decreasing regular increments. On the other hand, the case of city youths shows that there is a possibility of further growth acceleration for the entire child population. To explain this phenomenon, one must take into consideration many mutually correlated and uncorrelated factors which are associated with technological progress and the development of the medical sciences, better education, improved hygienic and cultural conditions.

REFERENCES

FETTER, V., M. PROKOPEC, J. SUCHÝ

1965 Developmental acceleration in children and youth according to anthropometrical investigations from the years 1951 and 1961. *Anthropologie* 2(3):45–47.

FETTER, V., M. PROKOPEC, J. SUCHÝ, S. TITLBACHOVÁ

1967 *Antropologie*. Praha: Academia.

MATIEGKA, J.

1933 "Tělesná povaha dnešního lidu československé-ho," in *Československá vlastivěda*, volume 2. Edited by V. Dedina. 193–240. Praha: Sfinx, B. Janda.

PROKOPEC, M.

1955 Body development of children in the city of Prague and in the Pardubice region with special reference to the environmental conditions (in Czech). *Acta Hygienica* 3(6). Prague: Institute of Hygiene.

1956 *Somatometrical survey of youth (regional analysis)*. Report for the Ministry of Health. Prague: Institute of Hygiene.

1959 "The impact of birth order on the child's body development," in *Proceedings of the first meeting of Czechoslovak anthropologists in 1958*, 219–223. Opava.

1964 Studies of child growth in Czechoslovakia. *Indian Pediatrics* 1(3):100–110.

1970 "Growth and socioeconomic environment," in *Physical growth and body composition*. Edited by J. Brožek, 55–60. Chicago: University of Chicago Press.

1976 New anthropometric data of the ČSSR Population. *Acta Facultatis Rerum Naturalium Universitatis Comenianae Anthropologia* XXII:187–194.

PROKOPEC, M., J. SUCHÝ, S. TITLBACHOVÁ
1973 Výsledky třetího celostátního výzkumu mládeže 1971 (české kraje) [Results of the third state-wide survey of youths 1971]. *Československá pediatrie* 28(7):341–346.

TANNER, J. M.
1955 *Growth and adolescence.* Oxford: Blackwell Scientific Publications.

Some Biosocial Determinants of the Growth, Health, and Nutritional Status of Papua New Guinean Preschool Children

L. A. MALCOLM

Within primitive populations, social and cultural factors determine to a major extent the biological interrelationships between growth, health, fertility, and morbidity patterns. Interwoven with these variables is the impact of poor nutrition that interacts with the infective load in a community to modify and distort the capacity of populations to achieve their potential size, fertility, and life span (Scrimshaw, Taylor, and Gordon 1968; Jelliffe 1966; Scrimshaw 1964). Adaptive mechanisms developed by populations to deal with the stresses are both biological and social. I will examine here these interrelationships within the rapidly changing context of the human biology of Papua New Guinea.

The study of growth and development in Papua New Guinea has revealed as diverse a spectrum of growth patterns as has been reported from any country (Malcolm 1970a, 1970b, n.d.). Figure 1 shows the variation in growth rate noted in children aged up to three years from various populations. The most rapid rates are achieved in urban children (Malcolm n.d.), and especially from families of the members of the Papua New Guinea Defense Force, who are provided with a reasonable income, family allowances, permanent housing, and adequate health services. In these children, weight at three months actually exceeds that of the Harvard standards. By contrast, the growth rate observed in the Asai children of the highlands (Malcolm 1970b) is slower than in any other reported population and seems to be related to a deprived nutritional environment, with a low protein intake, poor health services, and limited cultural contact with social changes which have occurred in other parts of the country. The curves between these extremes are from populations whose environmental characteristics exhibit a corresponding spectrum of variation and change.

It will be noted that the shape of the growth curve in all populations

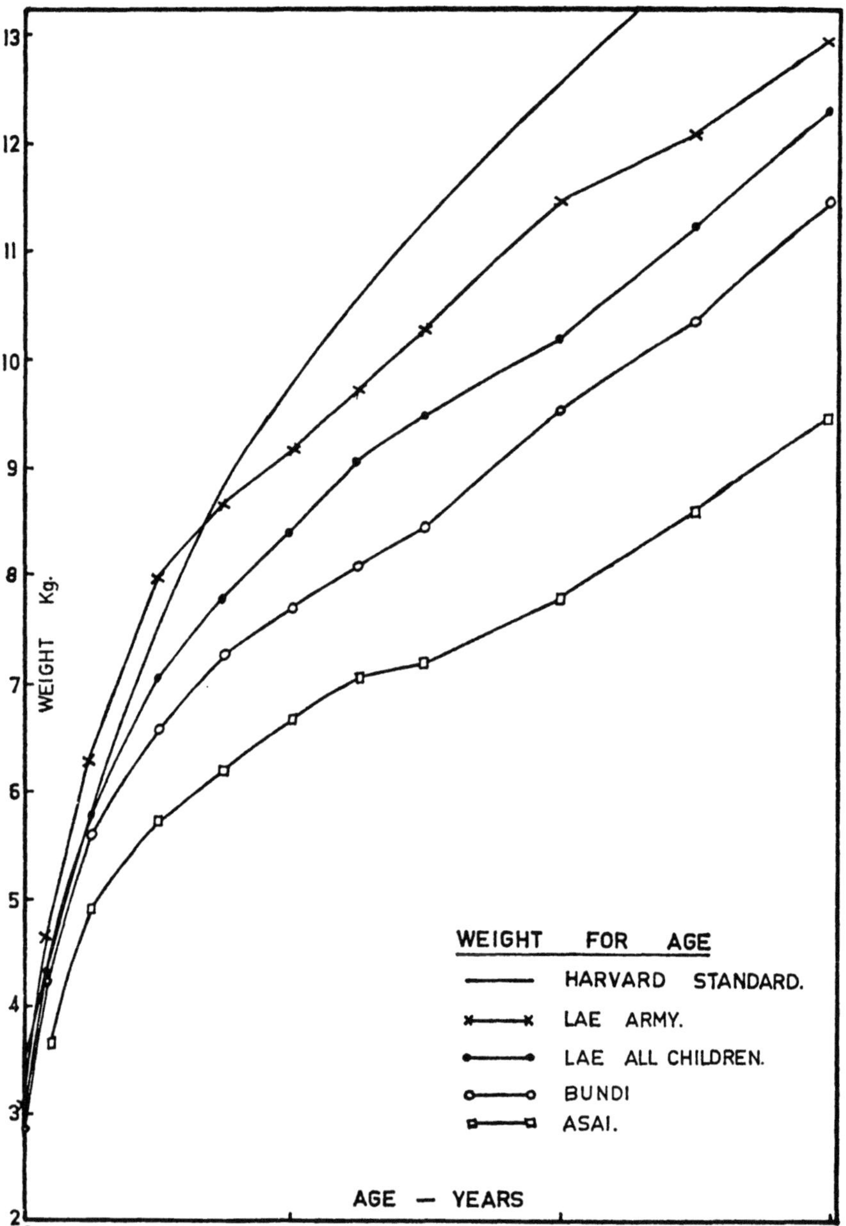

Figure 1. Weight by age of children from some Papua New Guinean populations compared with Harvard standards

follows a similar pattern, but at different levels, with reasonably satisfactory increments for the first few months, followed by a progressive departure from the trend of the European child. Nutritional opinion in the past has tended to regard this pattern, which is typical for developing countries (Jelliffe 1966; Guzman et al. 1968), as being due to a poor quality, protein-deficient weaning diet. A close examination, however, suggests that deficient total food intake is the more probable explanation. Evidence for this conclusion comes from the comparatively less affected range of height variation, which is a more specific measure of protein intake (Malcolm 1972) and from the changes observed in the skinfold thickness, a more specific indicator of total energy intake. Figures 2 and 3 show the changes in this latter parameter in two Papua New Guinean populations (Malcolm n) compared with that of European children. The New Guinean children from both urban Lae and a typical rural population, Kaiapit, reach a peak prior to six months, following which there is a steady fall until the age of one and one-half to two years. By this time a considerable gap has developed, especially in the triceps area, between European and Papua New Guinean values. Although European norms may not be the most appropriate standards, the changes strongly suggest that total food intake is inadequate and is a major factor determining impaired growth. It cannot be the case that the relatively small needs of the child at this age cannot be met by the available food in any society. A possible explanation, however, may be found in the social and cultural attitudes to child care and feeding. Few observers appear to have asked the obvious but perhaps important questions: why does a mother feed her child; what determines whether a mother feeds her child at a particular time; and what motivating factors operate in child feeding in different cultures?

The Western observer, examining the cultural patterns of primitive people, too often interprets his findings, including child care practices, in concepts familiar to himself and his own cultural background. The Western mother normally feeds her child to satisfy her own preconceptions of the child's needs. In many situations, this concept of need is in excess of the actual need. Feeding is based on a timetable, and success is measured by satisfactory weight gain. Failure to attain certain predetermined goals, either in food intake or weight gain, may lead to distress and anxiety in the Western mother. To the mother in a primitive culture, however, factors determining feeding practice may be quite different. Child rearing in most Papua New Guinean societies is permissive and unrestricted, and much of a child's activity such as sleeping, eating, excreting, etc., is determined by the personal choice of the child.

Studies of the mother–child relationship using prolonged, unobtrusive, and unstructured film sequences (Sorenson, personal communication) suggest that child feeding in Papua New Guinea is based almost entirely,

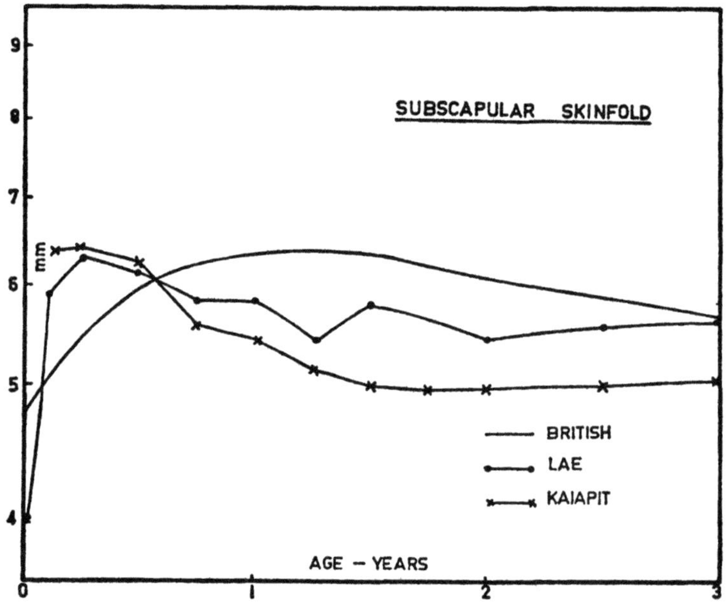

Figure 2. Subscapular skinfold thickness by age of Lae and Kaiapit children compared with that of British children

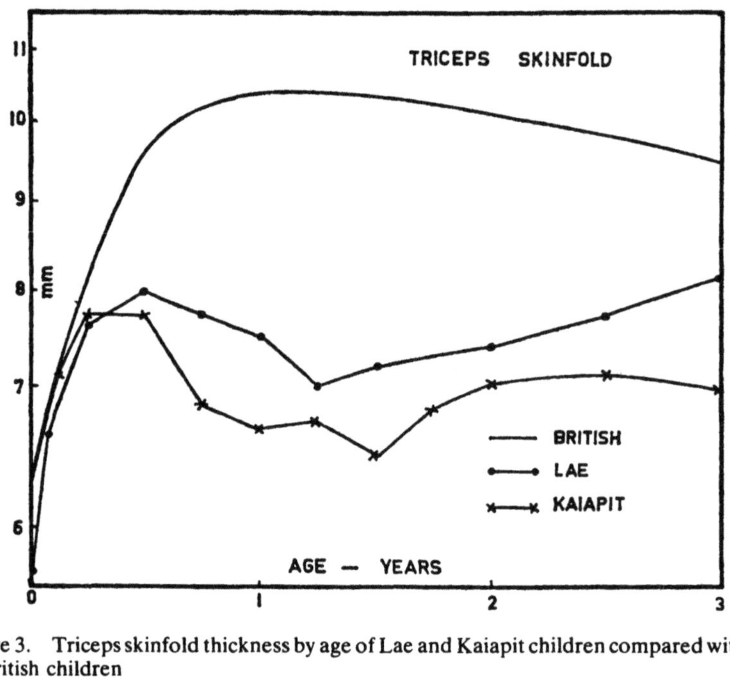

Figure 3. Triceps skinfold thickness by age of Lae and Kaiapit children compared with that of British children

and at all ages, even from birth, on the demands of the child. It is the child who determines when and with what he is fed. In the early weeks of life, the breast is immediately available when he cries and is given without hesitation at any time of the day or night. As soon as the infant is old enough to grasp the breast, he will take the nipple himself. The infant is thus capable of satisfying his own needs by his demands with the breast until the age of six months. However, beyond this time or in the event of lactational failure, the maternal response to his demands for the food resources immediately available to him appears insufficient to maintain an adequate nutritional intake. It is only later in the second year of life that he becomes sufficiently mobile and vocal to satisfy his needs through his own efforts. A gap therefore exists between the age of six months and one and one-half to two years, during which this demand/permissive type feeding regime fails to satisfy the nutritional requirements of the child.

While this pattern may be interpreted as maternal neglect, it must be emphasized that it is of physical needs only, not of emotional needs. The converse may well be true, however, of Western type feeding and child care patterns. These findings suggest that the quantitative deficit described above may be largely due to a maternal failure to appreciate this demand gap and to fill this gap from whatever food might be available. The different levels within this pattern may be attributable to the different quantitative and qualitative success of continued breast-feeding under different nutritional circumstances (Malcolm n.d.) as well as a better weaning diet and increased awareness of the needs of the child, with socioeconomic improvement.

The importance of the social microenvironment of the child has recently been stressed by Cravioto and DeLicardie (i.p.) as a determinant of development of clinical malnutrition. The quality of maternal care influences, to a major extent, the growth, health, and nutritional status of the preschool child rather than the macroenvironmental factors, such as food availability and quality, which have been the targets of traditional nutrition improvement programs.

The prolonged breast feeding pattern of Papua New Guinean mothers has been noted with considerable interest by many researchers (Malcolm 1970a, n.d.; Bailey 1964; Table 1). The duration varies from a mean of one-and-a-half years in urban areas (Malcolm n.d.) to over three years in the New Guinean highlands (Malcolm 1970a). Related to this prolonged period is a consequent variation in the length of the birth interval between successively born children. Studies of urban families indicate that, following pregnancy in the average mother, amenorrhea, presumably lactational, ensues for nearly a year, that conception occurs at one-and-a-quarter years, breast-feeding consequently ceases at one-and-a-half years, and delivery occurs at two years. A situation of

Table 1. Birth interval, duration of breast feeding, and fertility in Papua New Guinean populations

Population	Breast feeding duration (years)	Birth interval (years)	Birth rate/1000
Bundi[a]	3.5	4.0	30
Sepik[b]	2.5	3.3	30–26
New Guinea Is.[c]	—	2.6→2.1	40–55
Urban[d]			
Migrant	2.0	2.4	60–80
Permanent	1.5	2.0	

[a] Malcolm (1970a).
[b] Sturt and Sturt (1973).
[c] Ring and Scragg (1973).
[d] Malcolm (n.d.).

maximum biological fertility therefore exists, restrained only by the prolonged lactational amenorrhea.

By contrast, the prolonged birth intervals of four years reported for Bundi mothers (Malcolm 1970a) are also medicated through prolonged breast feeding, but culturally, not biologically. Throughout much of Papua New Guinea there is, but to a diminishing extent, a prohibition on sexual intercourse during the breast feeding period (Malcolm 1970a; Ring and Scragg 1973), due to the belief that the semen may poison the breast milk of the mother. This cultural belief may have had a basis in the biological wisdom of previous generations in that it acts to preserve the health of the mother from the stress of too frequent child bearing, as well as giving the valuable support of the breast milk for the crucial first three years of life.

In a recent study of a Sepik population, Sturt and Sturt (1973) report that this belief is still strongly held to the extent that only sixteen births in 512 (3 percent) occurred with a birth interval of less than one-and-a-half years between surviving children. Of these sixteen, eight were accepted into the family and of the other eight, one was starved to death, three were adopted out, and in four the previous child was either killed or starved to death. Following the early death of a child, however, the average interval was much shorter and was closely correlated with the age at death of a child, an observation also noted by Ring and Scragg (1973).

Throughout many areas in Papua New Guinea, however, this belief is having a diminishing influence with a consequent increase in fertility (Ring and Scragg 1973) and, as indicated above (Malcolm n.d.), has already been abandoned by the urban mother, whose fertility is now determined biologically rather than culturally. The impact of these changes on the health of both the mother and child, however, is apparently not serious at this stage, as the changes have coincided with equivalent socioeconomic improvements which have minimized the potential consequences. Of some concern, however, is the impact of change and increase in the traditional family size of five to six children, which, particularly in urban areas (Malcolm n.d.) is becoming an insupportable burden within the confines of a nonexpanding home and income, and in the rural areas (Ring and Scragg 1973) where population expansion, due to both decreased mortality and increased fertility, is already leading to accelerated urban drift. Change in the attitude toward intercourse during the breast feeding period in such areas as the New Guinean highlands, which may be anticipated and indeed is already occurring, could have a major impact on family size, health, and the limited land resources of people, whose adaptation within the current ecosystem is only just being maintained.

A further factor in the changing health and fertility patterns is the biological length of the menstrual span. In all traditional New Guinean populations the occurrence of the menarche, which is related to growth and protein intake, is delayed, the mean age in most populations ranging from sixteen to nineteen years (Malcolm 1970a). Marriage occurs some two years later than menarche, with the first child being born at a mean age for the mother of twenty to twenty-four (Sturt and Sturt 1973; Scragg 1973). Recent studies (Scragg 1973) indicate that late menarche is also associated with both early menopause and earlier cessation of child bearing, presumably due to the deprived nutritional circumstances of the slow-maturing groups.

Chow (personal communication) has shown from animal experiments that protein deprivation markedly delays the onset of sexual maturity and shortens the total duration of the reproductive life. A shortened menstrual span, therefore, may be considered as a possible adaptive mechanism to the deprived nutritional environment and the restriction of fertility, with its protein demands for fetal tissue and lactation, to within tolerable limits. On the other hand, improved nutrition and socioeconomic circumstances result in an earlier menarche and later menopause (Wolanski 1972) with a consequent extension of menstrual span and potential fertility.

It is possible that protein malnutrition may also reduce sexual drive and hence make tolerable both the restriction on intercourse during the breast feeding period and the reported restriction on intercourse for the early months of a new marriage for a Bundi couple (Malcolm 1970a). Reduction in sexual activity has also been noted in animal experiments (Chow, personal communication); but the measurement of such a complex, but perhaps important, biosocial factor in fertility would be most difficult in human populations.

CONCLUSION AND SUMMARY

Biosocial interrelations determine to a considerable extent the patterns of growth, health, and fertility of primitive populations, such as those in Papua New Guinea. The typical growth in the weight pattern of the child under three years old is related to a quantitative deficit in the total food intake, which in turn seems to be due to a pattern of feeding care determined more by the demand of the child than any maternal appreciation of the child's food needs. The spacing of children, so important for the maintenance of the health of the mother and child, is determined in traditional societies by social constraints on sexual intercourse during a prolonged breast feeding period, a constraint which is rapidly disappearing under the impact of cultural change and has entirely disappeared in the urban family, where spacing and fertility are now determined biologically and limited only by prolonged lactational amenorrhea. Fertility is further restricted biologically by protein malnutrition, which both delays menarche and leads to an earlier menopause and, hence, to a shortening of the menstrual span and reproductive life. Malnutrition may also have led to a reduction in sexual drive, making more tolerable the cultural constraints on sexual activity both before and early in marriage as well as after child bearing.

REFERENCES

BAILEY, K. V.

1964 Growth of Chimbu infants in the New Guinean highlands. *Journal of Tropical Pediatrics* 10:3–16.

CRAVIOTO, JOAQUIN, ELSA R. DeLICARDIE

i.p. *Ecology of malnutrition: environmental variables associated with clinical severe malnutrition.*

GUZMAN, M.A., N. S. SCRIMSHAW, H. A. BRUCH, J. E. GORDON

1968 Nutrition and infection field study in Guatemalan villages, 1959–1964, VII: Physical growth and development of preschool children. *Archives of Environmental Health* 17:107–118.

JELLIFFE, D. B.

1966 *The assessment of the nutritional status of the community.* WHO Monograph Series 59. Geneva.

MALCOLM, L. A.

1970a *Growth and development in New Guinea: a study of the Bundi people of the Madang district.* Institute of Human Biology Madang, Monograph Series 1.

1970b Growth, malnutrition and mortality of the infant and toddler of the Asai valley of the New Guinea highlands. *American Journal of Clinical Nutrition* 23:1090–1095.

1972 "Anthropometric, biochemical and immunological effects of protein supplements in a New Guinean highland boarding school." Paper delivered at the Ninth International Congress of Nutrition, Mexico.

n.d. "Health and nutrition, growth and development of children in urban Lae, Papua New Guinea." In preparation.

RING, A., R. F. R. SCRAGG

1973 A demographic and social study of fertility in rural New Guinea. *Journal of Biosocial Science* 5:89–121.

SCRAGG, R. F. R.

1973 "Menopause and reproductive span in rural Nuigini," in *Proceedings of the Ninth Annual Symposium of the Medical Society of Papua New Guinea,* Port Moresby, July 1973.

SCRIMSHAW, N. S.

1964 Ecological factors in nutritional disease. *American Journal of Clinical Nutrition* 14:112–122.

SCRIMSHAW, N. S., C. E. TAYLOR, J. E. GORDON

1968 *Interactions of nutrition and infection.* WHO Monograph Series 57. Geneva.

STURT, R. J., A. E. STURT

1973 "A study of family health in a Sepik population," in *Proceedings of the Ninth Annual Symposium of the Medical Society of Papua New Guinea,* Port Moresby, July 1973.

WOLANSKI, NAPOLEON

1972 Comment on "Anthropology and population problems," by S. Polgar, et al. *Current Anthropology* 13:255–258.

Processes of Human Growth in High Altitudes

N. N. MIKLASHEVSKAYA, V. S. SOLOVYEVA,
and E. Z. GODINA

Processes of human growth at high altitudes have been insufficiently studied and are described in few anthropological works. One should mention here the research done by Baker (Baker, Frisancho, and Thomas 1966; Frisancho and Baker 1970) in the Andes among the Quechua Indians from the region of Nunoa, Peru, at the altitude of 4,000–5,000 meters above sea level. Their total sample numbered 1,100 children and teenagers. Analysis of this material showed an extremely retarded process of physical growth at a high altitude. Sexual differences in body dimensions were found only in samples above sixteen years of age. Statural growth ends by the age of twenty in women and by the age of twenty-two in men. Nutritional analyses in the Nunoa population have revealed that protein and iron consumption is high, and, on the whole, their diet could be regarded as just as adequate and balanced as that of inhabitants of low altitude regions. This suggests that it is not undernourishment or lack of calories that retard or prolong growth in high altitude populations. It is likely that other factors are involved, such as hypoxia and genetics (Frisancho and Baker 1970).

Similar results were obtained by Turusbekov (1970) who studied the Kirgizian children in the Tyan-Shan mountains (2,020 meters above sea level), where poor physical development, retarded pubescence, and much longer periods of growth were also found.

The data on the growth and development of a high altitude population in the mountains of Ethiopia (Harrison et al. 1969) seem to be somewhat contradictory to the above. As noted above, people living at a high altitude (the village of Debarek, 3,048 meters above sea level) were compared to those from a middle altitude region (the village of Adi-Arkai, 1,524 meters above sea level). In these studies, it was found that children and adults living at the higher altitude were larger both in height and weight than those from the medium altitude region. As no differences in skinfold thickness were found, it does not appear that weight differences are due to differences in distribution of subcutaneous fat. No significant differences in skeleton maturity or appearance of permanent teeth were discovered.

The researchers Harrison et al. (1969) attribute the differences in growth and development of children in Peru and Ethiopia primarily to different race compositions of the studied populations and to differences in the length of time these populations have resided at high altitudes. The Peruvians have lived longer at high altitudes than the Ethiopians and, consequently, seem better adapted.

Investigations into the nature of growth and development at high altitudes have so far been limited to those made in Peru, Ethiopia, and the Soviet Union (in the Tyan-Shan mountains). Problems of human adaptation have become of increasing interest to scientists in recent years in part because of the wide range of habitats they occupy today.

As a result of the limited number and contradictory nature of the available materials there is a need for further investigation of the effects of high altitudes on human populations.

In 1968 and 1969 the Institute of Anthropology at the University of Moscow in conjunction with the Kirgizian State Medical Institute set up an expedition to the high altitude regions of the Pamir-Altai mountains. Under examination were the settlements of Kyzyl-Djar and Sagyndyk and the Sidorov collective farm[1] in the Soviet district (former Uzgen district) of the Osh region, Kirgizian SSR situated at 2,300–2,800 meters above sea level and isolated from the main transport routes of the republic by the Pamir-Altai mountain range. The inhabitants of these settlements number about 5,000, chiefly engaged in cattle breeding. In summer they migrate with their herds to high altitude pastures sometimes as high as 3,500 meters above sea level.

According to one ethnographic account (Vinnikov 1956) these Kirgizians entered the region from northern Kirgizia some 350 years ago.

In 1970, we studied a control sample of Kirgizians from several villages of the Kirov district of the Frunze region (700 meters above sea level).

The investigation was carried out, at both high and low altitudes of Kirgizia, by anthropologists, pediatricians, neuropathologists, psychiatrists, physiologists, and geneticists.

The extensive anthropological examination included measurements, evaluation of pubescence and constitutional peculiarities, definition of teething stages, dermatoglyphics, and collection of blood samples, which were analyzed for ABO, MN, Rh, P, and Lewis specificities.

One hundred adult males and 900 schoolboys of eight to eighteen years of age were studied in Kyzyl-Djar. In the Kirov district the sample was

[1] For the sake of brevity, all three settlements will hereinafter be referred to as "Kyzyl-Djar".

made up of 1,200 children and teenagers of the same age range. The group studied is typical of the south Siberia racial type. In facial height and width, head size, tertiary hair cover, hair and eye pigmentation, the population of Kyzyl-Djar is similar to that of other regions of Kirgizia, especially to southern Tyan-Shan (Miklashevskaya, Solovyeva, Godina, and Kondik 1972).

Among the racially diagnostic traits we determined was taste sensitivity to PTC. Under study were seventy subjects, of whom 88.6 percent were tasters, 11.4 percent nontasters, which seems to be within the expected range of frequencies in Mongoloid groups. The dermatoglyphics of the subjects also supported the assumption of close similarity of the population of Kyzyl-Djar to other Kirgizian groups.

Therefore, the Kirgizians of the studied isolate do not differ racially from other ethnoterritorial groups of Kirgizia and the growth differences found at high altitude cannot be attributed to racial differences.

Of particular interest are regularities in the attainment of body size and pubescence level in children and teenagers of the high altitude isolate (Miklashevskaya, Solovyeva, and Godina 1972). Some results of the investigation are shown in Tables 1–5 and Figures 1–12.

As seen from the tables, the average body dimensions in the schoolboys of Kyzyl-Djar are smaller than those of Kirgizians from low altitudes. Thus, the stature of Kyzyl-Djar boys, except those of twelve years of age, averages two centimeters shorter than that of children from the Kirov district. The difference in chest circumference is 1.5 centimeters. The pubescence of high altitude adolescents is also retarded. Thus, breast development of girls at high altitudes is on the average some 3.5 months later and menarche occurs nine months later than in girls from the low altitude regions. Change of voice is 6.5 months later and axillary hair eight months later in high altitude boys.

Our further investigations in the western Pamirs (Miklashevskaya et al. 1974) have also shown retarded growth and sexual maturation in high altitude Tadjics.

Comparison of growth curves for stature in both sexes at Kyzyl-Djar reveals a single crossover of the curves at the age of fourteen years, three months, the same age at which the growth curves for Andean Indians cross (Frisancho and Baker 1970). Before that age the heights of the Kyzyl-Djar boys and girls are nearly identical, but thereafter the boys begin to leave the girls behind. The Kirgizian schoolchildren at the low altitude exhibit crossovers of statural growth curves at ten years, ten months and at fourteen years, four months. At high altitudes, shoulder width also exhibits a single crossover at fourteen years, eight months, after which the boys leave the girls behind. In low altitude Kirgizians the crossover of growth curves for shoulder width occurs at ten years, eight months and at fourteen years, five months.

It can be seen that the processes of growth and pubescence at high altitudes is much retarded compared to that of inhabitants of lower ones.

What are the reasons for retarded physical development in the high altitude Kirgizians? Among the principal reasons, we may cite nutrition. The diet of high altitude dwellers contains less protein and vitamins. Also of potential significance is the degree of isolation which makes marriages between close relatives unavoidable. There are cases of polydactilia among the Kyzyl-Djarians. But we do not believe inbreeding to be a decisive factor in the retarded physical development of this population. This is because similar retardation also occurs in unisolated high altitude groups. According to Turusbekov (1970), who studied the physical development of schoolchildren in the town of Naryn (2,020 meters above sea level), these children also revealed poor physical growth and retarded pubescence. Baker, Frisancho, and Thomas (1966), working in the Peruvian Andes at 4,000 meters, found the population to be one of the most retarded in the world.

Physiological data may be of great help in the interpretation of these observations. The medical staff of Kyzyl-Djar examined the functional state of the cardiovascular systems mainly in adults. Their research showed that Kyzyl-Djarians are characterized by a comparatively low pulse rate, delayed circulation time, a somewhat lower arterial pressure, and a lower basal metabolism (Mirrakhimov 1968). We believe that the latter is particularly significant. The level of basal metabolism is known to depend upon the function of the thyroid gland. In all cases when this function decreases there is also a decrease in basal metabolism, while an increase in thyroid hormone concentration in the body causes higher metabolic exchange. Therefore, it is reasonable to suggest that at high altitudes the function of the thyroid gland decreases. And, indeed, as proved by Kalyuzhny and Belekova (1968) with the help of radioisotope diagnostics, the healthy aboriginal populations of high altitude Tyan-Shan (2,020 meters) and Pamir (3,760 meters) show a hypofunction of the thyroid gland. These authors hold that a moderate hypofunction of the thyroid gland is of an adaptive and compensatory nature. It facilitates economy of oxygen utilization which, by reducing the strain on the respiratory and vascular systems, make adaptation to conditions of hypoxia easier. Therefore, hypofunction of the thyroid gland at high altitudes is adaptive.

Thyroid hormones not only regulate basal metabolism in man but are also of vital importance in the processes of physical growth and development. According to Tanner (1962), even individual differences in growth rates are associated with differences in the activity of the thyroid. It is therefore possible that retarded growth and development at high altitudes is a side effect of adaptation to hypofunction of the thyroid causing both decreased basal metabolism and retarded growth and development.

It is reasonable to think that nutrition and hypoxia are not the only factors affecting growth and development in the children and teenagers of Kyzyl-Djar. It is necessary to continue research in order to evaluate the extent to which all these different factors retard growth.

Table 1. Means and standard deviations of some anthropometric characters in the high altitude Kirgizians (boys)

Age in years	*n*	Height		Chest circum-ference		Shoulder breadth		Bicrystal diameter		Chest breadth		Chest depth	
		m	σ	m	σ	m	σ	m	σ	m	σ	m	σ
8	24	118.53	4.21	60.42	3.51	25.85	1.21	19.48	0.97	18.77	1.11	13.73	0.89
9	41	123.23	5.13	61.50	2.54	26.80	1.04	20.27	1.04	19.32	0.90	14.05	0.88
10	36	126.45	4.64	63.45	2.26	27.45	1.46	21.01	1.20	19.84	0.88	14.31	1.03
11	40	130.45	5.44	64.95	3.43	28.38	1.45	21.85	1.93	20.28	1.15	14.53	1.05
12	49	136.65	4.94	67.58	2.81	29.56	1.31	21.99	0.83	20.85	0.91	14.85	0.95
13	51	139.39	6.80	68.82	2.95	30.07	1.56	22.87	1.28	21.19	1.02	14.91	0.99
14	53	146.45	7.46	72.39	3.72	32.04	1.86	24.06	1.45	22.57	1.32	15.72	1.25
15	43	153.33	8.38	75.73	5.06	33.07	1.98	24.95	1.75	23.32	1.73	16.11	1.33
16	28	156.29	6.13	78.36	4.52	34.25	2.13	25.59	1.37	23.82	1.39	16.57	1.08
17	30	162.12	4.79	83.05	5.33	35.42	2.00	26.18	1.20	25.02	1.88	17.15	1.22

Table 2. Means and standard deviations of some anthropological characters in the high altitude Kirgizians (girls)

Age in years	n	Height		Chest circum-ference		Shoulder breadth		Bicrystal diameter		Chest breadth		Chest depth	
		m	σ	m	σ	m	σ	m	σ	m	σ	m	σ
8	38	117.96	4.73	58.42	2.31	26.34	1.24	19.26	1.04	18.56	0.97	12.98	0.78
9	39	124.04	5.52	59.82	2.57	27.21	1.40	20.37	1.12	19.08	1.07	13.31	0.85
10	33	126.93	5.28	61.87	3.22	28.01	1.37	20.98	1.03	19.83	1.15	13.19	0.89
11	36	130.12	5.74	62.73	2.71	28.67	1.53	21.64	1.00	19.89	0.97	13.69	0.94
12	39	135.68	7.63	65.24	3.74	29.62	1.92	22.36	1.63	20.69	1.24	13.95	0.93
13	33	139.84	6.10	67.58	3.10	30.68	1.64	23.25	1.47	21.25	1.17	14.31	1.20
14	53	147.25	5.90	72.00	4.10	32.36	1.47	24.98	1.61	22.43	1.39	15.12	1.11
15	32	148.76	6.35	73.31	4.65	32.78	1.56	25.19	1.65	22.41	1.46	15.09	1.10
16	27	150.97	6.17	76.71	4.87	33.70	1.98	26.19	1.63	23.19	1.39	15.81	1.66
17	25	155.93	4.58	78.64	3.47	34.13	1.72	27.18	1.59	23.86	1.17	15.84	1.11

Table 3. Means and standard deviations of some anthropological characters in the low altitude Kirgizians (boys)

Age in years	n	Height		Chest circum-ference		Shoulder breadth		Bicrystal diameter		Chest breadth		Chest depth	
		m	σ	m	σ	m	σ	m	σ	m	σ	m	σ
8	61	120.52	4.32	61.13	2.24	26.44	0.96	19.69	0.94	19.31	0.83	13.88	0.74
9	60	123.76	5.09	62.58	2.85	27.14	1.44	20.28	1.09	19.69	0.99	14.13	0.79
10	62	128.94	5.99	64.90	3.03	28.03	1.47	21.00	0.92	20.31	1.03	14.56	0.98
11	59	132.90	4.80	66.15	2.82	28.86	1.48	21.80	1.62	20.66	1.01	14.54	0.78
12	63	136.38	6.35	68.02	3.23	29.45	1.54	22.28	1.10	21.29	1.06	14.86	0.94
13	60	141.50	6.48	70.49	3.44	30.79	1.75	23.06	1.33	22.12	1.26	15.23	0.94
14	60	148.11	6.75	74.22	4.29	32.29	1.91	23.99	1.37	22.81	1.23	15.88	1.09
15	60	155.98	7.18	77.59	5.34	33.95	2.05	25.32	1.73	23.91	1.59	16.32	1.17
16	58	160.47	6.97	81.57	5.03	35.43	1.89	26.33	1.40	24.85	1.41	16.92	1.22
17	56	163.88	6.27	84.69	5.87	36.25	2.13	26.83	1.63	25.56	1.84	17.65	1.46

Table 4. Means and standard deviations of some anthropological characters in the low altitude Kirgizians (girls)

Age in years	*n*	Height		Chest circum-ference		Shoulder breadth		Bicrystal diameter		Chest breadth		Chest depth	
		m	σ	m	σ	m	σ	m	σ	m	σ	m	σ
8	60	118.36	4.61	59.23	2.75	26.11	1.16	19.71	1.02	18.88	0.87	13.24	0.79
9	61	122.18	4.88	60.36	2.59	26.71	1.08	20.32	0.97	19.11	0.93	13.44	0.66
10	59	125.88	4.80	61.43	2.64	27.42	1.16	20.55	0.97	19.49	0.99	13.51	0.73
11	61	133.76	5.42	64.75	3.09	29.24	1.32	21.98	1.20	20.40	1.06	14.06	0.87
12	57	137.75	6.67	66.44	3.38	29.93	1.60	22.84	1.21	20.88	1.11	14.17	0.94
13	62	146.05	6.80	70.27	4.16	31.61	1.75	23.99	1.64	21.99	1.23	14.77	0.96
14	60	148.90	5.92	72.64	5.11	32.62	1.72	25.18	1.59	22.81	1.38	15.01	1.19
15	62	153.98	4.95	76.27	4.43	33.46	1.55	26.34	1.58	23.43	1.41	15.72	1.03
16	61	153.86	5.12	78.17	3.63	33.83	1.46	27.04	1.59	23.86	1.05	15.85	1.07
17	57	154.97	4.65	79.48	4.71	34.19	1.62	26.99	1.44	24.17	1.31	15.99	1.19

Table 5. Average age at which secondary sexual traits appear irrespective of their development level in high and low altitude Kirgizians

Boys			Girls		
Traits	High altitude	Low altitude	Traits	High altitude	Low altitude
Change of voice	13 yr. 3 mo.	12 yr. 8.5 mo.	Breast gland	11 yr. 4.5 mo.	11 yr. 7 mo.
Male breast	13 yr. 10 mo.	14 yr. 2 mo.	Pubic hair	13 yr. 5 mo.	13 yr. 7 mo.
Adam's apple	15 yr. 8 mo.	16 yr. 3.5 mo.	Axillary hair	14 yr. 0 mo.	14 yr. 2 mo.
Pubic hair	14 yr. 7 mo.	14 yr. 6 mo.	Menarche	15 yr. 2 mo.	14 yr. 5 mo.
Axillary hair	16 yr. 6 mo.	15 yr. 10 mo.			
Moustache	16 yr. 1 mo.	14 yr. 10 mo.			

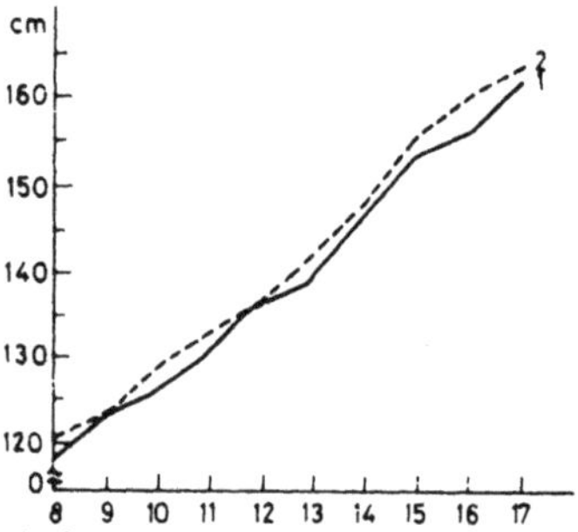

Fig. 1.

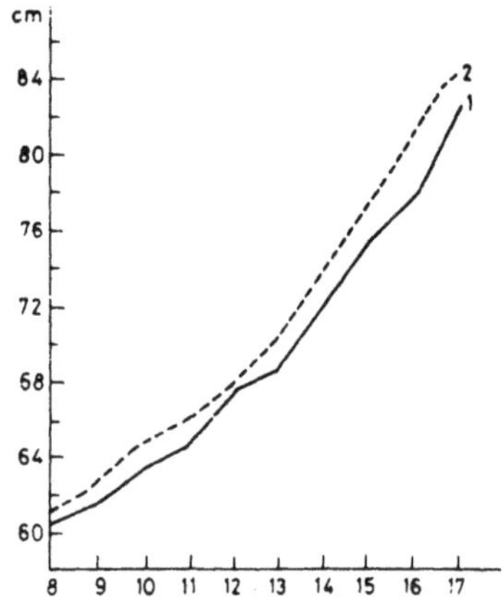

Fig. 2.

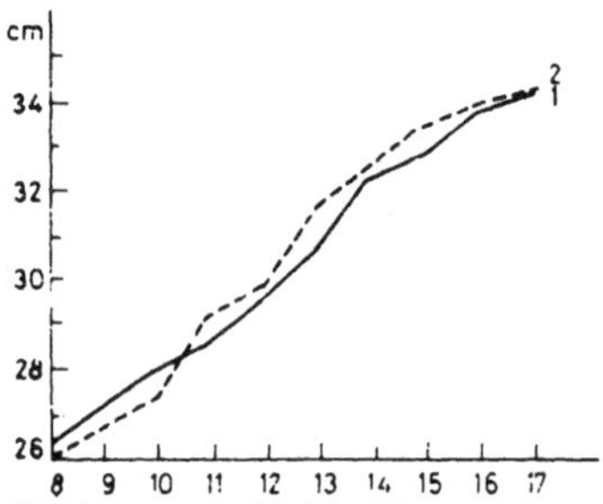

Fig. 3.

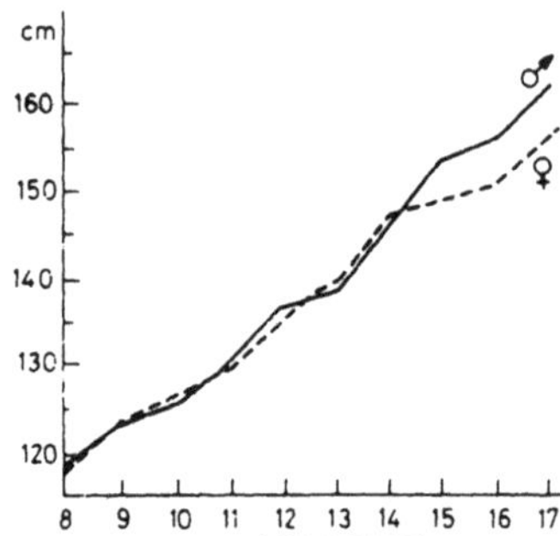

Fig. 4.

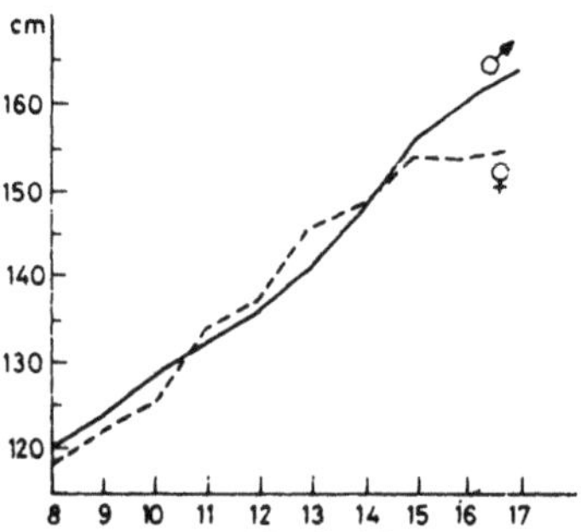

Fig. 5.

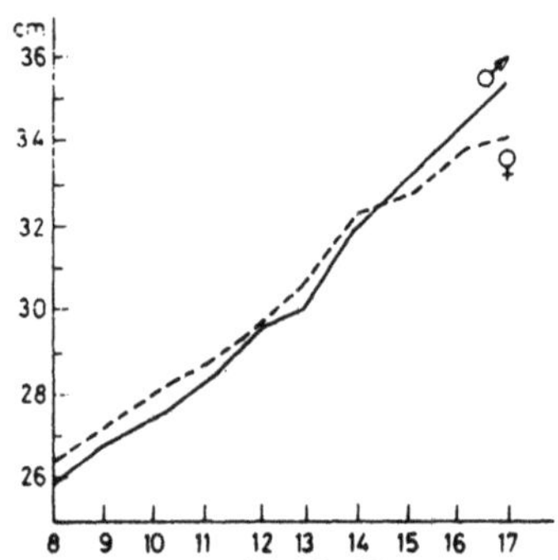

Fig. 6.

Figure 1. Growth curves for stature of boys: (1) high altitude; (2) low altitude
Figure 2. Growth curves for chest circumferences of boys: (1) high altitude; (2) low altitude
Figure 3. Growth curves for biacromial diameter for girls: (1) high altitude; (2) low altitude
Figure 4. Growth curves for stature of boys and girls in high altitude
Figure 5. Growth curves for stature of boys and girls in low altitude
Figure 6. Growth curves for biacromial diameter of boys and girls in high altitude

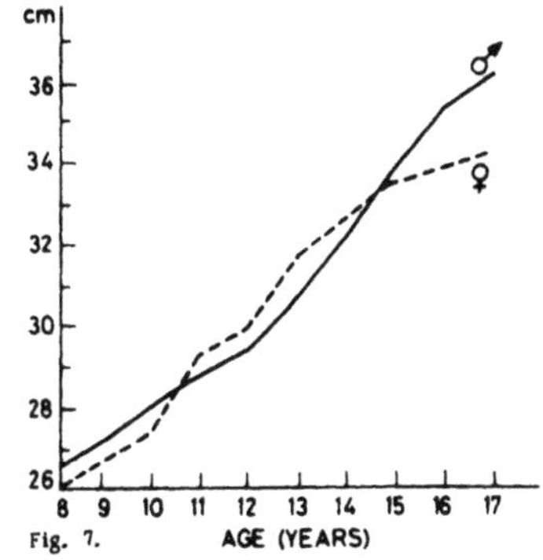

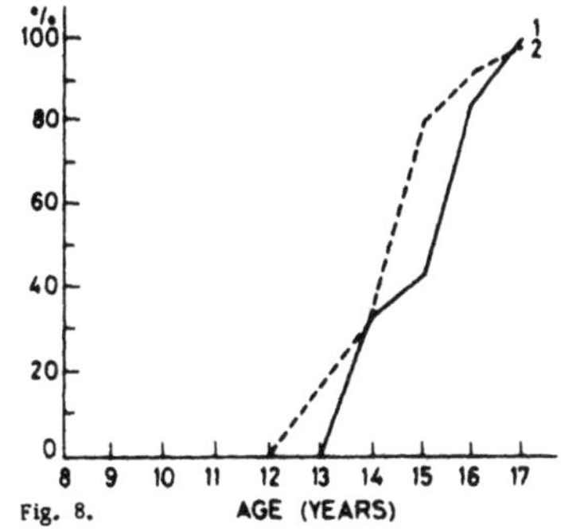

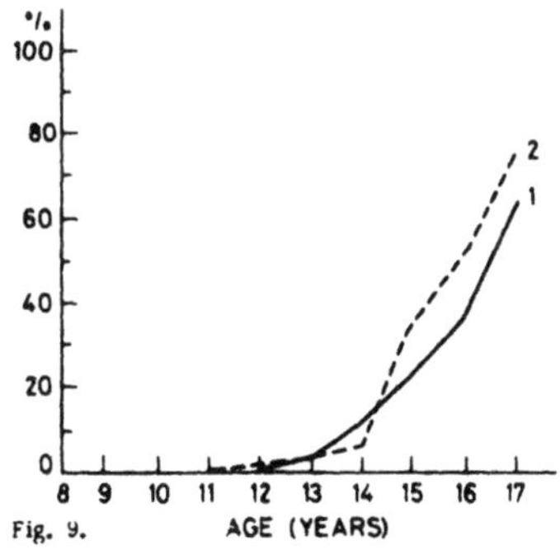

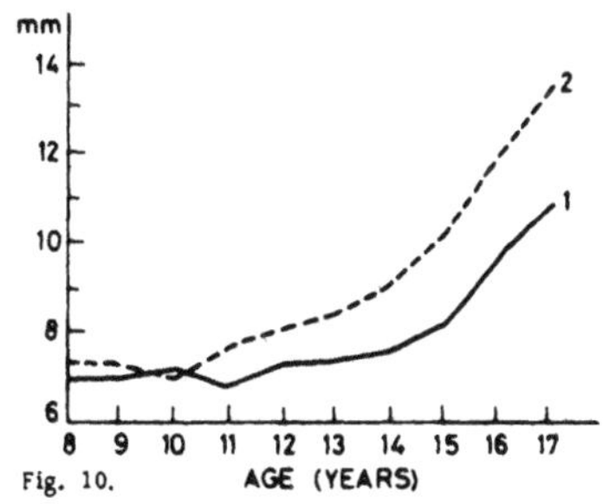

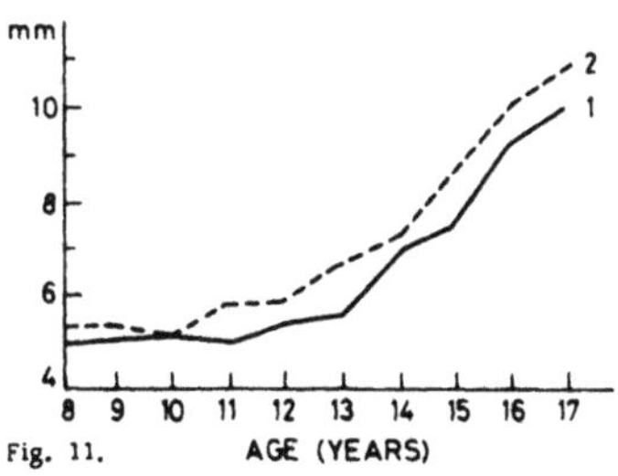

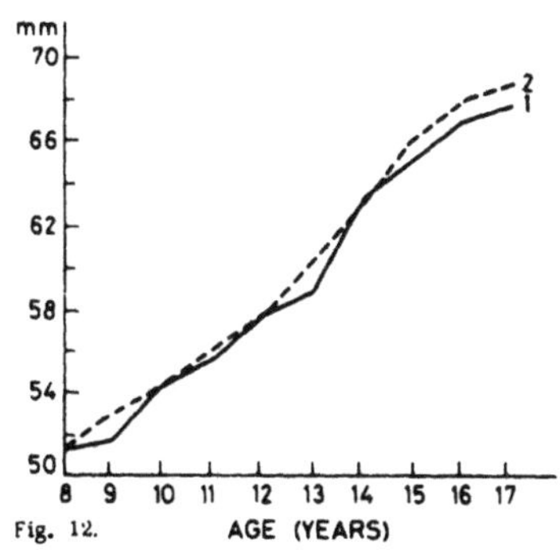

Figure 7. Growth curves for biacromial diameter of boys and girls in low altitude
Figure 8. Percentage of girls who have already menstruated, by age: (1) high altitude; (2) low altitude
Figure 9. Percentage of boys having already axillary hair (irrespective of the stage of development), by age: (1) high altitude; (2) low altitude
Figure 10. Growth curves for triceps skinfold values for girls: (1) high altitude; (2) low altitude
Figure 11. Growth curves for subscapular skinfold values for girls: (1) high altitude; (2) low altitude
Figure 12. Growth curves for diameter of the bicondylar humerus for boys: (1) high altitude; (2) low altitude

REFERENCES

BAKER, P. T., A. R. FRISANCHO, R. B. THOMAS

1966 "A preliminary analysis of human growth in the Peruvian Andes," in *Human adaptability to environment and physical fitness*. Edited by M. S. Malhotra. Madras.

FRISANCHO, A. R., P. T. BAKER

1970 Altitude and growth: a study of the patterns of physical growth of a high altitude Peruvian Quechua population. *American Journal of Physical Anthropology* 32(2).

HARRISON, G. A., C. F. KUCHEMAN, M. A. S. MOORE *et al.*

1969 The effects of altitudinal variation in Ethiopian population. *Philosophical Transactions of the Royal Society of London* 256.

KALYUZHNY, I. T., R. B. BELEKOVA

1968 *O prisposobitel 'no-kompencatornoy roli shchitovidnoy zhelezy u zdorodykh prozhivayushchix v usloviyx vysokogor'ya* [The accomodative-compensatory role of the thyroid gland in healthy and sick persons living at high altitudes]. Mat-li Symposiuma "Visokogorye i Krasnaya Krov" i Konferentzii "Visokogorye i Lekarstvo". Frunze.

MIKLASHEVSKAYA, N. N., E. Z. GODINA, V. S. SOLOVYEVA

1974 *New anthropological investigations on the Pamirs.* Rassovogeneticheskie Protzessy v Etnicheskoi Istorii. Moscow: Nauka. (Published in Russian.)

MIKLASHEVSKAYA, N. N., V. S. SOLOVYEVA, E. Z. GODINA

1972 Rose u razvitie detey v vnsokogornyx rayonax yuzhnoy Kirgizii [Growth and development of children in high altitude regions of southern Kirghizia]. *Voprosi antropologii* 40.

MIKLASHEVSKAYA, N. N., V. S. SOLOVYEVA, E. Z. GODINA, V. M. KONDIK

1972 Rostovne protsessn u chepoveka v uspovnyax vysokogor'ya [Growth processes in people living at high altitudes]. *Chelovek.*

MIRRAKHIMOV, M. M.

1968 *Serdechno-sosudistaya sistema v uspoviyax vysokogor'ya [The cardiovascular system in high altitude conditions]*. Leningrad: Meditzina.

TANNER, J. M.

1962 *Growth at adolescence* (second edition). Oxford: Blackwell Scientific Publications.

TURUSBEKOV, B. T.

1970 *Osobennosti vegetativnyx funktsiy u zhiteley gornoy Kirgizii [Peculiarities of vegetative function in mountain peoples of Kirghizia]*. Frunze: Kirghizstan.

VINNIKOV, YA. F.

1956 Rodo-plemeiioy sostav i rasselenie Kirgizov na territorii yuzhnoy Kirgizii [*Tribal composition and settlement of the Kirghizs on the territory of southern Kirghizia*]. Trudi Kirgizskoi Archeologo-etnographicheskoi Ekspeditzii. Moscow.

Research on Puberty in Haiti

J. B. ROMAIN

Puberty, which is considered to be central to processes of physical development, is the phase of organic maturation of the human body which corresponds to the onset of potential for reproductive functions. It is most easily discernible in women because of the onset of the menses. The occurrence of puberty is influenced by several factors which can counterbalance, complement, or be superimposed upon one another. Among these factors are race and heredity, nutrition and lifestyle, and climate and social class (Millot 1952: 6–10). The hormones secreted by the endocrine glands are of overriding importance.

In analyzing the influence of these different factors, Benoist (1963: 333–335) has stated that within the population of the racially mixed island of Martinique, the variations dependent upon race are indeed small when compared with those which are of social or climatic origin; he has suggested that if race were the only variable, puberty would occur at a relatively mature age (14.06 years).

In Haiti,[1] an island of the Caribbean group with five million inhabitants (St. Surin 1962: 21–22), mixture of the races has not appeared to be very extensive — 15 percent nonblack and 85 percent black. On this island we observed the onset of puberty in 1,729 individuals of the female sex to determine at what age the menses generally begins.

The sample group questioned included 1,350 blacks and 379 non-blacks; 517 subjects from the peasant class, 712 from the middle class, and 500 from the upper class. All are natives of the western department. The peasants are from the regions of Bourg-Champagne, Marlique, Nan-Citron, Mariman, and Bargout Croix-des-Missions. The middle

[1] The island of Haiti is situated in the archipelago of the Antilles in the Carribean, east of Cuba and it consists of the Dominican Republic on the east and the Republic of Haiti on the west. This study concerns the latter.

class girls come from Port-au-Prince and its suburbs, Petit Goâve, Arcahaie, and Mirebalais. The upper class subjects are from Pétionville, Port-au-Prince, and its wealthy neighborhoods.

The sample includes school girls whose ages varied from ten to eighteen years. These school girls were recruited from the following primary, middle, secondary, and professional schools: the Soeurs de la Sagesse in the church of St. Joseph du Sacré-Coeur in Turgeau, Christ-Roi in Bourdon, Elie Dubois in the rue du Centre, and the Ecole République de l'Argentine et République du Brésil, as well as the rural schools and the farm schools of the peasant towns.

The girls were divided into age groups, each of which contained seventy-five individuals. Each subject was asked the question, based on the formula used by Jean Benoist (1963) in Martinique, "Have you begun to menstruate?" The girl could answer simply "yes" or "no."[2]

The study produced the following results:

Table 1. Peasant class: 517 black subjects

Age in years	Percent of menstruating females
10	2.5
11	10
12	34
13	68
14	85
15	90
16	95
17	100

Table 2. Middle class: 443 black subjects

Age in years	Percent of menstruating females
10	7
11	12
12	55
13	70
14	90
15	95
16	99
17	100

[2] This question prevents any imprecision and elicits a "yes" or "no" response from every age group.

Table 3. Middle class: 279 nonblacks

Age in years	Percent of menstruating females
10	2
11	11
12	50
13	60
14	80
15	90
16	94
17	100

Table 4. Upper class: 500 blacks and nonblacks

Age in years	Percent of menstruating females
10	7.5
11	13
12	75
13	85
14	98
15	100
16	100

Table 5. Comparison of social classes

Age in years	Peasant class	Middle class	Upper class	Nonblack females
10	2.5	7	7.5	2
11	10	12	13	11
12	34	55	75	50
13	68	70	85	60
14	85	90	98	80
15	90	95	100	94
17	100	100	100	100

At the age of ten, there is a more significant percentage of menstruating girls within the upper and middle classes (7.5 percent) than there is in the peasant class and in the group of nonblacks (2 percent).

At eleven and twelve years, there is a clear increase in the percentages with respect to social class.

At thirteen years, the trend begins to equalize. But the girls from the upper class remain in the lead with 85 percent; the nonblacks stay behind (60 percent).

At fourteen, more precisely at about thirteen years and seven months, puberty is practically universal in the upper class, reaching a proportion of 98 percent. It does not reach similar proportions until fifteen years in the middle class and sixteen in the peasant class and in the group of nonblacks.

If we place the frequency curves for the four groups on the same graph with age on the abscissa and percentages of menstruating girls on the ordinate, these curves will appear from right to left as follows: nonblacks, peasant class, middle classes, upper class. Such an arrangement suggests a convergence of factors and their influence (see Figure 1).

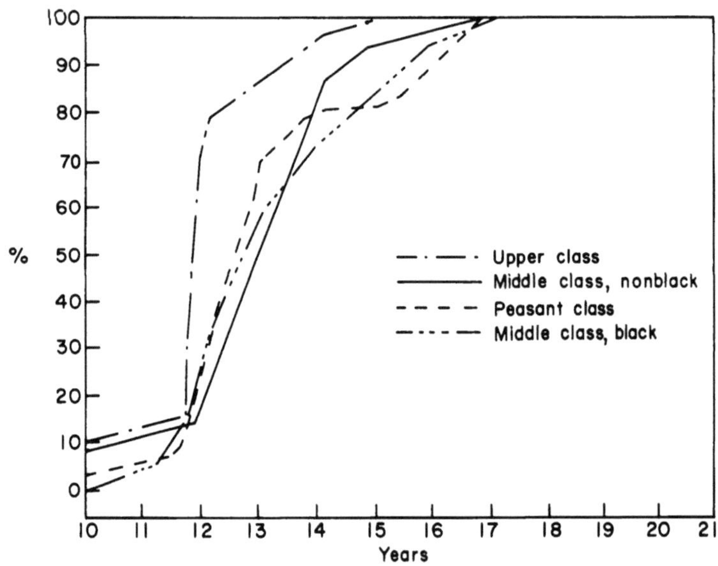

Figure 1. Menstruating girls according to age

Social and economic conditions, for example, contribute greatly to the general shape of the curves in the Figure. The early onset of puberty (thirteen years, seven months) — earlier than that apparent in Benoist's study in Martinique (14.06 years) and closer to that attested in black Americans (thirteen years) (Michelson in Benoist 1963: 418) — is exhibited by the upper class of all races, including girls from both moderate income and wealthy families.

In that upper class, the monthly income is at least $300 per family, as far as we could determine from our study; some resources, however, may have escaped our notice. Highly regarded in many ways, this class enjoys both European and American lifestyles, incorporating as much as possible the values and symbols of those societies. Their diet is balanced according to rational, proven formulas. Their education regarding nutrition seems to be well established; at any rate it is organized so as to be easily transmissible from generation to generation, as are other similar social amenities.

The middle class is spread over many sectors of the population. It includes small businessmen, teachers, intellectuals, professionals, and public school officials whose monthly income ranges from $150 to $200. It also contains a heterogeneous group characterized by the tendency to satisfy its aspirations and resolve its conflicts both by trying to enter the bourgeoisie, which rejects it, and by trying to reject the masses, which feeds it. This mixed group is plagued by complexes inhibiting its slightest advances. An appreciable portion of this social class is malnourished (OEA 1965:88–119). The state of nutrition of school children of this class parallels that of their school mistress as portrayed in a popularized image (Department of National Education 1967); both are undernourished.

Other material and psychosocial conditions of existence also have some effect upon physical development and the onset of puberty. For instance, the peasant class has existed for a long time apart from other social classes. For this reason, it has retained more of the physical characteristics associated with the black race than have the other groups, which have been more or less affected by miscegenation. It scarcely seems to have benefited from this at all, however, for if it had, one would expect a larger number to attain puberty at a young age (Romain 1971:354).

The 517 subjects representing this group rank, for the most part, third with regard to the age of the menses. We suspect that nutrition has played a large part in this. Both foreign and native nutritionists who have studied the problem of the peasant diet have agreed that they see not only a problem "of hunger, of undernourishment, of insufficient caloric consumption, but also a problem of malnutrition, of insufficient quality and quantity of foodstuffs" (Despeignes n.d.: 87).

In fact, the average daily intake of the peasant hardly ever exceeds 1,800 calories — indeed it is often only 1,500 — although it should be at least 1,950 calories according to the estimations by the CNRS (Moral 1959: 42), or even 2,600 according to the FAO (Lebret 1958: 58, 573). Moreover, this diet is also considered insufficient because of the excessive proportion of carbohydrates in comparison to proteins, and because of the taboos against certain slightly acidic foods such as *sponnias dulcia forts, magnifera indicas, citrus grandis osbeak, saturus shultz* (pineapple), whose absorption would bring a delay of puberty, or even a suppression of the menses.

The nonblacks, used here as a control group, can be placed parallel to the peasants. They begin menstruating later than the subjects from the middle class, with whom they share socioeconomic conditions but from whom they differ in skin color. Could their retarded development be due to miscegenation? Whatever the cause, Table 6 compares the various classes.

Table 6. Middle classes

Age in years	Percent of black females menstruating	Percent of nonblacks menstruating
10	7	2
11	12	11
12	55	50
13	70	60
14	90	80
15	96	90
16	97	94
17	100	100

In summary, puberty was studied in a population of 1,729 female individuals, taken from four different groups of Haitian society: the upper, middle, and peasant classes, as well as a group of nonblacks. From this study and from the schema presented in Table 5 it appears that in Haiti the average age of this very important event varies between thirteen years, seven months and fifteen years. Near the fourteenth year, in fact, at least 80 to 98 percent of the girls have begun menstruating, regardless of their social group. This average age has been determined by the interplay of diverse factors such as race, economic status, nutrition and social status. Socioeconomic conditions and nutrition seem to have had a marked effect. There is some temptation to explain the later development of the nonblacks on the basis of racial factors.

REFERENCES

BENOIST, JEAN
1963 "Les Martiniquais. Anthropologie d'une population metissée" [The people of Martinique. Anthropology of a mixed population]. *Bulletins et Mémoires de la Société d'Anthropologie de Paris* 4(11):241–432.

DEPARTMENT OF NATIONAL EDUCATION
1967 Unpublished manuscripts from a seminar on the improvement of education in Haiti.

DESPEIGNES, MIMOSE
n.d. *Service social et le problème de la nutrition.* Social service and the problem of nutrition in Haiti. Port-au-Prince: Imprimerie des Antilles.

HOUSE, G. A.
1957 *Biologie du noir* [Biology of the black race]. Brussels: Les Editions Problèmes d'Afrique Central.

LEBRET, L. J.
1958 *Suicide ou survie de l'occident* [Suicide or survival of the West]. Les Editions ouvrières.

MILLOT, JACQUES
1952 *Biologie des races humaines* [Biology of the human races] Paris: Armand Colin.
MORAL, PAUL
1959 *L'Economie haitienne* [Haitian economy]. Port-au-Prince: Imprimerie de l'Etat.
ROMAIN, J. B.
1971 *L'anthropologie physique das Haitians* [Physical anthropology of the Haitian people]. Port-au-Prince: Seminaire adventiste.
OEA (*Institut Interamericain de l'Enfant*)
1965 Recommendations et travaux du seminaire national de nutrition d'Haiti [Recommendations and reports from the national conference on nutrition of Haiti]. Montevideo.
OLIVIER, GEORGES
1960 Pratique anthropologique [Anthropological methodology]. Paris: Vigot Frères.
ST. SURIN, JACQUES
1962 Demographic indices and perspectives on Haitian population from 1950 to 1980.
TWIESSLEMANN, F.
n.d. Contibution à l'étude de la croissance pubertaire chez l'homme. [Contributions to the study of the development of puberty in man]. Pamphlet 35. Belgium: l'Institut Royal des sciences naturelles de Belgique.
VALLOIS, H. V.
1948 L'intérêt de l'anthropométrie et de la morphologie en biologie et en médecine [The importance of anthropology and of morphology in biology and medicine]. *La Semaine des Hôpitaux de Paris* 13.
VANDERVAEL, F.
1943 *Biometrie humaine* [Human biometry]. Paris: Masson.

The Development of Gypsy Children in Changing Living Conditions

JAROSLAV SUCHÝ

The differentiation of mankind into races and ethnic types took place most probably in conditions of isolation, under the influence of the natural environment. It was a natural process of structural development that could not be influenced by man. On the other hand, socially conditioned isolation is a product of human activity. In the interest of the fruitful development of mankind, it is desirable to remove social obstacles that contribute to maintaining isolation. States with progressive legal systems have eliminated these obstacles, but many social barriers still exist in people's minds and in human relationships.

In developing this theme, we shall examine the case of the Gypsies. The first migratory Gypsies (*Zigeuner* in German) came to Greece and Spain in approximately the tenth century A.D. Their migration started in India — probably in the seventh or eighth century — and from there they travelled to Europe. In the Middle Ages they reached all the European countries. Their history, including their voyages and various events during the nineteenth and twentieth centuries, has already been described (Suchý 1968a).

From an anthropological point of view, the history of the Gypsies is instructive. It shows how the structure of the European population was enriched by new elements. While this phenomenon had taken place several times before, the Gypsies represent the most recent influx of population into Central Europe, one which took place at a time when most European countries were already in their historical period. While it is difficult to shed light on older migrations (for which we find evidence only in excavations), the study of the most recent Indo-Europeans to enter Europe should lead to a more exact and critical evaluation of various ethnogenetic theories.

More than 200,000 Gypsies (or *Romi*, as they call themselves) live in

Czechoslovakia. They have their own social and economic history and ethnic character. Their integration with other segments of the population is a slow process, involving the interaction of genetic and environmental factors. Gypsies, like the populations of developing countries, are in a state of demographic transition because of their consistently high birth rate in comparison to other Central European populations. Thus, their presence makes particular demands on educational and health care institutions. Pediatricians, for instance, must be aware of, and apply, different health, growth, and maturation standards when examining Gypsy children.

At the Pedagogical Faculty of Charles University in Prague we have, since 1960, studied the physical development of Gypsy youth under

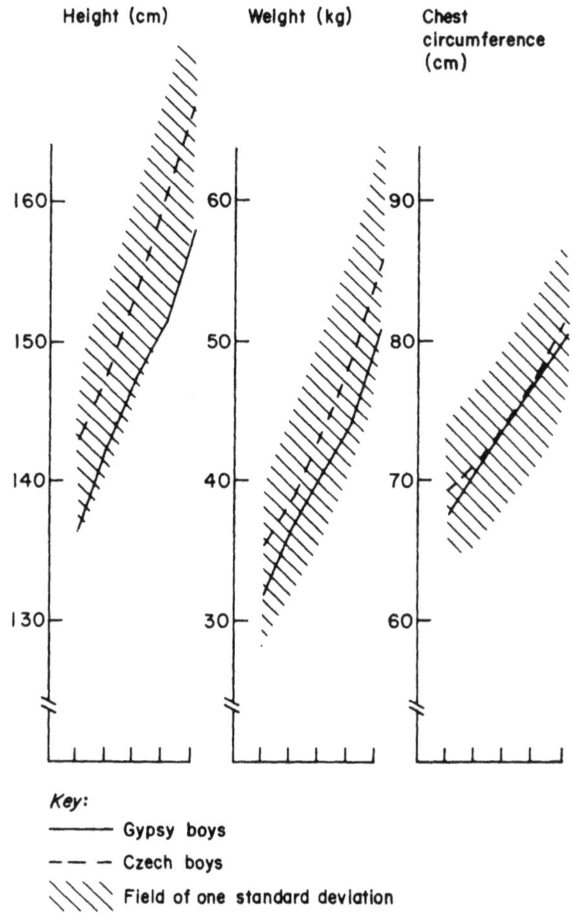

Figure 1. Comparison of the physical development of Gypsy and Czech boys aged eleven to fifteen

present conditions, namely at a time, when the Gypsies are abandoning their traditional nomadic way of life. Our goal has been to assist in their transition from isolation to integration. In 1967 this topic was included in the complex of problems considered by the International Biological Program. Both Malá and myself have worked toward a resolution of the problems presented by this shift in lifestyle (Malá and Suchý 1970).

From the results we achieved it is clear that the ethnic characteristics of this group are still very apparent. Figures 1 and 2 show comparisons in the development of the basic physical features of Gypsy children in comparison with Czech children. These graphs give the average values and standard deviations for height, weight, and circumference of chest. We also examined the usual morphological indices (head, face, and nose). The heads and faces of Gypsy children are narrow in comparison to those of Czech children, and their noses are broader.

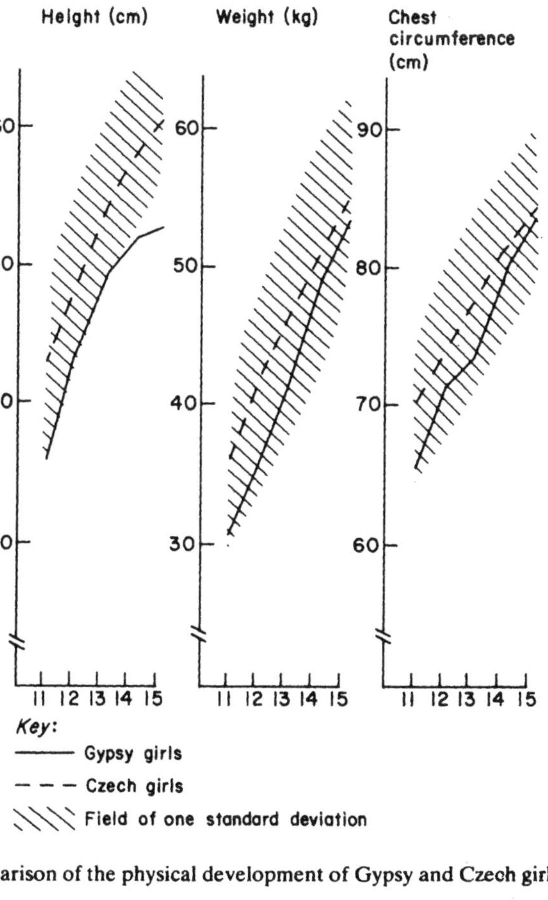

Figure 2. Comparison of the physical development of Gypsy and Czech girls aged eleven to fifteen

These conclusions are based on the examination of 163 Gypsy children in western and northern Bohemia from 1960 and 1963. The development curves derived from examination of 545 Gypsy children in Slovakia in 1961 and 216 Gypsy children in southern Bohemia (1968–1970) are similar to the first. All three groups are composed of children living in Gypsy families and attending basic nine-year schools or special schools along with other Czech children.

A long-term research study of a group of boarding school children gave somewhat different results. Three hundred and twenty-two observations were carried out at the Bořislav Home for Children (district of Teplice) from 1964 to 1967 (Suchý and Malá 1969). The averages for height and weight of these children were higher than those of other groups. Children living in the boarding school receive adequate hygiene and medical care. They usually come from neglected families. The results of these investigations are illustrated in Tables 1 through 9.

The results can be summed up in this manner:

1. The physical development of Gypsy children has a specific character which should be respected in the evaluation of their state of health.

2. Anthropological research of Gypsy children shows that environmental factors play an important role in influencing their development.

Table 1. Age, origin, and time of observation of subjects

		Age in years												
		6	7	8	9	10	11	12	13	14	15	16	17	18
1960–1963 Western and Northern Bohemia	M = 89	–	8	4	11	10	17	10	14	10	4	1	–	–
	F = 74	1	2	7	14	18	14	6	4	5	2	1	–	–
1961 Slovakia	M = 312	–	28	35	33	36	44	29	34	35	24	11	3	–
	F = 233	–	30	34	27	27	30	27	24	19	13	1	–	1
1968–1970 Southern Bohemia	M = 109	2	11	16	11	16	11	22	11	6	3	–	–	–
	F = 107	–	11	12	10	16	13	14	7	11	13	–	–	–
1964–1967 Boarding school Bořislav	M = 234				8	10	26	42	45	56	47			
	F = 88				1	4	10	10	14	24	25			

Key: M = males; F = females

Table 2. Height of Gypsy boys (centimeters)

Age	1960–1963 Western and Northern Bohemia	1961 Slovakia		1968–1970 Southern Bohemia		1964–1967 Boarding school Bořislav	
	x'	$\bar{x}$	s.d	$\bar{x}$	s.d.	$\bar{x}$	s.d.
7	114.0	115.0	4.9	116.4	8.3		
8	119.0	119.6	4.8	120.6	4.8		
9	124.0	125.0	6.1	124.2	7.3	127.9	–
10	129.0	128.9	5.9	127.0	5.5	130.8	3.42
11	133.0	134.0	7.1	133.5	6.4	136.7	5.01
12	138.0	138.1	7.1	139.1	7.3	142.4	4.35
13	143.0	143.7	8.3	140.1	7.8	147.3	6.96
14	148.0	147.6	8.4	(151.3)	–	151.6	8.79
15	154.0	154.6	7.4	(162.9)	–	158.1	7.47
16	161.0	(166.4)	6.2	–	–	–	–
17	–	156.3	5.5	–	–	–	–

Key: x' = mean (derivated from empirical curve of growth)
$\bar{x}$ = average
s.d. = standard deviation

Table 3. Height of Gypsy girls (centimeters)

Age	1960–1963 Western and Northern Bohemia	1961 Slovakia		1968–1970 Southern Bohemia		1964–1967 Boarding school Bořislav	
	x'	$\bar{x}$	s.d.	$\bar{x}$	s.d.	$\bar{x}$	s.d.
7	114.0	114.1	4.6	115.7	5.3		
8	199.0	120.5	5.0	120.9	4.4		
9	124.0	125.3	4.3	122.7	5.0	125.3	–
10	128.0	128.9	5.6	133.3	7.2	127.3	–
11	134.0	134.7	7.2	131.4	7.2	136.2	7.77
12	140.0	140.7	6.9	142.4	10.9	144.0	3.78
13	146.0	148.1	5.6	146.9	5.8	149.4	4.35
14	151.0	152.9	6.9	151.8	4.6	152.1	4.56
15	153.0	152.7	6.4	152.1	5.0	152.9	4.74
16	154.0	(159.0)	–	–	–	–	–

Table 4. Weight of Gypsy boys (kilograms)

Age	1960–1963 Western and Northern Bohemia	1961 Slovakia		1968–1970 Southern Bohemia		1964–1967 Boarding school Bořislav	
	x'	$\bar{x}$	s.d.	$\bar{x}$	s.d.	$\bar{x}$	s.d.
7	21.0	21.0	2.8	20.6	3.3		
8	23.5	21.9	2.5	22.7	3.0		
9	26.9	25.8	4.2	24.7	3.2	26.4	–
10	28.5	27.0	4.0	26.6	3.9	29.1	2.31
11	31.0	31.1	6.5	29.5	3.9	31.9	3.78
12	33.5	33.2	5.1	34.9	5.6	36.6	4.23
13	38.0	35.9	5.6	33.2	4.6	40.1	6.72
14	43.0	41.8	6.4	(43.8)	–	43.8	7.56
15	50.0	46.9	8.1	(60.3)	–	50.9	7.53
16	55.0	(54.0)	6.7	–	–	–	–
17	–	49.0	7.3	–	–	–	–

Table 5. Weight of Gypsy girls (kilograms)

Age	1960–1963 Western and Northern Bohemia	1961 Slovakia		1968–1970 Southern Bohemia		1964–1967 Boarding school Bořislav	
	x'	$\bar{x}$	s.d.	$\bar{x}$	s.d.	$\bar{x}$	s.d.
7	20.5	20.4	2.0	20.8	3.3		
8	24.0	23.0	3.3	23.1	3.0		
9	25.0	25.3	3.7	23.8	3.7	23.5	–
10	28.0	26.1	4.0	33.1	11.6	27.1	–
11	31.0	31.1	6.9	32.8	5.4	30.6	4.92
12	35.5	34.5	4.2	35.5	9.6	35.6	4.23
13	40.0	42.0	7.6	45.5	9.1	41.6	4.83
14	45.0	49.4	7.8	48.3	15.5	48.8	4.83
15	49.5	46.8	8.2	47.2	6.1	53.5	8.22
16	51.5	(54.0)	–	–	–	–	–
17	–	–	–	–	–	–	–

Table 6. Chest circumference of Gypsy boys (centimeters)

Age	1960–1963 Western and Northern Bohemia	1961 Slovakia		1968–1970 Southern Bohemia		1964–1967 Boarding school Bořislav	
	x'	$\bar{x}$	s.d.	$\bar{x}$	s.d.	$\bar{x}$	s.d.
7	59.0	59.0	2.4	57.2	3.0		
8	60.5	59.7	3.5	59.0	2.2		
9	62.5	62.2	2.6	61.5	2.8	62.3	–
10	65.0	63.1	3.8	62.0	3.9	65.1	–
11	67.0	66.5	5.1	64.2	2.0	67.5	2.64
12	68.5	68.6	3.7	68.0	3.4	70.5	3.80
13	72.0	70.0	4.7	67.2	4.2	73.7	4.20
14	75.0	73.5	5.2	(74.2)	–	77.0	6.20
15	–	78.2	5.4	(84.0)	–	79.0	5.80
16	–	(80.0)	3.8	(84.0)	–	–	–
17	–	77.0	–	–	–	–	–

Table 7. Chest circumference of Gypsy girls (centimeters)

Age	1960–1963 Western and Northern Bohemia	1961 Slovakia		1968–1970 Southern Bohemia		1964–1967 Boarding school Bořislav	
	x'	$\bar{x}$	s.d.	$\bar{x}$	s.d.	$\bar{x}$	s.d.
7	58.0	57.2	2.6	55.6			
8	60.0	59.8	3.2	58.3			
9	62.5	61.2	3.6	58.2		61.0	–
10	65.0	61.9	3.0	66.1		66.5	–
11	67.0	67.6	6.9	66.4		65.6	–
12	68.5	69.4	5.0	67.1		71.4	–
13	74.0	77.2	7.7	77.7		73.4	–
14	78.0	80.3	7.6	77.3		79.9	4.56
15	78.0	80.3	7.6	76.9		83.7	4.20
16	–	81.0	–	–		–	–

Table 8. Indices for Gypsy boys

Age	1960–1963 Western and Northern Bohemia	1968–1970 Southern Bohemia		1964–1967 Boarding school Bořislav	
	x'	$\bar{x}$	s.d.	$\bar{x}$	s.d.
Cephalic index					
7	85.4	85.1	3.6		
8	84.0	85.1	2.4		
9	85.5	85.7	3.7	85.3	–
10	85.9	83.9	2.7	84.6	–
11	83.9	84.3	2.6	83.8	3.4
12	84.7	84.9	3.2	84.9	4.1
13	82.3	85.9	3.0	84.4	3.3
14	83.8	83.5	–	84.8	3.4
15	83.7	81.8	–	84.9	3.3
Facial index					
7	81.5	80.5	4.9		
8	83.0	80.7	4.0		
9	81.4	80.1	3.8	80.9	–
10	83.0	80.9	5.4	80.2	–
11	83.8	83.0	4.5	82.0	3.4
12	82.6	81.9	5.7	82.7	4.2
13	86.2	83.6	4.1	84.3	4.2
14	86.4	85.5	–	84.7	4.0
15	84.5	86.0	–	85.8	4.0
Nasal index					
7	72.1	71.0	3.8		
8	72.0	72.8	7.1		
9	72.5	72.4	5.2	73.4	–
10	70.9	71.9	8.6	75.6	–
11	72.3	70.3	4.7	71.8	6.0
12	79.2	72.0	7.8	73.6	4.9
13	79.9	69.1	4.6	72.1	5.0
14	77.6	67.5	–	71.9	5.4
15	78.0	70.1	–	70.5	5.6

Table 9. Indices for Gypsy girls

Age	1960–1963 Western and Northern Bohemia	1968–1970 Southern Bohemia		1964–1967 Boarding school Bořislav	
	x'	$\bar{x}$	s.d.	$\bar{x}$	s.d.
Cephalic index					
7	83.0	85.0	3.2		
8	86.8	87.3	5.3		
9	86.1	83.6	3.1	89.5	–
10	83.7	85.1	3.0	81.8	–
11	83.6	83.4	2.8	85.1	–
12	84.0	83.7	3.5	85.7	–
13	84.5	83.0	3.0	86.1	–
14	84.2	82.7	3.9	83.9	3.8
15	86.0	84.3	2.6	83.4	4.9
Facial index					
7	83.0	78.9	4.0		
8	76.7	78.9	5.8		
9	83.5	82.0	5.5	88.8	–
10	83.8	79.5	3.1	86.6	–
11	83.9	82.0	4.7	86.5	–
12	81.5	82.9	3.6	84.9	–
13	84.7	82.6	6.9	86.1	–
14	84.6	80.9	3.9	83.0	3.9
15	86.5	85.7	5.6	83.8	4.5
Nasal index					
7	72.5	72.4	6.5		
8	75.4	71.1	4.2		
9	70.0	70.5	5.8	70.7	–
10	69.2	72.9	6.8	69.0	–
11	68.6	69.4	4.6	70.2	–
12	73.5	70.0	5.9	68.0	–
13	68.5	72.1	7.4	68.4	–
14	69.6	71.9	4.2	69.2	7.2
15	62.5	67.7	7.9	70.6	7.0

REFERENCES

MALÁ, H., J. SUCHÝ
1970 Anthropological research on Gypsy-children and youth in Czechoslovakia. *Glasnik Antropologie društva Jugoslavije* 7:39–63.

SUCHÝ, J.
1968a "Die Zigeuner," in *Rassengeschichte der Menschheit* ["The Gypsies" in *A racial history of mankind*]. Edited by K. Saller, 185–221. Munich and Vienna: R. Oldenbourg Verlag.
1968b Bodily development of Gypsy children in Czechoslovakia. *Anthropologie* 6:2, 73–78.

SUCHÝ, J., H. MALÁ
1969 Physical features of Gypsy youth. *Rivista di antropologia* 56:32–43.

The Relationship Between Human Body Constitution, Longevity, Disease, and Causes of Death

PETER RÂMNEANTZU

For a long time studies have demonstrated that individuals are not equally susceptible to disease and that they die from different causes at different ages. Pende (1955) and Sheldon (1950), among others, have brought to this overall concept some concrete observations. They have revealed that correlations exist, some of them quite close, between somatotypes and many disorders. As a matter of fact, medical treatises have shown the prediposition of certain somatotypes to certain diseases. The results published so far have been based on examinations of apparently healthy patients or population groups of mixed ages. But these investigations have had no time depth. There are few longitudinal studies based on repeated observations of the same individuals (Knussmann 1968).

The somatotypes of the population that we studied are the outcome of an interaction between genetic and environmental factors, characteristic of a region that has little human migration but also is not isolated. The first medical and biotypological examinations were performed in 1939 on 1,177 males and 975 females (Râmneantzu 1972).

The working methods and the classification of somatotypes used were those described by Viola (1932) and Pende (1955). Initial screening was followed by periodic medical examinations, and in 1971 and 1972, the individuals still alive were re-examined. Front and profile photos of the subjects were taken and the anthroposcopic procedure developed by Sheldon and his school (1950) was employed. In 1972, we recorded both the age and the cause of death of the individuals based on their death certificates.

The hypothesis being tested in the present investigation is that there is a firm relationship between somatotype and age. We have demonstrated the reliability of this relationship on the basis of two elements: the frequency distribution by five-year groups, from twenty to fifty years of age, of the individuals examined at the first biotypological examination; and the observations carried out during thirty-three years (from 1939 to 1972) in 473 males and 451 females still living. The biotypes were classified as follows: normosthenic and normoasthenic; mixed sthenic and asthenic; brevitype sthenic and asthenic; and longitype sthenic and asthenic. The relationship between somatotypes and age is shown in Table 1.

We have found that somatotypes do not change with age. The sthenic brevitype and sthenic longitype, obesity, gynandromorphism, in male and female, and even endocrine deviations, remain within similar limits, regardless of age. Individuals of asthenic brevitype or longitype undergo mild changes without passing from one group to another. Their adipose tissues get thicker at the age of about forty years, but after five to ten years they return to the original condition. Near the age of fifty, all somatotypes become more flexible, except the sthenic longitype called also chondroblast, which remains rather supple. Individuals of asthenic brevitype accumulate fat, but not exceeding ten percent of standard weight. Much of this appears to be genetically determined. At the age of sixty, both sexes show a tendency to lose weight through dehydration. Owing to these changes, the morphotype at adulthood and in old age may differ markedly.

As to the relation between somatotype and disease, shown in Table 2, 13.9 percent to 19.4 percent of those surviving have never resorted to medical treatment over the entire thirty-three year period. Many individuals affected by disease have shown close correlations between somatotype and disease.

Exogenous infectious diseases, especially tuberculosis, take a higher toll among linear, asthenic types. Females of this biotype were ill about twice as often as males, and females of the brevitype were more susceptible to influenza and viral hepatitis.

Endogenous illness was ten times more frequent than exogenous illness. This ratio can be explained by the fact that since 1955, many of the infectious diseases have been or are being eradicated in Romania. Females more often contracted endogenous diseases, especially disorders of the respiratory tract (pneumonia), arterial hypertension, rheumatoid polyarthritis, and diseases of the bile duct, chiefly gall bladder lithiasis. Males of the brevitype group were susceptible to arterial hypertension and heart disorders. Males of the asthenic longitype group displayed high frequencies of gastroduodenal ulcer, neuritis, and rheumatoid polyarthritis, while females of the asthenic brevitype group appeared to be susceptible to rheumatoid polyarthritis. Males of medium or mixed type were most free from endogenous disease. However, the occurrence of angina pectoris and myocardial infarction in males of normal type is significant.

Table 1. Relationship between somatotype and age frequencies

Age groups in years	Males					Females				
	Total number	Normo-types	Mixed	Brevi-types	Longi-types	Total number	Normo-types	Mixed	Brevi-types	Longi-types
20–24.9	123	25.8	45.9	13.1	15.2	169	26.3	36.7	14.3	22.7
25–29.9	252	27.4	44.0	12.3	16.3	220	26.6	34.3	15.0	24.1
30–34.9	214	27.6	43.3	12.9	16.2	174	30.1	31.2	14.1	24.6
35–39.9	203	26.9	43.8	13.5	15.8	175	26.5	33.7	15.7	24.1
40–44.9	212	27.5	43.3	14.1	15.1	119	26.4	35.5	15.3	22.8
45–49.9	173	26.8	44.4	13.5	15.3	118	26.9	34.4	15.8	22.9
Total	1,177	27.0	44.2	13.2	15.6	975	27.1	34.3	15.0	23.6

Table 2. Relationship between somatotype and morbididity on the basis of the illnesses between 1939 and 1972

	Number of examined subjects						
				Cause of illness[b]			
		In good health		Exogenous		Endogenous	
Somatotypes	Total[a]	Number	Percent	Number	Percent	Number	Percent
Males							
Normotypes	129	18	13.9	211	1.9	2198	19.8
Mixed	205	30	14.7	368	2.1	3728	21.3
Brevitypes	66	12	18.7	140	2.6	1447	26.8
Longitypes	73	11	15.1	192	3.1	1320	21.3
Females							
Normotypes	124	23	18.6	242	2.4	2505	24.8
Mixed	155	29	18.7	290	2.3	3339	26.5
Brevitypes	67	11	19.4	179	3.2	1926	34.4
Longitypes	105	19	18.1	482	5.6	2391	27.8

[a] The deaths from the war of 1941–1944 are not included.
[b] In the group of exogenous causes are included the code numbers 0–136 (WHO 1965). *Manual of the international statistical classification of diseases, injuries, and causes of death* (revised edition, 1965).

The age at death, that is, longevity or life span, should be viewed in light of two phenomena which took place in Romania. One is the decline of general mortality which, in rural environments varied from 22.2 to 16.1 percent between 1939 and 1948, but remained close to 9.0 percent from 1948 to 1971. The second important biostatistical event was the increase in the average life span, which in the years 1964 to 1967 reached 66.45 years for males and 70.51 for females.

From 1939 to 1972, the longevity and causes of death of individuals who died in the villages where they lived varied according to somatotypes (see Table 3).

Longevity in males fluctuated between 65.8 and 71.6 years. The latter mean age was attained by the brevitype. The average longevity of females was greater than that of males, fluctuating between 68.8 and 75.6 years. The latter average was that of individuals of average somatotype.

Biotypes show association with cause of death as well. Exogenous causes accounted for 3.5 percent of all deaths. The age of death from exogenous causes was much lower than that from endogenous causes. The most frequent exogenous cause of death among females and males of the asthenic longitypes was pulmonary tuberculosis. Some males of mixed longitype were accident victims.

When cause of death was endogenous, the life duration ranged from 68.7 to 76.2 years with brevitype, normotype, and mixed type females

Table 3. Somatotype, longevity, and cause of death from 1939 to 1972 in subjects aged 20 to 83 years

Somatotypes	Cause of death: Total[a] Number	Total[a] Age in years	Exogenous[b] Number	Exogenous[b] Age in years	Endogenous[c] Number	Endogenous[c] Age in years
Males						
Normotypes	102	68.8	4	35.8	98	70.1
Mixed	161	70.0	7	37.9	154	71.5
Brevitypes	49	71.6	3	40.2	46	73.6
Longitypes	60	65.8	5	34.0	55	68.7
Females						
Normotypes	98	75.6	1	32.3	97	76.1
Mixed	122	75.4	1	34.2	121	75.8
Brevitypes	58	74.9	2	38.1	56	76.2
Longitypes	83	68.8	3	29.7	80	70.3

[a] Endogenous causes of death: 140; 239; 240–279; 280–289; 390–458; 480–486; 520–577; 580–584; 780–794 (WHO 1965).
[b] Of the persons examined in 1939: 372 males and 361 females died between 1939–1972; 332 males and 163 females changed residence and 473 males and 451 females remained available for observation.
[c] We considered C1–C19 as exogenous causes of diseases and C20–25; 34–55 were classified as endogenous (WHO 1965).

living longest. Normotype males of mixed and brevitype constitution exhibited a shorter life span while male and female longitypes show the least longevity. Death as a result of cerebral hemorrhage occurred at an earlier than average age in asthenic longitype males. Asthenic longitypes also experienced a higher than average rate of arteriosclerosis, heart disorder, and gastric cancer. Heart disease shortened the life span in other male groups. Senility, associated with long life, was more frequent in brevitypes of both sexes. Increased longevity of brevitype females was due chiefly to their asthenic constitution. Cancer was not selective for somatotype, but was probably responsible for a general five to six year reduction in life span.

Forty percent of the 733 individuals studied might have been considered malnourished through the mid-1950's. Their fat intake consisted of approximately five-sixths animal fat and one-sixth vegetable oil. Females worked in agriculture along with males.

The present investigation of the same individuals, conducted from 1939 to 1972, led to the following conclusions: the constitutional type of adults twenty to fifty years of age, and of old people, does not change with age up to eighty-three years; somatotypes play a role in disease susceptibility; a connection exists between certain causes of death and biotype; and life span is likewise dependent upon the cause of death.

REFERENCES

KNUSSMANN, RAINER
1968 "Entwicklung, Konstitution, Geschlecht," in *Humangenetik,* volume 1. Edited by P. E. Becker, 280–437. Stuttgart: G. Thieme.

PENDE, NICOLA
1955 *Traité de médicine biotypologique.* Paris: G. Doin.

RÂMNEANTZU, P.
1972 Constituţia umană somato-funcţională a populaţiei adulte din sate de pe Valea Arieşului din judeţul Turda. *Studii şi Cercetări de Antropologie* 9:129–138.

SHELDON, H. WILLIAM
1950 *Les variétés de la constitution physique de l'homme.* Paris: Presses Universitaires de France.
1954 *Atlas of men.* New York: Harper.

VIOLA, GIACINTO
1932 *La costituzione individuale,* two volumes. Bologna: Cappelli.

WHO (WORLD HEALTH ORGANIZATION)
1965 *Manual of the international statistical classification of diseases, injuries, and causes of death,* volume 1. Geneva: World Health Organization.

One of the Factors of Brachycephalization over the Centuries: The Susceptibility of Dolichocephals to Tuberculosis

G. OLIVIER and M. E. DE CASTRO E ALMEIDA, with the collaboration of H. TISSIER

Two of the major evolutionary trends which have developed since the appearance of *Homo sapiens* are gracilization of the skeleton and rounding of the head. The latter phenomenon appears to be due to the action of several selective factors. Bielicki and Welon (1964) have shown that long-headed individuals had fewer surviving children than other people. Cessation of such differential mortality, according to Huizinga and Slob (1965), has led to what has been termed "de-brachycephalization."

The present study is concerned with correlations between cranial dimensions and causes of death. We have used the data of Macedo (1892),[1] who measured 1,000 skulls exhumed in Lisbon in 1880. These skulls belonged to citizens born toward the middle of the last century. For each subject Macedo noted the sex, profession, and cause of death. Although diagnoses of that period were often inaccurate, certain illnesses were well-known. Particularly notable was the great number of deaths due to pulmonary tuberculosis (one-third of the cases, a figure possibly exaggerated by the nonrepresentative social status of the sample).

ANALYSIS OF RESULTS

In our sample, tuberculous patients died an average of ten years earlier than other subjects. Thus, their anatomical characteristics were less likely to be transmitted to the next generation. The difference in age would probably be even greater if Macedo had not limited his sample to subjects

[1] We are grateful to Pr. G. da Fonseca Sacarrao, Director of the Bocage Museum, Lisbon, for authorizing us to use Macedo's unpublished data, and to Chabcuf Hiernaux, Jacquard and Thoma for their criticism of this text before publication.

from eighteen to seventy years of age (since old people rarely die of tuberculosis).

When we compare the cranial measurements and indices of individuals exhibiting symptoms of tuberculosis with those of subjects who died of other illnesses, we note that there is almost no difference among women. Therefore, among females, we found no particular type associated with greater susceptibility to tuberculosis. This is one of several instances where selection does not appear to affect females to the extent it affects males (the same is true for the weight/surface relationship of the body).

Among males, there are marked differences between those suffering from tuberculosis and the rest. Tuberculous males exhibit longer skulls and longer faces (with corresponding effects on the indices) than non-tuberculous males. They also share with tuberculous women the characteristic of a narrow face.

If selection is at work we would expect evidence of differential reproductive success. We do not have direct information on numbers of children, but many of the subjects were unmarried: more than half the men, and more than two-fifths of the women. Evidence of this is that their bodies had not been claimed by families. The percentage of unmarried people is higher in consumptives, probably due to the fact that they die at a younger age. Out of the people whose marital status is known (488 men and 505 women), there were only 232 married men (of whom 24.8 percent were consumptives) and 290 married women (of whom 27.6 percent were consumptives). Fifty-two and one-half percent of the non-tuberculous males were married, while only 37 percent of those with tuberculosis were married. In women these percentages are 63 and 45 percent, respectively.

Next, we addressed the question of whether those who did have children transmitted the traits in question to them. We did this indirectly, comparing the unmarried and married in terms of cause of death, assuming that only the latter had children. Thus, we find a curious fact: the subjects who died of tuberculosis after marriage scarcely differ from those who were married and died of other illnesses. On the other hand, unmarried consumptives (who were younger) present the greatest number of differences from nonconsumptives, whether married or unmarried.

We could hypothesize that death due to tuberculosis among subjects without descendants was associated with a reduction in the genetic disposition toward lengthening of the face and head in the general population. The opposite is seen in the non-tuberculous unmarried subjects, who present no unique characteristics, except, of course, their younger age. Their absence of descendants does not alter the genetic heritage of the population. We find a similar situation in unmarried women, as can be seen from the measurements in Table 1 (and from the *t*-tests which accompany the latter).

DISCUSSION

As is often the case, our task of interpretation would be simpler if we had less information, particularly if we did not know the marital status of the subjects. However, the importance of the subdivision into married and unmarried categories is evident.

It is necessary also to take into account the effects of age. It is known that cranial dimensions increase slightly with age, in part through thickening of the bones. So it is reasonable to inquire whether the smaller dimensions of tuberculous patients do not quite simply reflect the fact that they died younger. This necessitates a separate study correlating dimensions with age in Macedo's series. We found some correlation, though of little significance. Some correlations involve characteristics which do not differentiate skulls of tuberculous patients from others.

We also compared the averages for subjects from eighteen to twenty-nine years old, to equalize age in the tuberculous and nontuberculous categories. When this is done, we find three of the differences shown in Table 1 do indeed appear to be correlated with age, for they no longer appear at the same age. This is true for biauricular breadth in men and biasterial and minimum frontal breadths in women. On the other hand, narrow bizygomatic breadth in both sexes is more marked in young consumptives ($t = 2.57$), although the number of subjects is smaller. Therefore, it would be possible to say that the overall correlation with age is due to the presence of the consumptives.

Finally, two dimensions which present marked differences in men, the length of the head and total facial height (the latter measured in accordance with Broca's old technique) have no relation to age. Furthermore, these dimensions are greater in consumptives, even though they are younger, and correlations with age are usually positive.

Explanations based on differences in socioeconomic environment (probably lower in our subjects) also seem unlikely. Our subjects were studied by profession, and scarcely any correlation between cranial dimensions and trades as found.

We have no data on the post-cranial morphology of these subjects, for whom we have only skulls. Koch had pointed out differences of pigmentation associated with resistance to tuberculosis. Some large studies have shown that tuberculous individuals are taller, leaner, and have a smaller thoracic diameter than the general population. Thus, there is reason to think that the tuberculous subjects studied were taller. As a rule, taller stature is accompanied by a longer head and more elongated face, since

Table 1. Craniometry of subjects

	Tuberculous		Non-tuberculous		*t*	Married tuberculous		Married non-tuberculous		*t*	Unmarried tuberculous		Unmarried non-tuberculous		*t*
Males															
$n=$	158		336			60		173			98		163		
Age	34.75	11.75	45.13	14.59	*8.45*	40.98	11.07	51.16	12.36	*5.95*	30.93	10.49	38.73	14.08	*5.10*
Anterior-posterior diameter	134.78	6.23	184.42	6.37	*2.97*	186.08	5.60	184.60	6.42	1.70	186.26	6.61	184.23	6.33	*2.45*
Vertical diameter (Ba.-Br.)	134.78	5.35	133.79	5.11	*1.96*	135.42	5.16	134.10	4.94	1.73	134.40	5.45	133.46	5.28	1.36
Maximum transverse diameter	137.59	5.69	137.97	5.48	0.70	137.93	5.43	138.24	5.70	0.38	137.39	5.86	137.67	5.24	0.39
Biasterial diameter	109.59	4.64	109.33	4.90	0.57	109.73	5.03	109.74	4.86	0.01	109.50	4.42	108.90	4.92	1.02
Biauricular diameter (base)	120.01	4.79	121.07	5.48	*2.21*	120.10	4.29	120.91	5.32	1.18	119.96	5.09	121.24	5.66	1.89
Minimum frontal diameter	95.75	4.55	96.23	4.64	1.09	95.88	4.78	96.02	4.50	0.20	95.66	4.43	96.45	4.78	1.36
Bigonial breadth	110.42	4.83	111.44	4.68	*2.21*	111.12	4.27	111.32	4.41	0.31	109.99	5.12	111.56	4.97	*2.43*
Total facial height (Broca)	92.61	5.45	91.23	5.51	*2.61*	92.92	6.14	91.67	5.56	1.39	92.42	5.01	90.77	5.43	*2.50*
Bizygomatic breadth	126.93	5.70	128.08	5.58	*2.10*	127.85	5.17	127.95	5.33	0.13	126.37	5.97	128.23	5.86	*2.46*
Cranial index	73.95	3.36	74.88	3.51	*2.83*	74.16	3.11	74.96	3.62	1.64	73.82	3.51	74.80	3.40	*2.21*
Facial index	73.05	4.66	71.32	4.68	*3.85*	72.76	5.25	71.73	4.71	1.34	73.23	4.29	70.88	4.63	*4.16*

Table 1. *(continued)*

Females

$n=$	172		334			80		210			92		124		
Age	34.94	13.07	46.53	14.59	*9.08*	39.20	12.19	48.47	13.91	*5.56*	31.24	12.73	43.24	15.16	*6.32*
Anterior-posterior diameter	176.56	5.76	176.53	5.94	0.05	177.74	5.71	176.87	5.78	1.15	175.54	5.64	175.94	6.18	0.49
Vertical diameter (Ba.–Br.)	127.52	4.42	127.51	4.96	0.05	128.21	4.55	127.67	4.82	0.89	126.92	4.24	127.28	5.20	0.56
Maximum transverse diameter	133.35	5.14	133.32	4.84	0.06	133.41	4.96	133.35	4.76	0.09	133.29	5.31	133.27	4.98	0.03
Biasterial diameter	104.13	4.55	105.15	4.04	*2.48*	104.56	4.65	105.37	4.07	1.37	103.75	4.46	104.80	3.98	1.79
Biauricular diameter (base)	115.10	4.77	115.58	4.33	1.10	115.14	4.79	115.83	4.20	1.13	115.06	4.78	115.14	4.53	0.12
Minimum frontal diameter	91.47	3.90	92.45	4.21	*2.61*	91.09	3.39	92.75	4.12	*3.43*	91.80	4.22	91.93	4.33	0.22
Bigonial breadth	103.71	5.02	104.90	4.40	*2.62*	103.79	4.80	105.22	4.39	*2.32*	103.65	5.23	104.35	4.40	1.04
Bizygomatic breadth	118.51	5.02	119.24	4.70	1.51	118.94	4.87	119.52	4.44	0.93	118.14	5.15	118.76	5.09	0.88
Total facial height (Broca)	86.06	5.19	86.59	5.52	1.06	86.51	5.52	86.88	5.58	0.51	85.66	4.89	86.08	5.40	0.60
Cranial index	75.58	3.34	75.58	3.22	0.00	75.13	3.48	75.45	3.17	0.72	75.98	3.17	75.81	3.29	0.38
Facial index	72.67	4.17	72.69	4.94	0.05	72.77	4.19	72.76	4.90	0.02	72.58	4.18	72.58	5.02	0.00

Broca's measurement techniques. The italicized t-tests are significant.

transverse dimensions do not increase as much as sagittal (Pittard's law). It is also known that selection has favored small size in unfavorable economic conditions. Thus, we may ask whether the phenomenon which we have described in skulls is not simply the consequence of selection against large size, and only secondarily, against elongated heads and dolichocephals.

Finally, consumptives represent 31.4 percent of the total male sample, but the percentage falls to 24.8 percent of the married males. The difference between these two percentages (6.6 percent) could represent selective pressure. It is true that this difference only exists in men. But half of these 6.6 percent in each generation (i.e., 3.3. percent) lose one cranial index unit and gain one and one-half facial index units, assuming that the two categories had had as many children.

Such drastic selection should have caused dolichocephalism to disappear where tuberculosis has selectively affected human populations. Since this has not happened, it appears that the foregoing observations arise from fluctuations between two percentages and between two generations. It may reflect a purely local, or accidental and temporary, result. However, the few cases of plague and smallpox found in our series (forty-nine men, thirty-nine women) yield results similar to those of tuberculosis, giving indication of the effects of endemic and epidemic diseases elsewhere, even when occurring in much lower frequencies. Some other selective factors act similarly, while still others have an opposite effect (for example, by handicapping brachycephalics in still unknown ways). We may never know the role of infant mortality or which morphological type of children died early. This may well be the key to the interpretation of the diachronic evolution involving fluctuations in the gracilization of the skeleton, changes in racial characteristics in a region, etc.

Brachycephalization, which began in the Mesolithic era, has ceased since the beginning of this century (Billy 1968). It is logical to argue that present-day de-brachycephalization, as well as increase in stature, are associated with better hygiene and medical care. A low fitness for a certain type (tall dolichocephals) has been replaced by selective pressure in their favor, possibly until a new equilibrium is attained.

If confirmed, this explanation is important to anthropologists. If true, it means that the advantage of certain characteristics will decrease. In the face of fluctuations over time, or between contemporary populations, we may still think that certain aspects of variation respond to changes in health care. There is a tendency to explain everything through differential mortality, as has been done with polymorphism of the ABO blood group. But we should not minimize the role of population movements and changes in the environment in producing the diversity of populations.

The underlying problem is of a genetic nature: what is the association between the predisposition to certain illnesses and body measurements which are apparently independent? There may well be other conditions of life and environment, even certain medications having a similar effect, making possible unpredictable alterations in our species through the operation of selection.

CONCLUSIONS

The study of the skulls of subjects who died of pulmonary tuberculosis reveals that the men (but not women) are characterized by long heads and faces, narrow faces, and probably were tall. Subjects who died of the plague and smallpox present similar characteristics. They were ordinarily younger, more often unmarried, contributing less than others to the composition of the following generation. Brachycephalization over the centuries may be explained, in part, by differential mortality associated with a lesser resistance of tall dolichocephals to epidemics and endemic diseases. Reduction of this selective pressure is being reflected in de-brachycephalization and increased stature.

REFERENCES

BIELICKI, T., Z. WELON

1964 The operation of natural selection on human head form in an east European population. *Homo* 15:22–30.

BILLY, G.

1966 Nouvelles données sur l'evolution contemporaine des dimensions cephaliques. *L'Anthropologie* 70:283–308.

1968 Nouvelles données sur l'evolution contemporaine des parametres raciaux: II, La Stature. *L'Anthropologie* 72:41–63.

HUIZINGA, J., A. SLOB

1965 Progressive brachycephalization: reproduction and head form in the Netherlands. *Kon. Ned. Akad.* 68:297–301.

MACEDO, F. DE

1892 *Crime et criminel.* Lisbon.

OLIVIER, G., M. E. DE CASTRO E ALMEIDA

1972 Forme du crane et mortalité differentielle par tuberculose. *L'Anthropologie* 76:471–499.

A Study of Morphological Variables and Genetic Markers in Urban and Rural South African Negro Male Populations

HERTHA DE VILLIERS

The object of this study — part of the South African Multidisciplinary Human Adaptability Project of the International Biological Program — is to determine the morphologic and genetic differences, if any, between urban and rural South African Negro male populations.

Transition from a rural peasant environment to an urban industrial environment is accompanied by changes in climate, nutrition, levels of mental and physical activity, and culture. It is to be expected that these changes would affect not only the rates of physical growth in children, but also the adult physique, and that the adaptations made by different populations to the first three changes would be influenced by differences in genetic constitution.

POPULATIONS SAMPLED

This study is based on two South African Negro ethnic groups — the Venda and the Pedi — and is restricted to males between the ages of seventeen and sixty. Ethnological requirements were fully satisfied in both the urban and rural subsamples. All subjects were fully identified in terms of their tribal affiliations to either the Venda or the Pedi. Random sampling models proved to be impractical in all but the urban Venda subsample. Coverage of the four populations, however, was such that

I wish to thank Professor P. V. Tobias, head of the Department of Anatomy, University of the Witwatersrand, Johannesburg, for his assistance and for providing the facilities for this study; the International Biological Program: Human Adaptability Committee, the South African Council for Scientific and Industrial Research, and the University of the Witwatersrand for their financial support in this undertaking; my student assistants, and our subjects, whose help and cooperation made this study possible.

there could be little doubt about the representativeness of the samples (Vorster 1972). The ideal of identifying completely rural and completely urbanized subsamples of similar age distribution and with long continuous residence in one area proved impossible to attain. South African Negro male populations are highly mobile and would appear to be in transition from a rural to an urban industrialized life, though at different levels of urbanization. Further, the middle age groups are less well represented in the rural areas. There is thus a preponderance of men in the thirty-one to fifty year age groups in the urban areas and a corresponding deficiency of these age groups in the rural areas. However, the samples retained their essential characteristic of being "in the labour market" (Vorster 1972: 11).

Urban Venda

This subsample was drawn from the Venda population of Tshiawelo township, southwest of Johannesburg, Witwatersrand. It consisted of 148 males, each of whom had resided for a period of at least ten years under urban conditions. None of the individuals in the sample was related to any other as far as could be ascertained.

Rural Venda

The rural subsample was drawn from the chieftaincy of Lwamondo, Vendaland, in the northeastern Transvaal, and consisted of 201 subjects who had been living in this district for at least three months. Of these 201 subjects, thirty-five were related.

Urban Pedi

This subsample consisted of 227 males, residing in the Katlchang township (originally from Sekhukhunil proper in the central Transvaal) who had been employed in the Germiston area of the Witwatersrand for a period of at least six months prior to the study. None of them was related to any other as far as could be determined.

Rural Pedi

This subsample of 210 males resided in the chieftaincy of the Paramount chieftainess Mankopodi Thulare Sekhukhune, at Mohlaletsi, near Paradys, Gelukslocations in Sekhukhuniland proper, district Lydenburg, in the central Transvaal.

Age Distribution

The age distribution in the four subsamples was as follows:

Table 1.

Years	Urban				Rural			
	Venda		Pedi		Venda		Pedi	
	No.	%	No.	%	No.	%	No.	%
17–30	24	16.2	99	43.5	91	45.2	114	54.3
31–40	29	19.5	71	31.4	18	8.9	43	20.5
41–50	65	43.9	36	15.8	52	26.0	28	13.3
51 plus	30	20.4	21	9.3	40	19.9	25	11.9
Total	148	100.0	227	100.0	201	100.0	210	100.0
Mean age	41.5	(s.d. 8.3)	35.1	(s.d. 10.4)	35.1	(s.d. 9.1)	32.4	(s.d. 13.7)

There were no significant differences ($P < 0.05$) in the skeletal measurements of the four age groups in each of the subsamples. The age groups, therefore, have not been considered separately.

METRICAL VARIABLES

The metrical analyses are based on thirty-five absolute body measurements, six derived measurements, and two indices detailed in Tables 2 and 3. All measurements are given in millimeters and are standard measurements determined according to internationally accepted definitions. These measurements concern absolute size and are designed to give a description of the size and shape of the body and its principal parts.

The mean, the standard error of the mean, and the standard deviation for each measurement have been calculated. To test for the significance of the differences between pairs of mean values for absolute measurements in the four subsamples, the student or t-test has been used.

RESULTS AND DISCUSSION

Table 2 sets out the mean values and standard deviations of the absolute measurements, and Table 3 the derived measurements and indices for the Venda and Pedi urban and rural subsamples.

Table 2. Mean body measurements for urban and rural Venda and Pedi males

Variables	Urban						Rural				
	Venda			Pedi			Venda			Pedi	
	No.	Mean±s.e.	s.d.	No.	Mean±s.e.	s.d.	No.	Mean±s.e.	s.d.	No.	Mean±s.e.
Weight (in kg)*	247	*64.0*± 0.5	9.1	223	*60.2*±0.5	8.18	200	56.8±0.4	6.7	202	56.2±0.6
Stature	148	1668.9± 5.1	63.1	226	1668.2±3.6	54.8	199	1664.7±3.9	56.1	204	1659.3±4.7
Span	144	1783.6± 6.6	79.3	225	1762.1±8.5	88.1	191	1785.2±4.9	67.9	203	1742.0±9.5
Supra sternale height	148	1372.4± 4.6	56.6	226	1366.8±3.4	51.4	199	1370.8±3.6	51.6	204	1359.7±4.2
Tragion height	148	1545.0± 5.1	62.9	226	1545.3±3.6	54.4	199	1546.0±3.8	54.3	204	1533.4±4.6
Acromiale height	148	1370.8± 4.8	58.9	226	1365.4±3.6	54.1	199	1371.7±4.1	59.1	204	1359.3±4.2
Radiale height	148	1049.7± 3.7	45.8	226	1054.8±2.8	42.8	199	1052.2±4.0	57.5	204	1044.3±3.4
Stylion height	148	791.6± 3.9	48.0	226	812.8±2.3	35.3	199	797.4±6.4	40.9	204	809.8±2.9
Dactylion height	148	599.3± 3.9	48.4	226	618.3±2.2	33.7	199	600.5±3.1	43.9	204	609.0±3.5
Cristale height	148	1023.1± 4.0	49.3	225	1024.8±4.0	60.4	199	1020.4±3.1	44.8	204	1011.7±5.7
Trochanterion height	148	893.1± 3.7	45.3	225	901.7±4.1	62.6	199	896.4±2.9	40.9	204	898.5±3.2
Sitting height	148	846.1± 4.3	52.2	226	853.7±1.9	30.0	196	847.0±2.1	30.9	203	844.7±3.5
Tibiale height	148	469.7± 2.0	25.4	226	468.7±2.3	34.0	199	468.6±1.5	21.2	204	464.1±2.0
Malleolare height	148	75.2± 4.4	5.3	226	65.1±0.4	5.9	199	71.5±1.4	5.7	204	65.4±0.5
Biacromial breadth*	147	*387.7*± 1.4	17.8	226	*379.4*±1.5	22.0	199	373.3±1.2	17.7	204	369.1±2.1
Bicristal breadth	148	259.2± 1.4	17.2	225	251.4±1.1	16.1	199	256.0±1.0	14.5	203	247.3±0.9
Chest breadth*	147	*272.9*± 1.4	17.7	226	*269.8*±1.2	18.5	199	256.0±1.0	14.3	204	261.8±1.4
Chest depth*	147	*212.2*± 1.4	17.8	226	*205.8*±1.1	17.1	198	201.2±1.0	14.3	204	199.8±1.0

Table 2. *(continued)*

Variables	Urban						Rural				
	Venda			Pedi			Venda			Pedi	
	No.	Mean±s.e.	s.d.	No.	Mean±s.e.	s.d.	No.	Mean±s.e.	s.d.	No.	Mean±s.e.
Thoracic circumference*	147	*873.8*±10.9	82.7	226	*851.3*±4.0	61.1	199	834.3±5.6	80.0	204	835.8±3.3
Abdomen circumference*	148	*789.4*± 8.0	98.1	226	*770.9*±5.3	79.9	199	732.1±6.0	85.6	204	730.7±5.8
Arm circumference*	148	*270.2*± 2.3	28.8	225	*269.3*±1.7	26.2	199	240.7±1.7	24.2	204	249.2±1.5
Forearm circumference*	148	*223.6*± 2.1	26.5	225	*235.9*±1.7	26.0	199	196.8±1.1	15.9	204	209.1±1.1
Thigh circumference*	148	*491.3*± 3.4	42.4	225	*503.7*±3.6	54.3	199	460.3±2.8	39.4	204	480.1±3.0
Leg circumference*	148	*349.9*± 2.4	28.9	225	*342.2*±1.6	25.0	199	324.4±1.9	26.8	204	325.1±1.9
Bicondylar femur	148	92.2± 0.6	8.2	226	92.0±0.4	5.4	199	91.3±0.3	5.4	204	92.0±0.4
Bi-epicondylar humerus	148	67.8± 0.5	6.3	226	69.2±0.3	5.5	199	66.6±0.3	5.4	204	69.3±0.3
Hand length	148	189.3± 1.3	16.5	226	183.8±1.1	16.8	199	183.8±1.2	17.2	204	185.4±0.8
Hand breadth*	148	83.3± 0.3	4.5	226	*82.9*±0.3	4.5	199	81.5±0.3	4.9	204	79.5±0.4
Skinfold measurements:											
Juxta nipple*	148	*9.8*± 0.5	6.5	208	*7.7*±0.3	4.8	198	4.3±0.1	1.7	203	4.5±0.2
Juxta umbilical*	147	*14.1*± 0.7	8.7	207	*11.39*±0.4	6.9	199	7.2±0.2	3.1	203	8.3±0.2
Subscapular*	146	*11.6*± 0.5	6.4	208	*10.0*±0.3	4.7	199	7.7±0.1	2.1	201	8.2±0.2
Triceps*	148	*7.8*± 0.3	4.3	208	*7.4*±0.2	3.1	198	6.2±0.2	3.7	204	5.5±0.2
Biceps	148	4.0± 0.1	2.0	208	4.1±0.2	3.9	198	3.2±0.6	0.8	204	4.0±0.1
Thigh*	146	*8.8*± 0.4	5.1	203	*8.1*±0.2	3.7	196	6.0±0.2	2.3	201	5.2±0.1
Calf*	137	*8.6*± 0.2	3.1	185	*6.2*±0.2	2.7	198	6.8±0.2	3.1	162	4.0±0.2

* Statistically significant differences between urban and rural mean values for both the Venda and Pedi ($P<0.05$).
Key: s.e. = standard error; s.d. = standard deviation.

Table 3. Derived limb measurements (in millimeters) and body indices (percent) for urban and rural Venda and Pedi males

Variable	Urban				Rural			
	Venda		Pedi		Venda		Pedi	
	No.	Mean	No.	Mean	No.	Mean	No.	Mean
Upper limb length (acromiale–dactylion)	148	771.5	226	747.1	199	771.2	204	750.3
Arm length (acromiale–radiale)	148	321.1	226	310.6	199	319.5	204	314.9
Forearm length (radiale–stylion)	148	258.1	226	242.0	199	254.8	204	234.5
Lower limb length (stature–sitting height)	148	822.8	226	814.5	196	817.7	204	812.6
Thigh length (cristale–tibiale)	148	553.4	225	556.0	199	551.8	204	547.5
Leg length (tibiale–malleolare)	148	394.5	226	412.7	199	397.1	204	406.7
Body Indices (percentage)								
Relative shoulder breadth (biacromial breadth/stature)	147	23.2	226	22.7	199	22.4	204	22.6
Cormic index (sitting height/stature)	148	50.69	226	51.2	196	50.88	203	51.0

Characteristics of the Body (Tables 2 and 3)

The mean values indicate that the urban male can be distinguished from the rural male by weight, breadth, and depth measurements. The trunk diameters — biacromial breadth, chest breadth, and depth — are significantly larger in the urban than in the rural populations. This significant increase in breadth and depth dimensions is also apparent in the chest and abdominal circumferences. The limb circumferences, skinfold measures (except the biceps skinfold), and hand breadth are likewise significantly greater in the urban male ($P < 0.05$). On the other hand, the urban–rural differences in height and length dimensions — stature, sitting height, supra-sternal height, tragion height, span, upper and lower limb length, arm and forearm length, thigh and leg length — are not significant by the t-test ($P < 0.05$).

Urban-rural differentiation, particularly with reference to increased urban stature and body weight, has been observed in a variety of different populations (e.g., Acheson and Fowler 1964; Chamla, Marquer, and Vacher 1959; Grobbelaar 1963; Hiernaux 1972; Schlaginhaufen 1956; Schreider 1966, 1967).

However, in the South African Negro male, urban-rural differentiation is confined to body weight, breadth, depth, circumference, and skinfold measures; that is, to bulk and girth only. Stature and length dimensions, on the other hand, show no such differentiation. A possible explanation for the latter finding may lie in the fact that most of the urban subjects had been born in rural areas and had migrated to the urban areas only after full stature had been attained.

Studies on urban and rural Zulu male populations in Natal have provided parallel findings; that is, body weight and skinfold measures, but not stature, show significant increases in the urban Zulu populations (Slome, Gampel, and Abramson 1960; Abramson, Slome, and Ward 1960). Most of the urban Zulu subjects studied were born in rural areas and had probably migrated to the urban areas after full stature had been attained.

The urban–rural differences in body weight, breadth, depth, and circumference measures may be ascribed to differences in muscularity or percentage of body fat or both.

A significantly higher percentage ($P < 0.05$) of urban males of both the Venda and Pedi are capable of hard work with maximum oxygen intakes of more than forty-five millimeters per minute, per kilogram of weight (van Graan et al. 1969; Greyson et al. 1971). In urban subjects the mean cross-sectional muscle area of the arm was shown to be significantly greater than in their rural counterparts (Loots and Lamprecht 1971). The mean calf circumferences are also significantly greater in the urban than in the rural male.

Of the rural males, 82 percent of the Venda and 80 percent of the Pedi were unemployed, and of the unemployed 18.8 percent and 8 percent respectively had never been employed. The rural male is thus, on the whole, relatively inactive, although no increase in the degree of fatness was observed with age, in contrast to findings in other populations (Brozek 1960; Khosla and Lowe 1968). Of the urban males, on the other hand, 48.8 percent of the Venda and 63.9 percent of the Pedi were employed in heavy industry. Of the Pedi a further 13 percent were employed in the mining industry (Hall and Harris 1970; Hall 1971). These figures indicate that the urban males are physically more active than the rural males. The increased urban breadth, depth, and circumference measures would appear, at least in part, to reflect an adaptation to greater muscular activity.

Skinfold measures provide a good estimate of the quantity of fat in the body (Brozek and Keys 1956), and it had been demonstrated that there is a significant correlation between body weight and skinfold measurements (Slome et al. 1960). This correlation indicates that the urban–rural differences in body weight and skinfold thickness are, in part, a reflection of a greater degree of adiposity among urban males. An increase in the percentage of body fat with age was demonstrated in urban males (Loots and Lamprecht 1971). The body fat increase in urban males indicates that, although physically more active than their rural counterparts, their calorie intakes, in general, are more than adequate. The nutritional status of the urban Venda, with respect to the main nutrients, was shown to be superior to that of the rural Venda (Lubbe 1970).

Further, urban subjects had a higher animal protein intake (mean 66.4 grams) than the rural subjects (mean 17.7 grams), and in general the protein nutritional status of the urban males was substantially superior to that of the rural males (Lubbe 1970). These factors may have contributed toward the better muscular development of the urban males. The study of the nutritional status of the Pedi was limited to biochemical analyses which showed a significant, though small, difference between the serum cholesterol values obtained for the urban (mean 169) and rural (mean 152) males and a nicotinic acid deficiency in the rural males (Louw, Du Plessis, and Laubscher 1971).

Hiernaux (1972), in a study of the adult physique of urban and rural Sara males from the Moyen-Chari Province of Chad, demonstrated that the urban male, resident since early childhood in Fort Archambault, is taller, less muscular, has smaller biacromial and biiliac dimensions, and more fat than the rural male. Hiernaux did not consider this urban–rural differentiation to be related to nutritional factors, but interpreted the observed differences "as resulting from manual labour during the growth period in the rural areas but absent in urban Fort Archambault" (Hiernaux 1972:93). In Chad, the rural male populations would seem to be physically more active than the corresponding urban populations.

The observed urban–rural differentiation in South African Negro male populations would appear to be due to environmental differences and related to nutritional status and physical activity. However, a genetic differential may also be involved. A selective factor may be in operation which could result in the urban population being a nonrandom sample of the rural parental group, thus accounting for their differences.

Similar but more marked differentiation has been observed between sedentary and migrant populations (Thieme 1957; Greulich 1957). Data obtained in such studies suggests that a selected — that is, nonrandom — group is involved in migration. Schreider (1966) suggested that there is selection on the basis of mental factors, and the gradient of physical characters is a secondary correlated consequence of a selective pressure exerted on psychological, rather than physical, traits. Stature and weight differences observed between urban and rural groups have also, in part, been ascribed to heterosis and the breaking up of isolates (Grobbelaar 1963; Schull 1965).

GENETIC MARKERS

In an attempt to determine whether or not the urban and rural Venda and Pedi populations are genetically comparable, a wide range of sero-genetic markers was included in the multidisciplinary study, i.e., fourteen sero-genetic systems for the Pedi subsamples, and eleven for the Venda subsamples (Jenkins 1972). In addition, the following four gene markers were included in the present study:

1. Frequency of "Nontaster" t *Gene (Table 4)*

The "taste blindness" trait — that is, the inability to taste phenylthiocarbamide — is inherited as an autosomal recessive. The incidence of the "taste blindness" trait was recorded, using the sorting technique of Harris and Kalmus (1949), and the *t* gene frequency calculated. The frequency of

Table 4. Frequency of nontaster *t* gene in urban and rural Venda and Pendi

	Venda		Pedi	
	Urban	Rural	Urban	Rural
Number tested	148	201	225	204
Gene frequency (*t*)	0.218	0.255	0.230	0.249

the *t* gene shows no significant differences between the corresponding urban and rural populations ($P < 0.05$).

2. *Frequency of Mutant Allele for Red-green Color Blindness (Table 5)*

Red–green color blindness is inherited as an X-linked recessive. For detecting, classifying, and estimating the degree of defective color vision, the A–O pseudo-isochromatic plates were used. The frequency of the mutant allele is the same in the two Venda populations (0.020), 0.017 in the urban Pedi and 0.014 in the rural Pedi.

Table 5. Frequency of mutant allele for red-green color blindness in urban and rural Venda and Pedi

	Venda		Pedi	
	Urban	Rural	Urban	Rural
Number tested	148	201	227	204
Gene frequency	0.020	0.020	0.017	0.014

3. *Distribution of Midphalangeal Hair (Table 6)*

It has been suggested that the distribution of midphalangeal hair on the four fingers is controlled by a set of five alleles of the A gene (Bernstein 1949). These have been named A_1, A_2, A_3, A_4, A_0 and arranged in decreasing order of dominance, the subscript corresponding with the number of fingers with midphalangeal hair, A_0 lacking hair on all four fingers. The phenotypic frequencies in the four subsamples are given in Table 6.

The urban and rural Venda and Pedi males follow the general African pattern in that hair on the middle phalanx is an unusual feature. The intergroup differences in the phenotypic frequencies are not significant ($P < 0.05$).

4. *Digital Dermatoglyphics (Tables 7 and 8)*

Comparative procedures in fingerprint studies are based on the fact that, while all populations thus far examined share all characteristics studied, the frequencies of these characteristics vary among populations. These phenotypic variations in all probability reflect variations in gene frequency. The total ridge count provides a good basis for statistical analysis

Table 6. Distribution of midphalangeal hair in urban and rural Venda and Pedi

Phenotype	Venda				Pedi			
	Urban		Rural		Urban		Rural	
	No.	%	No.	%	No.	%	No.	%
A_0	142	96.5	177	89.9	171	77.0	167	82.7
A_1	2	1.3	8	4.0	17	7.7	13	6.4
A_2	2	1.3	7	3.6	17	7.7	11	5.4
A_3	1	0.9	1	0.5	15	6.7	10	5.0
A_4	–	–	4	2.0	2	0.9	1	0.5

Table 7. Pattern frequency in urban and rural Venda and Pedi

Pattern frequency	Arches		Loops		Whorls	
	No.	%	No.	%	No.	%
Venda: Urban						
Right hand	25	3.7	438	64.9	212	31.4
Left hand	45	6.5	435	63.1	210	30.4
Pairs: right and left	67	5.1	437	64.4	396	30.5
Venda: Rural						
Right hand	35	3.9	515	57.2	350	38.9
Left hand	57	6.1	550	59.2	323	34.7
Pairs: right and left	83	4.6	1048	58.2	669	37.2
Pedi: Urban						
Right hand	53	4.9	710	65.7	317	29.4
Left hand	77	7.1	704	65.2	299	27.7
Pairs: right and left	130	6.0	1414	65.4	616	28.6
Pedi: Rural						
Right hand	61	6.3	608	63.4	291	30.3
Left hand	82	8.5	615	64.0	263	27.5
Pairs: right and left	143	7.4	1223	63.7	554	28.9

and appears to be an inherited metrical character, controlled by a number of additive genes, the prenatal environment playing a comparatively small part (Holt 1961).

The dominant fingerprint pattern in African Negro populations is the loop and then the whorl. Arches are uncommon. The fingerprint patterns of both the urban and rural Venda and Pedi follow the same trends — their fingerprints rarely show arches and frequently exhibit loops. The urban–rural differences in pattern frequencies are not significant ($P < 0.05$).

The mean total ridge count for the urban Venda is 126.3 and for the rural 129.8. The corresponding values for the Pedi are urban 128.2 and

Table 8. Ridge counts in urban and rural Venda and Pedi

	Right hand		Left hand		Total Ridge Count	
	Mean s.e.	s.d.	Mean s.e.	s.d.	Mean s.e.	s.d.
Venda						
Urban	64.6±1.8	21.8	62.2±2.0	24.0	126.3±3.9	45.2
Rural	66.1±1.6	20.9	63.6±1.6	22.2	129.8±3.4	42.4
Pedi						
Urban	65.3±1.6	23.9	62.9±1.7	24.9	128.2±3.3	48.0
Rural	61.5±1.8	25.8	58.1±1.8	25.7	119.6±3.6	50.1

rural 119.6. The urban–rural differences in the total ridge count are not significant ($P < 0.05$).

On the basis of the study of the red cell and serum protein polymorphisms Jenkins (1972) concluded that the respective urban and rural Venda and Pedi subsamples "may be considered as constituting homogeneous populations" (Jenkins 1972:36). If it is accepted that the polymorphisms examined are representative of the overall genetic constitution of the Venda and Pedi respectively, the urban–rural differentiation observed may then be attributable to environmental factors. The urban-rural differences are largely confined to bulk and girth and appear to be related to nutritional status, physical activity, and age at urbanization.

CONCLUSION

The evidence from both metrical and genetic data suggests an absence of genetic selectivity at least with respect to the characters studied. It has not been possible to determine the presence of a heterotic effect. Dietary amelioration, physical activity, and age at urbanization would appear to be the main factors that contributed to urban–rural differentiation in the South African Negro male populations studied.

REFERENCES

ABRAMSON, J. H., B. GAMPEL, C. SLOME, N. SCOTCH

1961 Sex variation in the nutritional state of urban Zulu adults. *American Journal of Clinical Nutrition* 9:217–228.

ABRAMSON, J. H., C. SLOME, N. WARD

1960 Diet and health of a group of agricultural workers in South Africa. *American Journal of Clinical Nutrition* 8:875–884.

ACHESON, R. M., G. B. FOWLER

1964 Sex, socio-economic status and secular increase in stature. *British Journal of Preventative Social Medicine* 18:25–34.

BERNSTEIN, M.
1949 The middigital hair genes. *Journal of Heredity* 40:127–131.

BROZEK, A.
1960 "The measurement of body composition," in *An introduction to physical anthropology*. Edited by Ashley Montagu, 637–679. Springfield, Illinois: Charles C. Thomas.

BROZEK, A., A. KEYS
1956 Body measurements and estimation of human nutrition. *Nutrition Review* 14:289–294.

CHAMLA, M. C., P. MARQUER, J. VACHER
1959 Les variations de la stature en fonction des milieux socioprofessionnels. *Anthropologie* 63:37–61.

DE VILLIERS, H.
1972 "A study of morphological variables in urban and rural Venda male populations," in *Human biology of environmental change*. Edited by D. J. M. Vorster, 110–113. London: International Biological Program.

GREULICH, W. W.
1957 A comparison of physical growth and development of American-born and native Japanese children. *American Journal of Physical Anthropology* 15:489–515.

GREYSON, J. S., C. H. VAN GRAAN, J. RIJK, C. H. WYNDAM, N. B. STRYDOM
1971 The determination of physical work capacities of urban and rural Pedi males. *Chamber of Mines of South Africa Research Report* 52(71):1–8.

GROBBELAAR, C. S.
1963 Observations of the differences between body measurements in urban and rural European boys of the age ten to eighteen years. *South African Journal of Science* 59:565–572.

HALL, S. K. P.
1971 Motivation among a rural and an urban employed group of adult Pedi males. *Council for Scientific and Industrial Research Special Report* 154:1–33.

HALL, S. K. P., J. HARRIS
1970 Motivational pattern of a rural and an urban group of adult male Vendas. *Council for Scientific and Industrial Research Special Report* 154:1–33.

HARRIS, H., H. KALMUS
1949 The measurement of taste sensitivity to phenylthiourea (PTC). *Annals Eugenics,* London 15:24–31.

HIERNAUX, J.
1972 "A comparison of growth and physique in rural, urban and industrial groups of similar ethnic origin: A few case studies from the Congo and Chad," in *Human biology of environmental change.* Edited by D. J. M. Vorster, 93–95. London: International Biological Program.

HOLT, S.B.
1961 "Dermatoglyphic patterns," in *Genetical variations in human populations.* Edited by G. H. Harrison. Oxford: Pergamon Press.

JENKINS, T.
1972 "Sero-genetic considerations," in *A multidisciplinary study of two ethnic groups in transition*. Final report to South African National Committee for International Biological Program (Human Adaptability Section). *Council for Scientific and Industrial Research Special Report* 159. Edited by D. J. M. Vorster, 36–40.

KHOSLA, T., C. R. LOWE
1968 Height and weight of British men. *Lancet* 1:742–745.
LOOTS, J. M., DE V. LAMPRECHT
1971 A comparative study of rural and urban Venda males. *South African Medical Journal* 45:1285–1288.
LOUW, M. E. J., J. P. DU PLESSIS, N. F. LAUBSCHER
1971 A biochemical evaluation of the nutritional status of rural and urban Pedi males. *National Institute for Nutritional Diseases. South African Medical Research Council Special Report*, 1–16.
LUBBE, A. M.
1970 The nutritional status of urban and rural Venda males as determined by dietary surveys. *National Institute for Nutritional Diseases. South African Medical Research Council Special Report*, 1–15.
SCHLAGINHAUFEN, O.
1956 *Verteilung anthropologischer Merkmale in Stadt und Land.* Zürich: Art Institut Ovell Füssli.
SCHREIDER, E.
1966 Investigations in social stratification of biological traits. *Year Book of Physical Anthropology* 12:184–209.
1967 Possible mechanism of social differentiation in biological traits. *Human Biology* 39:14–20.
SCHULL, W. J., J. V. NEEL
1965 *The effects of inbreeding on Japanese children.* New York: Harper and Row.
SLOME, C., B. GAMPEL, J. H. ABRAMSON
1960 Weight, height and skinfold thickness of Zulu adults in Durban. *South African Medical Journal* 34:505–507.
THIEME, F. P.
1957 *A comparison of Puerto Rican migrants and sedentes.* Michigan Academy of Science, Arts and Letters 42:249–267.
VAN GRAAN, C. H., C. H. WYNDAM, N. B. STRYDOM, J. M. VILJOEN, J. S. GREYSON, G. G. ROGERS
1969 Determination of the physical work capacities of urban and rural Venda males. *Chamber of Mines of South Africa Research Report* 47(69):1–8.
VORSTER, D. J. M.
1972 *A multidisciplinary study of two ethnic groups in transition.* Final Report to South African National Committee for International Biological Program (Human Adaptability Section). *Council for Scientific and Industrial Research Special Report* 159:1–17.

Physical Growth and the Social Environment: A West African Example

D. K. FIAWOO

The hypothesis that Euro-American children have better growth performance than their African counterparts and that such a difference is due primarily to nutrition is no longer new to educators and clinicians in Africa. The few scattered investigations on the continent seem to prove the hypothesis.[1] But beyond this, we know very little about the physical growth patterns of the African child. Only a few figures, even of heights and weights, can be found in the published literature that go beyond the formative period. We need to know what local factors promote and retard growth. The specific objectives of this paper are as follows:

1. to give figures on heights and weights of Ghanaian children;
2. to examine the relative growth patterns of boys and girls; and
3. to compare the heights and weights of children from different environmental situations in Ghana.

METHODS

The height and weight measurements of 4,585 primary–middle school children and 715 Achimota (secondary) school students were collected between 1966 and 1968. The children, wearing uniforms but without shoes, were measured with the help of trained undergraduates of the Department of Sociology of the University of Ghana at Legon. Data were

[1] See the references, especially Waddy (1956), Deane and Geber (1964), and Welbourne (1951). The only published material on height and weight growth in Ghana comes from Waddy's survey of children from the Northern Territories. Although the sample is quite large, the limited age range of six to twelve for boys and girls tends to hamper analysis in terms of the development cycles and the relative growth of boys and girls.

regrouped into bivariate frequency tables and analysis was by the regression technique.[2]

By far the most important factor in the analysis of heights and weights of African children is the problem of accuracy of age. Exact age was calculated for each child from the date of birth and the date of measurement. Unusual care was taken to ensure accuracy. The schedules for the measurements required details of information on age, including evidence of age. Such evidence included inspection of birth certificates, school records, and the interviewing of teachers and of parents where necessary. In the villages where recorded ages were rare, all the available techniques were used in the assessment of age.

Subjects and their Social Environment

Five groups of Ghanaian children, plus a small sample of expatriate non-African children, representing different social environments in Ghana, were selected for study. With the exception of the international schools, all the Ghanaian children in the selected schools were weighed and measured. Of the former, only half the total number in school were randomly selected.

ACHIMOTA SCHOOL STUDENTS. This was the only group of secondary school students. Measurements were taken by the medical department of the school.[3] Included in this sample were 227 girls and 488 boys.

Very high standards of secondary school education exist in the country generally; nevertheless, Achimota School, founded in 1925, has since its foundation enjoyed the reputation of an elite school. Being a residential school with a tradition of good discipline, it was expected that the students would have more than average growth performance.

"PRIVILEGED" URBAN CHILDREN. These were Ghanaian children attending various expensive international schools at Accra during the period of this study. They included the Accra International School at the Cantonments, Christ the King, and the Ridge Church School. These are highly rated schools and very much in demand among the elite Ghanaian families. They are mostly operated by expatriates, and the large majority of teachers are expatriate graduates or holders of high professional certificates.

[2] The author is grateful to Dr. Kpedekpo of the Institute of Statistics, Legon, for statistical advice.

[3] I am grateful to the headmaster and the medical officer-in-charge for their permission to analyze these data and to incorporate them in this study.

The results of their Common Entrance Examination[4] from year to year and their high standard of discipline are said to justify the exorbitant fees paid. The children who attend these schools tend to come from relatively wealthy or highly educated Ghanaian homes. Included in this sample were 223 girls and 213 boys.

OTHER URBAN CHILDREN. These children were drawn from various government and City Council primary–middle schools in the heart of Accra. Their parents live in Accra and the children enjoy the amenities of West African urban life. There were 735 girls and 756 boys in this group.

SUBURBAN CHILDREN. Like the group above, these children were attending various government-approved primary–middle schools, but at the outskirts of Accra or in suburban areas. All the settlements featured in this study were within a radius of five to ten miles from the heart of Accra and included Apenkwa, Anumle, Achimota (village), and Alogboshie on the Accra–Nsawam Road.

Large numbers of children attending these schools were drawn from the nearby villages of Kisseihma and Christian's Village situated between Legon and Achimota. Apart from Anumle, which represents a planned settlement,[5] all these are very primitive communities with the barest minimum of social amenities. The parents of the children are mostly wage earners who commute to Accra, Legon, and Achimota School. Only a tiny fraction are full-time farmers. There were 408 girls and 593 boys in this sample.

RURAL SCHOOL CHILDREN. They were attending various Local Authority and Mission Schools in the Tonu (Tongu) Lower Volta District. The selected villages included Tefle, Sogakofe, Sokpoe, Vume, and Kpotame. All the villages lie on the main trunk road from Accra to Lome in the Togo Republic. The men are mostly farmers and fishermen. The women assist the men but are also petty traders in their own right (Fiawoo 1961). This group consisted of 651 girls and 1,005 boys.

EXPATRIATE CHILDREN. This group was made up mostly of Euro-American children attending the same international schools as the privileged urban children discussed above. Their parents held key positions in international organizations, in government, education, commerce, and the diplomatic fields. They included children of university professors and lecturers, and Syrian-Lebanese children whose parents were very successful traders in Ghana. The sample included 146 girls and 156 boys.

[4] This is an examination conducted by the West African Examinations Council for admission to approved secondary schools.

[5] Anumle represents the staff village of Achimota School.

GROUP CHARACTERISTICS. From these group characterizations, the following deductions were made:

1. Families in the urban area are probably wealthier than those in the rural area and the suburbs.

2. Urban parents who send their children to expensive international schools are probably better off economically than those who send their children to normal government schools.

If it is further postulated that if the physical growth of the child is closely related to parental socioeconomic status, then the following hypotheses are relevant:

1. Privileged urban children will grow taller and heavier than their normal counterparts in the city.

2. Other urban children will have better growth performance than their rural and suburban counterparts.

RESULTS

Among 2,568 primary–middle school boys aged five to eighteen, there is a marked degree of correlation between age and height, the correlation coefficient being 0.99; the older the boy, the taller he grows. The highest point of acceleration is at age fourteen to fifteen which falls within the Ghanaian pubertal age range of thirteen to fifteen.[6] At age eighteen boys are still growing; the mean is 65.6 inches and the range between 57 and 73 inches.

Among 2,017 primary–middle school girls there is almost perfect correlation between age and height (r = 0.98), the highest point of acceleration being at age ten to eleven. The curve rises sharply between ages ten and thirteen (see Figure 1). This is not different from the Euro-American pattern, where the highest point of acceleration for girls is between eleven and thirteen (Cole and Morgan 1961). At age fifteen growth declines appreciably and is hardly noticeable at seventeen to eighteen. The mean height at eighteen is 63.4 inches and the range is between fifty-nine and sixty-eight inches.

Relationship between Age and Weight

Table 1 shows the distribution of weight according to age for boys. There is a positive correlation between age and weight (r = 0.98). The highest point of acceleration is at age fourteen to fifteen, the same as for height. The annual gain remains significant at seventeen, and at eighteen the

[6] From various studies conducted at Accra and the Sogakofe district of the Volta Region (unpublished materials).

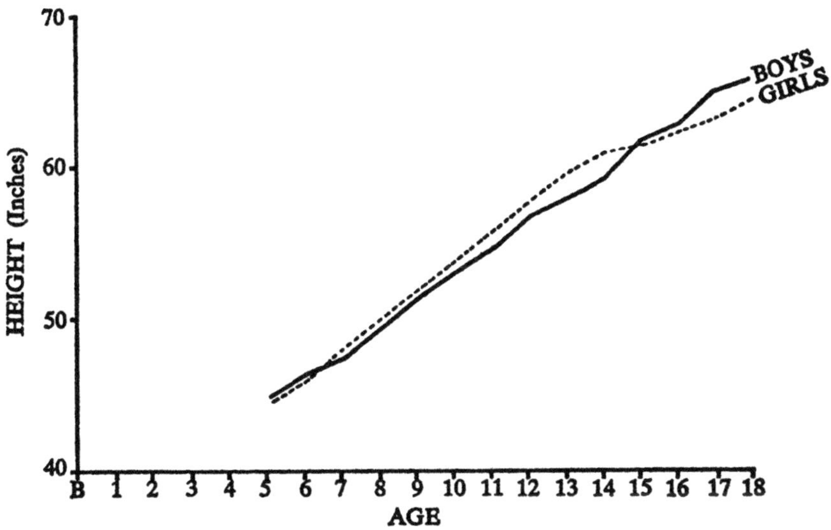

Figure 1. Relative rates of growth of boys and girls in primary–middle schools

Table 1. Primary-middle schools: distribution of weight according to age—boys

Age	Mean weight	Standard deviation	Standard error of mean	Total number of boys
5	43.4	3.3	0.63	27
6	43.6	5.1	0.50	104
7	46.0	7.1	0.46	242
8	50.9	7.4	0.47	248
9	55.7	8.4	0.52	261
10	61.0	10.5	0.69	235
11	64.8	8.4	0.57	220
12	71.7	10.4	0.71	214
13	76.3	11.9	0.82	214
14	81.1	10.2	0.72	198
15	92.6	15.0	1.05	203
16	97.9	16.7	1.28	169
17	109.0	15.7	1.32	142
18	114.8	14.4	1.51	91

boys are still growing. At eighteen the mean is 114.8 pounds and the range between 79 and 148 pounds. This wide range is probably suggestive of the varying socioeconomic groups of children featured in this study.

The growth distribution for girls is expressed in Table 2. Once again, there is a marked degree of correlation between age and weight (r = 0.99). Except for age fourteen, the curve rises sharply between twelve and sixteen years. At eighteen the weight curve is still rising. There is no marked difference between growth in height and growth in weight for

Table 2. Primary–middle schools: distribution of weight according to age—girls

Age	Mean weight	Standard deviation	Standard error of mean	Total number of girls
5	44.8	4.5	0.90	25
6	45.2	6.9	0.71	95
7	48.0	5.3	0.36	215
8	52.0	9.8	0.59	272
9	57.2	11.1	0.71	243
10	63.5	12.1	0.74	268
11	69.5	10.5	0.70	224
12	78.0	15.4	1.25	151
13	86.6	15.1	1.25	146
14	91.8	13.5	1.22	123
15	100.8	13.2	1.25	113
16	108.2	9.4	1.08	76
17	113.4	13.2	1.86	51
18	117.4	8.0	2.07	15

girls except that ages fourteen and fifteen (followed closely by age thirteen) stand out as the highest point of acceleration in weight growth, whereas in height, growth begins to decline at this age. This might suggest that growth in weight declines less rapidly than growth in height. The mean weight at eighteen years is 117.4 pounds and the range between 96 and 140 pounds (see Figure 2).

Relative Rates of Development for Boys and Girls

Contrary to the Euro-American pattern, Ghanaian girls are heavier and taller than the boys for a considerable part of their growing period, including the elementary school years. In height, boys and girls are almost at par between the ages of five and six. The boys lead by only a tiny fraction of an inch. Between six and seven years the girls catch up with the boys and outdistance them at seven. They maintain the lead from seven until fourteen. The boys overtake them at fifteen and maintain the lead for the rest of adolescence.

In weight, the female superiority is even more striking; girls are heavier than boys throughout the growing period, i.e., from five to eighteen.

These represent significant departures from the Euro-American norms where the growth trend shows a tendency for boys to exceed girls in height and weight from birth through the preschool and elementary school years. The girls become physiologically mature two years sooner than boys and thus attain adolescence earlier. This has the effect of carrying them ahead of the boys between eleven and thirteen. By fifteen, however, the boys are back in the lead "when their greater and probably

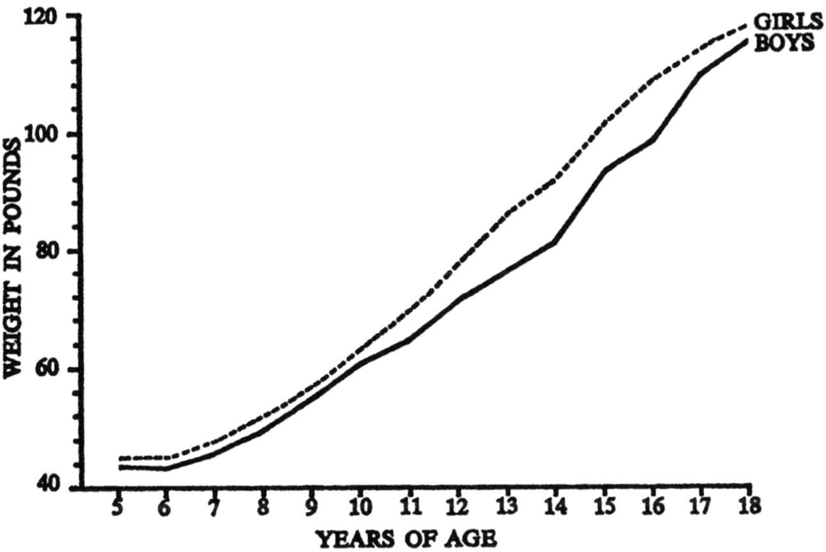

Figure 2. Relative rates of weight increase of boys and girls in primary–middle schools

more sustained adolescent spurt begins to take effect" (Tanner 1967:18). This male superiority is true of height and weight growth. It is significant that the Ghanaian expatriate group is parallel to the Euro-American standard in the relative growth of boys and girls. The Ghanaian rural, suburban, and urban groups are well behind this standard.

While there is some physiological explanation for the superiority of girls over boys between preadolescence and early adolescence, there seems to be no such explanation for the overall superior growth performance of boys in the Euro-American pattern. Similarly, one can find no biological basis for the superiority of girls over boys in Ghana. Perhaps the answers may be found in the Ghanaian social and cultural setting.

Since nutrition is the prime factor in the superior growth performance of Euro-Americans over Africans, one may conclude that Ghanaian girls have more appetite than boys. In traditional Ghanaian homes, it is customary for girls to assist with the domestic chores of their mothers, including the preparation of food.

It is not improbable that the nearness to the hearth affords them an advantage over boys in feeding. This desire to eat more, nurtured from the home, is probably carried over to the residential schools. This hypothesis is supported by the extreme obesity of market women in Accra, who apparently eat more than their male counterparts, but exert themselves less, and thus burn less of the stored fat. Hence, a combination of quantitative difference in diet and occupational roles of the sexes appears to tip the scales in favor of the females. But the whole question of the superior performances of females needs further study.

Individual Group Performance

The Ghanaian growth pattern delineated above represents the average performance of the groups of primary–middle school children discussed above in the introduction. It does not include Achimota School. To a large extent, each group typifies this growth pattern, especially in the case of the relative growth of boys and girls. In the following paragraphs, I examine each group in relation to the others and to the total pattern. It should be borne in mind, however, that each part of the whole is relatively small and that statistical errors are possible in the reckoning of an individual group mean.

Tables 3 and 4 show the mean height growth for boys and girls. In the former, Achimota shows a trend of marginal superiority to rural, suburban and urban groups, the differences being mostly between one and two inches. Compared to privileged urban and the expatriate groups, however, Achimota shows consistently inferior performance at eleven to thirteen years, the only age groups where data are comparable. The height growth for girls shows a fairly similar trend.

In weight growth, the pattern is strikingly similar to height: Achimota is generally inferior to the privileged urban/expatriate group, but is clearly superior to the rural/suburban/urban group. In weight, however, the gap

Table 3. Achimota and primary–middle schools: distribution of height according to age—boys (mean height, inches)

Age	Rural	Suburban	Urban	Privileged urban	Achimota	Total boys	Expatriate
5	–	–	42.0	45.0	–	27	44.8
6	45.2	43.5	44.6	48.1	–	104	46.3
7	47.1	45.5	46.5	49.2	–	242	48.8
8	49.2	48.0	48.4	51.2	–	248	52.1
9	51.0	49.4	50.3	53.7	–	261	54.4
10	52.0	51.0	52.3	55.4	–	235	55.2
11	53.6	53.7	53.6	57.1	55.5	226	58.1
12	55.9	55.1	56.0	58.4	58.2	242	58.9
13	57.2	56.7	57.4	59.0	58.9	268	59.0
14	58.4	57.9	59.1	–	61.5	278	–
15	60.5	61.9	60.9	–	62.9	291	–
16	62.4	62.1	61.8	–	65.1	239	–
17	64.1	64.4	63.7	–	65.9	201	–
18	65.2	64.7	65.1	–	65.3	147	–

Table 4. Achimota and primary–middle schools: distribution of height according to age—girls (mean height, inches)

Age	Rural	Suburban	Urban	Privilegd urban	Achimota	Total girls	Expatriate
5	44.5	43.0	–	44.5	–	25	44.7
6	44.5	43.9	45.9	46.7	–	95	45.9
7	47.1	46.0	48.2	48.4	–	215	49.1
8	49.7	47.4	49.7	51.4	–	272	51.7
9	51.1	48.9	51.3	53.7	–	243	53.3
10	52.7	51.3	53.5	56.0	52.5	269	55.3
11	55.0	53.8	55.6	58.2	58.9	235	59.1
12	56.7	55.0	58.2	61.5	59.7	176	60.7
13	58.8	57.7	59.7	62.7	61.3	181	–
14	59.4	59.0	61.9	–	61.5	182	–
15	61.0	60.8	61.3	–	62.4	140	–
16	62.7	61.5	62.1	–	61.8	104	–
17	61.8	62.3	63.4	–	62.1	74	–
18	63.9	61.5	64.5	–	62.7	25	–

between Achimota and the rural/suburban/urban group is very wide compared to the privileged urban/expatriate group.

When the privileged urban group is compared with the expatriate group, the results are very close indeed. In height growth for girls aged five, seven, eight, and eleven the expatriate group has the edge; the Ghanaian privileged urban group is dominant at ages six, nine, ten, and twelve. For the boys, the privileged urban group takes the lead from five to ten. The expatriate group catches up and outdistances it at eleven and maintains the lead up to age twelve. At thirteen the privileged urban group catches up and both groups remain at par.

The same dramatic performance characterizes weight development (Tables 5 and 6). For the boys, the Ghanaian group takes the lead at five and six; it is overtaken by the expatriate group at seven and eight. The privileged urban group recovers the lead at nine and maintains it until ten, when it is again overtaken by the expatriate group which maintains the lead from eleven to thirteen. The same general features characterize the growth for girls.

In height and weight growth for boys and girls, the urban group shows complete superiority over the suburban. But there is an interesting development when urban is compared with rural. While the urban girls are consistently taller and heavier than the rural girls, the rural boys are taller than the urban boys, except for the postpubertal period of seventeen and eighteen. In weight, the rural boys are superior between six and eleven; the urban boys seize the lead from twelve to eighteen.

In effect, urban superiority over rural is not a *fait accompli*; it is

Table 5. Achimota and primary–middle schools: distribution of weight according to age—boys (mean weight, pounds)

Age	Rural	Suburban	Urban	Privileged urban	Achimota	Total boys	Expatriate
5	–	–	35.0	45.9	–	27	45.5
6	40.7	37.9	40.6	51.9	–	104	50.4
7	45.4	42.2	44.0	57.7	–	242	59.4
8	50.5	47.7	48.4	62.2	–	248	66.0
9	55.0	51.1	53.9	69.5	–	261	69.4
10	57.8	54.4	60.2	75.0	–	235	71.7
11	63.5	63.2	63.1	83.2	76.4	226	87.7
12	71.2	68.9	71.3	86.1	85.6	242	91.5
13	76.1	73.5	77.1	88.0	90.4	268	91.0
14	81.1	78.0	83.8	–	102.6	278	–
15	88.7	99.2	92.0	–	109.4	291	–
16	99.2	96.7	102.5	–	122.5	239	–
17	109.9	108.5	111.3	–	125.5	201	–
18	114.8	118.2	121.0	–	124.9	147	–

Table 6. Achimota and primary–middle schools: distribution of weight according to age—girls (mean weight, pounds)

Age	Rural	Suburban	Urban	Privileged urban	Achimota	Total girls	Expatriate
5	41.5	40.0	43.0	45.7	–	25	44.4
6	40.7	38.2	45.7	49.6	–	95	48.7
7	45.0	41.3	49.7	55.8	–	215	56.6
8	51.6	41.3	52.1	64.1	–	272	65.8
9	53.6	45.9	57.2	72.7	–	243	67.1
10	60.8	54.9	62.6	78.2	67.0	269	74.5
11	67.2	62.9	69.8	88.6	87.9	235	89.7
12	73.4	68.3	83.0	95.5	94.4	176	94.3
13	78.1	81.3	91.1	99.0	108.7	181	–
14	85.0	87.3	99.2	–	111.7	182	–
15	96.3	99.0	103.8	–	116.3	140	–
16	104.5	106.2	109.8	–	121.8	104	–
17	106.8	116.0	118.5	–	120.5	74	–
18	114.7	115.0	127.0	–	122.0	25	–

questionable in places. The superior development in height among the rural boys is probably due to open-air activities which contribute to a better growth of the long bones.[7] Correspondingly, strenuous farm activities on the part of the rural boys may lead to the burning of much of the stored fat.

[7] Personal communication from Dr. Oracca-Tetteh, a biochemist at Legon.

Factors Promoting and Retarding Growth

In terms of aggregate growth performance, the six groups featured in this study may be ranked as follows: (1) expatriate, (2) urban privileged, (3) Achimota, (4) urban, (5) rural, and (6) suburban.

The first three groups show better-than-average growth performance. The children are taller and heavier than the last three.

There is considerable evidence that children from both expatriate and privileged urban groups belong to the highest socioeconomic strata. Their parents are highly educated and hold status positions with corresponding earning power. Such well-to-do parents are better able to afford the balanced diet necessary for growth than their less affluent urban counterparts.

Achimota school has long ceased to be the preserve of the elite as a result of the free educational policy of the government of Ghana. The selection of students from the "common entrance" pool ostensibly gives a fair chance to all Ghanaians; nevertheless, an unobtrusive selective process operates whereby the students — usually the cream of the pool — tend to come from the well-to-do homes with a relatively long tradition of education.[8] Secondly, Achimota is a residential school with a tradition of good feeding and good discipline. These combined factors tend to encourage better-than-average growth performance.

The urban group represents a heterogeneous body. It includes the middle-range stratum of Ghanaian society and the "working classes." Obviously, some of the children in this group would have excellent growth opportunities, and others would be far less fortunate because of varying parental socioeconomic status. It would be desirable in future research to break this group down into two or three discrete components. The results would probably show that the higher the parental socioeconomic status, the better the growth performance in the urban area.

Rural performance in this study reveals that the natural environment, the geographical setting, as well as the occupation dictated by these surroundings may contribute to physical growth.

Finally, the poor suburban growth performance is a clear reflection of the effects of unsanitary surroundings, economic poverty, and gross illiteracy.

SUMMARY AND CONCLUSIONS

The following major conclusions have emerged from this study:

1. The highest point of acceleration in height and weight growth for boys is at age fourteen to fifteen, and for girls at ten to thirteen.

[8] See Forster (1965; Chap. 7) for a fuller discussion on the relationship between school attendance and parental socioeconomic status.

2. Girls are bigger and taller than boys for a considerable part of their growing period.

3. In terms of individual group performance, it has been shown that the higher the socioeconomic group of the child's parent, the better the growth performance. For example:

a. Privileged urban children perform as well as expatriate children in Ghana and are taller and heavier than the average Ghanaian urban child;
b. urban children have better growth performance than their suburban counterparts; and
c. rural children perform better than suburban children and sometimes as well as urban children.

Some of these conclusions, especially the differential growth performance of children of different socioeconomic groups, are already familiar in the Euro-American context where periodic measurement of the height and weight of school children has become a regular feature of educational research. While the social strata are less clearly defined in Ghana — and in Africa generally — than in Europe or America, one can safely distinguish between the urban elite and the average urban dweller in terms of education and/or earning power.

Similarly, the distinctions between urban and rural, urban and suburban are almost clearcut in the African context. One is likely to find a higher standard of education with a stronger earning power in the urban area. It is the relationship between rural and suburban which needs further study. The relatively permanent village settlement adjacent to a major road and with clearly defined means of sustenance is certainly superior to a suburban community where members have to eke out an existence; it may even offer a challenge to the urban settlement as this study reveals.

Thus, the various groups in this study represent defined socioeconomic groups in the African context. The study further reveals that there is some relationship between these groups and the growth performance of the children.

The conclusions relating to the growth spurt as well as the superior growth performance of girls over boys need further study.[9] These conclusions need to be treated as hypotheses which must be tested and retested by successive investigations. The final picture will probably emerge only after years of painstaking field work.

Finally, it is important to stress that the data analyzed in this paper have been collected only from southern Ghana. How far they reflect Ghana as a whole will depend upon future investigations.

[9] Note that in his study of Baganda children in the vicinity of Kampala, Welbourne (1951) observed that girls up to nine years were taller and heavier than the boys, but from twelve to fourteen years they were shorter and lighter. He was unable to comment on the significance of this pattern because of an alleged uncertainty concerning the ages of his subjects.

REFERENCES

COLE, L., J. J. B. MORGAN

1961 *Psychology of childhood and adolescence.* New York: Holt, Rinehart and Winston.

DEANE, R., M. GEBER

1964 The development of the African child. *Discovery* 25:1, 14–19.

FIAWOO, D. K.

1961 *Social survey of Tefle, Ghana.* Child Development Monograph (mimeograph).

1972 "Puberty and adolescence in Ghana." Unpublished manuscript.

FOSTER, PHILIP

1965 *Education and social change in Ghana.* Chicago: University of Chicago Press.

GRANT, M. W.

1951 "The technique for the analysis of heights and weights." Mimeographed manuscript. Applied Nutrition Unit, London School of Hygiene and Tropical Medicine.

GRANT, M. W., G. R. WADSWORTH

1959 The height, weight, and physical maturity of Liverpool schoolgirls. *The Medical Officer* 102:303–306.

JELLIFE, D. B.

1962 The African child. *Transactions of the Royal Society for Tropical Medicine and Hygiene* 46:13.

ROBERTS, D. F.

1960 "Effects of race and climate on human growth as exemplified by studies on African children," in *Human growth.* Edited by J. M. Tanner. Oxford: Pergamon Press.

TANNER, J. M.

1967 *Education and physical growth.* London: University of London Press.

WADDY, B. B.

1956 Heights and weights of children in the northern territories of the Gold Coast. *The Journal of Tropical Medicine and Hygiene* 59:1, 1–4.

WELBOURNE, H. F.

1951 The growth of Baganda children in the vicinity of Kampala. *East African Medical Journal* 28:428.

Progressive Decline in Stature in India: A Study of Sixty Population Groups[1]

PRANAB GANGULY

During the last hundred years or so, trends toward accelerated physical maturation and greater adult height have been observed in many parts of the world. Continuing increases in certain bodily dimensions reflecting a general increase in body size have been reported in several countries of Europe, North America, and Japan, in men, women, and particularly in children. These observations have prompted some scholars to presume that the progressive increase in stature is an almost universal phenomenon. The presumption, however, seems unwarranted because the information derived from a few technologically advanced countries cannot serve as the basis for a generalization about the entire species. There are, in fact, three distinct logical possibilities: the average height of a population may decrease, increase, or remain unchanged over the generations. To these possibilities Meredith in 1941 has added a fourth one (Kimura and Yamazaki 1961). Instead of being unidirectional, the secular change may be cyclical; that is, the average height of a population may rise and fall in succession during certain decades.

GENERAL TRENDS IN STATURE

An excellent example of stability in height is furnished by the Micronesian

I am most grateful to my teachers, Professor M. N. Basu and Mr. D. Sen of Calcutta University, for their keen interest in the work and for their encouragement. I am particularly indebted to Mr. H. K. Rakshit of the Anthropological Survey of India for providing me with some hitherto unpublished data relating to the peoples of Deccan and for his advice and help in statistical matters. The suggestions and criticisms made by Dr. D. P. Mukherjee and Mr. A. Pal (my collaborator on a brief communication on the subject of this paper, presented at Calcutta University in 1972) helped me a great deal in analyzing and drafting the material.

[1] This essay was awarded the Bengal Immunity Research Prize in Science in 1975.

people of Yap Island. They were measured by six anthropologists at different times from 1876 to 1948. During this period the island suffered severe depopulation and was successively occupied by Spain, Germany, Japan, and the United States. But through these vicissitudes, the culture and dietary habits of the Yapese underwent little modification, and the height of the people remained virtually unaltered (Hunt 1958). Another island community, namely the people of Tristan Da Cunha, showed no evidence over a span of thirty years of a trend toward either increased adult stature or earlier physical maturation (Marshall et al. 1971).

Short-term but intense environmental adversities have often had stunting effects on the heights of populations, but such effects are usually obliterated soon after the conditions have returned to normal. In Japan, an increase of approximately 1 centimeter in average height was observed in each decade since the beginning of the present century. The rising trend was interrupted by the food shortage during World War II, and there was a conspicuous decline in stature which was most apparent among adolescents. But the trend toward decline was soon reversed, and the average height of the Japanese in the late 1950's surpassed its prewar level (Suzuki 1959). The stunting effects of wartime deprivations were evident in Madrid children during the Spanish Civil War and in children of Guam growing during the Japanese occupation (Acheson 1960). The consequences of famine on body size and height were shown by Ivanovsky (1923) in his study of the Russian population under famine. The diminution of stature due to some specific cause like war or famine has been reported from time to time from different parts of the world; but regular decrease, that is, decrease not attributable to a specific acute condition, has rarely been reported. As an example of the trend to diminution, I may mention the study of Kenntner (1969) of the population of Chile in South America. Over the span of forty years, from 1920 to 1960, the decrease in average height of the male general population of Chile amounted to almost 5.5 centimeters, that is, more than 1 centimeter in each decade. On the other hand, the decrease in the stature of Chilean military recruits in this period was only 1.4 centimeters. The difference is presumably due to the selection procedure adopted for military recruitment which prescribed a minimum body size.

Reports of increases in adult height are, of course, far more numerous than reports of regression or stability in stature. A possible reason for this is that secular bodily changes have been researched almost exclusively in a few developed countries where the environment generally favors greater realization of hereditary growth and development potentials. Of the studies demonstrating increases, that of Bowles (1932) is perhaps the best known. Important changes in adult American male size were observed by Bowles in two generations of Harvard college students. Three hundred and ninety-nine Harvard fathers had an average stature of

173.95 centimeters while at college, whereas 480 Harvard sons yielded an average of 177.5 centimeters at a comparable age. This is an increase of about 3.5 centimeters in one generation. Bowles's samples of mothers and daughters from four women's colleges in the eastern United States revealed similar increases in stature in American women.

Bakwin and McLaughlin (1964) compared the heights of 536 Harvard freshmen entering in the 1930's with 500 entering in 1957–1958. The students were almost equally divided between public and private school graduates, and the latter generally came from economically better-off families. It was found that the public school graduates had increased in height over the two decades, but the private school graduates had not.

The practice of measuring and weighing entrants to Harvard University was discontinued in 1965. Damon (1968) examined the Harvard records from 1870 to 1965 and was able to retrieve the height and weight data and other relevant information for eighty-five members of twelve "Old American" families for four generations. An "Old American" was operationally defined as a person whose four grandparents had been born in the United States. Among the eighty-five individuals there were twelve fathers, twenty-four sons, thirty grandsons, and nineteen great-grandsons. The birthdates of the four generations centered around 1858, 1888, 1918, and 1941, and their mean ages at the time of measurement were 21.9, 19.0, 18.9, and 18.8 years, respectively. The average height increased 2.63 centimeters between the first and second generations, and 1.14 centimeters between the second and third. No increase was observed in the fourth generation. The "Old American" families studied by Damon belonged to the economically favored sector of society. The findings of Bakwin and McLaughlin and Damon are believed to be indications that the secular increase in height has come to an end among the economically favored Americans. It is held that the well-to-do Americans, being stimulated by environmental amelioration, have already reached the upper limits of plasticity.

The cessation of the increase in height has been observed in immigrants also. Among the Japanese immigrants in Hawaii the increase in height, more rapid in males, ended after one American-born generation, suggesting that a maximization of hereditary growth potential had been reached. Increase in height was more gradual in women in whom it extended into the third generation (Froehlich 1970). Cultural conservatism of Japanese women and male dominance in Japanese families are believed to have brought about this sex difference.

In some countries, the increase in the economically better-off classes may still be continuing, but the rate of increase appears to be higher in the lower classes. In France and Scandinavia the change in height has been greater in the lower social classes (Boyne in Kimura and Yamazaki 1961). Kimura and Yamazaki have nonetheless observed that in Japan the increase in stature is extremely noticeable, even in the upper social class. Among young upper class Japanese women, namely the female medical students of Toho University in Tokyo, the increase has been more than 2 centimeters during the period from 1933–1935 to 1959–1960.

An example of the increase in stature in the American Indian population is provided by the Western Apache (Miller 1970). Three hundred and fifty-five Western Apache men were measured by Gabel in 1940, and 143 sons of these men were measured by Miller in 1967. Between 1940 and 1967 there was a 1.3-centimeter increase in stature, and the head length and body weight also increased. Similar increases have been reported among the inhabitants of Bougainville Island in the Territory of New Guinea (Friedlaender 1971). Reports of secular trends in stature have also been received from Holland, France, Sweden, Norway, England, and Russia (Miller 1970), but it appears that much of the material on which such reports are based was obtained either from military recruits or from school children. The usefulness of the material is consequently rather limited. Anthropometric data are obtainable from young recruits and school students through institutional channels with comparative ease, and this is possibly the reason why so much data has been amassed from recruits and children.

There is, however, another side of the coin to be taken into consideration. Statistically speaking, samples wholly constituted of men, enlisted or drafted for military service, cannot truly represent the general population. Owing to the selection procedure followed at the time of recruitment, such men tend to be significantly taller than the general population and cannot, therefore, be considered as a cross section of the general population. Furthermore, any change in the selection procedure, particularly the raising or lowering of the minimum physical standards, is likely to influence the mean stature of the recruits. One can seldom be certain whether the recruitment procedure remained exactly the same over the decades or whether it underwent some change. Udjus (1965) compared 5,765 Norwegian recruits of 1962, aged 19.5 to 20.5 years, with the recruits of 1920–1922, aged twenty-one years, and found that the mean stature had increased 5.12 centimeters in about four decades (177.49 centimeters in 1962). But it is difficult to draw any inference from this observation about trends for the general male Norwegian population.

The studies on children pose difficulties of another sort. These studies amply demonstrate that the general trend is toward accelerated growth and earlier maturation. For example, in the Soviet Union there has been a trend over the last fifty years toward greater height and weight in children at all ages, earlier attainment of adult stature, and an earlier menarche. In 1923 Moscow boys continued to grow beyond the age of eighteen, but in 1960 their growth was usually completed at the age of seventeen

(Vlastovsky 1966). In Japan, both menarche and the period of most rapid growth now occur about a year earlier than they did in the first decade of the present century (Kimura 1967). Further data on height and weight for children and college students demonstrating more rapid growth will be found in Krogman (1941). There is also some evidence for increased prenatal growth and increased body length of newborn babies (Cantalini 1967, Oster 1967). It must, however, be borne in mind that accelerated growth and greater adult height are altogether different phenomena, and it is still debated whether or not speedier growth has meant increase in adult stature or merely earlier maturity (Boyne, Aitken, and Leitch 1957). Accelerated maturation need not, in fact, increase eventual body size. Evidence from animal growth studies indicates that the rate of growth and the final body size may not be controlled by the same genes (Hunt 1958). Tanner (1971) has summed up the position by observing that the increase in adult size has not been so marked as that in children's and that most of the children's increased size is due to earlier maturity.

From the foregoing survey of the relevant literature, it appears that the dominant trend is toward increased prenatal growth, increased body length of neonates, accelerated growth in children, and earlier attainment of adult size. The absolute increase in eventual adult size has been less striking. It should nevertheless be noted that these changes have been known to be taking place almost exclusively in a few Occidental populations and in Japan; we have very little information about the direction of human microevolution for the rest of the world. Our knowledge about the trends for the vast majority of mankind inhabiting Asia (excluding Japan), Africa, South America, eastern Europe, Australia, and Oceania is woefully inadequate. The scantiness of basic data prevents us from arriving at any generalization regarding the human species.

TRENDS IN STATURE IN INDIA

With a population of some 550 million (1971 census) and a net annual increase of more than ten million, India is the second most populous country in the world. China, of course, ranks first in the world with regard to absolute population size. Since China and India contain between them more than one-third of the present-day humanity, some knowledge of the temporal shifts in stature in these two countries would be valuable for determining the global trends and the general direction of human microevolution. We know regrettably little of the trends in China, but as far as India is concerned, some useful information can no doubt be retrieved from early and more recent anthropometric records. Thanks to the painstaking efforts of a few European and Indian scholars, two or more series of comparable anthropometric data are available for some Indian population groups. One has simply to compare these measurements to determine the increase or decrease, if any, occurring over time.

An almost unavoidable difficulty in this kind of study is the scantiness or absence of information about the background or setting of certain investigations. Some authors tell us little about the actual period of investigation, geographical areas covered, techniques followed, and instruments used. Statistical details beyond the arithmetic mean and the sample size are quite often not available for the earlier data, and the size of the sample is often too small to permit valid conclusions. But, even if we discard the materials which are deficient in one respect or another, there still remains a substantial body of data that may throw some light on the questions relating to trends.

Before proceeding further, I would point out an interesting characteristic of the Indian population. No human population is known in which mating between its members takes place at random. Large populations are usually composed of smaller segments with varying degrees of reproductive isolation. In India this segmentation has been carried to an extreme. The breeding structure of the Indian population has been quite complex because the elaborate caste hierarchy has created almost permanent reproductive barriers. The word "caste" is derived from the Latin *castus* and implies purity of breed (Risley 1915). More often than not a caste is divided into subcastes, and a subcaste, unless it is very small, tends to consist of subsubcastes. This fissile tendency is most marked in the Hindu religious community, but it is present in the Indian Christian and Muslim communities as well in a more attenuated form. In this context, it would not be quite correct to talk about a "general population" of India. The general population as such does not exist; it is a mere abstraction. There are communities segmented into castes and smaller groups by religious, social, linguistic, and geographical barriers. These groups practice endogamy to varying degrees. In addition, there are the aboriginal tribes which are, from a demogenetic point of view, relatively small and isolated Mendelian populations.

While considering the temporal shifts in stature in India, it would be more realistic to treat these population groups as discrete. Fortunately, data are available separately for quite a number of castes, tribes, and communities.

SOURCES OF DATA

The earliest available anthropometric record in India is that of J. Shortt who measured several population groups in the Nilgiri Hills in South India (Shortt and Ouchterlony in Gupta and Dutta 1966). But since Shortt's samples are small in size, the data have not been used in the present work. The earliest material that I have been able to use was collected by the distinguished Hungarian anthropologist, Charles de Ujfalvy, in Kashmir in 1881 and 1882 (Gupta and Dutta 1966).

Sir Herbert Risley must be mentioned as one of the pioneers who contributed most to anthropometry in India in the late nineteenth and the early twentieth century. Volumes I and II of his work (1891) contain valuable anthropometric data collected from extensive regions of eastern India from 1886 to 1888. Risley's survey was later extended to other parts of India as well as to Baluchistan, Burma, and Ceylon. Madras and Assam were excluded from the scope of his survey, possibly to avoid duplication with that of Thurston and Waddell who had already begun anthropometric investigations in Madras Presidency and Assam, respectively. Edgar Thurston, then Superintendent of the Madras Government Museum, obtained on loan a set of anthropometric instruments from the Asiatic Society of Bengal and commenced the collection of data in 1894. He obtained measurements from a large number of castes and tribes of the four southern states, viz., Mysore, Andhra Pradesh, Kerala, and Madras (Thurston 1909).

Risley's and Thurston's data have been extensively used in the present study, but it has not been possible to utilize Colonel Waddell's data collected from the Brahmaputra Valley in Assam for two reasons. First, most of the samples of Waddell consist of very few individuals and are, therefore, inadequate for making valid comparisons. Second, there are no corresponding series of data lately collected from the Brahmaputra Valley with which the remaining samples of Waddell could be compared.

Crooke's work (1896) is an important source from which I have culled some data for North Indian peoples. Crooke's measurements were actually taken by Surgeon-Captain Drake Brockman who wrote, "In order to obtain as large a number as possible of representative castes, long distances have been travelled; only males of the age of twenty-five years and upwards have been selected as subjects for measurements on account of their mature physical development" (Crooke 1896: xxvii). I also use other published works (see Appendix), and unpublished data for South Indian peoples collected under the All India Anthropometric Survey Project (South Zone) by H. K. Rakshit and associates during 1961 to 1963.

CRITERIA FOR DATA

Criteria used in sorting out useful data are as follows:

1. The time period between the first and subsequent investigations should be at least one generation. In India that is approximately twenty-five years. Intervals between investigations that are too short risk not

revealing long-term trends, whereas those that are too long may lead to inconclusive results based on inadequate comparable data.

2. The sample size ideally should be 100; however, given the generally small size of samples used in early investigations, we will lower the bottom limit in this study to thirty.

3. Both samples must be drawn from the same population group or a smaller subdivision of said population group.

4. The geographical distribution of samples must be considered. Although it would be more desirable to limit comparisons to the district level — in the descending hierarchy of state, district, and locality — the origin of too few samples is specified at this level. Thus, we shall permit comparisons between samples taken at the state level.

A diligent search was made of the relevant literature, and sixty population groups were found to satisfy these conditions. Great care was taken to keep possible biases from entering the sorting procedure. The sixty groups may be divided into two categories according to the availability of geographical information for samples. For thirty-six groups there are records showing the localities, districts, or zones (comprised of more than one district) in which population samples were obtained. For the remaining twenty-four groups, the information is very scanty and is usually limited to the names of the states. The data for the thirty-six groups are presented in Table 1 and the remainder in Table 2. The data, as here given, were mostly culled from the original works listed in the bibliography. When the original work could not be consulted, information was obtained from other sources, particularly *Anthropometry in India* (Gupta and Dutta 1966), a handy compilation of anthropometric data.

RESULTS

The size of the sample and the mean stature derived from earlier and later investigations are presented in Tables 1 and 2. We have subtracted the later mean from the earlier mean of each population and the fourth column shows the differences. The final column shows the actual or estimated time intervals between investigations. To avoid crowding the tables with too much detail, some of the relevant items of information have been listed alphabetically in the appendix. The ethnic status of the groups (caste, tribe, or minority community status), geographical area, earlier and later authors, the years of investigation, and publication are also given there. Where the years of investigation are not known, the years of publication of the earlier and later materials have been used for estimating the time interval. Where three or more comparable measurements are available, the first and last ones have been considered because the time interval between them is the longest.

Table 1. Difference in stature between pairs of samples drawn from same district or zone

Population group	Earlier work n_1	m_1	Later work n_2	m_2	m_2–m_1	Time gap (years)
1 Abor	84	1586.1	335	1583.9	− 2.2	34
2 Badaga	40	1641.0	104	1649.6	+ 8.6	68
3 Bagdi	99	1603.0	100	1585.8	−17.2	64
4 Bhil (M.P.)	50	1622.5	100	1600.9	−21.6	29
5 Billava	50	1632.0	131	1635.8	+ 3.8	68
6 Brahman Chitpavan	100	1655.0	104	1633.7	−21.3	47
7 Brahman Paschatya Vaidik	50	1662.4	114	1658.2	− 4.2	36
8 Brahman Shashana	52	1635.0	143	1642.9	+ 7.9	29
9 Brahman Tamil	50	1640.7	83	1641.9	+ 1.2	31
10 Bunt	40	1657.0	123	1683.0	+26.0	68
11 Chero	90	1626.0	100	1620.3	− 5.7	45
12 Cheruman	60	1566.0	64	1570.9	+ 4.9	61
13 Chetty	50	1623.0	34	1625.3	+ 2.3	31
14 Chhatri	100	1661.0	139	1633.3	−27.7	54
15 Coorg	32	1687.0	287	1671.0	−16.0	34
16 Ho	165	1600.0	100	1600.0	0.0	37
17 Kharwar	100	1617.0	197	1607.9	− 9.1	45
18 Korku	51	1636.4	50	1620.6	−15.8	25
19 Kurmi (Bihar)	100	1608.0	51	1599.6	− 8.4	64
20 Kurmi (U.P.)	100	1642.0	94	1613.5	−28.5	54
21 Ladaki	36	1630.0	34	1634.0	+ 4.0	30
22 Mahar	100	1634.0	102	1622.0	−12.0	47
23 Mal Paharia	100	1577.0	54	1573.6	− 3.4	48
24 Mappila Muslim	40	1648.0	152	1636.4	−11.6	68
25 Maratha	100	1632.0	55	1646.4	+14.4	47
26 Munda	100	1589.0	250	1581.5	− 7.5	45
27 Nayar	186	1656.0	50	1650.3	− 5.7	61
28 Oraon	100	1621.0	100	1604.0	−17.0	69
29 Pallan	50	1643.0	94	1624.8	−18.2	68
30 Paniyan	101	1565.0	100	1547.0	−18.0	27
31 Pod	100	1625.0	100	1618.9	− 6.1	64
32 Santal	100	1614.0	100	1593.0	−21.0	46
33 Shanan	130	1701.0	155	1640.0	−61.0	61
34 Syrian Christian	70	1650.8	333	1640.9	− 9.9	61
35 Tiyar	40	1642.0	63	1609.0	−33.0	32
36 Toda	82	1698.0	73	1719.0	+21.0	32

Key: n_1 and m_1 indicate the earlier sample size and mean stature (in millimeters), and n_2 and m_2 the later sample size and mean stature (in millimeters).

An examination of the data presented in Table 1 will show that the average stature has remained unchanged only in one group, viz., the Hos of Seraikela. The average stature has shown some increase in ten groups and a falling off in the remaining twenty-five. Table 2 likewise shows

Table 2. Difference in stature between pairs of samples drawn from same state

Population group	Earlier work n_1	m_1	Later work n_2	m_2	m_2–m_1	Time gap (years)
1 Agamudaian	40	1658.0	102	1635.7	−22.3	68
2 Ahir	350	1648.0	67	1613.7	−34.3	45
3 Bhantu	30	1640.0	150	1633.8	− 6.2	45
4 Bhil (Rajasthan)	200	1629.0	56	1588.0	−41.0	38
5 Brahman Deshastha	100	1642.0	107	1636.9	− 5.1	31
6 Brahman Nagar	100	1643.0	105	1648.6	+ 5.6	31
7 Brahman Nambudiri	55	1635.6	106	1619.8	−15.8	31
8 Chakkiliyan	50	1622.0	202	1610.1	−11.9	68
9 Chamar	333	1648.0	158	1618.8	−29.2	45
10 Chenchu	40	1625.0	100	1626.3	+ 1.3	68
11 Dom	100	1655.0	113	1665.3	+10.3	54
12 Idaiyan	42	1643.0	155	1630.9	−12.1	68
13 Jangam	30	1651.0	33	1643.8	− 7.2	49
14 Kahar	80	1636.0	56	1605.3	−30.7	45
15 Khandayat	40	1645.0	30	1632.8	−12.2	49
16 Kuruba	50	1636.0	176	1643.3	+ 7.3	68
17 Mukkuvan	40	1631.0	115	1632.9	+ 1.9	68
18 Muslim (Madras)	80	1645.0	177	1641.3	− 3.7	68
19 Muslim (U.P.)	441	1660.0	166	1624.5	−35.5	45
20 Palli	40	1625.0	214	1624.7	− 0.3	68
21 Pana	40	1607.0	50	1598.8	− 8.2	49
22 Paraiyan	40	1621.0	265	1617.6	− 3.4	68
23 Vellala	40	1624.0	225	1642.5	+18.5	68
24 Vokkaliga	50	1672.0	186	1666.9	− 5.1	68

Key: n_1 and m_1 indicate the earlier sample size and mean stature (in millimeters), and n_2 and m_2 the later sample size and mean stature (in millimeters).

increase in stature in six groups and decline in stature in as many as eighteen groups. Considering all sixty populations together, we find that sixteen groups have become taller on the average, one has remained stable, and the remaining forty-three have actually become shorter to varying degrees. Roughly speaking, three groups out of every four have become shorter.[3]

[3] It is apparent that two aspects of temporal change must be considered: the direction of the change and its magnitude. We omit, for the time being, the magnitude, and examine the statistical significance of the direction by applying the sign test, which is one of the easiest and most useful of nonparametric tests. In the tables, the increases in average stature have been given a plus sign and the decreases a minus sign. As one of the differences is zero, half of it is to be counted as positive and half as negative. Let n denote the number of comparisons and r denote the number of times the less frequent sign occurs. Then, for the data given in Table 1, $n = 36$ and $r = 10.5$; for the data in Table 2, $n = 24$ and $r = 6$; and for the combined data, $n = 60$ and $r = 16.5$. The sign test actually measures the significance of deviations from a 50-50 distribution, and for this purpose, the critical value of r for different

We must bear in mind that the sixty pairs of means used in the present analysis were derived from as many as 120 investigations. The investigators had different backgrounds, and their data were obtained from fourteen states from samples of unequal size and of different ethnic composition over a period extending from 1881 to 1963. Notwithstanding such heterogeneity in the basic material, a clear trend toward gradual decline in stature emerges.

The mean of the thirty-six differences given in Table 1 is −8.5 millimeters and the mean of the twenty-four differences given in Table 2 is −9.97 millimeters (see fn. 3). The mean time interval between the earlier and the later investigations has also been computed. It comes to about forty-eight years for the groups cited in Table 1 and fifty-four years for those cited in Table 2. Hence, roughly speaking, it appears that the average stature has declined in India at the rate of nearly 1 centimeter in fifty years. The rate of decline, of course, has not been uniform all over the country. For instance, in comparison with the Deccan, the decrease has been more marked in the north and northeast India. On the whole, however, the decline has been slow and steady rather than rapid and transitory.

CAUSAL PROPOSITIONS

Change in stature is a complex phenomenon to which more than one factor seems to bear a causal relationship. Kaplan has raised the interesting question whether the increases in bodily dimensions, presumed by many to be a worldwide phenomenon, are due to the operation of a single

---

levels of significance have been computed by Dixon and Mood (1946). A reference to this work shows at once that r is significant at the 5 percent level for both the series of data given in Tables 1 and 2. For the combined data, however, r is significant at a still higher level, that is, at the 1 percent level. Having thus established that a *prima facie* case exists for assuming that significant changes in stature have taken place in India since the 1880's, we proceed to apply the t-test to our data which takes into consideration the magnitude of the changes as well.

Evidently the thirty-six differences given in the fourth column of Table 1 are independently distributed; that is, one is in no way influenced by any other. It seems also reasonable to assume that their distribution is more or less normal or, at least, that it does not deviate far from the normal pattern so as to make the t-test invalid. We compute the mean and the standard deviation of the differences which come to −8.5 millimeters and 15.82 millimeters respectively, and use the t-test to verify the hypothesis that the mean difference is zero. We apply Bessel's correction for small samples and calculate t by using the formula cited by Moroney (1957: 229). The value of t is observed to be 3.18 with thirty-five degrees of freedom, which is significant indeed at the 1 percent level of probability. Similarly, for the twenty-four differences given in Table 2, the mean is found to be −9.97 millimeters and the standard deviation 15.12 millimeters. The observed value of t is 3.16 which, with twenty-three degrees of freedom, is definitely significant at the 1 percent level of probability. One cannot but be impressed by the remarkable agreement in the results yielded by two totally different series of data presented in Tables 1 and 2.

evolutionary force (Kaplan 1954). But we have already seen that such increases are really not worldwide; the instances of metric stability and diminution in stature, albeit seldom reported, are not entirely absent. We also have seen that the secular increments in stature tend to diminish and disappear within two or three generations, as evinced by Harvard students and Japanese immigrants in Hawaii. The concept of a single evolutionary force leading the species to gigantism therefore is not tenable.

The degree of inbreeding in a population has been found to have a notable bearing on the variability of stature. Dahlberg's hypothesis that heterosis is a major factor contributing to the secular increase in stature has been supported by research conducted lately in France, Poland, and elsewhere. It has been observed that the degree of inbreeding is negatively correlated with average body height (Schreider 1967, Wolanski, Jarosz, and Pyzuk 1970). This may lead the unwary to speculate that the falling off in stature in India might be due to a rise in the level of inbreeding. But there is no evidence whatsoever for postulating that the practice of inbreeding of late has increased in India. If anything, the frequency of consanguineous marriages may have actually decreased in recent times with the spread of urbanization and literacy, both of which have been found to be negatively related to high consanguinity rate (Dronamraju 1967). In the present context, therefore, inbreeding cannot be considered a cause of the secular decrease in stature.

Of other pertinent observations, Sir Francis Galton's statement that tall stature correlates to some extent with decreased fertility intrigues us because its implication is that the relative proportions of short individuals would gradually increase, thereby lowering the mean stature of the population (Galton 1886). But, if such a relationship between stature and fertility actually existed, the secular decrease in stature would have been a universal trend. In reality, however, the trend is far from universal.

Of all the factors that may be causally related to human plasticity in body size, the per capita consumption expenditure and nutritional status, which are closely related phenomena, appear to be most important. Severe malnutrition retards growth and delays the attainment of adult stature. If malnutrition is of short duration, the child is often capable of a compensatory growth spurt after slowing. If it is chronic, even the prolongation of growth period does not fully compensate for the slow rate of growth, and the result is stunted adult stature (Acheson 1960, Frisancho, Garn, and Ascoli 1970). In Japan, growth in height by prefecture was observed to be correlated with economic indices and particularly with the expenditure incurred for milk and eggs (Takahasi 1966).

In a survey I recently carried out in northern India, the mean difference in stature between the better-off and the worse-off sections of the same population was found to be well over 2 centimeters. This difference was persistently observed in different populations in urban and rural areas. The dietary surveys conducted by the Indian Council of Medical Research and the Anthropological Survey of India amply demonstrate that the diet of the common people of India is grossly deficient in calories, proteins, and various nutrients; large sections of the Indian population, tribal as well as nontribal, are living in a state of undernutrition (Malhotra 1967). Throughout the country, grains are the principal sources of protein. But these do not contain all the essential amino acids. Meat, fish, and dairy products are considered luxuries by common people.

Certain questions obviously follow. Were the economic conditions of the people, i.e., their standard of living and nutrition, appreciably better in the past than now? Has the deterioration in these standards, if it exists, been sufficient to induce decreases in stature?

For an answer to the first question, I shall use per capita figures of national income (see Table 3) as furnished in a recent work on the national income of India (Mukherjee 1969).[4]

Table 3. Average per capita national income of India at 1948–1949 prices

Years	Per capita income at 1948–1949 prices
1860	169
1865	169
1870	172
1875	177
1880	197
1885	216
1890	204
1895	201
1900	199
1905	203
1910	220
1915	241
1920	253
1925	261
1930	260
1935	260
1940	265
1945	255
1950	253
1955	275

[4] The figures are for overlapping nine-year periods from 1865 to 1950. But the first and the last figures centering on 1860 are averages over seven years. The averages have been calculated at the constant prices prevailing in 1948 and 1949 in order to make the figures, pertaining to different periods, comparable. I need scarcely mention that since 1860 the Indian rupee has depreciated greatly to a very small fraction of its original gold value.

From the data in Table 3 it emerges that, notwithstanding some fluctuations, a slow but sustained growth of per capita income has taken place. This rules out the possibility of ascribing secular decrease in stature to economic deterioration. For assessing the standard of living, however, the per capita consumption expenditure appears to be more useful than the per capita national income. In India, the per capita consumption expenditure at the constant 1948–1949 prices was Rs.260.80 in 1962–1963 as against Rs.234.10 in the base year, 1948–1949. The per capita level of consumption, as indicated by the expenditure in real terms, has shown a slow improvement in recent times. Reliable information on consumption levels for earlier periods is not readily available.

Concerning consumption habits of the early nineteenth century people of eastern India — namely in Bihar, Bengal, and Assam — a series of findings (Hamilton 1838) are interesting even at this distance in time. They relate to the frequencies of consumption rather than the levels of consumption. From this material Mukherjee (1969) prepared an estimate of the percentages of families consuming specified items of food at least once every month. These percentages, together with the corresponding percentages for 1953–1954 obtained from the seventh round of the National Sample Survey, are presented in Table 4. It will be apparent that there has been some fall in recent times in the percentages of families consuming items such as meat, fish, milk, and pulses which are rich in proteins. On the other hand, the percentages relating to certain necessities, such as rice, vegetables, and oil, have shown some increases. From this, one might generalize that in terms of protein intake, the standard of diet has to some extent deteriorated. But for a variety of reasons it is difficult to accept such a generalization. First, this method of comparison, though it relates to two distant points of time, is too imperfect to be relied upon. Second, the data relate only to eastern India and not the whole or

Table 4. Percentages of families consuming specified items of food at least once a month

Items of food	Early nineteenth century	1953–54
Meat	39	34
Fish	86	73
Ghee (clarified butter)	9	9
Milk	92	67
Sugar	100	62
Pulse (curry)	80	52
Vegetables	59	77
Spices	44	85
Oil	74	90
Salt	96	98
Rice	81	93
Wheat	41	44

Source: Mukherjee (1969: 448–449)

greater parts of India. Third, we have no information about the populations covered by Hamilton and by the later survey. It must be borne in mind that ethnic groups differ considerably from one another with regard to food habits. Finally, Hamilton's material is too old; the present study is concerned with the changes since the 1880's and not with those for earlier periods. We have to agree with what Aykroyd wrote years ago, "There is no convincing evidence that the national diet deteriorated during the century up to the outbreak of the Japanese War" (Aykroyd 1946:22).

Before proceeding further, one point should be clarified. We are not contending with the fact that chronic malnutrition causes stunting in adult stature. Malnutrition does prevail in India and it interferes with the maximum realization of the hereditary growth potentials. What we want to point out is that the food situation in India has not worsened since the 1880's or, at least, there are no reliable empirical observations indicating that it has become any worse. It would, therfore, be rather rash to assume that malnutrition is the principal cause of decrease in stature.

Conterio and Cavalli-Sforza (1967) conducted a highly interesting follow-up study on 773 men who were born in the years 1890 and 1891 and were subjected to medical and anthropometric examinations at the age of twenty. The subsequent demographic records for these individuals were also available. It was found that the men who died earlier were significantly smaller than the men who survived. In other words, taller stature had a higher survival value. Now, if such a relationship between stature and survival really exists, the Indian situation may lend itself to a fresh interpretation. Over the past few decades, the mortality rates in India steadily declined due to improved public health measures and other reasons. The death rate per thousand people per annum was forty-four from 1891 to 1901; it decreased to twenty-seven from 1941 to 1951, and has declined farther in recent years. The infant mortality rate (deaths of infants under one year per thousand live births) was 215 from 1900 to 1905; it fell to 134 from 1946 to 1950, and has dropped farther since then (Chandrasekhar 1959). Assuming that taller individuals tend to survive longer, we find that the decline in mortality rates in the younger age groups actually means *a relaxation of selection against shorter stature*. Shorter individuals, who would die at younger ages in the past, now survive, bringing about a slow but continued shift in the relative proportions of taller and shorter individuals. The survival of undersized individuals in larger numbers, due to a lessening of the intensity of selection against them, may partly account for the falling off in average stature in India. But, let me reiterate, changes in body size are very complex phenomena which are possibly causally related to a number of factors. We cannot satisfactorily explain them by postulating a single cause.

CONCLUDING REMARKS

In this study I have attempted to show that, contrary to the general impression, the trend toward greater body height is not a universal phenomenon. Its occurrence is mainly limited to some Occidental populations and Japan. For the vast majority of mankind living in the Afro-Asian countries, the available data are too fragmentary to draw general conclusions. I have culled from various sources almost all the comparable data for stature of different population groups living in India and have demonstrated that three groups out of every four have become shorter in varying degrees in the course of one or two generations. The dominant trend in India is toward diminution in height; the actual rate of decrease — about one centimeter in fifty years — is slow but sustained. The trend cannot be ascribed to any recent increase in the severity of malnutrition or in the degree of inbreeding because there is no evidence for either of them which stands up under scrutiny. A hypothesis of relaxation of selection against undersized individuals has been advanced that can explain the observed trend only partly. It has been stressed that secular changes are too complex to be assignable to a single cause.

Moroney (1957:123) suggests that the conclusions obtained from an investigation can be classified into three categories: "(a) the ones that were well known to start with, (b) the ones that confirm or refute previous 'hunches', (c) the ones that no one ever thought of (and possibly cannot be believed, in some cases)." Some of the findings of the present study fall, perhaps, in the third category and are susceptible to the treatment parenthetically suggested by Moroney. If those who disagree with the findings proceed to verify them by further research, it will be a most gratifying and encouraging response.

APPENDIX

The data for each population group are arranged in the following manner:

(1) name of the caste (C), tribe (T), and community (Cm) measured; (2) locality and/or the name of the state; (3) name of the earlier author with the year of publication; (4) year of the earlier investigation; (5) name of the more recent author with the year of publication; (6) year of the subsequent investigation.

1. Abor (T); Abor Hills, North-East Frontier Agency; Coggin Brown and Kemp 1915; Roy Gupta and Dutta 1966); 1948–1949.
2. Agamudaian (C); Madras; Thurston 1909; 1894; Rakshit (personal communication); 1961–1963.
3. Ahir (C); Uttar Pradesh; Crooke 1896; Mahalanobis, Majumdar, and Rao 1949; 1941.

4. Badaga (T); Nilgiri Hills, Madras; Thurston 1909; 1894; Rakshit (personal communication); 1961–1963.
5. Bagdi (C); West Bengal; Risley 1891; 1886–1888; Raychoudhuri 1952; 1951.
6. Bhantu (C); Uttar Pradesh; Crooke 1896; Mahalanobis, Majumdar, and Rao 1949; 1941.
7. Bhil (T); Rajasthan; Risley 1903; 1901; Weninger 1952; 1939.
8. Bhil (T); Western Vindhya Hills, Madhya Pradesh; Guha 1935; 1931; Bhargava and Kher 1960.
9. Billava (C); South Kanara District, Mysore; Thurston 1909; 1894; Rakshit (personal communication); 1961–1963.
10. Brahman Chitpavan (C); Western Maharashtra; Gupte 1907; 1901; Karve and Dandekar 1951; 1946–1950.
11. Brahman Deshastha (C); Maharashtra; Gupte 1907; 1901; Guha 1935; 1930–1933.
12. Brahman Nagar (C); Gujarat; Gupte 1907; 1901; Guha 1935; 1930–1933.
13. Brahman Nambudiri (C); Kerala; Guha 1935; 1931; Rakshit (personal communication); 1961–1963.
14. Brahman Paschatya Vaidik (C); West Bengal; Chanda 1916; Raychaudhuri 1952.
15. Brahman Shashana (C); Puri District, Orissa; Risley 1903; 1901; Guha 1935; 1930.
16. Brahman Tamil (C); Madurai, Tanjore, and Trichi Districts, Madras; Guha 1935; 1931; Rakshit (personal communication); 1961–1963.
17. Bunt (C); South Kanara District, Mysore; Thurston 1909; 1894; Rakshit (personal communication); 1961–1963.
18. Chakkiliyan (C); Madras; Thurston 1909; 1894; Rakshit (personal communication); 1961–1963.
19. Chamar (C); Uttar Pradesh; Crooke 1896; Mahalanobis, Majumdar, and Rao 1949; 1941.
20. Chenchu (T); Andhra Pradesh; Thurston 1909; 1894; Rakshit (personal communication); 1961–1963.
21. Chero (T); Mirzapur District, Uttar Pradesh; Crooke 1896; Mahalanobis, Majumdar, and Rao 1949; 1941.
22. Cheruman (T); Malabar, Kerala; Risley 1903; 1901; Rakshit (personal communication); 1961–1963.
23. Chetty (C); Madurai, Trichi, and Tanjore Districts, Madras; Guha 1935; 1931; Rakshit (personal communication); 1961–1963.
24. Chhatri (C); North and East Uttar Pradesh; Risley 1891; 1886–1888; Mahalanobis et al. 1949; 1941.
25. Coorg (T); Coorg District, Mysore; Holland 1901; Cipriani 1936; 1935.
26. Dom (C); Uttar Pradesh; Risley 1891; 1886–1888; Mahalanobis, Majumdar, and Rao 1949; 1941.
27. Ho (T); Seraikella, Bihar; Chatterjee and Das 1927; 1925; Ghosh 1965; 1962.
28. Idaiyan (C); Madras; Thurston 1909; 1894; Rakshit (personal communication); 1961–1963.
29. Jangam (C); Mysore; Risley 1903; 1901; Karve 1954; 1949–1950.
30. Kahar (C); Uttar Pradesh; Crooke 1896; Mahalanobis, Majumdar, and Rao 1949; 1941.
31. Khandayat (C); Orissa; Risley 1903; 1901; Karve 1954; 1949–1950.
32. Kharwar (C); Mirzapur District, Uttar Pradesh; Crooke 1896; Mahalanobis, Majumdar, and Rao 1949; 1941.

33. Korku (T); Melghat Forest, Madhya Pradesh; Chattopadhyay 1952; 1938; Basu 1970; 1963.
34. Kurmi (C); South Bihar; Risley 1891; 1886–1888; Mitra (Gupta and Dutta 1966); 1950–1951.
35. Kurmi (C); West and East Uttar Pradesh; Risley 1891; 1886–1888; Mahalanobis, Majumdar, and Rao 1949; 1941.
36. Kuruba (C); Mysore; Thurston 1909; 1894; Rakshit (personal communication); 1961–1963.
37. Ladaki (T); Ladak, Kashmir; Ujfalvy 1884 (Gupta and Dutta 1966); 1881–1882; Chakladar 1921; 1911.
38. Mahar (C); Western Maharashtra; Gupte 1907; 1901; Karve and Dandekar 1951; 1946–1950.
39. Mal Paharia (T); Santal Parganas, Bihar; Risley 1891; 1886–1888; Sarkar 1935–1936.
40. Mappila Muslim (Cm); Malabar, Kerala; Thurston 1909; 1894; Rakshit (personal communication); 1961–1963.
41. Maratha (C); Poona, Maharashtra; Risley 1903; 1901; Karve and Dandekar 1951; 1946–1950.
42. Mukkuvan (C); Kerala; Thurston 1909; 1894; Rakshit (personal communication); 1961–1963.
43. Munda (T); Ranchi District, Bihar; Risley 1891; 1886–1888; Basu 1932–1933.
44. Muslim (Cm); Madras; Thurston 1909; 1894; Rakshit (personal communication); 1961–1963.
45. Muslim (Cm); Uttar Pradesh; Crooke 1896; Mahalanobis, Majumdar, and Rao 1949; 1941.
46. Nayar (C); Malabar, Kerala; Fawcett 1901; Rakshit (personal communication); 1961–1963.
47. Oraon (T); Ranchi District, Bihar; Risley 1891; 1886–1888; Banerjee 1960; 1956.
48. Pallan (C); Tanjore, Trichi, and Madurai Districts, Madras; Thurston 1909; 1894; Rakshit (personal communication); 1961–1963.
49. Palli (C); Madras; Thurston 1909; 1894; Rakshit (personal communication); 1961–1963.
50. Pana (C); Orissa; Risley 1903; 1901; Karve 1954; 1949–1950.
51. Paniyan (T); Wynaad Taluk, Malabar, Kerala; Eickstedt 1934; 1926(?); Das 1955; 1953.
52. Paraiyan (C); Madras; Thurston 1909; 1894; Rakshit (personal communication); 1961–1963.
53. Pod (C); 24 Parganas District, West Bengal; Risley 1891; 1886–1888; Raychaudhuri 1952; 1951.
54. Santal (T); Santal Parganas District, Bihar; Risley 1891; 1886–1888; Biswas 1956; 1931–1934.
55. Shanan (C); Tinnevelly, Kanyakumari, and Ramanathpuram, Madras; Risley 1903; 1901; Rakshit (personal communication); 1961–1963.
56. Syrian Christian (Cm); Travancore, Kerala; Risley 1903; 1901; Rakshit (personal communication); 1961–1963.
57. Tiyer (T); Malabar, Kerala; Thurston 1909; 1894; Eickstedt 1934; 1926(?).
58. Toda (T); Nilgiri Hills, Madras; Thurston 1909; 1894; Eickstedt 1934; 1926(?).
59. Vellala (C); Madras; Thurston 1909; 1894; Rakshit (personal communication); 1961–1963.

60. Vokkaliga (C); Mysore; Thurston 1909; 1894; Rakshit (personal communication); 1961–1963.

REFERENCES

ACHESON, ROY M.
1960 The environment and the growth of children. *Excerpta Medica I, Anatomy* 14:254.
1961 Effects of nutrition and disease on human growth. *Biological Abstracts* 36:2680.

AYKROYD, W. R.
1946 *Nutrition*. Bombay: Oxford University Press.

BAKWIN, H., S. D. MCLAUGHLIN
1964 Secular increase in height. Is the end in sight? *Lancet* 2:1195–1196.

BANERJEE, PAPIA
1960 A comparative study of the Oraon physical characters. *Bulletin of the Anthropological Survey of India* 9:89–100.

BASU, A.
1970 Anthropometry of the Korkus of Melghat Forest. *Man in India* 50:149–156.

BASU, P. C.
1932–1933 Racial affinities of Mundas. *Transactions of the Bose Research Institute* 8:211–247.

BHARGAVA, I., G. A. KHER
1960 An anthropometric study of Central India Bhils of Dhar District of Madhya Pradesh. *Journal of the Anatomical Society of India* 9:14–19.

BISWAS, P. C.
1956 *Santals of the Santal Parganas*. Delhi: Bharatiya Adimjati Sevak Sangha.

BOWLES, G.
1932 *New types of Old Americans at Harvard*. Cambridge: Harvard University Press.

BOYNE, A. W., F. C. AITKEN, I. LEITCH
1957 Secular change in height and weight of British children, including an analysis of measurements of English children in Primary Schools: 1911–1953. *Nutrition Abstracts and Reviews* 27:1–18.

CANTALINI, C.
1967 Infants in the Province of L'Aquila. *Excerpta Medica I, Anatomy* 21:750–751.

CHAKLADAR, H. C.
1921 Translation of the first outlines of a systematic anthropology of Asia. *Anthropological Papers, University of Calcutta* 6. Translated by V. Guiffrida-Ruggeri.

CHANDA, R.
1916 *The Indo-Aryan races*. Rajshahi: Varendra Research Society.

CHANDRASEKHAR, S.
1959 *Infant mortality in India 1901–1955*. London: George Allen and Unwin.

CHATTERJEE, A. N., T. C. DAS
1927 The Hos of Seraikella. *Anthropological Papers, University of Calcutta*, part one, n.s. 1.

CHATTOPADHYAY, K. P.
1952 Korku physical types and racial affinities. *Journal of the Asiatic Society of Bengal: Science* 18:23–24.

CIPRIANI, LIDIO
1936 Su alcuni gruppi umani del Curg. *Archivio per l'Antropologia e la Etnologia* 45:87–124.

COGGIN BROWN, J., S. W. KEMP
1915 "Anthropometrical section", in *Abors and Galongs*. Edited by G. D. Sutherland-Dunber, 88–91. Calcutta: Asiatic Society of Bengal.

CONTERIO, F., L. L. CAVALLI-SFORZA
1967 Selections for quantitative characters in man. *Biological Abstracts* 48:8370.

CROOKE, W.
1896 *The tribes and castes of the North-West Provinces and Oudh*, volume one. Calcutta: Government Press.

DAMON, A.
1968 Secular trend in height and weight within Old American families at Harvard, 1870–1965. *American Journal of Physical Anthropology* 29:45–50.

DAS. S. R.
1955 A somatological study of the Paniyans of Wynaad. *Bulletin of the Department of Anthropology* 4:51–61.

DIXON, W. J., A. M. MOOD
1946 The statistical sign test. *Journal of American Statistical Association* 41:557–566.

DRONAMRAJU, K. R.
1967 Caste and consanguinity in Andhra Pradesh. *Eugenics Quarterly* 14:238–240.

EICKSTEDT, EGON F. VON
1934 *Rassenkunde und rassengeschichte der menschheit.* Stuttgart: Ferdinand Enke.

FAWCETT, F. W.
1901 Anthropology: Nayars of Malabar. *Bulletin of the Madras Government Museum* 3:36–191.

FRIEDLAENDER, J. S.
1971 Secular trend and aging in Bougainville, Territory of New Guinea. *American Journal of Physical Anthropology* 35:278.

FRISANCHO, A. R., S. M. GARN, W. ASCOLI
1970 Childhood retardation resulting in reduction of adult body size due to lesser adolescent skeletal delay. *American Journal of Physical Anthropology* 33:325–336.

FROEHLICH, J. W.
1970 Migration and the plasticity of physique in the Japanese-Americans of Hawaii. *American Journal of Physical Anthropology* 32:429–442.

GALTON, SIR FRANCIS
1886 Family likeness in stature. *Proceedings of the Royal Society* 40:42–73.

GHOSH, G. C.
1965 An anthropometric study of the Hos of Seraikella, Singhbhum. *Man in India* 45:306–310.

GUHA, B. S.
1935 "Racial affinities of the peoples of India," in *Census of India, 1931*,

volume one, part three, A:i–116. Simla: Government of India Press.

GUPTA, P., P.C. DUTTA
1966 *Anthropometry in India (1868–1961)*. Calcutta: Anthropological Survey of India.

GUPTE, B. A.
1907 *Anthropometric data from Bombay*. Ethnographic Survey of India. Calcutta: Government Press.

HAMILTON, BUCHANON
1838 *History, antiquities, topography and statistics of eastern India*, volume two.

HARRISON, G. A.
1968 The biological structure of human populations. *Proceedings of the Royal Anthropological Institute of Great Britain and Ireland for 1967*: 29–36.

HOLLAND, T. H.
1901 The Coorgs and Yeruvas, an ethnological contrast. *Journal of the Asiatic Society of Bengal* 70:59–98.

HUNT, E. E., JR.
1958 Human growth and body form in recent generations. *American Anthropologist* 60:118–131.

IVANOVSKY, A.
1923 Physical modifications of the population of Russia under famine. *American Journal of Physical Anthropology* 4:331–353.

KAPLAN, B. A.
1954 Environment and human plasticity. *American Anthropologist* 56:780–800.

KARVE, I.
1954 Anthropometric measurements in Karnatak and Orissa and a comparison of these two regions with Maharashtra. *Journal of the Anthropological Society of Bombay* n.s. 8:45–75.

KARVE, I., V. M. DANDEKAR.
1951 *Anthropometric measurements of Maharashtra*. Poona: Deccan College.

KENNTNER, G.
1969 Absolute differences and changes of the average stature among the Chileans. *Biological Abstracts* 50:3248.

KIMURA, K.
1967 A consideration of the secular trend in Japanese for height and weight by a graphic method. *American Journal of Physical Anthropology* 27:88–94.

KIMURA, K., C. YAMAZAKI
1961 Secular changes in the physique of Japanese woman medical students. *Journal of Anthropological Society of Nippon* 69(1):46–50.

KROGMAN, W. M.
1941 *Growth of man*. The Hague: W. Junk.

MAHALANOBIS, P. C., D. N. MAJUMDAR, C. R. RAO
1949 Anthropometric survey of the United Provinces, 1941: A statistical study. *Sankhya* 9:90–324.

MALHOTRA, M. S.
1967 "People of India including primitive tribes — A survey on physiological adaptation, physical fitness, and nutrition," in *The biology of human*

adaptability. Edited by P. T. Baker and J. S. Weiner, 329–355. Oxford: Clarendon Press.

MARSHALL, W. A., J. M. TANNER, H. E. LEWIS, MARGARET A. RICHARDSON
1971 Anthropometric measurements of the Tristan Da Cunha Islanders 1962–1968. *Human Biology* 43:112–139.

MILLER, PETER S.
1970 Secular changes among the Western Apache. *American Journal of Physical Anthropology* 33:197–206.

MORONEY, M. J.
1957 *Facts from figures.* Harmondsworth: Penguin.

MUKHERJEE, M.
1969 *National income of India: trends and structure.* Calcutta: Statistical Publishing Society.

OSTER, H.
1967 Has the body length of neonates changed in the past decade? *Excerpta Medica I, Anatomy* 21:487.

RAYCHOUDHURI, T. C.
1952 The racial problem of Bengal. Presidential address, anthropology and archaeology section. *Proceedings of the 39th Indian Science Congress*: 169–188.

RISLEY, SIR HERBERT H.
1891 *The tribes and castes of Bengal*, volumes one and two. Anthropometric measurements. Calcutta: Government Press.
1903 *Census of India, 1901*, part one, volume one. Report of ethnographic appendices. Calcutta: Government Press.
1915 *The people of India.* Calcutta: Thacker, Spink and Company.

SARKAR, S. S.
1935–1936 Racial affinities of the inhabitants of Rajmahal Hills. *Transactions of the Bose Research Institute* 11:123–153.

SCHREIDER, E.
1967 Body height and inbreeding in France. *American Journal of Physical Anthropology* 26:1–3.

SUZUKI, S.
1959 "Basal metabolism in the Japanese population," in *World review of nutrition and dietetics*. Edited by G. H. Bourne, 102–124. London: Pitman Medical Publishing Co.

TAKAHASI, E.
1966 Growth and environmental factors in Japan. *Human Biology* 38:122–130.

TANNER, J. M.
1971 Growing up faster. *Medical Times, Bombay* 1:4–8.

THURSTON, EDGAR
1909 *Castes and tribes of southern India*, volume one. Madras: Government Press.

UDJUS, L. G.
1965 *Anthropometrical changes in Norwegian men in the twentieth century.* Oslo: Norwegian Research Council.

VLASTOVSKY, V. G.
1966 The secular trend in the growth and development of children and young persons in the Soviet Union. *Human Biology* 38:219–230.

WENINGER, M.
1952 Physisch-anthropologische untersuchungen an einigen stammen Zentralindiens. *Acta Ethnologica et Linguistica*. 3:1–156.
WOLANSKI, N., E. JAROSZ, M. PYZUK
1970 Heterosis in man: growth in offspring and distance between parents' birthplaces. *Social Biology* 17:1–16.

A Study of Height, Weight, and Arm Girth Measurements of Punjabi Males

RAGHBIR SINGH

Information on even the simplest nutritional anthropometric measurements, such as height, weight, and upper arm girth, of adults of North Indian populations is scanty. Earlier somatometric investigations on adults in North India were mainly concerned with ethnic studies (Karve 1948; Karve and Dandekar 1951; Mahalanobis and Bose 1941; Sen and Gupta 1966; Majumdar and Rao 1958), and more emphasis was placed on various head and face measurements or those measurements that were useful for racial classification. Body weight and arm girths were often ignored.

Upper arm girth is gaining importance in nutritional anthropometric investigations. It is easily practicable and shows high positive correlations with muscle thickness as well as subcutaneous fat measurements (Singh 1967a, 1967b). Since arm girth is a valuable nutritional anthropometric parameter, its relationships with body weight and weight/height index would also be of great significance in nutritional studies. The object of the present study, therefore, is twofold: (1) to collect information about the means and standard deviations of somatometric measurements, including height, weight, and arm girth measurements of adult, white-collar Punjabi Hindu Khatris (a selected occupation, caste, and population group in Delhi); and (2) to indicate the usefulness of the arm girth measurement as a nutritional index by studying its relationships with body weight and weight/height index.

MATERIAL AND METHODS

Two hundred and twenty-three Punjabi-speaking Hindus between the ages of twenty and fifty-five from one endogamous caste named Khatri constitute the sample of the present study. They were all white-collar clerks of the Ministry of Railways, New Delhi and their monthly incomes ranged between Rs. 200 to Rs. 600 per month.

Stature, body weight, upper arm girth, wrist girth (maximum), and sitting height measurements were obtained for each subject. The measurement of lower limbs was obtained by subtracting the sitting height measurement from stature. Body weight was taken to the nearest pound with appropriate reduction for clothing. All other measurements were recorded to the nearest 0.1 centimeter. Martin's (1928) techniques for measurement were followed. Arm girth measurements were taken with a flexible steel tape and at a level mid-way between the tip of the acromial process of the scapula and olecranon process of the ulna while the arm was hanging down freely, as suggested by the Committee on Nutritional Anthropometry (1956).

RESULTS

Somatometric Measurements

In Table 1, arm girth and body weight measurements showed high values for coefficient of variation compared to the other measurements.

When the somatometric data were divided (Table 2) into three broad age groups of twenty to twenty-nine, thirty to thirty-nine, and forty to fifty-five years, there were no significant differences (Table 3) between the means of age groups except in arm girth measurements, which were significantly higher in the thirty to thirty-nine year group than in the twenty to twenty-nine year group. But there was no significant difference between the thirty to thirty-nine and the forty to fifty-five year age groups even for the arm girth measurements.

Correlations

High positive correlations existed between arm girth and body weight. The values of the correlation coefficient "*r*" in the three age groups — twenty to twenty-nine, thirty to thirty-nine, and forty to fifty-five years — were 0.918, 0.883, and 0.900, respectively. The value for the whole sample (all observations) was 0.901. The correlation was linear.

High positive correlations existed also between arm girth and weight/height index. The values of "*r*" in the three age groups (same order as above) were 0.946, 0.905, and 0.924, respectively. The value for the whole sample was 0.925. The relationship between the two parameters was linear.

Table 1. Mean values and statistical constants of somatometric measurements of 223 Punjabi adult males

Measurements	Mean s.e.	s.d. s.e.	coeff. v. s.e.	Range
Stature (cm)	167.95±0.39	5.89±0.28	3.51±0.17	153.8 -183.3
Sitting height (cm)	87.59±0.19	2.90±0.14	3.31±0.16	76.8 - 96.0
Height–sitting height (cm)	80.36±0.27	4.05±0.19	5.04±0.24	70.1 - 90.0
Body weight (kg)	60.70±0.70	10.51±0.50	7.85±0.37	39.91- 90.72
Arm girth (cm)	25.39±0.19	2.78±0.13	10.95±0.52	18.9 - 34.2
Wrist girth (cm)	15.91±0.06	0.88±0.04	5.53±0.26	13.9 - 18.2

Key: s.e. = standard error; s.d. = standard deviation; coeff. v. = coefficient of variation

Table 2. Mean values and standard deviations of somatometric measurements in three age groups of Punjabi adult males

	Age groups					
	20–29 years (n = 74)		30–39 years (n = 88)		40–55 years (n = 61)	
Measurements	Mean	s.d.	Mean	s.d.	Mean	s.d.
Stature (cm)	168.10	6.10	167.86	5.77	167.88	5.90
Sitting height (cm)	87.76	3.00	87.64	2.76	87.29	2.98
Height–sitting height (cm)	80.34	4.10	80.22	4.19	80.59	3.81
Body weight (kg)	58.64	10.44	61.76	10.28	61.69	10.72
Arm girth (cm)	24.78	2.78	25.79	2.72	25.54	2.77
Wrist girth (cm)	15.76	0.93	15.99	0.76	15.98	0.98

Table 3. Values of t for somatometric measurements between three age groups of Punjabi adult males

Measurements	Values of t between age groups					
	(20–29 years) and (30–39 years)		(20–29 years) and (40–55 years)		(30–39 years) and (40–55 years)	
	t	Significance	t	Significance	t	Significance
Height	0.255	n.s.	0.212	n.s.	0.020	n.s.
Sitting height	0.264	n.s.	0.910	n.s.	0.732	n.s.
Height–sitting height	0.182	n.s.	0.364	n.s.	0.556	n.s.
Body weight	1.909	n.s.	1.669	n.s.	0.361	n.s.
Arm girth	2.339	significant	1.603	n.s.	0.550	n.s.
Wrist girth	1.691	n.s.	1.292	n.s.	0.655	n.s.

Key: n.s. = non-significant.

Significant positive correlations were also observed between height and weight measurements. The values of "r" in the three age groups (same order) were 0.438, 0.420, and 0.494, respectively. The value of "r" for the whole sample was 0.440.

DISCUSSION

Body weight tended to increase up to about forty years of age, but not between that and fifty-five. Arm girth measurements were significantly higher in the thirty to thirty-nine age group than in the younger age group, and this may be due to additional deposition of subcutaneous tissue in the older group.

The finding of positive correlation between height and weight is in line with expectations in view of their interrelatedness.

High positive correlation between arm girth and body weight in the present sample of adults is in keeping with the author's earlier observations (1970) in adolescents.

Extremely high positive correlations between arm girth and body weight may be explained on the basis of a battery of correlations and intercorrelations reported in the literature. Arm girth shows a high positive correlation with muscle width of the upper arm (Singh 1967a) as well as with subcutaneous fat measurements of the upper arm (Singh 1967b), and muscle thickness of different parts of the body are interrelated (Tanner 1962, 1964, 1965). Subcutaneous fat measurements at different sites of the body are also highly interrelated (Garn 1954; Tanner 1962, 1964, 1965; Singh 1964). Further, it has also been reported that muscle width as well as subcutaneous fat measurements are related to body weight (Singh 1967a). Arm girth shows a high correlation with muscle thickness as well as with subcutaneous fat of the upper arm. And muscle and subcutaneous tissue at different sites of the body are correlated with corresponding tissues at other parts of the body. Also, since muscle and subcutaneous fat both are related to body weight, the high correlations between arm girth and body weight in the present study may be due to the above-mentioned correlations and inter-correlations. It may also be mentioned here that high correlations between arm girth and body weight may also be due to other unkown causes which can be identified only through further research.

Body weight is a well-known nutritional anthropometric parameter. High positive correlation between arm girth and body weight indicates that arm girth may be considered an important nutritional anthropometric measurement. But it must be emphasized here that both body weight and arm girth measurements are valuable nutritional anthropometric parameters only if the subjects measured are free from oedema, a clinical sign of acute protein calorie malnutrition. The author stressed earlier (1963) that in any anthropometric investigations, subjects must be free from nutritional deficiency signs or nutritional somatoscopic traits.

REFERENCES

COMMITTEE ON NUTRITIONAL ANTHROPOMETRY
1956 Food and Nutrition Board, National Research Council recommendations concerning body measurements for the characterization of nutritional status. *Human Biology* 26:115–123.

GARN, S. M.
1954 Fat patterning and fat intercorrelations in the adult male. *Human Biology* 26:59–69.

KARVE, I.
1948 *Anthropometric measurements of the Marathas.* Deccan College Monograph 2. Poona.

KARVE, I., V. M. DANDEKAR
1951 *Anthropometric measurements of Maharashtra.* Deccan College Monograph 8. Poona.

MAHALANOBIS, P. C., C. BOSE
1941 Correlation between anthropometric characters in some Bengal castes and tribes. *Sankhya* 5:249–260.

MAJUMDAR, D. N., C. R. RAO
1958 Bengal anthropometric survey, 1945: statistical study. *Sankhya* 18: 203–408.

MARTIN, R.
1928 *Lehrbuch der Anthropologie.* Jena: G. Fischer Verlag.

SEN, D. K., P. GUPTA
1966 Physical characteristics of rural Kashmiri Muslims. *Journal of Indian Anthropology* 1:81–88.

SINGH, RAGHBIR
1963 A short note on nutritional somatoscopy. *Anthropologist* 10:45–48.
1964 "Nutritional somatometric study of adult males with special reference to skinfold measurements and inter-correlations," in *Aspects of Indian anthropology*. Edited by P. C. Biswas and I. P. Singh, 173–184. Delhi: Department of Anthropology, University of Delhi.
1967a A study of fat, muscle, bone and comparison between roentgenogrammetric and somatometric methods for assessing subcutaneous fat and muscle thickness in the upper arm of adult males. *Zeitschrift für Morphologie und Anthropologie* 58:308–317.
1967b Relationship of arm girth measurements with roentgenogrammetric and caliper measurements of subcutaneous fat at triceps and biceps in adult male volunteers. *Zeitschrift für Morphologie und Anthropologie* 59:85–91.
1970 Relationship of arm girth measurements with body weight and weight/ height index in Punjabi boys aged eleven to eighteen years. *Acta Medica Auxologica* 2:55–58

TANNER, J. M.
1962 *Growth of adolescence.* Oxford: Blackwell Scientific Publications.

1964 "Human growth and constitution," in *Human biology*. Edited by G. A. Harrison, J. S. Weiner, J. M. Tanner and N. A. Barnicot, 299–397. Oxford: Oxford University Press.
1965 "Radiographic study of body composition in children and adults. Symposia of the society for the study of human biology," in *Human body composition*, volume six, 211–236. New York: Pergamon.

Nutrition and Pregnancy

JOHN R. K. ROBSON

Restrictions or prohibitions on dietary components during pregnancy are well known. The most appropriate diet during this important period is still largely a matter of conjecture (Committee on Maternal Nutrition 1970). Perfectly normal, fully developed, healthy infants have been born to mothers consuming a wide variety of diets during pregnancy. Despite this, there is evidence that physical growth and development are adversely affected by inadequate nutrition. A grossly inadequate maternal diet has been associated with infertility. Lesser degrees of nutritional deficiency have been associated with abortion, prematurity, and low birth weight. Fatal abnormality has been attributed to specific nutritional deficiencies (Smith 1947; Antonov 1947; Osofsky 1969).

In animals, there is evidence that deprivation of nutrients results in a reduction of the number of cells in fetal organs including the brain (Winick 1968). The timing of the deficiency is important. Thus, malnutrition imposed on animals and humans when the cells are increasing in size rather than number results in a reduction in the cell mass (Winick and Rosso 1969), although recovery is possible on refeeding (Winick and Noble 1966). Anatomical changes and reduction of cell numbers do not necessarily mean that function is impaired, but reduction in brain lipids and enzymes (Chase et al. 1971) suggests such dysfunction may occur.

In human infants, the neurons of the nervous system are organized very early in pregnancy, and by the end of the sixth week following conception, reflex function has been established. Development and differentiation of the various parts of the brain, including the myelin sheath of nerves and supporting tissue, takes place from the fourth month of pregnancy through the second year of postnatal life. Thus, dietary deprivation, both early in pregnancy, when the foundation of the nervous system is being laid, and later, when the higher centers are being formed, may have deleterious effects on the nervous system. These may include disturbance in functions such as impaired learning ability. Whereas a group may be physically handicapped if a large proportion of its members have received an inadequate diet, these disadvantages will be compounded if they also suffer from impaired learning ability. Thus, the successful development of a society must begin with adequate maternal nutrition.

Unfortunately, little is known of the nutritional status of mothers during pregnancy or of the effect of inadequate nutrition on her offspring and, ultimately, on the society as a whole. Field studies are needed to evaluate the effects of nutrition on reproductive performance. Studies should include: observations of the mother before, during, and after pregnancy; descriptions of her food habits; measurements of intake and the utilization of nutrients; and an assessment of the nutritive value of the foods.

MEASUREMENTS OF REPRODUCTIVE PERFORMANCE

Indices of effective reproductive performance might include the prevalence of amenorrhea not associated with pregnancy, or infrequent scanty menses. This would be especially significant between the ages of twenty and thirty-five years of age. Age of menarche and marriage are important, too. In pregnant women, an obstetrical history including details of prior fetal loss through unintentional abortion should be collected. Successful and unsuccessful attempts at intentional abortion should also be noted. The description and recording of fetal abnormalities, physical deformities, and development anomalies in the infant are important as they may be related to maternal nutrition. Birth order and birth weights should be ascertained and correlated to duration of gestation. Body length, head, arm and chest circumference and the thickness of subcutaneous fat should be measured (Jelliffe 1966). Signs of intrauterine growth failure should be noted. These include decreased weight for length and peeling and wrinkling of the skin. Where sophisticated facilities for biochemical studies exist, diagnosis of fetal malnutrition may be made from an examination of the cord blood (*Chemical diagnosis* . . . 1971).

OBSERVATIONS OF THE MOTHER

Early fetal development also depends on the nutritional status of the mother before she becomes pregnant. Therefore, the health of the mother and her diet prior to pregnancy should be reported. Debilitating diseases, such as malaria, severe upper respiratory tract infections, intestinal parasites, especially hookworm, may all influence the reproductive performance. Body weight, the thickness of the subcutaneous fat over the triceps and subscapular regions, and careful records of physical activity may be used to evaluate energy balance. If the mother expends more energy than her food intake provides, there is likely to be interference with proper fetal growth; this is especially important at adolescence. Thus, estimates of energy intake from the food intake, and energy output from physical activity and increases or decreases in subcutaneous fat, and body weight during and after pregnancy (Taggart et al. 1967) will indicate whether the mother is maintaining her energy balance. If possible, measurements should be made of blood pressure, pulse rates, and biochemical indices including serum transferrin, hemoglobin, and changes in hematocrit. All of these measurements could be made by anthropologists, provided they were given the necessary training. Such data would provide valuable information on the circulatory and hemapoeitic system on which fetal respiration and nutrient transfer depends.

FOOD HABITS

Quantities of foods consumed, and condiments which may have an important role in protein sparing (Townsend et al. 1973) and in providing trace elements and vitamins, should be carefully measured. The method of selection of the food from the plant in the field or in the market, as well as the method of preparation and cooking should be described in detail. Analysis of the foods in the raw state and as consumed is vital to the understanding of the nutritive value of diets, and overcomes many of the inadequacies of Food Composition Tables (Robson et al. 1972). In addition to recording food restrictions and prohibitions, the mother and relatives should be questioned on the diet in pregnancy. Supplemental foods may be reserved specially for the mother who eats them "to promote the prenatal growth of the baby," to ensure that the child will be "intelligent," or to reduce "the risk of obstetrical difficulties or physical abnormality" (Ferro–Luzzi. i.p.).

In evaluating the data, note should be taken of other nutrients that may affect the specific nutrient being measured. Thus, protein adequacy can not be estimated without taking into account energy intake. Similarly, vitamin A levels in the blood are dependent on protein and thiamine needs on the carbohydrate source of energy. Consideration of a single nutrient in isolation is likely to overlook important side effects resulting from synergism or antagonism between nutrients or even the food supplying those nutrients.

Anthropologists should be able to make significant contributions to nutritional knowledge, and collaboration between researchers in nutrition and anthropology should provide much needed information on maternal nutrition and its influence on development.

REFERENCES

ANTONOV, A. M.
1974 Children born during the siege of Leningrad in 1942. *Journal of Pediatrics* 30:250.

CHASE, H. P., C. S. DABIERE, N. NOREEN WELCH, D. O'BRIEN
1971 Intrauterine undernutrition and brain development. *Pediatrics* 47:491.

Chemical diagnosis . . .
1971 Chemical diagnosis of fetal malnutrition. *Nutrition Review* 29:160.

COMMITTEE ON MATERNAL NUTRITION
1970 Maternal nutrition and the course of pregnancy. Washington, D.C.: National Academy of Sciences.

FERRO-LUZZI, G. E.
i.p. Food avoidances of pregnant women in Tamilnad. *Ecology of Food and Nutrition.*

JELLIFFE, D. B.
1966 The assessment of nutritional status of the community. World Health Organization, Monograph 53. Geneva: WHO.

OSOFSKY, H. J.
1969 Antenatal malnutrition, its relationship to subsequent infant and child development. *American Journal of Obstetrics and Gynecology* 105:1150.

ROBSON, J. R. K., F. A. LARKIN, A. M. SANDRETTO, G. TADAYYON
1972 Malnutrition — its causation and control. New York: Gordon and Breach.

SMITH, C. A.
1947 Effects of maternal undernutrition upon newborn infants in Holland (1944–1945). *Journal of Pediatrics* 30:229.

TAGGART, N. R., R. M. HOLLIDAY. W. Z. BILLEWICZ, F. E. HYTTEN, A. M. THOMSON
1967 Changes in skinfolds during pregnancy. *British Journal of Nutrition* 21:439.

TOWNSEND, P. K., J. E. KONLANDE, SHU-CHUNG LIAO
1973 Nutritive contributions of sago ash used as a native salt in Papua, New Guinea. *Ecology of Food and Nutrition* 3:65.

WINICK, M.
1968 Retardation in cell division in organs of fetal rats after severe maternal undernutrition. International Congress of Pediatrics, Mexico City.

WINICK, M., A. NOBLE
1966 Cellular responses in rats during malnutrition at varying ages. *Journal of Nutrition* 89:300.

WINICK, M., P. ROSSO
1969 Head circumference and cellular growth of the brain in normal and marasmic infants. *Journal of Pediatrics* 74:774.

The Age at Menarche in Venezuelan Students Residing in Metropolitan Caracas

FRITZI KOHN DE BRIEF

OBJECTIVES OF THE INVESTIGATION AND DEFINITION OF THE PROBLEM

This investigation is concerned with the determination of the average age of menarche and the possible environmental influences on menarche in young Venezuelan students residing in the metropolitan Caracas, as well as their weight and height.

This subject is of interest because of the absence of anthropological research on this topic in Venezuela.[1] The results of this study will help determine the level of sexual maturity of the local female school population and contribute data for comparative genetic studies of populations.

It has been shown that the occurrence of menarche marks the culmination of hormonal changes leading to physical, sexual, and psychological maturity in the female (Tanner 1962). Menarche (the appearance of the first menstrual period) is not an independent phenomenon but is linked to environmental factors. The most important environmental agents are

I wish to express my deep appreciation to the persons and institutions which contributed to the development of this work: to Gisela González Orihuela, sociologist, researcher on the Guidance Commission for Medical Education — Faculty of Medicine; Luis Montero, statistician, Professor in the Schools of Statistics and Social Work; Lydia Abramovici and Beatriz Wolferman, students of economics and sociology at the Central University of Venezuela; and to the teachers and students of the various institutes. I am especially grateful to Dr. Adelaida de Díaz Ungría, anthropologist and head of the Physical Anthropology Section of the Institute for Economic and Social Research, for her continuous and excellent collaboration during the investigation.

[1] Few investigations have been carried out in Venezuela on this topic, and the majority of these have been based on data obtained from reviewing medical files of patients who have come to welfare centres in the country. Figures for Venezuela have been given by Doctors Joel Valencia Parpacén in 1936, Gabriel Trompiz in 1937, Ceferino Alegría in 1942 (see Bergher and Shelly 1965), as well as Bergher and Shelly themselves (1965).

socioeconomic status and nutrition (Levine 1953; Fluchmann 1958; Burell and Tanner 1961; Oettlé and Higginson 1961; Valsik 1965; Bojlen and Bentzon 1968). Studies carried out with regard to nutritional status show that a high-protein diet accelerates menarche, while a diet low in proteins and based on carbohydrates delays it.

Quality of the diet is known to be related to a family's economic status. It is also known that public schools recruit their students from more humble socioeconomic strata than private schools and that education in Venezuela is selective.[2] Thus, if the nutritional factor is taken into consideration, menarche should occur on the average later in students at public schools and their secondary branches (see below) than in girls who attend private schools.

MATERIALS AND METHODS

This study is based on a sample of 1,260 female students, attending institutions located in the Federal District (Libertador y Vargas Department) and the Sucre District of Miranda State in Venezuela. Of this sample, the majority are mestizo; 94 percent are natives of Venezuela; 86 percent are of urban origin; and 80 percent were born in Caracas. The size of the sample was determined by the total female student population attending institutions of primary and middle education (primary schools starting from the third grade), with an interest in the possibility of continuing to work with the sample chosen. We proceeded in this manner because no statistics on the ages of female students in Caracas were available.

The sample was chosen to reflect socioeconomic stratification by examining two criteria: type of education and type of institution. Type of education refers to primary and middle education, with its secondary branches — normal and industrial, crafts and technical. Type of institution refers to public (or "official") institutions and private ones. The sample has been divided into both categories for each stratum.

Institutions were chosen systematically in ratios proportional to their total enrollment, as well as the enrollment within each school. The starting point for selection was a listing of all pupils from the third grade on, and the proportion of that number in the institutions chosen.

In this manner, sixty-three institutions were selected for investigation: forty schools of primary education (twenty-eight public and twelve private) and twenty-three institutions of middle education (fourteen secondary, including seven public and seven private; and nine for industrial, crafts, technical, and normal education, including six public and three

[2] Those students with the most economic means continue middle education studies, especially in the secondary branch (Crema 1970).

private). Eight hundred primary and 460 middle students were interviewed.

The students were interviewed individually in private by three trained interviewers in the period between mid-January and mid-May 1969. Their ages ranged from eight to 23.6 years. These age limits were used because of the possibility of a great range of variation in the age at menarche.[3]

The survey asked for the interviewee's date and place of birth,[4] whether or not she had menstruated,[5] and her recollection of her age at its onset. The students did not show any embarrassment in answering these questions.

The average age at menarche was calculated by the statistical technique known as probit[6] (Finney 1962) and the retrospective method, or that of recalled age (Livson and McNeill 1962). The probit technique is based on the proportion of girls in the different age groups who are menstruating and is determined by using two questions: actual age and whether or not menstrual periods had started. The method used is one now being employed by doctors and anthropologists (Wilson and Sutherland 1960; Burrell, Healy, and Tanner 1961; Valsik 1965; Milicer and Sczotka 1966; Milicer 1968; Shakir 1971; Frisch and Revelle 1971).

The state of nutrition was evaluated by the twenty-four-hour-recall method with subsequent calculation of dietary adequacy (Krehl 1964), and a series of anthropometric measurements was taken. The Table of Nutritional Requirements,[7] as revised in 1963 by the Department of Food and Nutrition (Goldsmith 1964), was used to calculate nutritional intake.

ANALYSIS AND DISCUSSION

Determination of Age at Menarche

Table 1 shows the various age groups and the number and proportion of girls for whom menarche had or had not occurred. The average age at menarche, estimated by means of probit analysis, is 12.7 ± 0.07 years, with a standard deviation of 1.36.

The line of regression (probit) is given by the equation: $Y = 5.29 + 0.73\,(X - 13.10)$, where Y is the probit of the proportion having experi-

[3] In a study of 1,000 adolescent mothers younger than fifteen in the Maternidad Concepción Palacios in Caracas, carried out by R. Castellanos, O. Aguero, G. Franco, and L. F. Moreno, occurrence of the menarche at age seven is mentioned (personal communication).

[4] These figures were verified in the registration files in doubtful cases.

[5] Date of reference: May, 1969.

[6] In the terminology of statistics, probit is a measurement of probability which is based on deviation from the average.

[7] Food and Nutrition Board, National Academy of Sciences, National Research Council.

Table 1. Frequency of menarche by age group for students from the metropolitan area of Caracas (1969)

Age groups	Number of girls	Menarche has occurred		Menarche has not occurred	
		Number	Percentage	Number	Percentage
8.1– 9.99	52	0	0.0	52	100.0
10.0–10.5	35	1	2.9	34	97.1
10.6–10.99	63	4	6.3	59	93.7
11.0–11.5	79	8	10.1	71	89.9
11.6–11.99	116	21	18.1	95	81.9
12.0–12.5	115	44	38.3	71	61.7
12.6–12.99	98	58	59.2	40	40.8
13.0–13.5	103	69	67.0	34	33.0
13.6–13.99	75	63	84.0	12	16.0
14.0–14.5	85	77	90.6	8	9.4
14.6–14.99	80	75	93.8	5	6.3
15.0–15.5	83	80	96.4	3	3.6
15.6–23.99	276	275	99.6	1	0.4
Total	1260	775	62.0	485	38.0

enced menarche and X is the age in years. A probit value of 5.0 corresponds to a cumulative frequency of 50 percent, i.e., the mean age of menarche. A probit value of 3.0 corresponds to a cumulative frequency of 2.27 percent, and a probit value of 6.0 corresponds to a cumulative frequency of 84.13 percent.

Table 2 illustrates the arithmetic valuations obtained for the total sample and the different types of institutes.

The chi-square values reveal that the sample from private schools is more heterogeneous than that from public ones, and that the difference in average age at menarche is not significant. However, it should be noted that the sample from private schools is smaller.

It was not possible to evaluate the probit in institutions of technical education because the distribution is not normal. This is possibly because the sample is quite heterogeneous. It should be pointed out that in the age group from 12.6 to 12.99 years, menarche had occurred in all the students, while in the following group a quarter had still not menstruated (Table 3). This makes it impossible to plot the empirical line of regression, thereby making it impossible also to calculate the arithmetic probit. The sample of group 6 is homogeneous, and the statistics are quite similar to those of the total sample. This is an indication that group 7 has little influence in the total probit. The chi-square for group 6 is closer to its expected value (ten) than that of the total sample. The average menarche is one month older than that of the total sample. Ages at menarche in relation to the type and level of education are shown in Table 4.

Table 2. Results by probit analysis of the age at menarche in students from the metropolitan area of Caracas for the total sample and for various socioeducational groups (1969)

Material	No.	Average±s.e.	s.d.	Line of regression (Probit)	χ^2	d.f.
1 Total sample	775	12.70±0.07	1.36	$Y=5.29+0.73\ (X-13.10)$	6.98	10
2 Public schools	512	12.71±0.83	1.36	$Y=5.20+0.73\ (X-12.99)$	5.42	9
3 Private schools	263	12.64±0.12	1.52	$Y=5.12+0.65\ (X-12.83)$	18.78	10
4 Primary education	330	12.94±0.08	1.44	$Y=4.74+0.68\ (X-12.50)$	10.83	10
5 Secondary education	267	12.31±0.43	1.31	$Y=6.26+0.76\ (X-13.97)$	2.64	7
6 General education	597	12.80±0.06	1.37	$Y=5.02+0.72\ (X-12.83)$	10.94	10
7 Technical education	178	–	–	–	–	–

Key: General education = primary and secondary
Technical education = industrial, trade, technical, and normal
s.d. = standard deviation; d.f. = degrees of freedom

Table 3. Distribution by type of education related to age at menarche (expressed in absolute and relative numbers). Students from the metropolitan area of Caracas (1969)

		Primary				Secondary				Technical			
		Yes		No		Yes		No		Yes		No	
Age groups	Average age	No.	%	No.	%	No.	%	No.	%	No.	%	No.	%
8.0– 9.99	9.00	0	0.0	52	100.0	0	0.0	0	0.0	0	0.0	0	0.0
10.0–11.5	10.75	13	7.3	164	92.7	0	0.0	0	0.0	0	0.0	0	0.0
11.6–11.99	11.75	21	18.2	94	81.8	0	0.0	1	0.0	0	0.0	0	0.0
12.0–12.5	12.25	43	38.4	69	61.6	1	33.3	2	66.7	0	0.0	0	0.0
12.6–12.99	12.75	50	56.8	38	43.2	4	66.7	2	33.3	4	100.0	0	0.0
13.0–13.5	13.25	56	65.9	29	34.1	10	71.5	4	28.5	3	75.0	1	25.0
13.6–13.99	13.75	36	76.8	11	23.2	17	94.9	1	5.1	10	100.0	0	0.0
14.0–14.5	14.25	43	87.7	6	12.3	29	96.7	1	3.3	5	83.4	1	16.6
14.6–14.99	14.75	22	84.8	4	15.2	37	97.3	1	2.7	16	100.0	0	0.0
15.0–15.5	15.25	21	91.8	2	8.2	42	97.7	1	2.3	17	100.0	0	0.0
15.6–23.99	19.83	25	96.7	1	3.3	127	100.0	0	0.0	123	100.0	0	0.0
Total		330	26.2	470	37.3	267	21.2	13	1.0	178	14.1	2	0.2

Key: Technical education = industrial, trade, technical, and normal

These results are comparable to those obtained in other countries in recent decades. For example, the average age at menarche in the United States in the years 1940 to 1955 was between 12.5 and 13.0 years of age, depending on socioeconomic status and geographic location (Tanner 1966). Younger averages have been reported by Laska-Mierze Jeroska in 1970 (Carfagna et al. 1971) for Cubans: Negroids (12.36 ± 0.07) and Caucasians (12.41 ± 0.03).

Recent European data, tabulated by Tanner (1966), yield average ages ranging from 12.8 years for girls in Budapest to 13.1 years for girls in Copenhagen. The earliest menarcheal age in Europe, 12.49 years, occurs in students from Naples, Italy (Carfagna et al. 1971).

When socioeconomic conditions are considered, girls from the lowest stratum in Naples have an average menarcheal age later than those from the highest stratum (12.99 versus 12.23 years). The menarcheal age nearest that observed in Naples is the one estimated by Zukowski, Kmietowics-Zukorska, and Gruska (1964) in the Polish city of Wroclaw (12.6 years) and that estimated by Thoma (Tanner 1966) for girls from one of the residential districts of Budapest. Investigations carried out in cities in Great Britain gave the following results: Edinburgh in 1952, 13.4 years; Bristol in 1956, 13.2 years; and London in 1959, 13.06 years (Tanner 1962).

In Africa, Hautvast (1971) gives the figures from the Nyakyusa and Kisi tribes of Tanzania as 14.93 and 15.5 years, respectively. In Baghdad average menarcheal age ranges from 13.59 to 13.96 years and corresponds to socioeconomic factors, mainly nutrition (Shakir 1971: Table 1, Graph 1).

RECALL METHOD. All the girls in the present study over 15.8 years had menstruated and indicated that they recalled the age at menarche. It should also be mentioned that eighteen of these girls claimed that menstruation had begun before the lower limit determined in the probit analysis. It is possible that this discrepancy arises from errors of recollection. The average age of menarche calculated using these data is 12.33 ± 0.04 years with a standard deviation of 1.24. That this average is 0.37 years younger than the age calculated by probit analysis may be due to a tendency to brag about early sexual maturity. The average menarcheal age for this sample is 0.21 years younger than that calculated for the Andrés Bello and Aplicación secondary schools (Bergher and Shelly 1965).

Table 5 illustrates the statistics obtained for the total sample arranged by type of institution. For comparative purposes, replies from girls over fifteen were separated in recognition of the possibility that their replies would be more accurate. Table 6 shows average age at menarche, standard deviations, and variances for the over-fifteen group. From the

Table 4. Distribution by type of school and education related to age at menarche (expressed in Caracas (1969)

Age	Primary								Secondary			
	Public				Private				Public			
	Yes		No		Yes		No		Yes		No	
	No.	%	No.	%	No.	%	No.	%	No.	%	No.	%
8.0– 9.99	0	0.0	20	38.4	0	0.0	32	61.6	0	0.0	0	0.0
10.0–11.5	8	61.6	91	55.5	5	38.4	73	44.5	0	0.0	0	0.0
11.6–11.99	12	57.1	68	72.3	9	42.9	26	27.7	0	0.0	1	100.0
12.0–12.5	33	76.7	55	79.7	10	23.3	14	20.3	0	0.0	1	50.0
12.6–12.99	39	78.0	25	66.5	11	22.0	13	33.5	3	75.0	1	50.0
13.0–13.5	41	73.2	26	89.7	15	26.8	3	10.3	3	30.0	2	50.0
13.6–13.99	30	83.3	11	100.0	6	16.7	0	0.0	10	58.8	0	0.0
14.0–14.5	35	81.4	2	33.3	8	18.6	4	66.7	11	37.9	0	0.0
14.6–14.99	18	81.8	3	75.0	4	18.2	1	25.0	22	59.4	0	0.0
15.0–15.5	21	100.0	2	100.0	0	0.0	0	0.0	19	45.2	0	0.0
15.6–23.99	20	80.0	0	0.0	5	20.0	1	100.0	67	52.8	0	0.0
Total	257	45.9	303	54.1	73	30.4	167	69.6	135	96.4	5	3.6

statistics of Tables 5 and 6 we see that menarche occurs earlier in students who attend institutions of secondary education than in those attending institutions of industrial, craft, technical, and normal education. This is possibly due to the better living conditions of the former group. It will be noted that the mean ages calculated from the estimates of the over-fifteen group are later than those of the total sample. This may indicate some bragging on the part of the younger girls, as mentioned above.

Table 5. Results by retrospective analysis of age at menarche in students from the metropolitan area of Caracas for the total sample and in various socioeducational groups (1969)

Material	Number	Average±s.e.	s.d.	Asymmetry	Variance
1 Total sample	775	12.33±0.04	1.24	0.09	1.53
2 Public schools	512	12.39±0.05	1.21	0.16	1.48
3 Private schools	263	12.21±0.07	1.27	0.15	1.61
4 Primary education	330	12.11±0.06	1.19	0.32	1.41
5 Secondary education	267	12.40±0.07	1.20	0.16	1.44
6 General education	597	12.24±0.04	1.20	0.20	1.46
7 Technical education	178	12.65±0.09	1.29	−0.15	1.66

Key: General education = primary and secondary
Technical education = industrial, trade, technical, and normal

absolute and relative numbers). Students from the metropolitan area of

				Technical								
Private				Public				Private				
Yes		No		Yes		No		Yes		No		Age
No.	%	No.	%	No.	%	No.	%	No.	%	No.	%	
0	0.0	0	0.0	0	0.0	0	0.0	0	0.0	0	0.0	8.0– 9.99
0	0.0	0	0.0	0	0.0	0	0.0	0	0.0	0	0.0	10.0–11.5
0	0.0	0	0.0	0	0.0	0	0.0	0	0.0	0	0.0	11.6–11.99
1	100.0	1	50.0	0	0.0	0	0.0	0	0.0	0	0.0	12.0–12.5
1	25.0	1	50.0	2	50.0	0	0.0	2	50.0	0	0.0	12.6–12.99
7	70.0	2	50.0	1	33.3	0	0.0	2	66.7	1	100.0	13.0–13.5
7	41.2	1	100.0	2	20.0	0	0.0	8	80.0	0	0.0	13.6–13.99
18	62.1	1	100.0	3	60.0	0	0.0	2	40.0	1	100.0	14.0–14.5
15	40.6	1	100.0	8	50.0	0	0.0	8	50.0	0	0.0	14.6–14.99
23	54.8	1	100.0	9	52.9	0	0.0	8	47.1	0	0.0	15.0–15.5
60	47.2	0	0.0	95	77.1	0	0.0	28	22.9	0	0.0	15.6–23.99
132	94.4	8	5.6	120	100.0	0	0.0	58	96.6	2	3.4	Total

Socioeconomic Status

NUTRITION IN RELATION TO MENARCHE. Tables 7, 8, and 9 summarize the intake of proteins and carbohydrates in the various age groups as well as the associations between nutrition and age at menarche in the different socioeducational groups studied.

It was found that the diet with respect to carbohydrate intake was

Table 6. Results by retrospective analysis of the age at menarche in students (from the Caracas metropolitan area) fifteen years of age and older: total sample and by various socio-educational groups (1969)

Material	Number	Average age of menarche	s.d.	Variance
1 Total sample	351	12.80	1.20	1.44
2 Public schools	227	12.80	1.19	1.41
3 Private schools	124	12.70	1.16	1.34
4 Secondary education	169	12.60	1.05	1.10
5 Technical education	136	12.90	1.22	1.48

Key: Technical education = industrial, trade, technical, and normal

Table 7. Nutritional index according to age groups. Students in the metropolitan area of Caracas (1969)

	Carbohydrate intake				Protein intake			
	Satisfactory		Deficient		Satisfactory		Deficient	
Age groups	No.	%	No.	%	No.	%	No.	%
8.0– 9.99	52	100.0	0	0.0	52	100.0	0	0.0
10.0–10.5	35	100.0	0	0.0	35	100.0	0	0.0
10.6–10.99	63	100.0	0	0.0	63	100.0	0	0.0
11.0–11.5	79	100.0	0	0.0	79	100.0	0	0.0
11.6–11.99	116	100.0	0	0.0	0	0.0	116	100.0
12.0–12.5	115	100.0	0	0.0	0	0.0	115	100.0
12.6–12.99	98	100.0	0	0.0	0	0.0	98	100.0
13.0–13.5	103	100.0	0	0.0	0	0.0	103	100.0
13.6–13.99	75	100.0	0	0.0	0	0.0	75	100.0
14.0–14.5	85	100.0	0	0.0	0	0.0	85	100.0
14.6–14.99	80	100.0	0	0.0	0	0.0	80	100.0
15.0–15.5	83	100.0	0	0.0	0	0.0	83	100.0
15.6–23.99	276	100.0	0	0.0	219	79.3	57	20.7
Total	1260	100.0	0	0.0	448	35.6	812	64.4

Table 8. Index of relationships between nutrition, menarche, and type of school (expressed in absolute and relative values). Students from the metropolitan area of Caracas (1969)

		Satisfactory intake		Deficient intake		Total
		Menarche		Menarche		
Type of school		Yes	No	Yes	No	
Carbohydrates						
Public	No.	512	308	0	0	820
	%	62.4	37.6	0	0	100.0
Private	No.	246	176	17	1	440
	%	55.8	40.0	3.8	0.4	100.0
Protein						
Public	No.	160	4	352	304	820
	%	19.5	0.5	43.0	37.0	100.0
Private	No.	197	87	66	90	440
	%	45.0	20.0	15.0	20.0	100.0

Table 9. Index of relationships between nutrition, menarche, and type of education (expressed in absolute and relative values). Students from the metropolitan area of Caracas (1969)

Type of education		Satisfactory intake Menarche Yes	No	Deficient intake Menarche Yes	No	Total
Carbohydrates						
Primary	No.	330	470	0	0	800
	%	41.2	58.75	0.0	0.0	100.0
Secondary	No.	267	13	0	0	280
	%	95.35	4.64	0.0	0.0	100.0
Technical	No.	178	2	0	0	180
	%	98.88	1.11	0.0	0.0	100.0
Protein						
Primary	No.	21	167	309	303	800
	%	2.62	20.87	38.62	37.87	100.0
Secondary	No.	158	6	109	7	280
	%	56.42	2.14	38.92	2.5	100.0
Technical	No.	95	1	83	1	180
	%	52.77	0.55	46.1	0.55	100.0

satisfactory both in public schools (62.4 percent had menstruated) and in private ones (where menarche had occurred in 55.8 percent). On the other hand, protein intake among the public school students was not satisfactory; only 20 percent had abundant protein nutrition and of those, 19.5 percent had menstruated. In private schools, 65 percent attained a satisfactory protein level and of those, 45 percent had menstruated. Students at the secondary education level in whom the menarche had occurred had the best protein diet.

Comparing the average ages of menarche in each institution and the results obtained from the diet survey, it appears that secondary level students, because of their richer protein diet, resemble those attending private institutions, in experiencing menarche earlier.

The average heights and weights for each group are summarized in Table 10. Table 11 shows the values for menarcheal age, weight, and height attained in the various groups studied.

It is noteworthy that menarche occurred earlier in students from secondary schools and that they have greater average weight and height. In spite of the fact that there is little difference in weight and height between students in institutions of industrial, craft, technical, and normal

Table 10. Heights and weights of students in institutions located in the Federal and Sucre Districts of Miranda State (1969)

Age groups	Number	Average height (cm)	Average weight (kilograms)
8.0– 8.5	1	128.4	29.5
8.6– 8.99	5	132.6	28.7
9.0– 9.5	9	130.9	30.9
9.6– 9.99	37	133.8	31.1
10.0–10.5	35	136.0	32.4
10.6–10.99	63	138.6	34.6
11.0–11.5	79	141.6	36.3
11.6–11.99	116	146.5	37.2
12.0–12.5	115	146.4	40.3
12.6–12.99	98	148.9	42.0
13.0–13.5	103	150.5	42.7
13.6–13.99	75	153.0	45.0
14.0–14.5	85	153.4	46.8
14.6–14.99	80	154.0	46.8
15.0–15.5	83	155.5	49.6
15.6–15.99	57	155.2	50.1
16.0–16.5	44	156.8	51.9
16.6–16.99	38	156.0	50.5
17.0–23.6	137	157.1	51.4
Total	1260	146.1	40.9

Table 11. Index of relationships between menarche (probit), weight, and height in students from the metropolitan area of Caracas for the total sample and various socio-educational groups (1969)

Material	Number	Average age at menarche	Average weight (kg)	Average height (cm)
1 Total sample	775	12.70	48.28	154.37
2 Public schools	512	12.71	47.50	153.96
3 Private schools	263	12.64	49.82	155.18
4 Primary education	330	12.94	45.44	151.95
5 Secondary education	267	12.31	50.41	156.74
6 General education	597	12.80	47.67	154.09
7 Technical education	178	–	50.34	155.31

Key: General education = primary and secondary
Technical education = industrial, trade, technical, and normal

education, it could be hypothesized that this is in response to environmental influence.

Relationships between menarche, height, and educational levels are difficult to establish. Nevertheless, there are differences associated with type of institution attended, with earlier menarche, and with greater height in private institutions than in public ones. This could be an indication of a more adequate diet consumed by girls attending private schools. The variance in private schools with respect to weight (68.45) and height (35.99) is greater than in public schools (weight: 58.07; height: 32.64). This indicates that there is a smaller proportion of students concentrated around the average weight and height in private institutions. It could also indicate greater heterogeneity in type of diet.

It may be pointed out that twelve- to seventeen-year-old girls in the metropolitan area of Caracas are similar in height but lighter in weight than Nigerian girls (Tanner and O'Keefe 1962). They are also considerably shorter and lighter than girls of the same ages in southern England (Wilson and Sutherland 1950).

REFERENCES

BERGHER, E., R. SHELLY

1965 *Contribución al estudio de la menarquia en Venezuela.* Caracas: Ministry of Health and Public Assistance.

BOJLEN, K., M. W. BENTZON

1968 The influence of climate and nutrition on age at menarche, a historical review and a modern hypothesis. *Human Biology* 40 (1):69–85.

BURRELL, R. J. W., M. J. R. HEALY, J. M. TANNER

1961 Age at menarche in South African Bantu girls. *Human Biology* 33:250.

BURRELL, R. J. W., J. M. TANNER

1961 Age at menarche in South African Bantu schoolgirls living in the Transkei Reserve. *Human Biology* 33(3):250–261.

CARFAGNA, M., E. FIGURELLI, G. MATARESE, S. MATARESE

1971 Menarcheal age of schoolgirls in the district of Naples, Italy, in 1969–70. *Human Biology* 44(1):117–126.

CREMA, M.

1970 Aspectos sociales del estudiante de educación media en Caracas. *Estudio de Caracas* 6(2):133–324.

FINNEY, D. J.

1962 *Probit analysis: a statistical treatment of the sigmoid response curve,* second edition. London: Cambridge University Press.

FLUCHMANN, F.

1958 *Tratamiento de los trastornos menstruales.* Mexico: Editorial Interamericana S.A.

FRISCH, R., R. REVELLE

1971 The height and weight of girls and boys at the time of initiation of the adolescent growth spurt in height and weight and the relationship to menarche. *Human Biology* 43(1):140–159.

GOLDSMITH, G.
1964 "Diet Advice to the healthy patient," in *The medical clinics of North America*. Edited by S. L. Halpern, 1113–1128. Philadelphia: Saunders.
HAUTVAST, J.
1971 Physical growth and menarcheal age in Tanzanian schoolchildren and adults. *Human Biology* 44(1):117–126.
KREHL, W.
1964 "Diet advice to the healthy patient," in *The medical clinics of North America*. Edited by S. L. Halpern, 1129–1140. Philadelphia: Saunders.
LEVINE, V. E.
1953 Studies in physiological anthropology. The age of onset of menstruation of the Alaska Eskimo. *American Journal of Physical Anthropology* 2:252.
LIVSON, N., D. MCNEIL
1962 The accuracy of recalled age of menarche. *Human Biology* 34(3):212–221.
MILICER, H.
1968 Age at menarche of girls in Wroclaw, Poland in 1966. *Human Biology* 40(2):249–259.
MILICER, H., F. SCZOTKA
1966 Age at menarche in Warsaw girls in 1965. *Human Biology* 34(2):199–204.
OETTLÉ, A. G., J. HIGGINSON
1961 The age at menarche in South African Bantu (Negro) girls: with a comment on methods of determining mean age at menarche. *Human Biology* 3:181–190.
SHAKIR, A.
1971 The age at menarche in girls attending schools in Baghdad. *Human Biology* 43(2):265–270.
TANNER, J. M.
1962 *Growth at adolescence,* second edition. Oxford: Blackwell.
1966 *Educación y desarrollo físico,* first edition. Mexico: Siglo XXI Editores.
TANNER, J. M., B. O'KEEFE
1962 Age at menarche in Nigerian schoolgirls, with a note on their heights and weights from age 12 to 19. *Human Biology* 34:187–196.
VALENCIA, P. J.
1936 *Biología menstrual de la mujer venezolana.* Caracas: Tipografía La Nación.
VALSIK, J. A.
1965 The seasonal rhythm of menarche: a review. *Human Biology* 37(2):75–90.
WILSON, D. C., I. SUTHERLAND
1950 Further observations on the age of the menarche. *British Medical Journal* 2:862–866.
1960 The present age of the menarche in southern England. *Journal of Obstetrics and Gynaeocology of the British Empire* 67:320.
ZUKOWSKI, W., A. KMIETOWICS-ZUKORSKA, S. GRUSKA
1964 The age at menarche in Polish girls. *Human Biology* 36:223–234.

The Relevance of Nourishment to the Reproductive Cycle of the Female in India

JUDIT KATONA-APTE

This paper indicates that the low-income rural Indian female who represents the majority of the female population is undernourished, especially during the reproductive period of her life. Much of the literature surveyed, as well as my fieldwork in both Maharashtra and Tamilnad, support this statement. This can be seen in the following brief description of the reproductive period of such females.

A female infant in a low-income household in India begins life with a handicap. She is a financial liability in regions characterized by the dowry system. While female infanticide is not extensively practiced as it once was, female infants are not cared for as well as are male infants. A female infant is nursed, but not as long as the male. She is less likely to have supplements, such as powdered milk, added to her diet. She will not be taken to the hospital as quickly when ill. She is generally not fussed over as are her brothers. In a 1901 census report the following observation is made:

> But if the practice of doing away with female infants is now confined to a limited area . . . there is little reason to doubt that in most parts of India, female infants receive far less attention than males. It is almost universally the case that, whereas male offspring are ardently desired, the birth of a female child is unwelcome. It is especially so where the securing of a husband is a matter of difficulty and expense and where there are already several female children in the family. Consequently, even if there is no deliberate design of hastening a girl's death, there is no doubt that, as a rule, she receives less attention than would be bestowed on a son . . . she is also probably not fed as a boy would be and when ill her parents are not as likely to make the same strenuous efforts to insure her recovery (Risley and Gait 1903:115).

The situation does not seem to have changed much in the sixties.

> . . . while the risk of male deaths at most ages seems to have steadily decreased, . . .

> there has been no corresponding improvement in female deaths ... a larger proportion of boys than girls seem to be born in this country than in the West... girls, in the first few years of their life, still seem to suffer from greater neglect than boys, as a result of which Nature's balancing action of taking away more boys than girls in the first years of life does not properly come into play, and the survival of girls never seems to draw even with that of boys... (*Census of India* 1962:xvii).

There are some interesting statistics to support the above statement. In the first week of life, 61.9 of the male and 58.9 of the female infants per 1,000 die. But by the end of the first year, 172 male and 182 female babies per 1,000 have died (Agarwala 1970:122).

But the number of infant deaths under one year of age due to disease in the latest available year for the four major cities of India is 6,191 for males and 5,164 for females (Chandrasekhar 1972:173–175). These figures probably come from hospital records and may indicate that female infants do not receive medical attention as readily as males.

The ratio of females to males is as low as 719:1,000 in some areas (*Census of India* 1962:xviii). The figure for all of the U.S. is 1030:1000 (*United Nations* 1965). The life expectancy figures for females in India is 40.55 years, for males 41.89 years. In developed countries the figures for male life expectancy are considerably lower than the female United States figures, for example, are 66.90 (males) and 73.70 (females) (*United Nations* 1965).

Marriages in India are based more on socioeconomic factors than on companionship. Total devotion by the wife to her husband is demanded since according to the Laws of Manu — the ancient Indian code of behavior — eternal salvation for women is attained only through worshipping the husband while on earth. Thus, she does not look upon herself as her husband's equal, indeed does not address him directly or by name and does not question why her wishes and opinions may not be asked for when important decisions are made.

A woman generally becomes pregnant within a year of her marriage. For the first time she has an opportunity to justify her existence. If she gives birth to a male offspring, her status is elevated as befits the mother of a son. If she has only daughters, her status drops to just above that of a barren wife. Barrenness is the worst affliction for a female. Although the fathers of daughters are pitied, the mothers are blamed for "not having any sons in them."

As a wife and mother, a woman works hard. The preparation of food is complicated and time consuming. All grains and legumes (these are the staples) have to be picked, cleaned, and often ground by hand. Cooking is done on a smoky burner. Women sometimes cart water long distances. They collect fuel, such as cow dung or twigs, clean and wash cooking and eating utensils, clean the house, care for and worship at the "god's corner," and play a major part in raising children. They often work in the fields along with their husbands and tend the garden around the house. There is very little, if any, time for recreation for a lower class rural housewife.

At meals, the wife feeds her husband first, then the children, boys before girls; only then does she eat. Often she eats separately with her daughters. If available, the major portions of such nourishing foods as meat, fish, egg, or milk, and sweets are served to the males, because it is believed that they need it for strength or growth, and often not much is left over for the females. It is not unusual to find households where the women are vegetarian but the males are not. Vegetarianism among females may be rationalized on religious grounds, thus leaving more (or all) of the high protein foods for the males.

If a woman dies before her husband, she is considered fortunate. If her husband dies before she does, it is believed to be the result of her sins, either in this or a previous life. As a widow she is dependent upon a son's kindness for her support and can be relegated to the status of servant in her son's home. A widow eats sparingly, only one meal a day, because it is said to be unhealthy to eat much in this state of life. Even at the death feast the female is treated in an inferior manner. The food served does not consist of staples (as it is for the male) but only of snack items (lower status). For example, instead of clarified butter (very high status), water (low status) is used for the preparation of certain items.

The major part of an adult female's life in rural India is spent in some stage of the pregnancy, childbirth, and lactation cycle. Since infant mortality is still very high — in 1965 the number of registered deaths for infants was sixty-five per 1,000 as opposed to twenty-five per 1,000 in the United States (World Health Organization 1967) — often several pregnancies are necessary for each child who survives to adulthood.

In spite of the fact that the most important function for a female in India is reproduction, a girl is not given any formal sex education and is rarely aware of the connection between sexual intercourse and conception. Girls are seldom taught anything about puberty or sex. The first menstrual period is often a horrifying and frightening experience to them. Even after reaching menarche the connection between menstruation and conception is not explained, though they are given special foods for a few days "to strengthen the waist," presumably for future pregnancies. As they near marriage, girls are not prepared for the physical aspects and the first act of intercourse can be, and often is, a shocking and painful experience.

A woman learns sexual activity from her husband, who is often inexperienced himself. She is expected to provide sexual satisfaction for her husband. A pregnant woman traditionally goes to her father's house for her confinement. Sometimes she uses this occasion to stay away from her husband and his family as long as possible.

In Hinduism, wifehood equals motherhood. To be a mother gives meaning to the life of all women. To be pregnant is considered both honorable and auspicious. Women have little status until they have had a child. It is not surprising that motherhood — not womanhood — is considered to be the most sacred state.

The caloric intake of the pregnant women is far below Western standards. In the lower economic classes in India, it is between 1,300 and 1,700 calories (Chandrasekhar 1972:217). In Tamilnad, South India, a pregnant woman's nutrition was affected by a belief that too much food, as well as highly nutritious food, was not good for her and harmful to the fetus. A pregnant woman may be still nursing a previous child and thus have further restrictions on her diet.

The restrictions on newly delivered mothers were strict even sixty years ago: ". . . all nourishment is denied to her and for five days she is to live upon a stuff made up of molasses, gum acacia, and ajawan" (Bose 1913:324). The situation has not changed. Last year in South India I found women living *post partum* only on coffee for three days or more. After this initial period, they are still denied certain foods which are considered improper to eat while lactating. For as long as six weeks, the new mother eats only one instead of the accustomed two meals per day. Even that meal is lacking in foods considered important by Western standards, such as vegetables, milk products, meat, fish, eggs, etc. (Katona-Apte 1972). Lactating mothers are permitted all the coffee or tea and bread they want. Interestingly, these are the same foods that are considered proper for sick persons.

A mother's intake of nutrients is also restricted by the infant's state of health. When a child is ill, rather a common occurrence in developing nations, the illness is blamed on the mother's milk, and thus on the foods she consumes to produce the milk. Just as garlic is believed to help the production of milk, other foods are believed to be harmful to the infant, and she is expected to avoid them, further limiting her diet and reminding her that she may cause harm to the child by overindulgence or gluttony. The connection between nutrition and the physical well-being of the infant unfortunately is a negative one.

The nursing woman is the least well-fed member of the household. The number of restrictions on her diet is extensive. Yet the Indian mother:

> . . . continues to nurse her baby into the second and often into the third year . . . she continues to nurse the first baby until the new one arrives and occasionally even after the birth of the new baby. The Indian woman is thus in an almost continuous state of lactation — however meager — throughout the child-bearing period of her life (Chandrasekhar 1972:219).

A woman is considered impure during her menstrual periods and her confinement. During these times she is isolated and is expected not to touch anything or anyone, or to sleep near her husband. She is barred from the kitchen and can eat only what is given to her. Since she is not working during this period, she is often given less food, which affects her health in general.

After delivery, confinement may last from ten to forty days. The impure woman often stays within a small poorly lit and ventilated area in the house, reserved for such occasions. She is considered to be dirty and defiling.

Fasting is often a burden for the Hindu woman. Whether pregnant, lactating, healthy, or ill, a woman is expected to fast several times a week, either to appease certain selected gods, to bring about good events, or to avoid bad ones. Women fast to obtain male offspring, to gain long life for their husbands, and to assure themselves that they will get the same husband again in the next rebirth. They fast on the eves of holidays, and on days declared sacred by astrologers. In the lower economic classes even staples are in short supply. This, added to the fasting, means that much of the time women go without food (Katona-Apte 1976).

We can speculate that a chronic shortage of food forces people to make choices as to who is to get which foods and how much of them. In India, this choice favors those who are expending visible physical energy. Rationalization for not feeding a woman very much after delivery is that she is not expending energy and so would not be able to digest the food. The same is true while she is lactating. Since she is not visibly contributing any labor, she may be viewed as an economic burden on the family. It is likely that the dietary restrictions are related, amongst other things, to a lack of economic contributions to the family.

The Indian female has a difficult time. She is underfed, overworked, and her body is constantly supporting a fetus or breastfeeding infant:

> . . . a heavy toll of female lives is taken in the earlier reproductive age, i.e. between the ages of 15 and 34. The toll is so heavy that the difference between the male and female population grows remarkably rapid and wide and this gap is never made up in the middle or old age. What is more, proportionately greater deaths occur among females even between the ages of 35 and 54 than is commonly believed (*Census of India* 1962:xvii).

The Indian female's short life expectancy, 40.55 years (U.S. female 73.70 years) is largely due to malnutrition, especially during the strenuous reproductive years, and to various sociocultural restrictions just described. It is a vicious circle: poor maternal nutrition contributes to the unhealthy state of infants who often do not survive. This in turn causes repeated pregnancies which drain the mother. There is no question that improved nutrition would provide healthier infants and longer life for the female.

REFERENCES

AGARWALA, S. N.
1970 *A demographic study of six urban villages.* Bombay: Asia Publishing House.

BOSE, K. C.
1913 "Infantile mortality: its causes and prevention," in *The proceedings of the second all-Indian sanitary conference held in Madras, 1912*, volume two. Simla: Government Central Press.

Census of India
1962 *Census of India.* Final population totals, Paper Number 1.

CHANDRASEKHAR, S.
1972 *Infant mortality, population growth and family planning in India.* London: George Allen and Unwin.

KATONA-APTE, JUDIT
1972 "The relevance of socio-cultural factors to malnutrition." Paper presented at the meeting of the American Anthropological Association in Toronto, December.
1976 "Dietary aspects of acculturation: meals, feasts, and fasts in a minority community in South Asia," in *Gastronomy: the anthropology of food and food habits*. Edited by Margaret L. Arnolt. The Hague: Mouton.

RISLEY, H. H., E. A. GAIT
1903 *Census of India, 1901,* Volume one, part one: report Superintendent of Government Printing, Calcutta.

United Nations
1965 *United Nations Demographic Yearbook.* New York.

WORLD HEALTH ORGANIZATION
1967 *Epidemiological and vital statistical reports.* Geneva.

An Anatomical View of Demographic Data: Biomass, Fat Mass, and Lean Body Mass of the United States and Mexican Human Populations

GABRIEL LASKER and HENRY WOMACK

When anatomists are asked to look at human populations, they see them in a somewhat different light than do demographers. We cannot report definitive results of anatomical studies of populations here, but we will introduce some partial figures that will allow extrapolated conclusions useful chiefly for heuristic purposes. We had anticipated that the International Biological Program's Human Adaptability Project (IBP/HAP) would have developed representative data of just the kind needed, but if it has, the results are not readily available, and we have found it easier for the time being to rely on older and very limited data.

When demographers look at populations, they tend to emphasize numbers. The census is their most accurate tool. Even when population counts are broken down into age and sex subgroups, one purpose is still to help project future numbers. Without belittling other demographic objectives (actuarial, economic, and sociological considerations), we do note that the growing concern with human ecology has not led to much examination of the anatomy of populations by either demographers or human biologists. Those who have examined populations have all too easily accepted the postulate that peoples are demographically equivalent. On the other hand, those who have examined anatomical differences have usually been concerned with standards for judging individual health and physiological performance.

We do not know of published accounts of any efforts to estimate the biomass of the world population of *Homo sapiens* and to "dissect" it into the biomasses of his various tissues. We shall therefore indicate what one would try to do to evaluate the biomass and its components in two

We thank Dr. Bernice Kaplan for assistance with the collection of data in Peru, Dr. Marcus Goldstein for past access to his protocols, and Drs. Stanley Garn, A. R. Frisancho, and F. E. Johnston for their personal communications and helpful criticisms.

national populations, and we shall discuss some obligatory assumptions. The actual figures on the biomass of the populations of the United States and Mexico and on the mass of metabolically inactive (fat) and active tissues for these national populations remain tentative, and data on the whole human species are not yet available.

However, one may consider the world as made up of two kinds of populations: those having had reasonably ample food resources for the last several generations and those that have had relatively restricted amounts of food for their population. In a general way these populations correspond to "Western" versus "non-Western"; i.e., Europe and northern North America versus the rest of the world. It will be readily apparent that there are other areas where food may have been scarce but which now are prosperous (such as Japan), and others such as Israel, Australia, New Zealand, and Siberia which are, from this point of view, "Western." Data on the United States, on the one hand, and Mexico, Guatemala, and Peru, on the other, may stand for the two ecological types and hence be used to give at least a crude picture of the world population from an anatomical perspective.

The distinction between consistently well-fed populations and those which have often gone through "bottlenecks" of food shortages is not seen primarily in population density. The United States had an estimated population of 209,166,000 in 1972 for 3,675,911 square miles or fifty-seven persons per square mile. Mexico has an area of 761,602 square miles with a population estimated at 54,253,000 in 1972 or seventy-one persons per square mile. But if one discounts Alaska's 586,412 square miles, the population densities of the two countries are nearly equal and are about the world average (seventy-two per square mile). Prosperous areas of the world have densities ranging from four per square mile for Australia to 162 per square mile for Europe.

What we shall do is to take the average weight of Mexicans of various age and sex groups, multiply by the number of individuals in each group, respectively, and then sum all of the weights to get a figure for the total biomass of *Homo sapiens* in Mexico. When the same is done for the United States, the figure obtained will be compared with that of Mexico and the ratio will be compared with the corresponding ratio of population numbers.

An effort will then be made to estimate the amount of human fat in these populations. For our purposes we will use data from the 10-State Nutrition Survey (Garn, personal communication). These data include triceps skinfolds and age- and sex-specific constants which allow estimates of fat weight. Unfortunately, we know of no corresponding data for Mexico. However, one of us measured the triceps skinfolds of adult Peruvians of both sexes in three towns of the north coast of Peru. Like the national populations of Mexico, the origins of these Peruvians include a major indigenous element (usually in Mestizo form), a European element (rarely unmixed with Amerindian), and a very minor African element (Kaplan and Lasker 1953; Lasker and Kaplan 1965). For younger ages (one to eighteen years) data on triceps skinfolds are from the Instituto de Nutrición de Centro América y Panamá (INCAP) (1969) for six Latin American countries. We use those of rural Guatemala as being most similar ethnically and environmentally to Mexico. Once the biomass and fat mass of a population are known, the difference between them (body weight minus fat mass) gives the lean body mass (LBM), which is of great importance because of its relationship to the food needed by that population.

Since metabolically active tissues do not have equal levels of metabolism in different countries either in basal or habitual circumstances, one cannot directly equate the ratio of LBM of two countries with their relative food use, let alone their food needs. Of the edible portion of foods utilized, only part is consumed and, of that consumed, only part is absorbed and metabolized. The magnitude of both these fractions probably depends to some extent on the amount of food available, and there may well be human adaptations to the amount that has been available: (1) in the immediate past (influencing psychological hunger and some biological "hunger"); (2) during development (influencing body size and hence caloric needs); and (3) over generations (influencing the survival of those with genetically lower caloric needs or those with a genetically higher capacity to supply caloric needs under very complex biocultural conditions of natural selection).

RESULTS

The average body weights of the white population of each age and sex in the United States, as reported by Stoudt, Damon, and McFarland (1960), were converted into kilograms and accepted for the total population without correction for the minor nonwhite elements. These weights are given in Table 1. For the Mexican population, Table 2 shows weights at birth and one and two years of age averaged from two sources reported by Meredith and Goldstein (1952). Weights of Mexicans three to seventeen years old are given from one source by the same authors (and they cite similar figures for Mexicans in the United States). For ages eighteen and up, data for Mexicans measured by Goldstein and tabulated and graphed by Lasker (1953) have been utilized in Table 2. The Mexican population was measured some time ago (most of it in 1941), and it is possible that more recent data would indicate a secular trend to increased weights and earlier growth (as has been noted in most countries studied). For the United States, Hamill, Johnston, and Lemeshow (1972) have published

Table 1. Data on weight and triceps skinfold thickness of United States population samples by age, estimated body fat, lean body mass (LBM), and the constants by which these are derived

Age	Weight (kg)[a]	Triceps (mm)[b]	Constant[c]	Fat (kg)[d]	LBM (kg)[e]
Males					
1– 4	11.5	9.7	–	2.9	8.6
5– 9	24.4	9.3	–	6.1	18.3
10–14	41.2	11.7	0.869	10.2	31.0
15–19	64.4	11.5	0.971	11.2	53.2
20–24	71.7	12.0	0.971	11.7	60.0
25–34	74.7	13.4	0.829	11.1	63.3
35–44	75.5	14.1	0.887	12.5	63.0
45–59	75.1	14.2	0.924	13.1	62.0
60–64	73.5	13.2	0.962	12.7	60.8
65–74	72.4	12.7	0.962	12.2	60.2
75 up	69.8	12.1	0.962	11.6	58.2
Females					
1– 4	11.1	9.6	–	3.2	7.9
5– 9	23.4	10.8	–	6.8	16.6
10–14	41.8	14.8	0.815	12.1	29.7
15–19	55.0	17.5	0.911	15.9	39.1
20–24	56.7	18.0	0.911	16.4	40.3
25–34	58.3	20.3	0.984	20.0	38.3
35–44	63.5	22.9	1.057	24.2	39.3
45–59	65.8	24.0	0.965	23.2	42.6
60–64	66.2	23.1	0.873	20.2	46.0
65–74	65.8	21.8	0.873	19.0	46.8
75 up	65.3	19.0	0.873	16.6	48.7

[a] Data for U.S. whites from Stoudt, Damon, and McFarland (1960).
[b] Weighted averages of triceps fatfolds of U.S. white population from Garn (1972).
[c] Constants based on data from Garn (1972). The constants for calculating body fat from triceps fatfolds are for ages 10, 20, 30, 40, and 60, respectively. The age 20 constant is used for the 15–19 year old group. For ages 45–59 the constants for ages 40 and 60 are averaged.
[d] Body fat is determined by multiplying the triceps fatfold thickness by the constant. For ages 1–4 and 5–9 fat is taken to be in the same ratio as for 10–14 years of age.
[e] Lean body mass (LBM) is weight minus fat.

weight data for children aged six to seventeen and three-quarters years. These figures are slightly higher than those here, but being more recent they are even less comparable to the Mexican data.

The triceps skinfold measurements for the white population in the United States (Table 1) were provided by S. M. Garn (1972) from unpublished data collected for the 10-State Nutrition Survey. For the Latin American population (Table 2), ages one to eighteen are for Guatemalans from the INCAP studies (1969). Ages eighteen and over are for Peruvians from unpublished data of Lasker. In an effort to calculate the factors of metabolizing and nonmetabolizing tissues, we have estimated the total body fat from these skinfold measurements. This was done by multiplying the skinfold thickness by a specific constant for

Table 2. Data on weight and triceps skinfold thickness of Latin American samples by age, estimated body fat, lean body mass (LBM), and the constants by which these are derived

Age	Weight (kg)[a]	Triceps (mm)[b]	Constant[c]	Fat (kg)[d]	LBM (kg)[e]
Males					
1– 4	10.6	8.6	–	1.8	8.8
5– 9	21.2	7.0	–	3.7	17.5
10–14	33.9	6.7	0.869	5.9	28.0
15–19	53.2	8.8	0.971	8.5	44.7
20–24	57.5	11.3	0.971	11.0	46.5
25–34	59.8	10.0	0.829	8.3	51.5
35–44	63.6	11.9	0.887	10.6	53.0
45–59	63.6	8.6	0.924	7.9	55.7
60–64	62.0	8.3	0.962	8.0	54.0
65–74	62.0	6.8	0.962	6.5	55.5
75 up	62.0	4.9	0.962	4.7	57.3
Females					
1– 4	10.1	9.3	–	2.5	7.6
5– 9	20.3	8.7	–	5.1	15.2
10–14	35.2	10.8	0.815	8.8	26.4
15–19	48.0	15.7	0.911	14.3	23.7
20–24	51.6	14.0	0.911	13.6	38.0
25–34	52.7	15.2	0.984	15.0	37.7
35–44	55.5	16.3	1.057	17.2	38.3
45–59	59.1	16.4	0.965	15.8	43.3
60–64	57.3	12.8	0.873	11.2	46.1
65–74	57.3	16.1	0.873	14.1	43.2
75 up	57.3	11.7	0.873	10.2	47.1

[a] Ages 1–17 for Mexican children in Mexico from Meredith and Goldstein (1952); ages 18 and up from Mexican data of Goldstein (graphed by Lasker 1953).
[b] Ages 1–18 for Guatemala from INCAP (1969: vol. 25); ages 18 and up for Peru from unpublished data of Lasker (see Lasker [1962] for description of subjects).
[c] See Footnote c, Table 1.
[d] See Footnote d, Table 1.
[e] See Footnote e, Table 1.

each age-sex group. The constants were derived by Garn from data of the 10-State Nutrition Survey, and he has kindly permitted us to use these unpublished constants. The values for skinfold thickness and total body fat for the Latin American population (Table 2) are well below those for the United States (Table 1). A. R. Frisancho and S. M. Garn (personal communication) confirm the same finding for Latin American populations with which they have worked. The INCAP (1969) studies also show, in general, smaller triceps skinfolds in boys under eighteen in rural populations of six Central American countries, and similar or smaller values in girls, compared to English standards. Because the average skinfolds in England are below those in the United States, the same can be said of all six Central American countries studied.

Lean body mass was determined by substracting the total body fat from total body weight. The values for United States individuals are shown in Table 1 and those for Latin American individuals in Table 2. The values for total biomass, fat, and lean body mass for the entire United States and Mexican populations are shown in Tables 3 and 4, respectively. These values were obtained by multiplying the individual values by the total number of individuals as abstracted from United Nations demographic and statistical reports for the 1970 decennial censuses of the United States and Mexico (United Nations 1970).

Table 5 summarizes these total values and shows the various ratios between the two populations. The 1970 population of Mexico (48,377,200) is 23.8 percent of that of the United States (203,165,700). The biomass of the Mexican population is estimated here to be 17.2 percent of that of the United States. The fat mass of the people of Mexico as calculated here (409,058,800 kilograms) is 14.7 percent of the fat mass calculated for the people of the United States (2,781,028,100 kilograms). The LBM of the Mexican population (1,528,224,800 kilograms) is 18.0

Table 3. Estimates of the number, biomass, and constituency of that biomass in non-metabolizing (fat) and metabolizing lean body mass (LBM) fractions of the United States population by age and sex

Age	Number	Biomass (kg)	Fat (kg)	LBM (kg)
United States males				
1– 4	8,744,700	100,564,100	25,359,600	75,204,500
5– 9	10,166,300	248,057,700	62,014,400	186,043,300
10–14	10,589,200	436,275,000	108,009,800	328,265,200
15–19	9,633,900	620,423,200	107,899,700	512,523,500
20–24	7,920,100	567,871,200	92,665,200	475,206,000
25–34	12,217,000	908,944,800	135,608,700	773,336,100
35–44	11,223,100	847,344,100	140,288,800	707,055,300
45–59	15,952,700	1,198,047,800	208,980,400	989,067,400
60–64	4,027,000	295,984,500	51,142,900	244,841,600
65–74	5,431,400	393,233,400	66,263,100	326,970,300
75 up	2,976,500	207,759,700	34,527,400	173,232,300
United States females				
1– 4	8,422,300	93,487,500	26,951,400	66,536,100
5– 9	9,788,600	229,053,200	66,562,500	162,490,700
10–14	10,198,400	426,293,100	123,400,600	302,892,500
15–19	9,435,700	518,963,500	150,027,600	368,935,900
20–24	8,451,600	479,205,700	138,606,200	340,599,500
25–34	12,691,500	739,914,500	253,830,000	486,084,500
35–44	11,848,600	752,386,100	286,736,100	465,650,000
45–59	17,216,200	1,132,826,000	399,415,800	733,410,200
60–64	4,589,200	303,805,000	92,701,800	211,103,200
65–74	6,993,300	460,159,100	132,872,700	327,286,400
75 up	4,648,400	303,540,500	77,163,400	226,377,100
Totals	203,165,700	11,264,139,700	2,781,028,100	8,483,111,600

percent of the LBM estimated for the population of the United States (8,483,111,600 kilograms).

DISCUSSION

It is not easy to evaluate the quality of data on weight and fat until repeated estimates have been made. A preliminary (and independent) estimate of body fat in the United States was extrapolated on the basis of data from Brožek (1952), Škerlj, Brožek, and Hunt (1953), Newman (1956), Garn (1957), Lee and Lasker (1958), and Forbes and Amirhakimi (1970). The constants Garn has supplied yield slightly less body fat at ages one to twenty and slightly more at ages twenty-five and above. The total fat estimated is essentially the same, 2,802 versus the 2,781 million kilograms estimated in Table 3, a difference of 0.76 percent. The corresponding difference in LBM is 0.24 percent. Only further data from large samples of randomly selected and carefully measured

Table 4. Estimates of the number, biomass, and constituency of that biomass in non-metabolizing (fat) and metabolizing lean body mass (LBM) fractions of the Mexican population by age and sex (fat and LBM) estimated by analogy using data from Guatemala and Peru

Age	Number	Biomass (kg)	Fat (kg)	LBM (kg)
Mexican males				
1– 4	4,164,400	44,142,600	7,496,000	36,646,600
5– 9	3,947,300	83,682,800	14,605,000	69,077,800
10–14	3,281,400	111,239,500	19,360,300	91,879,200
15–19	2,497,900	132,888,300	21,232,200	111,656,100
20–24	1,935,100	111,268,300	21,286,100	89,982,200
25–34	2,870,200	171,638,000	23,822,700	147,815,300
35–44	2,202,000	140,047,200	23,341,200	116,706,000
45–59	1,927,000	122,557,200	15,223,300	107,333,900
60–64	452,600	28,061,200	3,620,800	24,440,400
65–74	589,300	36,536,600	3,830,500	32,706,100
75 up	273,100	16,932,200	1,283,600	15,648,600
Mexican females				
1– 4	4,029,800	40,701,000	10,074,500	30,626,500
5– 9	3,801,600	77,172,500	19,388,200	57,784,300
10–14	3,134,900	110,348,500	27,587,100	82,761,400
15–19	2,570,200	123,369,600	36,753,900	86,615,700
20–24	2,107,400	108,741,800	28,660,600	80,081,200
25–34	3,005,600	158,395,100	45,084,400	113,310,700
35–44	2,257,500	125,291,300	38,829,000	86,462,300
45–59	1,925,900	113,820,700	30,429,200	83,391,500
60–64	468,400	26,839,300	5,246,100	21,593,200
65–74	605,400	34,689,400	8,536,100	26,153,300
75 up	330,200	18,920,500	3,368,000	15,552,500
Totals	48,377,200	1,937,283,600	409,058,800	1,528,224,800

Table 5. The relative size of the human population of Mexico and the United States and the relative biomass, fat mass, and lean body mass of the two populations (in kilograms)

Population of Mexico (1970) / Population of United States	48,377,200 / 203,165,700	=0.238
Weight of people of Mexico / Weight of people of United States	1,937,283,600 / 11,264,139,700	=0.172
Fat mass of people of Mexico / Fat mass of people of United States	409,058,800 / 2,781,028,100	=0.147
Lean body mass of people of Mexico / Lean body mass of people of United States	1,528,224,800 / 8,483,111,600	=0.180

populations can provide really "good" estimates, however, especially for non-Western nations.

To the extent that one can accept the assumptions concerning the equivalency of Guatemalan, Peruvian, and Mexican populations with respect to triceps skinfolds as a measure of total body fat, the LBM of Mexicans is calculated as 18.0 percent of that of the population of the United States (Table 5). In fact, the lean body mass of Mexicans is slightly larger than that since Mexicans are, on average, smaller in body size (as measured by stature, for instance) than Americans, and the same triceps skinfold no doubt represents less total weight of fat in a Mexican. Since we used the same constants for Mexican and American populations we have no doubt overestimated the fat component in Mexico and hence underestimated lean body mass. In any case, both because of the greater proportion of children in the population and because of slower growth and maturation and smaller adult size, the average amount of metabolizing tissue per Mexican is much less than that per American.

Brožek and Grande (1955) have shown that components of the lean body do not metabolize equally. Liver is the most active tissue and we know next to nothing about the relative size of the liver and other organs in various parts of the world. During severe weight loss all tissues are known to be reduced in mass and the best assumption, for present purposes, is that various metabolizing tissues represent more or less constant proportions of lean body mass and that amount of mass is linearly correlated with measures of metabolism.

Data on metabolism are usually reported as Basal Metabolic Rate (BMR). Some early studies apparently indicated scores above the United States in at least some Latin American people. On the other hand, studies in China and other parts of the non-Western world usually show a BMR of less than one hundred. Unfortunately, the BMR is standardized for persons of equal surface area. If actual O_2 utilization or heat production were reported, inferences could be made directly relevant to the total populations as we have presented them. In any case, surface area (usually estimated from height and weight) is highly correlated with LBM since both reflect the size of the body frame. If the BMR is not elevated in such peoples as Mexicans (and there is every reason to believe that it is not), the bodily utilization of food as calories per individual must, like lean body mass, be considerably lower than in the United States. In life, of course, it is not basal conditions that count but the metabolism during the actual daily round of activities and at rest. Anecdotes about the long periods of siesta in Mexico have no basis in fact. Many Mexicans work through the day, and our best guess would be that the energy actually expended per individual in the United States and Mexico is probably more or less proportional to their respective mean lean body weights.

To the extent that this kind of finding can be generalized to Western versus non-Western countries, what are the policy implications of such studies? Do they demonstrate a lower per capita need for food in underdeveloped countries and hence justify less concern for their scarcity of food or less generous response to it? Concern and generosity are moral questions beyond the scope of a biological study; all one can say is that food needs at present may be less *per capita* in some countries than in others but that these differences may be the result of adaptation to past food shortages. The more abundant resources now available, through equality of opportunity for growth in the immediate future, open the possibility of equality of nutritional needs later.

The problems of hypercaloric nutrition, obesity, and their morbid consequences in the United States make it uncertain just what level of caloric intake and what type of adaptation are desirable. The fact remains, however, that without food surpluses there is no choice of response. With the access to food which wealth permits the individual can — to the extent his culture allows — make a voluntary choice of the level of caloric intake which he considers desirable for his immediate comfort and pleasure or for his long-term health.

The extension of studies on population anatomy may lead to an understanding of the probable effects of various population policies and may help eventually to inform decisions so as to extend "freedom from hunger" to a wider freedom of choice.

REFERENCES

BROŽEK, J.

1952 Changes of body composition in man during maturity and their nutritional implications. *Federation Proceedings* 11:784–793.

BROŽEK, J., F. GRANDE
1955 Body composition and basal metabolism in man: correlation analysis versus physiological approach. *Human Biology* 27:22–31.

FORBES, G., G. AMIRHAKIMI
1970 Skinfold thickness and body fat in children. *Human Biology* 42:401–418.

GARN, S. M.
1957 Roentgenogrammetric determinations of body composition. *Human Biology* 29:337–353.
1972 "A preliminary analysis of anthropometric, radiographic and dental data for the 10-State Survey." Unpublished manuscript.

HAMILL, P., F. JOHNSTON, S. LEMESHOW
1972 Height and weight of youths 12–17 years. *United States National Center for Health Statistics*, series 11, number 124. Washington, D.C.: U.S. Government Printing Office.

INSTITUTO DE NUTRICIÓN DE CENTRO AMÉRICA Y PANAMÁ
1969 *Evaluación Nutricional de la Población de Centro América y Panamá*, volumes twenty-five to thirty. Guatemala City, Guatemala.

KAPLAN, B., G. LASKER
1953 Ethnic identification in an Indian Mestizo community. *Phylum, the Atlanta University Review of Race and Culture* 14:179–190.

LASKER, G.
1953 The age factor in bodily measurements of adult male and female Mexicans. *Human Biology* 25:50–63.
1962 Differences in anthropometric measurements within and between three communities in Peru. *Human Biology* 34:63–70.

LASKER, G., B. KAPLAN
1965 The relation of anthroposcopic traits to the ascription of racial designation in Peru. Homenaje a Juan Comas en su 65 aniversario, Mexico, 2:189–220.

LEE, M., G. LASKER
1958 The thickness of subcutaneous fat in elderly men. *American Journal of Physical Anthropology* 16:125–134.

MEREDITH, H., M. GOLDSTEIN
1952 Studies on the body size of North American children of Mexican ancestry. *Child Development* 23:91–110.

NEWMAN, R.
1956 Skinfold measurements in young American males. *Human Biology* 28:154–164.

ŠKERLJ, B., J. BROŽEK, E. HUNT
1953 Subcutaneous fat and age changes in body build and body form in women. *American Journal of Physical Anthropology* 11:577–600.

STOUDT, H., A. DAMON, R. MCFARLAND
1960 Heights and weights of white Americans. *Human Biology* 32:331–341.

UNITED NATIONS
1970 *Demographic and statistical report, 1970 census.*

Individual Differences in Skeletal Muscle Fiber Composition

S. INOKUCHI

Skeletal musculature undoubtedly plays an important role in the organization of the body frame and motility of the individual. However, reports discussing individual variations in skeletal muscle itself are few, and only those concerning relations between the body frame and the total number of fibers of M. biceps brachii by Barin-Baum (1963) and Etemadi and Hosseini (1968) fall within this category. Variation not only of the M. biceps brachii, but of other muscles as well, should be studied extensively for comparison. To this end, a large number of cases must be examined to develop standards for the number of muscle fiber characteristic of each muscle.

MATERIALS

The materials studied were ten cases (four males, six females) of right biceps brachii muscle, eight cases (three males, five females) of right sternocleidomastoid muscle, twelve cases (five males, seven females) of right adductor longus muscle, and three cases (males) of sartorius muscle of humans, as well as ten cases (five males, five females) of right biceps brachii muscle of monkeys, obtained from dead subjects preserved with injection of a 10 percent formalin solution in water. Representative transverse sections of the muscle tissue, 1 centimeter approximately in thickness, from center portions of the muscles at their greatest widths were used.

METHODS

The transverse sections of muscle tissue were refixed in 10 percent formalin solution in water, processed as usual for rinsing, dehydration, and embedding in celloidin and were cut into thin sections (20~25 μm), followed by double staining with hematoxylin and eosin or by Van Gieson's stain.

The entire area of transverse section of each muscle was determined on ×10 magnified photomicrographs of the thin sections by planimeter. A count was made of muscle fibers located in individual square sections, with an area of $(100\,\mu m)^2$. One hundred sections were examined for each muscle section, and their sum taken as the number of muscle fibers per square millimeter of the transverse section of muscle. Then the total number of muscle fibers for that section was estimated by multiplying the sum by the cross-sectional area (1 square millimeter). In addition, measurement of the minor diameter and major diameter orthogonal to it of the individual fibers contained in the square sections and a cross-sectional area of each fiber was calculated by these diameters.

RESULTS AND DISCUSSION

Table 1 shows the mean values of the cross-sectional area of the center of muscles, total number of muscle fibers, number of muscle fibers per square millimeter, and diameter of muscle fibers in each muscle.

The total number of muscle fibers was approximately 200,000 in biceps brachii and adductor longus in man, and approximately 100,000 in the sternocleidomastoid and sartorius muscle in man and biceps brachii in monkey. Biceps brachii and adductor longus have a larger total number in males than in females, while sternocleidomastoid and sartorius are similar. The range of variation expressed by the ratio between maximum and minimum values was greater in the former two muscles than in the latter three. In the former two, the maximum value of total fibers was approximately equal in male, but in female larger in biceps brachii than in adductor longus. Minimum values were approximately equal in males, but larger in adductor longus than in biceps brachii in females. In the other three muscles, maximum and minimum values were approximately equal.

The cross-sectional area was larger in human biceps brachii and adductor longus muscles than in the other, and the range of variation was larger in these two muscles than in the others. This tendency was most pronounced in males. In these two muscles, the maximum value of the cross-sectional area was larger in adductor longus than in biceps brachii in males and approximately equal in females, while minimum values were equal in males and larger in adductor longus than in biceps brachii in females. Therefore, the range of variation was greater in adductor longus in males and in biceps brachii in females than in the other muscles. The

Table 1. Mean value of the cross-sectional area, total numbers of muscle fibers, number of the muscle fibers per square millimeter, and diameter of muscle fibers on each muscle

				Cross-section area (mm^2)			Total number of muscle fibers			Number of muscle fibers per (mm^2)			Thickness of muscle fibers (μm^2)		
Muscle	Species	Sex	No.	Mean	Max.	Min.	Mean	Max.	Min.	Mean	Max.	Min.	Mean	Max.	Min.
Biceps brachii	Human	M	4	399	540	170	251,051	345,667	170,850	696	1005	486	830.8	967	585
		F	6	147	305	85	172,543	341,295	114,297	1210	1423	859	534.9	759	301
		Total	10	248	540	85	203,947	345,665	114,297	1005	1423	486	653.2	967	301
	Monkey	M	5	83.3	110	53	110,438	122,243	97,997	1391	1849	982	476.1	615	333
		F	5	106.1	143	48	99,168	124,410	80,288	1022	1664	840	626.5	897	295
		Total	10	94.7	143	48	104,804	124,410	80,288	1198	1849	840	551.3	897	295
Sternocleidomastoid (50y)		M	3	146.5	179	89	104,122	111,070	94,512	862	1363	605	652	860	394
		F	5	119.5	165	77	111,042	136,703	81,205	916	1206	565	637	1250	394
		Total	8	129.6	179	77	108,442	136,703	81,205	896	1363	565	642	1250	394
Adductor longus		M	5	334.7	712	168	253,401	339,150	153,335	873	1203	476	616.3	1305	381
		F	7	251.2	337	141	212,936	246,201	156,505	884	1133	699	474.2	709	263
		Total	12	286.0	712	141	229,797	339,150	156,505	880	1203	476	533.4	1305	263
Sartorius		M	3	165.3	257	102	118,077	130,192	99,654	804	977	484	874	1520	498

mean values decrease slightly in human sartorius and sternocleidomastoid and monkey biceps brachii in that order. The range of variation was approximately equal in these muscles.

The largest number of muscle fibers per square millimeter in monkey biceps brachii occurred in males. The values decrease in human adductor longus, sternocleidomastoid, sartorius, and biceps brachii in that order. The value in human biceps brachii was approximately one half that of the monkey. In human females, the value for biceps brachii was larger than that for any other muscle, and that of monkey biceps brachii, human sternocleidomastoid, and adductor longus decrease in that order.

The diameter of muscle fibers in males was greater in human biceps brachii and sartorius, intermediate in adductor longus and sternocleidomastoid, and least in monkey biceps brachii. In females, it was greatest in sternocleidomastoid and the monkey biceps brachii, decreasing in human biceps brachii and adductor longus in that order. Males exhibit a larger diameter of muscle fiber in the human biceps brachii and adductor longus, while females had larger diameters in sternocleidomastoid and the monkey biceps brachii.

As to the relationship among cross-sectional area, total number of muscle fibers, number of fibers per square millimeter, and diameter of muscle fibers (Figures 1, 2, and 3), greater cross-sectional area is associated with larger total number, smaller number of muscle fibers per square millimeter, and larger fiber diameter. Smaller cross-sectional area is associated with smaller total number, larger number of fibers per square millimeter, and smaller fiber diameter. This tendency is pronounced in human biceps brachii and adductor longus, less so for the relationship of diameter or number of muscle fibers per square millimeter in sartorious or total fiber number in monkey biceps branchii, and is absent in sternocleidomastoid. These observations suggest that the development of muscle varies greatly depending upon the composition of muscle fibers.

It also follows that the muscles used in this study may be divided into three groups with respect to composition. First, in human biceps brachii and adductor longus, the range of variation of total fiber number is large, muscle hypertrophy pronounced, and the diameter of muscle fibers and number of muscle fiber per square millimeter follow the development of muscle. Second, in human sternocleidomastoid and monkey biceps brachii, the range of variation of total fiber number is small, muscle hypertrophy not pronounced, and the diameter does not follow the development of muscle, having a greater value in females. Third, human sartorius exhibits an intermediate condition. According to the description by Kohashi (1937) and Halban (1894) the first group is comprised of muscles with intense contraction, while the second group is made up of muscles with continuous stable strength.

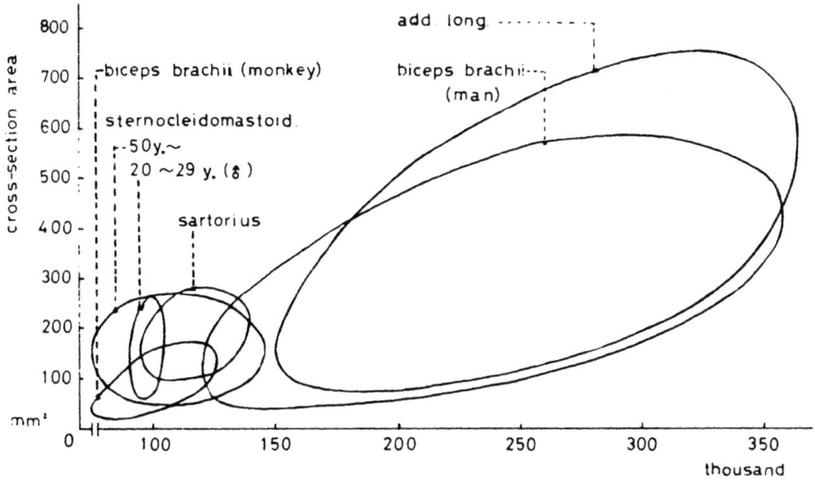

Figure 1. Correlation between the cross-sectional area of muscle belly and the total number of muscle fibers

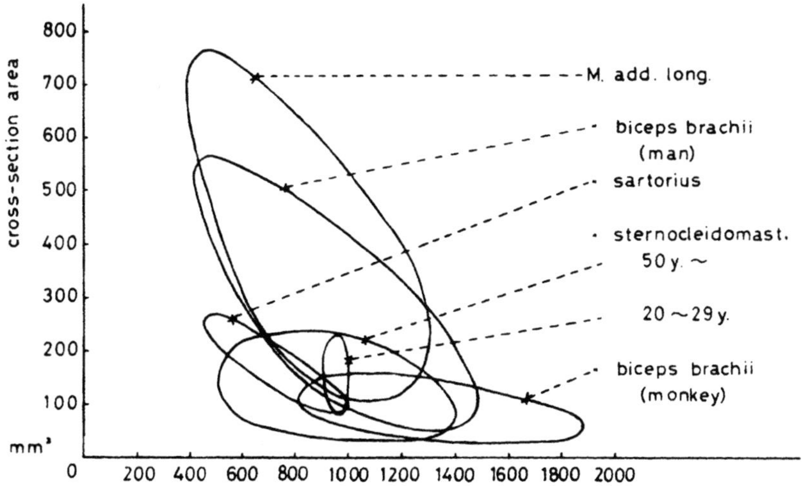

Figure 2. Correlation between the cross-sectional area of muscle belly and the number of muscle fibers per square millimeter

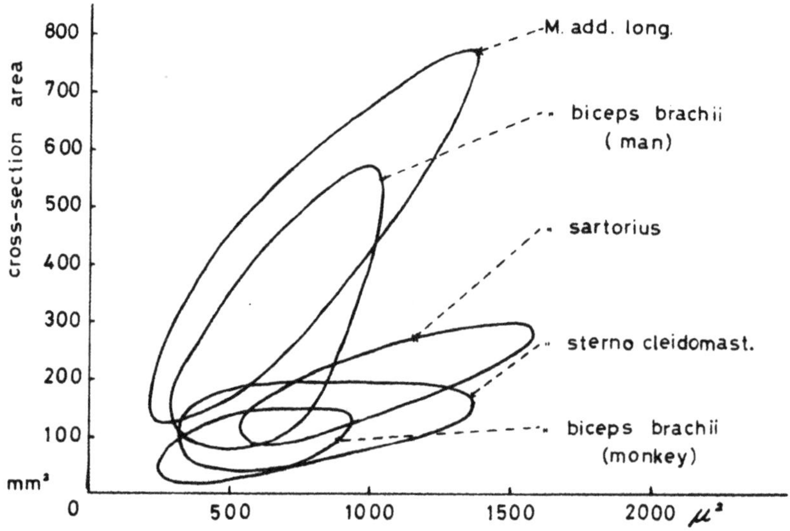

Figure 3. Correlation between the cross-sectional area of muscle belly and the size of muscle fibers

CONCLUSION

In human biceps brachii (ten cases), sternocleidomastoid (eight cases), adductor longus (twelve cases), and sartorius (three cases), and biceps brachii (ten cases) of crab-eating monkeys, total number of muscle fibers in maximal cross-sectional area, number of muscle fibers per square millimeter, and size of muscle fiber were studied to establish the individual differences in muscle fiber composition of skeletal muscle. The following were observed:

1. Total number of muscle fibers was largest in human biceps brachii and adductor longus, which showed a great range of variation.

2. The number of muscle fiber per square millimeter was largest in biceps brachii of male monkeys and human females, and tended to be inversely proportional to the cross-sectional area of the muscle belly. This tendency is pronounced in human biceps brachii and adductor longus, less so in monkey biceps brachii.

3. The diameter was greatest in human biceps brachii in males, in human sternocleidomastoid, and monkey biceps brachii in females, and was closely related to the number of muscle fibers per unit area.

4. The foregoing findings suggest that the features of muscle development are closely associated with muscle function.

REFERENCES

BARIN-BAUM, D. E.
1963 Über die Faswezahl in Muskulus biceps brachii des Menschen unter Berücksichtigung des Körper bautypus. *Acta Anatomica* 55:224–254.

ETEMADI, A. A., E. HOSSEINI
1968 Frequency and size of muscle fibers in athletic body build. *Anatomical Records* 162:269–274.

HALBAN, J.
1894 Die Dicke der quergestreiften Muskelfasern und ihre Bedeutung. *Anatomische Hefte* 3:267–308.

KOHASHI, Y.
1937 Histologische Untersuchungen der verschiedenen Skelettmuskeln beim Menschen, I. Untersuchungen beim Erwachsenen. *Okajimas Folia Anatomica Japonica* 15:175–188.

Adaptive Strategies of Human Populations under Nutritional Stress

WILLIAM A. STINI

It has long been recognized that average body size has been increasing in most human populations. This increase in stature and weight has been accompanied by a trend toward earlier maturation in these same populations (Tanner 1968). Interpretations of these concurrent events implicate improvement in nutritional state and in the quality of medical care as contributory factors. Support for this point of view is found in the variations in body size observed in different socioeconomic groups in contemporary (Mendez and Behrhorst 1963; Ashcroft, Heneage, and Lovell 1966; Sabharwal, Morales, and Mendez 1966; Villarejos et al. 1971; Dellaportas 1969) and past human populations (Tanner 1964). There appears to be a general pattern in which individuals respond to stress by altered growth trajectories. Because such alterations improve the chances for survival in the face of potentially lethal stresses, they may properly be called adaptive. The capacity to adjust individual metabolic demand avoids or defers the cost of natural selection and, as a result, retards the process of genetic change that would qualify as true adaptation. However, it must be kept in mind that the ability to make the necessary adjustment is a variable trait that will in itself be subject to natural selection. The existence of the ability to make nongenetic adjustments in body size may be inferred from current increases in body size which reveal the retention of genetic programs for larger body size although these programs have not been expressed to their maximum for many centuries. The most attractive explanation for this phenomenon is that body size *per se* is not being subjected to the full force of natural selection in most human populations. Instead, there exists a high degree of developmental plasticity. As a result, certain aspects of the genome of the population are "buffered." Slobodkin (1968) has argued for the existence of hierarchical response systems to environmental stressors.

Through such graded responses organisms undergo slow, deep-seated physiological changes that permit continued exploitation of more rapid, short-term adjustments in the most flexible manner. Slowing of growth and its termination before the attainment of full genetically programmed adult size would be an example of such a mechanism.

Living organisms are successful competitors in the game of evolution and, therefore, have accrued adaptations to a host of environmental challenges. A well-adapted population would be, as Slobodkin argues, one that enjoys a relatively high probability of survival under conditions highly likely to occur. Since man has become an agriculturist and a dweller in cities in increasing numbers, two regularly encountered stressors have been nutritional imbalance and disease. Adaptation to one of these conditions necessarily affects the degree of adaptation to the other. Diseases account for most mortalities of protein-deficient children. These children are unable to satisfy their anabolic amino acid requirements and simultaneously synthesize adequate quantities of antibodies essential to combat infection. Selection for adaptability in human populations would tend to favor physiological mechanisms that would deal with these problems synergistically. One such mechanism involves a shift in adrenocortical function that simultaneously affects anabolic processes and antibody synthesis. Such general stress responses produce profound changes in growth and maturation which, if suitably prolonged, will be reflected in late maturation and reduced adult body size. From the standpoint of the population's genetic constitution, such reductions relax the pressure on food resources available to each individual and thereby allow a larger number of individuals to coexist. Relaxation of one form of mortality permitting an increase in population size allows maintenance of genetic variability and reduces the likelihood of extinction. It may be convincingly argued that such nongenetic adaptations possess great importance in influencing the nature of genetic adaptations capable of surviving in a population and must therefore be taken into account whenever the process of adaptation is considered. It is in this context that the following discussion of human adaptive strategies under conditions of nutritional stress will be undertaken. It is hoped that the importance of the interaction of individual body size and population size will attain increased significance when viewed in this way.

METABOLIC DEMAND AND BODY SIZE

Although tissues vary in their metabolic demands, thereby making precise estimates of caloric requirements for a given total body weight difficult, there is a direct relationship between lean body mass as found in skeletal muscle and metabolic rate. Certain constants may be consulted when estimations of metabolic demand are attempted. For instance, speed of shortening of particular muscles is similar from species to species. When the generality of the mechanism of muscle contraction and sarcomere structure is considered, it can be seen why this similarity exists. Also, the power output of any muscle and the metabolic variables involved in maintaining energy flow to that muscle all depend upon its cross-sectional area (McMahon 1973). Of course, there are muscles with varying functions that contract at different rates and thereby differ in metabolic cost of maintenance. Broadly, differentiating between tonic and phasic muscles makes possible prediction of differing metabolic costs of contraction. Although the sharp distinction between red and white fibers has been criticized recently (Goldspink 1970), there is good reason to think that the energy requirement of white fibers is greater than that of red fibers, and the tonic muscles bear a greater proportion of red fibers than do the phasic muscles. It might be anticipated, then, that tonic muscles are metabolically less demanding than phasic ones. There is also evidence that fibers can shift from white to red characteristics during their lifetimes (Goldspink 1970). Such shifts may make possible the reduction of metabolic demand while allowing a full complement of fibers to survive during periods of nutritional deficiency. It also appears that metabolizing tissue may be replaced by inert collagen (Scrimshaw 1964).

While acknowledging the potential effects of the above-mentioned factors, it must be stressed that there is a close relationship between the amount of skeletal muscle present and the metabolic requirements of the individual. Activity levels will certainly affect caloric requirements as well, and the work performed by an organism includes the function of moving its own mass from one place to another. As that mass increases, the energy cost of work increases, creating additional caloric demand. If standards for metabolic cost of activity are consulted (Consolazio, Johnson, and Pecora 1963), it is possible to estimate the magnitude of difference in energy requirements for individuals of different body size. An example of such estimation may be seen in Table 1. The quantities used in

Table 1. Comparison of caloric requirements of males of mean body weight in United States and Colombian populations at various activity levels

	United States	Heliconia, Colombia
Mean body weight (kilograms)	70	60
Caloric costs (kilocalories)		
Resting (8 hours)	570	480
Sedentary activity (6 hours)	492	266
Light labor (8 hours)	1,527	1,123
Moderate labor (2 hours)	654	482
Total	3,243	2,351

Table 1 are drawn from United States standards (Consolazio, Johnson, and Pecora 1963) and data collected in a South American Caucasian population living in an area of endemic protein-caloric deficiency (Stini 1968).

As may be seen from comparison of the total energy requirement of the males shown in Table 1, the 70-kilogram man requires about 3,200 kilocalories per day in the performance of a normal round of activities. The 60-kilogram man performs the same round of activities at a cost of approximately 2,300 kilocalories. The difference between these two daily requirements, about 900 kilocalories per day per man, is about one-third of the daily requirement of the 60-kilogram man. Stated another way, four 60-kilogram males may be supported on a resource base capable of supporting three 70-kilogram men. Size differences between females in United States and South American populations are not as pronounced, a point that will be discussed below.

Differential protein requirements also emerge when individuals of different body size are compared. According to the somewhat generous daily recommendations published by the Food and Nutrition Board of the National Research Council (1958), a 70-kilogram man should ingest 70 grams of protein per day, while a 60-kilogram man would require 60 grams. A 58-kilogram woman would need 58 grams of protein, while a 51-kilogram woman would need 51 grams. All of the above figures apply to the adult segment of the population and take into account the lower metabolic rate prevailing in females. Energy but not protein requirements are reduced in later years in both sexes. Requirements for both protein and calories are higher during the adolescent growth spurt in well-nourished populations where the characteristic sigmoid curve of values for both attained height and weight makes its abrupt upward deflection. However, this pattern of growth and development may be altered so that both attained height and weight follow a more linear and prolonged trajectory. Later attainment of adult body size and less explosive growth both tend to be more economical in terms of caloric requirements of the individual. In this manner further reductions in caloric demand supplement those made possible through reduced adult body size.

When the yield of agricultural land is considered in the light of energy requirements of the human population, the significance of some of the reductions in individual caloric requirements becomes more apparent. For instance, maize culture as practiced in much of South America may be expected to yield 3,500 kilocalories per acre per day, while wheat yields 2,000 kilocalories per acre per day (Weiner 1972). By bringing additional land under cultivation, it is possible to support a larger population, and the benefit of a large family of immature, metabolically inexpensive workers is more practical than philosophical. If a population's division of labor is arranged to maximize the labor input of children under such circumstances, the biological advantages of a broad-based demographic profile, with many young people and relatively few in the higher age categories, can be seen (Thomas 1972). In terms of maximization of population number, this advantage holds even when no additional land remains to be brought under cultivation. However, when population density attains a level where all land is exploited to the limit, Malthusian factors will assert themselves, and the buffering effect of nongenetic adaptations will rapidly diminish under continued population pressure. Ultimately, the action of natural selection would favor adaptations of the genetic variety. More will be said on this point in another section of the present study.

MECHANISMS CONTROLLING BODY SIZE

Growth and development in man is a lengthy process made up of a number of stages. Intrauterine life is characterized by an early period of hyperplastic growth wherein the cell population increases from one to over two and one-half trillion. Before birth, hyperplastic growth has given way to hypertrophic growth in most tissues. Increase in cell size rather than number will characterize future growth. This is particularly true of functional cells of the nervous system and of skeletal muscle. Much has been written concerning the early termination of skeletal muscle hyperplasia. Some early discussions of this topic date from over a hundred years ago (Olivier 1869; Hayern 1875; Bompar 1887; MacCallum 1898; Durante 1902; Bramwell and Muir 1907; Bablet and Normet 1937), while experimental work pursuing the precise mechanisms involved continues today. Although there is some disagreement concerning the time when hyperplasia ceases (Cheek 1968), as well as evidence that in some species limited amounts of muscle fiber replacement occur (Cheek, Brasel, and Graystone 1968), there is abundant evidence that few if any new skeletal muscle fibers appear in humans after the sixth month postpartum. As a result of this early shift from hyperplastic to hypertrophic muscle growth, factors preventing the formation of a full complement of muscle fibers before the sixth month postpartum and those resulting in destruction of fibers after that time will permanently reduce the number of fibers present in the adult. Such reductions might be viewed as a "resetting" of the individual's future caloric demand. By what means are these reductions obtained? At this point all the evidence available must be drawn from nonhuman species, which may or may not react in a manner identical to human subjects.

It has been convincingly demonstrated by a number of researchers using several mammalian species that the stress of protein-caloric deficiency evokes a response from the adrenal cortex. The level of this response may be monitored by determination of the concentration of anti-inflammatory glucocorticoids in the serum (Kinsell et al. 1953; Castellanos and Arroyave 1961; Hamwi and Tzagournia 1970). It is believed that the presence of increased concentrations of these hormones effects a shift in liver function resulting in the cycling of essential amino acids through synthetic pathways which reduce the likelihood of damage to that vital organ (Waterlow 1955). The diversion of amino acids from anabolic pathways results in their loss to the maintenance and growth processes of skeletal muscle. Because skeletal muscle is constantly exchanging amino acids with the surrounding interstitial fluid, the cutoff of supply results in a net loss to the muscle, sometimes referred to as a reduction in "protein stores" (Clark 1953; Waterlow 1955; Leon 1966; Goldberg 1969; Korner 1960). Administration of adrenocorticotrophic hormone (ACTH) and hydrocortisone have both been shown to cause reductions in muscle mass in experimental animals (Sidransky, Wagle, and Verney 1969; Sidransky and Verney 1970). In addition, presumptive evidence of adrenocortical hyperactivity under prolonged stress of malnutrition may be found in the enlargement of the cortices and in some cases their breakdown under extended periods of stress (Newman 1973). The latter observation would be expected if the "exhaustion phase" of the general adaptive syndrome of Selye (1950) had been reached. It is significant that high serum concentrations of hydrocortisone have the additional effect of suppressing antibody synthesis (Dean 1964). This somewhat maladaptive concomitant of the general adaptation syndrome may be interpreted as furnishing additional evidence for the high priority of the preservation of liver function.

The foregoing discussion, though by no means exhaustive, provides a consistent explanation for several elements of the reduction in lean body mass occurring during prolonged protein–calorie malnutrition. Perhaps a fuller understanding of all the mechanisms involved will be possible in the near future, but many aspects of the process occurring in man are still unknown. Nevertheless, observations of reduction in human muscle fiber populations under severe protein–calorie malnutrition have been published (McFie and Welbourne 1962; Montgomery 1962). Detection of reduced muscle fiber population has been accomplished by means of total fiber count performed on a cross section of the sartorius muscle. Muscles from adults who had suffered severe malnutrition as children were compared with those of individuals having no record of malnutrition and were found to contain significantly fewer fibers. In an attempt to ascertain whether such early losses of muscle fiber could be induced in a relatively short time and still be permanent, the author examined sartorii from a sample of domestic hogs subjected to protein deficiency for three months after birth. These were compared with muscles from animals fed continuously high-protein diets. Despite the fact that all the animals were provided with high-protein diets for six months following the experimental period, the protein-deficient category showed significant reductions in body weight, sartorius cross section, and fiber count. The results of this study are shown in Tables 2 through 5. It is not known whether this permanent reduction in fiber population reflects an interruption in hyperplastic growth soon after birth or a destruction of fibers that had already formed. Whichever process (or possibly both) determined the final fiber population, the reduction in lean body mass and associated metabolic demand is impressive (Stini 1972b). Other studies have pointed out the permanence of size reductions associated with malnutrition during critical stages of growth (Sidransky and Verney 1970), while in some cases arguing that metabolic efficiency, at least in anabolic processes, may be somewhat impaired by early protein deficiency, particularly during nursing (Barnes et al. 1973).

Table 2. Weights of pigs in four nutritional categories

n	Category	Weight at 3 weeks (kg)	Slaughter weight (kg)	Dietary intake during experimental period
5	Control			
	11– 2	6.9	285	18 percent casein
	11– 4	6.5	319	protein diet
	10– 6	5.8	313	*ad libitum*
	16– 5	5.1	319	$\bar{x} = 307$
	16–13	3.3	299	
6	Restricted			
	9– 1	4.8	248	18 percent casein
	10– 3	4.8	272	protein limited
	11– 1	6.1	292	to 2 kilograms
	14– 1	6.0	263	total intake
	16– 3	4.8	271	per day
	16– 7	4.6	282	$\bar{x} = 271$
6	Intermediate protein			
	10– 1	6.9	287	5 percent casein
	9– 6	6.5	237	protein diet
	11– 7	5.8	272	*ad libitum*
	14– 6	5.5	302	$\bar{x} = 272$
	14–10	6.4	290	
	16– 9	3.4	245	
6	Low protein			
	9– 2	4.1	234	3 percent casein
	10– 7	5.1	255	protein diet
	11– 9	5.5	244	*ad libitum*
	14– 3	6.9	240	$\bar{x} = 243$
	16– 4	5.0	233	
	16–12	4.2	251	

Table 3. Analysis of variance: slaughter weights of four nutritional categories

	Category controls ($n = 5$)	Restricted ($n = 6$)	Intermediate ($n = 6$)	Low protein ($n = 6$)	Total
Σx	1,535	1,628	1,633	1,457	6,253
$\bar{x}$	307	271	272	243	
Σx^2	472,117	442,886	447,851	354,207	1,717,061
			Correction term		1,700,000
			Total sum of squares		17,061

Sources of variation	d.f.	Sums of squares	Mean square
Total	23	17,061	
Groups means	3	11,232	3,744
Within group	20	5,829	291

F = 3,744/291 = 12.86; P <0.001

Table 4. Mean corrected cross-sectional areas and fiber counts from sartorius muscles of four nutritional categories of pigs

	Category controls	Restricted	Intermediate	Low protein
Area (mm²)	200.5	206.9	190.0	176.4
Count (mm²)	29.7	31.2	32.6	32.0
Total fibers	5,934	6,395	6,155	5,562

Table 5. Analysis of variance for mean total muscle fiber populations within and between four nutritional categories of pigs

Sources of variation	d.f.	Sums of squares	Mean square
Total	89	103,644,166	
Group means	3	8,969, 257	2,989,752
Within group	86	94,674,969	1,100,872

F = 2,989,752/1,100,872 = 2.72, d.f. = 3.86; P ~0.05

Evidence of Lean Body Mass Reduction in Living Humans

There are several methods for estimating the lean body mass of human subjects. An underlying principle of most techniques involves differences in the specific gravity of muscle and that of fat. Most of the elaborate techniques, such as underwater weighing, are impractical under most field conditions. Field studies attempting to make such estimates generally rely on less precise but more feasible procedures. Among the latter category are such readily applicable procedures as measurement of muscle shadows on radiographic plates or the derivation of muscle circumference of the limb from measurement of its outer circumference minus skinfold thickness. Use of upper arm muscle circumference in the estimation of relative lean body mass has yielded much useful information in field studies and has been the subject of a long and active debate concerning its predictive value (for discussion and description of technique, see Stini [1972c]).

A reasonable position is that whatever the difficulties in extrapolating from upper arm muscle circumference to a precise estimate of total lean body mass, comparisons of population means for upper arm muscle circumference and mean body weight allow estimation of the relative amounts of lean body mass present in the respective populations. When a comparison of the values for upper arm muscle area is made between the adults of a protein-deficient Caucasoid population in South America and those of the United States, the figures shown in Table 6 are obtained.

The comparison shown in Table 6 clearly indicates the tendency for adults in Heliconia, Colombia, to attain significantly lower values for this measure of lean body mass. The observation lends support to the assertion that the difference in adult body weights observed in the two populations bears a sizable component of lean body mass and hence might reasonably be expected to affect relative caloric requirements to at least the extent calculated in Table 1. It is therefore possible that the difference in caloric requirements is greater than indicated by application of standard values, as was done in the preparation of Table 1. This observation would support the suspicion, held by many researchers, that the recommended values for both protein and caloric intake (National Research Council 1958), having been developed through the study of well (or excessively) fed populations of industrialized countries, may be highly

Table 6. Comparison of Heliconia, Colombia, and United States values for upper arm muscle area (square millimeters)

	Heliconia		United States		Heliconia percent of United States	
	area (mm^2)	weight (kg)	area (mm^2)	weight (kg)	area (mm^2)	weight (kg)
Male	4,267	60	6,464	70	66	(85)
Female	3,450	51	4,272	58	81	(88)
Percent female of male	81	85	66	83		

inaccurate when extrapolated to populations adapted to much lower nutrient intake. Additional support for this point of view may be found in comparison of actual versus recommended caloric and protein intakes in areas of endemic nutritional stress, as in Heliconia, Colombia (Vitale and Velez 1967). When such comparisons are made, it is found that actual nutritional intake is considerably lower than recommended with respect to both protein and calories.

Reductions in lean body mass (more accurately, total protein; for discussion, see Waterlow and Alleyne [1972]) are directly associated with reductions in caloric and protein requirements but cannot be sustained without other morphological adjustments. An important component of the adjustment is the alteration in the pattern of skeletal growth (Bartley 1968). The end result of the pattern observed in nutritionally stressed populations is a reduction in the size of the skeleton. The skeleton grows more slowly, particularly in males, with respect to both appearance of ossification centers and epiphyseal growth and closure. Termination of epiphyseal growth in males may not occur until the second half of the third decade of life. This attenuated and prolonged growth pattern under endemic nutritional stress, while associated with size reduction, appears to have little effect on allometric relationships. Generally proportionality is seldom affected to any appreciable degree. The most striking examples of this phenomenon may be seen in cases of "nutritional dwarfism" wherein statures 60 to 70 percent of normal for the local population occur in individuals whose body proportions are in every way normal (Stini 1968).

Despite the overall tendency to maintain normal proportions, one clearly discernible difference between populations living under conditions of nutritional stress and better-fed populations is the greater linearity of nutritionally stressed adults. This linearity, as measured by ponderal index scores, is most pronounced in males. As seen in Table 7, ponderal index scores are higher in males than in females of even well-fed populations, but the difference grows larger as the nutritional state becomes more marginal. Although fat is clearly implicated in the observed lower ponderal indices of well-fed populations as well as in those of the female members of poorly fed ones, reduction in lean body mass is the more important variable producing the greater linearity of poorly fed males. It is noteworthy that populations exhibiting this tendency toward reduction of sexual dimorphism for total body mass, and by implication protein and calorie requirements, show little if any difference with respect to sexual dimorphism for stature (Stini 1972a). The world range of male/female stature ratios is a fairly narrow one (approximately, from 1.04 to 1.09), and poorly fed populations show no systematic tendency to cluster near either end of this range. Values for a number of populations are shown in Table 8.

Table 7. Comparisons of adult mean values for stature, weight, and ponderal index in various parts of the world ranked by male ponderal index (from Stini 1972a)

	Male			Female			
	Stature	Weight	Ponderal index	Stature (cm)	Weight (kg)	Ponderal index	Rank
India	163.0	48.2	13.43	150.7	45.2	12.69	(1)
Burma	161.5	49.9	13.15	150.7	45.2	12.69	(2)
Thailand	161.0	53.2	12.83	151.0	45.8	12.65	(3)
Vietnam	157.6	49.1	12.82	147.6	63.5	12.59	(4)
Malaya	161.8	55.1	12.75	149.8	47.8	12.28	(9)
Japan	163.2	56.8	12.71	151.9	49.1	12.42	(6)
Nuñoa, Peru	160.7	55.6	12.68	148.8	52.6	11.89	(15)
Heliconia	162.0	57.1	12.59	151.3	50.7	12.24	(10)
Sweden	175.9	75.1	12.49	164.0	65.2	12.22	(11)
United States	175.5	77.6	12.45	161.8	61.7	12.39	(7)
Northeast Brazil	160.6	58.0	12.43	150.4	48.8	12.34	(8)
Colombia	160.1	59.6	12.28	152.2	49.4	12.44	(5)
Bolivia	161.3	61.0	12.28	151.0	54.0	11.97	(13)
Ecuador	157.7	57.5	12.25	146.3	51.3	11.81	(16)
Uruguay	168.1	70.0	12.23	155.5	60.0	11.90	(14)
Chile	162.8	64.3	12.18	151.9	56.4	11.98	(12)

Table 8. Male stature as percent of female stature (within populations) ranked in descending order (adapted from Stini 1972a)

Macedonia	109.0	Costa Rica (Urban)	106.9
Sweden	109.0	Bolivia	106.8
Lera (Africa)	108.8	Brazil	106.8
Venda (Africa)	108.8	Vietnam	106.8
Yugoslavia	108.5	Germany (North)	106.7
United States	108.5	Thailand	106.6
Bulgaria	108.4	Germany (South)	106.4
Portugal	108.2	Costa Rica (Rural)	106.3
India	108.2	Norway	106.2
Wales	108.2	Ukraine	106.2
Uruguay	108.1	Switzerland (Geneva)	106.1
Gypsy	108.0	Bassari (Africa)	106.1
Peru (Highland)	108.0	Switzerland (German)	106.0
Malaya	108.0	Poland	105.8
Ecuador	107.8	Highland Swiss	105.7
Czechoslovakia	107.8	Bisago (Africa)	105.5
Estonia	107.7	Gondar (Africa)	105.2
Baka-Bayaka (Africa)	107.6	Colombia (Lowland)	105.2
Japan	107.4	Mbuti (Africa)	105.1
Chile	107.2	Mangisi (Africa)	104.7
Burma	107.2	Lari (Africa)	104.6
Heliconia (Colombia)	107.1		

The Significance of Reductions in Sexual Dimorphism

The tendency for the male members of malnourished human populations to experience greater reductions in adult lean body mass than do females may well be adaptive for the population. Because most tasks are performed well below the maximum capacity of the smaller males and because division of labor may make possible the utilization of labor of slowly growing adolescents and young adults without serious loss of productivity, larger body size in males would raise the cost of their maintenance while providing little or no gain in terms of resources captured. It has been shown in several populations that smaller size may be combined with an equal (Viteri as cited by Waterlow and Alleyne [1972]) or superior functional efficiency in the performance of a number of tasks (Parizkova and Merhaustova 1970). There is a real advantage, then, in developmental adjustments producing smaller but well-proportioned and functionally unimpaired males. The problem is not as easily resolved in the case of females. While work performance is important for women in populations where endemic nutritional stress occurs, the capacity to reproduce successfully would be more directly subject to the action of natural selection.

From the standpoint of fitness, fertility and fecundity are all-important factors. Even so, success in conceiving, bearing, and giving birth to normal infants is genetically meaningless if the infants themselves do not mature and reproduce. One critical period in the formation of a viable organism occurs early in embryological development when differentiation gives rise to the organs and tissues essential to independent life. Damage occurring in this early period will result in a defective infant or, in many cases, early loss as an aborted fetus. The human mother is remarkably capable of shielding the developing embryo and fetus from nutritional stress during early gestation. It has often been said that the fetus parasitizes the maternal organism. Although there are differences of opinion on the degree to which the host-parasite analogy may be accepted, there can be little doubt that malnourished mothers provide essential nutrients to their fetuses even when they themselves receive inadequate quantities of them. This is particularly true with respect to protein. In most malnourished human populations, the recommended intake of essential amino acids, most readily obtained from animal protein, is not available. As mentioned earlier, adult protein requirements are frequently estimated as being on the order of one gram per kilogram body weight per day. To satisfy this requirement, a 50-kilogram woman would need 50 grams of protein every day, an amount she probably never receives. It is clear that physiological adjustments allow survival in the face of this persistent deficiency. The situation is severely exacerbated during the second half of pregnancy, when protein requirements are thought to rise by 20 grams per day. Even more severe is the stress occurring during lactation when a milk flow of 850 milliliters per day requires 40 grams of protein above the baseline requirement. Energy requirements also rise, by about 1,000 calories per day, during lactation. It is during the months following birth, when the mother is nursing her infant, that a second and possibly most critical threat to survival occurs. If the mother is incapable of providing sufficient milk, the infant will be unlikely to survive because of both nutritional and immunological factors. A reduction of the quantity of milk might, within limits, be tolerated and does appear to occur. Alterations in the quality of mother's milk are seldom if ever seen and would be more destructive to the infant's welfare if they did occur.

The maintenance of something near the normal quantity of flow of high quality mother's milk can only be accomplished at the expense of the maternal organism. Because the most serious imbalance occurs with respect to protein, it is carcass protein that must supply the difference. The mechanisms by which maternal tissue are caused to give up amino acids which ultimately supply the needs of a growing fetus and a nursing infant have been studied in experimental animals. It appears that the maternal adrenal cortex and placental hormones are capable of evoking this response (Aschkenasy-Lelu and Aschkenasy 1959; Kinzey and Srbnik 1963; Berg, Siggs, and Greengard 1967; Kinzey 1968). The possession of such a mechanism allows survival of the infant through depletion of the reserve of essential amino acids maintained in maternal tissue, particularly skeletal muscle. Such capability would be rendered useless if there were no maternal reserve. Therefore, reduction in the number of muscle fibers possessed by the female members of a malnourished human population to the degree occurring in males would be maladaptive.

The end result of this combination of selective forces is the tendency for males to reduce lean body mass to a greater extent than do females in the same population, thereby reducing the degree of sexual dimorphism for this trait. Here we see demonstrated the capacity of populations to exploit physiological adjustments to soften the impact of natural selection. Evidence that nonhuman primate mothers are capable of resorbing significant quantities of tissue to maintain their fetuses has recently been discussed by Watts and Riopelle (1973). The results observed in controlled experiments with rhesus macaques appear consistent with those observed in humans. It appears that delay of skeletal growth is not as great among females as among males in these same populations. This provides further evidence that females are better buffered against the effects of nutritional stress, with the likelihood that the preservation of the reproductive potential of the population is the biological factor involved.

The Limits of Physiological Adaptation

One view of the role of physiological adaptations in the process of evolution is that stated by Waddington (1957), i.e., such adaptations allow the population to survive and remain in an area characterized by a particular stress until such time as the proper genetic events occur to allow truly adaptive change. When more adaptive genes or gene combinations arise in the population, their superior fitness would result in a shift in gene frequencies producing a restructuring of the population's genome — an event that could properly be called adaptation *sensu strictu*.

What are the indications that natural selection might be altering gene frequencies in nutritionally stressed human populations? A number of studies have pointed out greater survivorship of smaller newborns in areas where maternal nutrition is inadequate. These studies include a sampling of lower socioeconomic categories in the city of Aberdeen, Scotland (Baird as cited by Drillien [1964]), where lower birth weights and a higher percentage of stillbirths occur than in higher socioeconomic categories. Other studies conducted in India reveal that survivorship of very small infants (1,600–2,000-gram birth weights) is higher among individuals of lower socioeconomic strata than among their more prosperous neighbors. This is true despite the fact that superior medical attention, including in-hospital births, is available to members of the higher socioeconomic categories (Jayant 1964). Similar observations have been made in Africa (Hollingsworth 1965) and England (Karn and Penrose 1951). Frisancho, in discussing Peruvian data (Frisancho et al. 1973), provides evidence of higher rates of survival for infants born of smaller parents. In all of these studies, however, it is difficult to ascertain whether higher survivorship of smaller infants may, in some cases, reflect early loss of less viable embryos and fetuses, excluding them from calculation in the categories of neonatal and infant mortalities. Future demographic studies of malnourished populations will gain considerable value through increased attention to these categories of conceptus loss.

One line of evidence concerning the mechanism of conceptus loss is revealed by the frequent observation of reduced adrenal weights in stillborn infants of poor parents in developed countries (Newman 1973). Such reductions may reflect increased adrenocortical activity on the part of stressed mothers. The genetic implications of whatever mechanisms affect the survivorship of infants could, over an extended period, be profound. It is unlikely, however, that simple genetic mechanisms will be sufficient to explain the form of selection occurring. The traits involved are polygenic and the interaction of the changes which might be anticipated under various forms of stress would make the most acceptable model the one that involved "multiple adaptive peaks" as proposed by Wright (1948), rather than "a single best type" model which would lend itself to straightforward deterministic treatment. In short, there is evidence that natural selection is at work in populations under nutritional stress. Although it is not possible at this time to produce a predictive model that could be tested in these populations, it seems warranted to assume that the cultural, behavioral, physiological, and developmental buffers that modify the intensity of natural selection under conditions of long-term nutritional stress have limits to their effectiveness. When the stress is severe, intense, and prolonged enough, the price of natural selection must be paid. The cost of the loss of a conceptus rises with each day it survives, so that early loss of embryos would be the most effective adaptive strategy for populations under severe stress. Loss of third trimester fetuses would be more costly, and loss of neonates and infants would, of course, be increasingly more maladaptive from the standpoint of the population's survival. Whether human populations employ adaptive strategies in a series that effectively minimizes the cost of natural selection during gestation and early postnatal life is not clear from the available evidence. It is unquestionably an important area for future research.

SUMMARY AND CONCLUSIONS

Using an analogy introduced by Slobodkin, it might be said that adaptation restores flexibility to physiological responses much as learning restores flexibility to behavior. Taking the analogy a step further, we might compare genetically determined response patterns to a particular stress to stereotypic behavior in a given circumstance. Both are extremely effective and economical in the absence of drastic change in conditions. But both bear the potential of being severely disadvantageous when conditions are sufficiently altered. Because the adaptedness of a population might be stated in terms of the probability of its survival under conditions that are very likely to occur, it might be anticipated that there would be a high value associated with the possession of a sequence of adaptive responses of increasing cost for frequently encountered stressors. The most costly response would involve the loss of an adult individual before that individual contributed either to the energy capture function or to the gene pool of the population.

Because nutritional stress is a frequently encountered and potentially lethal stressor, it might be anticipated that human populations would possess a variety of adaptive responses to it. Evidence for a number of physiological and developmental adjustments occurring in human populations makes it possible to discern the outline of a graded response sequence to this significant form of stress. There are obviously many gaps in the account presented here. There are also important lines of evidence excluded, such as a variety of behavioral alterations and cultural practices observed in areas of persistent nutritional deficiency. The reader is directed to such works as those by Jelliffe (1968) and Geertz (1970) for treatment of some of these factors. It is hoped that the present account will alert others to the importance of considering the variety of ways a population may adapt to a severe stress before genetic change occurs. Genetic adaptation alone would frequently result in a stereotypic and potentially maladaptive and rigid response.

The evolutionary importance of a multiplex and graded response system should not be underestimated merely because it is difficult to quantify. If a flexible array of adaptations allows populations to survive and reproduce in the face of environmental fluctuations of large magnitude, persistent exposure to stress will allow beneficial mutations and recombinations to occur and spread through the population. This process makes possible the sort of genetic change that can rightly be called evolution.

Interesting questions arise from any attempt to reconstruct human evolutionary history using adaptive strategies in response to nutritional stress as an example. One question involves the probable size of our ancestors. If the current trend of increase in body size is indeed revealing the genetic potential possessed by the species before agriculture and urbanization created conditions favoring reduced body size, is there not the possibility that selection favored larger size at some earlier date? There is little evidence for large body size in early *Homo sapiens* populations. However, the amount of evidence available is not adequate to exclude the possibility that early hunters and gatherers possessed mean body size resembling our own more than those of our more recent ancestors.

Another question arises from consideration of adaptive strategies of human populations under nutritional stress: is it desirable to induce larger adult body size by nutritional practices stressing excessively high animal-protein consumption? There is evidence that high-protein diets may have undesirable side effects, including greater susceptibility to cardiovascular diseases. From the standpoint of human populations throughout the world, it should be pointed out that the effects of the current population explosion are amplified by the human biomass explosion that is accompanying it. This is particulary true of the increased consumption of organisms of high trophic level, i.e., beef cattle and fattened hogs. Also, if the adaptive strategy of poorly fed populations involves reduced adult body size, it may well be that dietary supplementation stressing animal-protein intake in the early stages of life may inactivate one level of adaptation and force the population to fall back on its next line of defense. In most cases, the next line of defense involves the payment of the cost of natural selection through increased mortality rates. It is necessary, then, to be certain that early dietary supplementation can be followed by the supply of sufficient nutritional resources to satisfy the inevitable increase in both protein and calorie requirements.

REFERENCES

ASCHKENASY-LELU, P., A. ASCHKENASY
1959 "Effects of androgens and estrogens on the metabolism of protein and the growth of tissues," in *World review of nutrition and dietetics*. Edited by G. H. Bourne, 33–60. Philadelphia: Lippincott.

ASHCROFT, M. T., P. HENEAGE, H. G. LOVELL
1966 Heights and weights of Jamaican school children of various ethnic groups. *American Journal of Physical Anthropology* 24:35–44.

BABLET, J., L. NORMET
1937 Les lésions histopathologiques de la bouffissure d'Annam. *Bulletins of the Academy of Medicine* 117:242.

BARNES, R. H., E. KWONG, L. MORRISSEY, L. VILHJALMSDOTTIR, D. LEVITSKY
1973 Maternal protein deprivation during pregnancy or lactation in rats and the efficiency of food and nitrogen utilization of the progeny. *Journal of Nutrition* 103:273–284.

BARTLEY, M. H., JR.
1968 "Structural activities of the anti-inflammatory steroids and their relationship to osseous tissue." Unpublished doctoral dissertation, University of Utah.

BERG, B. N., E. B. SIGGS, P. GREENGARD
1967 Maintenance of pregnancy in protein-deficient rats by adrenocortical steroid or ACTH administration. *Endocrinology* 80:829–839.

BOMPAR, F.
1887 Etude sur l'atrophie des muscles thoraciques chez les tuberculeux. *Journal of Medicine of Bordeaux* 17:121–130.

BRAMWELL, F., A. MUIR
1907 A remarkable muscular lesion occurring in sprue with notes of a case of peculiar (myopathic) muscular atrophy in which somewhat similar changes were pictured. *Quarterly Journal of Medicine* 1:1.

CASTELLANOS, H., G. A. ARROYAVE
1961 Role of adrenal cortical system in response of children to severe protein malnutrition. *American Journal of Clinical Nutrition* 9:186–195.

CHEEK, D. B.
1968 "Muscle cell growth in normal children," in *Human growth, body composition, cell growth, energy and intelligence*. Edited by D. B. Cheek, 337–351. Philadelphia: Lea and Febiger.

CHEEK, D. B., J. BRASEL, J. E. GRAYSTONE
1968 "Muscle cell growth in rodents: sex difference and the role of hormones," in *Human growth, body composition, cell growth, energy and intelligence*. Edited by D. B. Cheek, 306–325. Philadelphia: Lea and Febiger.

CLARK, I.
1953 The effect of cortisone upon protein synthesis. *Journal of Biological Chemistry* 200:69–76.

CONSOLAZIO, C. F., R. E. JOHNSON, L. J. PECORA

1963 *Physiological measurements of metabolic functions in man*. New York: McGraw-Hill.

DEAN, R. F. A.

1964 "Kwashiorkor," in *Recent advances in pediatrics*. Edited by D. Gardner, 234–265. London: Churchill.

DELLAPORTAS, G. J.

1969 Growth of school children in Gondar area, Ethiopia. *Human Biology* 41:218–222.

DRILLIEN, C. M.

1964 The growth and development of the prematurely born infant. *Journal of Pediatrics* 7–8:73–95.

DURANTE, G.

1902 "Anatomie pathologique des muscles," in *Manuel d'histologie pathologique*. Edited by A. V. Cornil and L. A. Ranvier. Paris: Alcan.

FRISANCHO, A. R., J. SANCHEZ, D. PALLARDEL, L. YANEZ

1973 "Adaptive significance of small body size under poor socioeconomic conditions in Southern Peru." Paper delivered at the annual meeting of the American Association of Physical Anthropologists, Dallas, Texas, April 12–14.

GEERTZ, CLIFFORD

1970 *Agricultural involution: the processes of ecological change in Indonesia.* Berkeley: University of California Press.

GOLDBERG, A. L.

1969 Protein turnover in skeletal muscle, II: Effects of denervation and cortisone on protein catabolism in skeletal muscle. *Journal of Biological Chemistry* 244:3223.

GOLDSPINK, G.

1970 "Morphological adaptation due to growth and activity," in *Physiology and biochemistry of muscle as food.* Edited by D. Brisky, L. Casseus, and D. Marsh, 521–536. Madison: University of Wisconsin Press.

HAMWI, J. J., M. TZAGOURNIA

1970 Nutrition and diseases of the endocrine glands. *American Journal of Clinical Nutrition* 23:311–329.

HAYERN, M. G.

1875 Note sur les maladies musculaires qu'on observe dans les maladies chroniques. *Comptes Rendus Sociologiques et Biologiques* 1:69.

HOLLINGSWORTH, M. J.

1965 Observations on the birthweights and survival of African babies: single births. *Annals of Human Genetics, London* 28:291–300.

JAYANT, K.

1964 Birthweight and some other factors in relation to infant survival: a study in an Indian sample. *Annals of Human Genetics, London* 27:261–267.

JELLIFFE, D. B.

1968 *Infant nutrition in the subtropics and tropics.* Geneva: World Health Organization.

KARN, M. N., L. S. PENROSE

1951 Birthweight and gestation time in relation to maternal age, parity and infant survival. *Annals of Eugenics, London* 16:147–164.

KINSELL, L. W., J. W. PARTRIDGE, L. BOLERIG, S. MARGEN

1953 Dietary modification of the metabolic and clinical effects of ACTH and cortisone. *Annals of Internal Medicine* 37:921–929.

KINZEY, W. G.
1968 Hormonal activity of the rat placenta in the absence of dietary protein. *Endocrinology* 82:266–270.

KINZEY, W. G., H. H. SRBNIK
1963 Maintenance of pregnancy with short-term injections of ovarian hormones. *Proceedings of the Society for Experimental Biology and Medicine* 114:158–168.

KORNER, A.
1960 The role of the adrenal gland in the control of amino acid incorporation into protein of isolated rat liver microsomes. *Journal of Endocrinology* 21:177.

LEON, H. A.
1966 Early effects of corticosteroids on amino acid incorporation by rat liver systems subsequent to its in vitro injection. *Endocrinology* 78:481.

MACCALLUM, J. B.
1898 On the histogenesis of the striated muscle fiber, and the growth of the human sartorius muscle. *Johns Hopkins Hospital Bulletin* 90–91:208–215.

MCFIE, J., H. F. WELBOURNE
1962 Effect of malnutrition in infancy of the development of bone, muscle and fat. *Journal of Nutrition* 76:97.

MCMAHON, T.
1973 Size and shape in biology. *Science* 179:1201–1204.

MENDEZ, J., C. BEHRHORST
1963 The anthropometric characteristics of Indian and urban Guatemalans. *Human Biology* 35:457–469.

MONTGOMERY, R. D.
1962 Muscle morphology in infantile protein malnutrition. *Journal of Clinical Pathology* 15:511–521.

NATIONAL RESEARCH COUNCIL
1958 *Recommended daily dietary allowance*. National Academy of Sciences Food and Nutrition Board, Publication 589. Washington, D.C.

NEWMAN, M. T.
1973 "Unsolved problems in human nutrition." Paper delivered in symposium "What we need to know but do not know about ecology and adaptation in man" at the annual meeting of the American Association of Physical Anthropologists, Dallas, Texas, April 12–14.

OLIVIER, A.
1869 "Des atrophies musculaires." Unpublished thesis, Paris.

PARIZKOVA, J., J. MERHAUSTOVA
1970 The comparison of somatic development, body composition and functional characteristics in Tunisian and Czech boys of 11 and 12 years. *Human Biology* 42:391–400.

SABHARWAL, K. P., S. MORALES, J. MENDEZ
1966 Body measurements and creatine excretion among upper and lower socio-economic groups of girls in Guatemala. *Human Biology* 38:131–140.

SCRIMSHAW, N. S.
1964 Ecological factors in nutritional disease. *American Journal of Nutrition* 14:112.

SELYE, H.
1950 *The physiology and pathology of exposure to stress.* Montreal: Aetna Medical Publishers.

SIDRANSKY, H., E. VERNEY
1970 Skeletal muscle protein metabolism changes in rats force-fed a diet inducing an experimental kwashiorkor-like model. *American Journal of Clinical Nutrition* 23:1154–1159.

SIDRANSKY, H., D. S. WAGLE, E. VERNEY
1969 Hepatic protein synthesis in rats force-fed a threonine-devoid diet, and treated with cortisone acetate or threonine. *Laboratory Investigation* 20:364.

SLOBODKIN, L. B.
1968 "Toward a predictive theory of evolution," in *Population biology and evolution*. Edited by R. C. Lewontin. Syracuse: Syracuse University Press.

STINI, W. A.
1968 "Physiological and morphological correlates of kwashiorkor in Colombia." Unpublished doctoral dissertation, University of Wisconsin.
1969 Nutritional stress and growth: sex difference in adaptive response. *American Journal of Physical Anthropology* 31:417–426.
1972a Malnutrition, body size and proportion. *Ecology of Food and Nutrition* 1:121–126.
1972b Lean body mass reductions in domestic hogs subjected to protein-deficient diet. (Abstract.) *American Journal of Physical Anthropology* 37:451.
1972c Reduced sexual dimorphism in upper arm muscle circumference associated with protein-deficient diet in a South American population. *American Journal of Physical Anthropology* 36:341–352.

TANNER, J. M.
1964 "Hormonal, genetic and environmental factors controlling growth", in *Human Biology*. Edited by G. A. Harrison, J. S. Weiner, J. M. Tanner, and N. A. Barnicot, 340–354. New York: Oxford University Press.
1968 Earlier maturation in man. *Scientific American* 218:21–27.

THOMAS, R. B.
1972 "Human adaptation to a High Andean energy flow system." Unpublished doctoral dissertation, Pennsylvania State University.

THOMPSON, A. M., W. Z. BELEWICZ
1963 Nutritional status, maternal physique and reproductive efficiency. *Proceedings of the Nutrition Society* 22:55–60.

VILLAREJOS, V. M., J. A. OSBORNE, F. J. PAYNE, J. A. ARGUEDAS
1971 Heights and weights of children in urban and rural Costa Rica. *Journal of Tropical Pediatrics and Environmental Child Health* 17:32–43.

VITALE, J. J., H. VELEZ
1967 *Estudio Socio-econ6mico del Municipio de Heliconia*. Investigacion Upjohn. Medellin, Colombia, February 1967, 8–10.

WADDINGTON, CONRAD
1957 *The strategy of the genes*. London: Allen and Unwin.

WATERLOW, J. C.
1955 *Protein malnutrition*. A symposium held under the auspices of FAO, World Health Organization , and the Josiah Macy Foundation. London: Cambridge University Press.

WATERLOW, J. C., G. A. O. ALLEYNE
1972 Protein malnutrition in children — advances in knowledge in the last ten years. *Advances in Protein Chemistry* 26:117–241.

WATTS, E. S., A. J. RIOPELLE
1973 "The effects of maternal protein deprivation on size and skeletal matu-

rity of newborn Rhesus monkeys." Paper delivered at the annual meeting of the American Association of Physical Anthropologists, Dallas, Texas, April 12–14.

WEINER, J. S.
1972 "Tropical ecology and population structure," in *The structure of human populations*. Edited by G. A. Harrison and A. J. Boyce. London: Oxford University Press.

WRIGHT, S.
1948 On the roles of directed and random changes in gene frequency in the genetics of natural populations. *Evolution* 2:279–294.

SECTION THREE

Effects of Chromosomal Variation on Human Development

Comments on the Effects of Chromosomal Variation on Human Development

FATHI ABDEL-HAMEED

To exist, a population must adapt to its immediate local environment. But to persist, this population must be able to maintain adaptation by altering its genetic structure, i.e., gene pool, in response to environmental changes. This means that most of the individuals in the population must possess genotypes and phenotypes which permit them to survive their contemporary environment and also to produce and reproduce new genotypes and phenotypes that are adapted to, and meet the demands of, the new environment.

Such plasticity requires a wide base of genetic variability. In sexually reproducing diploid species, a large amount of this variability is usually maintained in the heterozygous condition as concealed, or hidden, genetic variation. While polyploids can maintain a larger amount of unexpressed genetic variation, they have a greater difficulty in producing new phenotypes than the diploids. For example, a recessive allele must be present in all four sets of chromosomes of a tetraploid in order to be expressed. Therefore, diploidy seems to represent the most adaptive ploidy level since it represents the best compromise between haploidy, where concealed variation is at a minimum, and polyploidy, with its excessively concealed but extensive amount of variation.

The genetic load manifested in the elimination of maladapted individuals represents the cost of evolution, or what Haldane (1957) termed "the price a population must pay for the privilege of evolution."

Sexual reproduction provides a means of continuous reshuffling of genetic variability. Much of this genetic variation is revealed as phenotypic variation, which is exposed to the action of natural selection. The adaptive value of each genotype is dependent on the action and interaction of all the genes of the individual. Therefore, it is difficult to adequately define the term "adaptive value." When we consider the whole genotype (instead of a single locus), we become more involved with the concepts of pleiotropy, polygeny, developmental canalization, homeostasis, allele and gene interactions, and nuclear-cytoplasmic interactions.

The term "adaptation" is frequently used in a different sense, to refer to the physiological or morphological changes experienced by individuals themselves. For example, we perspire in summer more than in winter, and if we spend several weeks at high altitudes where oxygen is scarce, our blood changes so that it contains a larger number of red blood cells. Under these conditions we "adapt" physiologically to cope with new environmental conditions. Adaptability of individuals may vary within a single population, and the adaptive properties of individuals are usually heritable. The ability of higher organisms to undergo individual physiological adaptations is undoubtedly an adaptive characteristic of the species. The genetic structure of a population underlies the range of physiological adaptations exhibited by its individual members, and this underlying genetic structure has arisen through the operation of natural selection in the past.

In spite of the large amount of genetic variation present in most natural populations, and the vast number of possible genotypes, we find that a single phenotype (so called "normal," or "wild type") is expressed uniformly. This simply means that an extremely large number of genotypes give rise to phenotypes which are indistinguishable, even under diverse environmental conditions. The explanation of this phenomenon is not fully understood; however, it is believed that it is due, at least in part, to developmental homeostasis and canalization.

According to Waddington (1962), canalization involves major controlling genes, sometimes called "switch genes," which stabilize the phenotype both against genetic variation and environmental variation. This seems to be true at the diploid level of a given species, and our observations on triploid chickens indicate that this may also be true for different ploidy levels (Abdel-Hameed 1979).

Several biologists have concluded that if a particular phenotype is advantageous, there will be strong selection for genes canalizing its development, so that the development pathways in virtually all organisms are highly canalized.

The diploid number of chromosomes in man was believed to be 48. But improvements in mitotic cell cultures, especially the application of colchicine and hypotonic treatments, have allowed cytogeneticists to obtain a more accurate discription of the normal karyotype. The correct chromosome number of 46 was finally established as late as 1956 by Tjio and Levan. Since then, deviations from the normal karyotype, either due to structural or numerical chromosome aberrations have been identified as a very important cause of fetal loss (i.e., in about one-third of spontaneous first-trimester abortions) and birth defects (i.e., in about one percent of live births). Many well known clinical syndromes such as Turner's, Klinefelter's, Down's syndromes have since been found to have their etiology in specific chromosomal abnormalities.

The first example of chromosomal anomaly was discovered in 1959 when Lejeune and his associates demonstrated that children with congenital malformations characteristic of Down's syndrome had 47 chromosomes and the extra chromosome is a small acrocentric known as chromosome 21. Nondisjunction of this acrocentric chromosome is believed to take place during meiotic divisions of parental gametogenesis, especially so in the oogenesis of older mothers. Such nondisjunction would lead to the formation of gametes having one extra acrocentric and thus forming the trisomic condition for this chromosome when fertilized by normal gametes. Another form of chromosomal aberration, i.e., translocation between chromosome 21 and chromosome 15 or 14, was found to be responsible for the manifestation of the so-called "translocation Down's syndrome." Such translocations are found in a small frequency, about 2.5 percent, of Down's syndrome. Patients with translocation Down's syndrome are identical phenotypically to those with the more frequent condition of trisomy 21. Both are genetically also similar in that the genetic material of chromosome 21 is essentially in triple dosage despite the fact that the total number of chromosomes is only 46 in the translocation condition versus 47 in the trisomy condition. The translocation genotype is transmitted from one parent only, usually the mother. In this case she is phenotypically normal, carries the translocation genotype as a "balanced translocation" with a total of only 45 chromosomes. The condition of balanced translocation entails having the normal dosage and balance of the genetic material but in an abnormal chromosomal arrangement, thus usually resulting in normal phenotype.

Undoubtedly the phenotypic anomalies observed in these clinical syndromes result from imbalance in the genetic material which in turn affect the early stages of fetal development and cause birth defects. Therefore such genetic defects due to numerical and structural aberrations of the chromosomes act as powerful teratogenic agents.

In higher animals, sexual differentiation occurs during a specific period in development. In early fetal development, certain major genes initiate the process of gonadal development. For example, the Y chromosome in mammals acts as a strong male determining element and it is believed that it carries genes that are necessary to initiate differentiation of the embryonic gonads into testes. In this case, the embryonic testis produces certain hormonal secretions that direct male differentiation; in their absence, a female develops. Such processes represent a good example of a highly canalized system that ensures a precise expression of sexual dimorphism in higher animals. Therefore, the chromosome complement of 46, XX is characteristic for normal females, while that of 46, XY is characteristic for normal males in man. However, exceptions to such a highly canalized system may be observed. Two recently recognized clinical syndromes, i.e., XX males and XY females, are cases in point.

The XX male and XY female in man represent instances in which the phenotypic sex is opposite to that which the individual should have had in accordance with his chromosomal constitution. The occurrence of these two phenomena is not clearly understood and appears to represent a genetic paradox. The XX male and XY female are normal in appearance, but are sterile individuals. In all cases some degree of gonadal dysgenesis, from partial to total, is characteristic. In XY females with testicular feminization, the production of hormones is normal or near that which is normal for males. The XX male is characteristically a case of hypogonadism, while the streak gonads in XY females with gonadal dysgenesis have no hormonal function. Several attempts have been made to explain the development of male primary and secondary sexual characteristics in the presence of a normal appearing female karyotype (XX male), and others to explain XY female syndromes in man. Both XY female with either gonadal dysgenesis or testicular feminization and the XY male represent certain clinical syndromes that have received a great deal of attention, but their etiology remains unexplained (see Abdel-Hameed and Gluesing 1979).

In many cases the study of abnormalities seems to be important in furthering our understanding of the so-called "normal condition," whether the abnormalities are *environmental,* as in the effect on development of high altitudes, or *genetic,* as in the case of polyploidy, or *both* as in the case of the XX male and XY female in man. The term "normal" simply represents the most common and widespread phenotypic expression within a given species.

REFERENCES

ABDEL-HAMEED, F.
1979 Polyploidy in humans and other vertebrates. In this volume.
ABDEL-HAMEED, F., L. M. GLUESING
1979 The XX male and XY female in humans. In this volume.
ABDEL-HAMEED, F., L. TANGEN
1979 "Position of male determining gene loci on the Y chromosome and possible explanation of XX male and XY female phenomena in man," in *Comparative karyology of primates*. Edited by Brunetto Chiarelli and Ann Koen. World Anthropology. The Hague: Mouton.
ATKIN, N. B., H. B. KLINGER
1962 Superfemale mole. *Lancet* 2:727.
HALDANE, J. B. S.
1957 The cost of natural selection. *Journal of Genetics* 55:511–524.

LEJEUNE, J., R. TURPIN, M. GAUTIER
1959 Le mongolisme, premier exemple d'aberration autosomique humaine. *Annales de Genetique* 1:41–49.
STENCHEVER, M. A., J. M. HEMPEL, M. N. MACINTYRE
1967 Cytogenetics of spontaneously aborted human fetuses. *Obstetrics and Gynecology* 30:683–691.
TJIO, J. H., A. LEVAN
1956 The chromosome number of man. *Hereditas* 42:1–6.
WADDINGTON, C. H.
1962 *New patterns in genetics and development*. New York: Colombia University Press.

The XX Male and the XY Female in Humans

FATHI ABDEL-HAMEED and LINDA M. GLUESING

In humans, the normal chromosome number is forty-six, consisting of twenty-two pairs and one pair of sex chromosomes. Males have the sex chromosome complement XY, receiving one X chromosome from the maternal parent and one Y chromosome from the paternal parent. Females have the sex chromosome complement XX, receiving one X chromosome from each parent. In humans, the Y chromosome is strongly male determining. Even in the presence of as many as four X chromosomes, the Y chromosome exerts its male determining effect and the result is a male. The sex chromosome complements XO, XX, XXX, and XXXX result in females while the sex chromosome complements XXY, XXXY, XYY, and XXXXY result in males. Deviation from this precise and effective sex determination in man occurs spontaneously and at a low frequency, producing the XX male and the XY female. In contrast to sex determination in man, the sex of certain species such as *Drosophila* is determined by the balance of X chromosomes in relation to the number of autosomal sets of chromosomes, rather than through the presence or absence of the Y chromosome (Abdel-Hameed and Shoffner 1971).

STRUCTURE AND FUNCTION OF THE Y CHROMOSOME

The Y chromosome has distinctive morphology and staining properties. The distal portion of the long arm of the Y chromosome exhibits an especially bright, fluorescent pattern when stained with fluorescing acridine derivatives. At interphase, the Y chromosome has been described as a small fluorescent single or double body (Manolov et al. 1971). Recent evidence indicates that the distal segment of the long arm consists of several fluorescent bands. Schwinger and Pera (1972) have observed a splitting of the Y chromosome fluorescence during prophase. At this time, the chromosomes are greatly extended and the Y chromosome appears as two fluorescent bodies, indicating that the fluorescent region of the long arm of the Y chromosome consists of two closely lying fluorochrome binding sections. Manolov et al. (1971) have observed several fluorescent bands. Observation of photographic negatives reveal a more complex pattern of fluorescence than that seen on the prints from the same negatives. In metaphase, the distal portion of the Y chromosome can be resolved into two subunits. The short arm exhibits a weakly fluorescent band and the long arm exhibits three fluorescent bands (Figure 1). Of these, p^1 and q^1 bands are located on each side of the centromere and fluoresce weakly, while q^2 and q^3 are brightly fluorescent, q^2 being more brightly fluorescent than q^3. Manolov and co-workers (1971) believe that the Y chromosome during interphase is composed of several fluorescent subunits which probably correspond to those they have observed at metaphase. They further hypothesize that the shape of the Y chromosome is approximately the same during interphase as during metaphase. The Y chromosome in man is distinctively smaller in size than the X chromosome. Morphologically the X and the Y chromosomes are distinguishable.

During prophase of the first meiotic division of the primary spermatocyte, the short arm of the X chromosome and the short arm of the Y chromosome associate terminally (Pearson and Bobrow 1970; Ford and Hamerton 1956). This association is necessary to prevent random

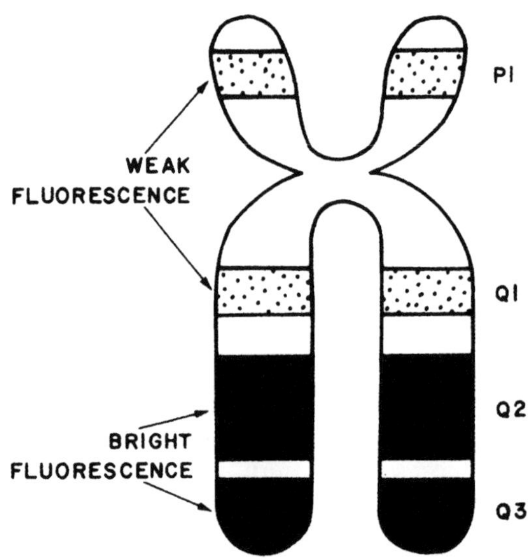

Figure 1. Characteristic fluorescent banding pattern of the Y chromosome

assortment of the X and Y chromosomes and the subsequent formation of zygotes with either XXY or XO sex chromosome complements (Ferguson-Smith 1966). The short arm of the X chromosome adjoins the short arm of the Y chromosome at diakinesis and metaphase-I without evidence of chiasma (Pearson and Bobrow 1970). A synaptinemal complex has been observed near or at the end-to-end association (Solari and Tres 1970).

The Y chromosome initiates the differentiation of the embryonic gonads into testes sometime early in fetal development (Money and Ehrhardt 1973). Before this stage, the gonads are indistinguishable and give no indication of whether they will develop into ovary or testis (Laurian et al. 1971). Once the embryonic gonads are developed, their hormonal secretions direct the differentiation of the genital ducts and external genitalia (Mittwoch 1967). In the absence of testicular secretion, a female develops. Jost (1947) found that if the gonads are removed early in fetal development, before differentiation has been initiated, both a genetic male and a genetic female will develop female external genitalia. Thus, the testis is necessary for male development, but the ovary is not necessary for female development.

The embryonic testis is believed to secrete two important substances. The first is a nonsteroidal, morphogenetic substance which suppresses the development of the Müllerian ducts (Figure 2). In the absence of this substance, the Müllerian ducts develop into female internal genitalia (Laurian et al. 1971). The second substance is a hormone analogous to testosterone which masculinizes the Wolffian ducts and the external genitalia. Thus the embryonic testis must stimulate the growth of the

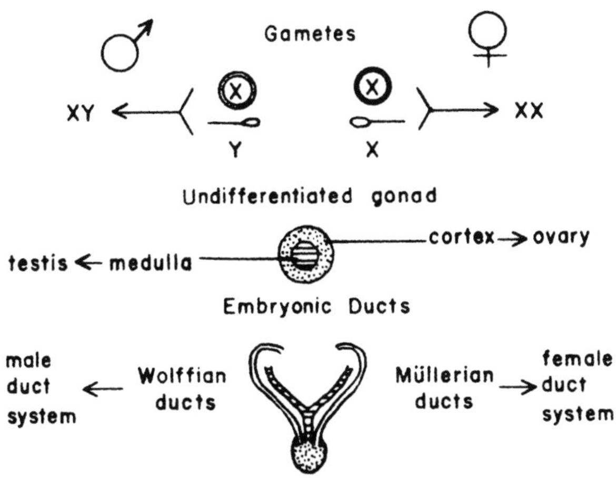

Figure 2. Diagrammatic illustration of sex determination in mammals

Wolffian system and male external genital organs and inhibit the Müllerian development in mammalian male embryos (Stenchever et al. 1969; Dewhurst 1971; Money and Ehrhardt 1973).

SEX CHROMOSOME ABNORMALITIES AND GONADAL DYSGENESIS

Generally, an alteration in the sex chromosome complement results in total to partial gonadal dysgenesis depending on the magnitude of the sex chromosome abnormality and the number of body cells affected (Eggen 1969). Gonadal dysgenesis is characterized by streak gonads, sexual infantilism and elevated urinary gonadotrophins (Goldberg et al. 1968; Christakos et al. 1969). If somatic abnormalities are present, the condition is usually referred to as Turner's syndrome. In Turner's syndrome, karyotyping reveals a modal number of forty-five chromosomes (Soltan 1968; Bishun 1968; Eggen 1969), or a recognizable structural alteration of one of the X chromosomes at least in some cell lines (Christakos et al. 1969). Individuals with true Turner's syndrome (45, X) develop as hypoplastic females. They have shieldlike broad chests, neck webbing, short stature, and are amenorrheic, fail to develop either pubic or axillary hair, and show no breast development. They have streak gonads which are represented by minute, amorphous, fibrous structures which show no microscopic evidence of germ cell function. Very often, gonadal morphology is indifferent and it is difficult to determine whether the gonad is an ovary or a testis (Eggen 1969). Buccal and vaginal epithelial cells fail to exhibit sex-chromatin mass of Barr (Barr-bodies) or drumstick appendages in the polymorphonuclear leukocytes (Nora and Sinha 1968; Bachman et al. 1971). The external genitalia are hypoplastic and the internal genitalia are represented by rudimentary cervix, uterus, and fallopian tubes.

Another common aneuploidy in man is XXY Klinefelter's syndrome, also known as chromatin-positive testicular dysgenesis. The extra X chromosome present in the cells of a genetic male results in the impairment of testicular differentiation. They are sterile and often eunuchoid in appearance (Eggen 1969). The incidence of gynecomastia is about 40 percent (de la Chapelle 1972) and mental retardation is common (Ferguson-Smith 1966). Buccal smears are chromatin-positive, drumstick appendages are present on the nuclei of polymorphonuclear leukocytes, and karyotyping reveals a modal count of forty-seven chromosomes (Soltan 1968; Eggen 1969). Testicular histology exhibits lack of spermatogonia, diminished diameter, hyalinization, and obliteration of the seminiferous tubules, peritubular fibrosis, and hyperplasia and polymorphism of the Leydig cells (de la Chapelle 1972).

XX MALES

General Appearance

The sex chromosome complement in normal human males is XY, yet in some reported cases, phenotypic males have been found in which the sex chromosome complement is XX. These males are sometimes considered a form of Klinefelter's syndrome, but definite differences exist between the two syndromes. The general appearance of the XX male is that of a normal male (de la Chapelle 1972; de la Chapelle et al. 1965; Powers et al. 1970; Therkelsen 1964). They have a male psychosexual orientation and frequently marry (de la Chapelle 1972). The pubic and axillary hair are of female type and distribution but may be typically male in distribution (de la Chapelle 1972; Boczkowski et al. 1968; de la Chapelle et al. 1965). Facial hair is usually decreased (de la Chapelle 1972; Therkelsen 1964; de la Chapelle et al. 1965), and the XX male may lack beard growth (Lindsten et al. 1966). The prostatic gland is of normal size and consistency (de la Chapelle 1972; Powers et al. 1970; de la Chapelle et al. 1971; Lindsten et al. 1966). The external genitalia is normal in appearance (de la Chapelle 1972; de la Chapelle et al. 1965; Therkelsen 1964). Although the testes are frequently undersized, they are occasionally normal in size (de la Chapelle et al. 1971). Macroscopically the testes, epididymis, and vasa deferentia are normal in appearance (Therkelsen 1964). The incidence of gynecomastia is approximately the same as in Klinefelter's syndrome, but mental deficiency is probably not higher in frequency than that recorded in newborn babies (de la Chapelle 1972). All XX males are sterile and exhibit azoospermia, the semen being absent of spermatozoa (de la Chapelle 1972; de la Chapelle et al. 1971; Powers et al. 1970; Andersson et al. 1972).

Histology of the Testes

Histologically, the testicular tissue of the XX male is distinctive in several aspects. A degree of abnormality is usually associated with the syndrome, but in some cases testicular histology is near normal (de la Chapelle 1972). The seminiferous tubules are usually sparse, exhibiting diminished diameter (de la Chapelle 1972; de la Chapelle et al. 1965, 1971; Boczkowski et al. 1968; Lindsten et al. 1966; Lisker et al. 1970) and thickening of the tubular membrane (de la Chapelle 1972). A slight degree of peritubular fibrosis is associated with the syndrome, and there is some occurrence of annular hyalinization of the tubules (de la Chapelle 1972; Boczkowski et al. 1968; Lindsten et al. 1966; Lisker et al. 1970). The Leydig cells, which produce testosterone, are numerous and hyperplastic (de la Chapelle et al. 1965; Boczkowski et al. 1968; Therkelsen 1964; Lindsten et al. 1966). They may be polymorphic and arranged as the cells are characteristically arranged in ademonas (Therkelsen 1964; Lindsten et al. 1966). Only Sertoli cells, mature or immature in appearance, are seen within the seminiferous tubules (de la Chapelle et al. 1965; de la Chapelle 1972; Therkelsen 1964; Boczkowski et al. 1968; Lindsten et al. 1966; de la Chapelle et al. 1971). Spermatogenic cells are usually absent, although they are sometimes observed (de la Chapelle 1972; de la Chapelle et al. 1965). In several reported cases, the testicular histology of the XX male resembles that of Klinefelter's syndrome.

Steroid and Pituitary Hormone Levels

Little information is available concerning hormone levels in XX males. Some general observations have been made and the clinical symptoms are referred to as hypogonadism (de la Chapelle et al. 1965; Powers et al. 1970). The response of XX males to administered pituitary gonadotrophins has been reported as poor and it is generally assumed that the Leydig cells are deficient in hormone production (de la Chapelle 1972). The biosynthetic pathways and metabolism of testosterones are summarized in Figures 3 and 4. Neuwirth and co-workers (1972) compared the plasma testosterone values of normal men with chromosome variants of Klinefelter's syndrome. The study included cases of 46, XY/47, XXY; 47, XXY; and XX males. The XX male was found to exhibit the lowest plasma testosterone values of those studied and their levels were very significantly lower than those in normal males. The reported plasma testosterone levels were 2.56 ng/ml ± 1.40 s.d.; the normal males tested having a value of 8.90 ng/ml ± 2.51 s.d. Urinary excretion of pituitary gonadotrophins are sometimes reported within normal male limits (de la Chapelle 1972; de la Chapelle et al. 1971; Therkelsen 1964; Boczkowski et al. 1968), but in the majority of cases, the secretion of gonadotrophins is elevated (de la Chapelle 1972; Lisker et al. 1970). The deficient Leydig cell hormone production is directly responsible for the increase in pituitary gonadotrophins (de la Chapelle 1972). Estrogen excretion is decreased or within the lower range for normal males (de la Chapelle 1972; Boczkowski et al. 1968; Therkelsen 1964). The 17-ketosteroids excreted in the urine are within the normal range for men (de la Chapelle et al. 1971; Lisker et al. 1970; Therkelsen 1964; Boczkowski et al. 1968; de la Chapelle 1972) and reflect metabolites or adrenal and testicular steroids. In normal adults, the adrenal cortex is responsible for two-thirds to three-fourths of the total 17-ketosteroids in man, and the testes are responsible for the remaining one-fourth. Thus, 17-ketosteroid measurement serves as a guide to androgenic activity (Henry and Krieg 1969).

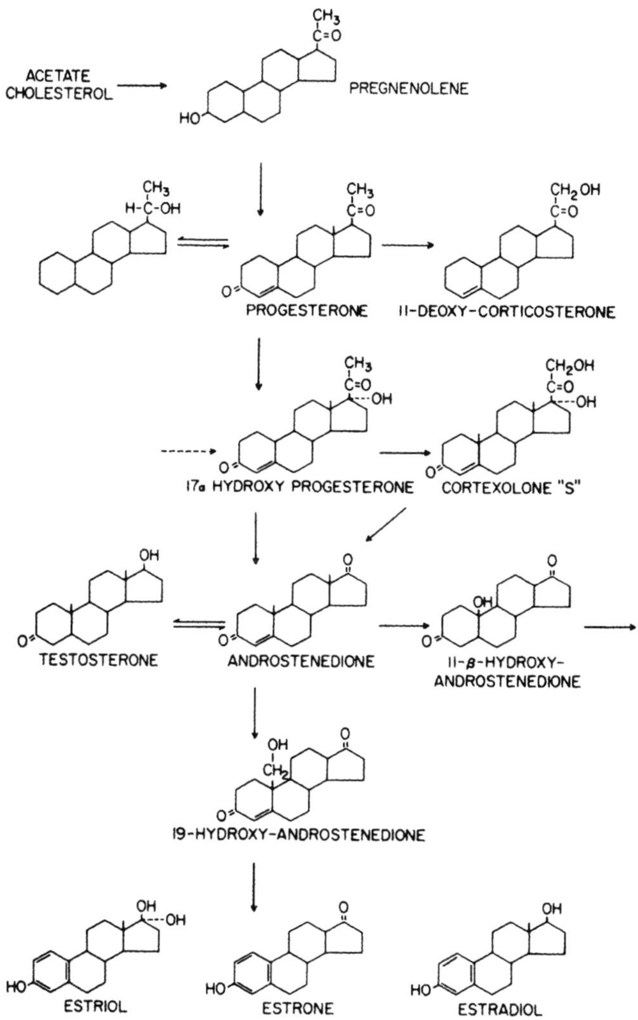

Figure 3. Synthetic pathway of steroid hormones used by the testicular interstitial cells

Frequency of XX Males

The incidence of XX males seems to be extremely low. In two studies cited by de la Chapelle (1972), one XX male was reported out of 9,232 newborn males, and in another study of 61,742 newborn males, two XX males were reported. Ratcliffe et al. (1970) have reported one XX male out of 3,500 newborn males tested. In a survey of 2,500 children, 1,264 of whom were male, only one XX male was reported (Stewart et al. 1969). Of 5,382 males in a study of 10,412 liveborn infants, no case of an XX male was observed (Pantelakis et al. 1970). The incidence of XX males

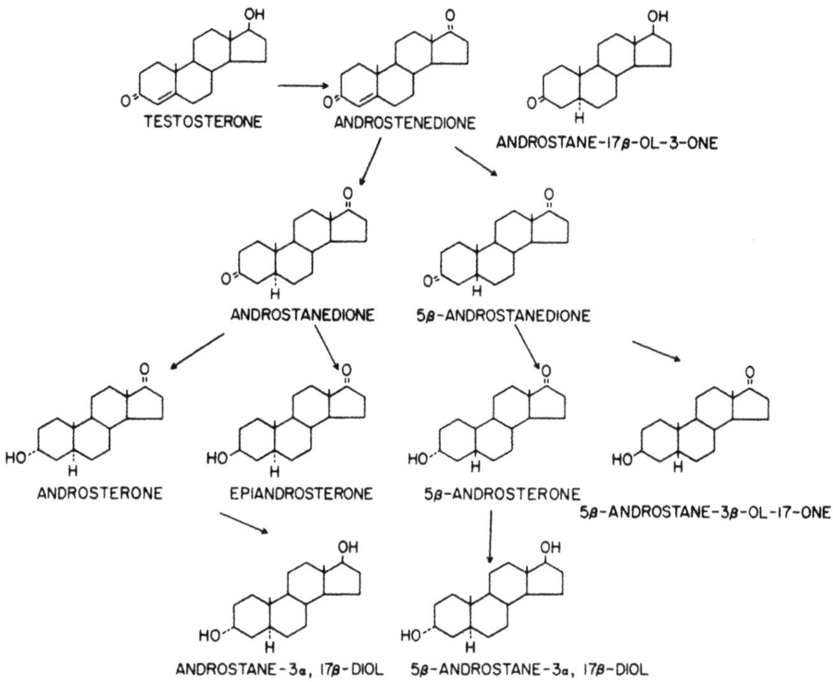

Figure 4. Metabolism of testosterone

would appear to be approximately 0.006 percent (5/81,120) of the total males reported (Table 1). The incidence of XX males in both newborn males and in spontaneous abortions is also very low. In one study of 100 spontaneous abortions, no case of an XX male was detected (Lauritsen et al. 1972), although in a similar study, one case of an XX male was reported (Stonova and Selezniova 1968).

Table 1. Incidence of XX male syndrome among liveborn male infants

Male infants tested	XX males reported	Percentage of XX males	Reported by
9,232	1	0.011	de la Chapelle (1972)
61,742	2	0.003	de la Chapelle (1972)
3,500	1	0.029	Ratcliffe et al. (1970)
1,264	1	0.079	Stewart et al. (1969)
5,382	0	0.000	Pantelakis et al. (1970)
Totals			
81,120	5	0.006	

XY FEMALE

Cases of phenotypic females with a 46, XY sex chromosome complement have been reported and these females can be grouped into two distinct clinical classes: those with gonadal dysgenesis and those with testicular feminization. Women who lack the somatic abnormalities characteristic of Turner's syndrome, but who possess streak gonads, are examples of "pure gonadal dysgenesis" (Cohen and Shaw 1965; Espiner et al. 1970; Perez-Ballister et al. 1970). Their chromosome complement may be 46, XX or 46, XY, although the majority of cases have the sex chromosome complement 46, XY (Wachstein 1951; Cohen and Shaw 1965).

XY Female with Gonadal Dysgenesis

Gonadal dysgenesis may vary in intensity but is usually described as either pure (complete) or partial gonadal dysgenesis. In 46, XY gonadal dysgenesis, the gonads are undifferentiated and the pelvic organs are usually typical of other forms of gonadal dysgenesis (Sternberg and Barclay 1967; Espiner et al. 1970). The XY gonadal dysgenesis females possess uterus, tubes, and female external genitalia including vagina, normal sized clitoris, and normal appearing vulva (Tetsuji et al. 1971; Hoffenberg and Jackson 1957; Smith and Pepler 1969). They are characterized by primary amenorrhea. In partial XY gonadal dysgenesis, some masculinization of the external genitalia may be present at birth, and vagina, uterus, and tubes may only be partially present (Dewhurst 1971).

HISTOLOGY AND HORMONE ACTIVITY OF THE DYSGENETIC GONADS. The gonads are composed of whitish streaks of tissue, lacking germ cell function (Dewhurst 1971; Espiner et al. 1970). Ova, follicles, and neoplastic tissues are absent (Chemke et al. 1970; Cohen and Shaw 1965). The tissue is nonspecific, consisting of cortical stroma and hilar structures (Cohen and Shaw 1965). The streak gonads are incapable of hormone production and spontaneous secondary sexual development fails to occur at puberty (Dewhurst 1971; Sternberg and Barclay 1967; Chemke et al. 1970). As in other cases of gonadal dysgenesis, urinary gonadotrophins are elevated, resulting from hormonal failure of the dysgenetic gonads (Christakos et al. 1969). The adrenal cortex of chromosomal males, whether phenotypically male or female, produce more androgens than those of chromosomal females (Chemke et al. 1970). As expected, serum and urinary testosterone values are raised in these XY females in comparison to normal females, but their values are low in comparison to normal males (Chemke et al. 1970) because their streak gonads are incapable of hormonal function. Therefore, their testosterone is, most likely, derived from their adrenal cortex (see our discussion of steroid and pituitary hormone levels of XX males).

Karyotyping reveals a normal appearing Y chromosome, but apparently this Y chromosome failed to direct differentiation of the gonads into testes during early stages of fetal development (Sternberg and Barclay 1967; Dewhurst 1971; Espiner et al. 1970; Miller 1964). When the embryonic gonads fail to differentiate into testis, neither Mullerian duct inhibition nor Wolffian system stimulation occurs. Under such conditions, uterus and tubes differentiate and female external genitalia result.

ABNORMAL DEVELOPMENT AND TUMORS OF THE DYSGENETIC GONADS. Gonadoblastomas arise more frequently in females and less frequently in males with dysgenetic gonads (Ferenczy et al. 1971). The term gonadoblastoma designates a specific type of gonadal tumor composed of a mixture of germ cells and other cells resembling immature granulosa or Sertoli cells. Leydig or lutein type cells may also be present (Kariminejad et al. 1972; Frasier et al. 1964; Miller 1964). A relatively high incidence of neoplasms occur in the dysgenetic gonads (Ferenczy et al. 1971; Sternberg and Barclay 1967; Mukerjee et al. 1972; Espiner et al. 1970; Miller 1964), and are particularly frequent in females with gonadal streaks, in females with an XY karyotype, or in mosaics carrying a Y chromosome or fragment (Sternberg and Barclay 1967; Ferenczy et al. 1971; Mukerjee et al. 1972; Miller 1964). This is not true of XO, XX, or mosaics lacking a Y chromosome with rudimentary gonads (Ferenczy et al. 1971; Sternberg and Barclay 1967). Gonadal dysgenetic XY females are particularly prone to the development of tumors. In some cases, these tumors develop very early in life (Dewhurst 1971; Sternberg and Barclay 1967). The incidence of malignancy is about 26 percent of the cases reported (Dewhurst 1971). Therefore, it is recommended that the gonads should be removed as soon as the patient's condition is realized. The gonads have no hormonal function and represent a detriment to the well-being of the patient (Dewhurst 1971).

The mode of inheritance of the XY female with gonadal dysgenesis and females with testicular feminization will be discussed in a later section.

Testicular Feminization

The second category of XY female is that of testicular feminization, existing in two variants: the complete form and the incomplete form (Morris and Mahesh 1963). The XY females with complete testicular feminization, unlike those with pure gonadal dysgenesis, do not exhibit sexual infantilism. They possess well-developed breasts (Boczkowski and Teter 1969; Dewhurst 1971; Sternberg and Barclay 1967; Morris 1953;

Rivarola et al. 1967), their bodies are typically very feminine, and they fail to show traces of masculinization (Laurian et al. 1971; Sternberg and Barclay 1967; Dewhurst 1971; Morris and Mahesh 1963). They lack a uterus, have inguinally or abdominally located testes, normal stature uncharacteristic of Turner's syndrome, scanty or absent pubic and axillary hair, short blind vaginas, normal sized clitoris, normal appearing vulva, primary amenorrhea due to the lack of uterus, and a normal female behavior pattern (Dewhurst 1971; Sternberg and Barclay 1967; Laurian et al. 1971; Rosenfield et al. 1971; French et al. 1965; Morris 1953; Ismail et al. 1972). The testes are usually considered normal in appearance except for their undescended position (Dewhurst 1971; Morris 1953). The presence of a uterus is one of the main distinguishing points between gonadal dysgenesis and testicular feminization.

The incomplete form of testicular feminization resembles the complete form in the presence of primary amenorrhea, normal or slightly less than normal breast development, short blind vagina, lack of uterus, and inguinally or abdominally located testes. It differs from the complete form in that varying degrees of masculinization are present (Saez et al. 1971). The clitoris is usually enlarged and some pubic hair growth develops (French et al. 1965; Rosenfield et al. 1971).

The syndrome of testicular feminization is thought to result from androgen insensitivity. In the complete form, large amounts of testosterone administered induce neither masculinization nor an anabolic response (Rosenfield et al. 1971; Laurian et al. 1971; Rivarola et al. 1967; Boczkowski 1968). In the incomplete form, some evidence of androgen response is evident from the presence of pubic hair and enlarged clitoris, but the difference of response to androgens between the two forms is frequently considered one of degree (French et al. 1965).

HISTOLOGY OF THE FEMINIZED TESTES. Histologically, the feminized testes are distinctive. The seminiferous tubules are lined with Sertoli cells which frequently appear immature (French et al. 1965; Sternberg and Barclay 1967; Laurian et al. 1971; Saez et al. 1971; Morris 1953). The tubules are thickened and hyalinized (Saez et al. 1971; French et al. 1965; Morris 1953). Increased fibrous stroma is present, the foci being nonspecific (French et al. 1965; Sternberg and Barclay 1967; Morris 1953; Morris and Mahesh 1963). Germ cell development is not evident and the testis usually lacks evidence of spermatogenesis (Laurian et al. 1971; Saez et al. 1971; Morris 1953). The testes exhibit marked Leydig cell hyperplasia and these cells are increased in number, grouped in large or small aggregates (Laurian et al. 1971; Saez et al. 1971; Sternberg and Barclay 1967; French et al. 1965; Rosenfield et al. 1971). Electron micrographs of these tissues reveal fine structures similar to that of fetal testis (Morris and Mahesh 1963). The feminized testis appears similar to that of normal males with cryptorchid testes, the testes being undescended (Morris 1953; Sternberg and Barclay 1967), but definite morphological differences exist (Sternberg and Barclay 1967; Morris and Mahesh 1963). The seminiferous tubules are less mature in testicular feminization, considering the age of the patient, than of normal males with cryptorchid testes (Sternberg and Barclay 1967; Morris and Mahesh 1963), and the tubules are often indistinguishable from the ones found in tubular adenomas (Morris and Mahesh 1963). Some degree of Leydig cell hyperplasia is present in cryptorchid testes of otherwise normal XX males, but not to the degree exhibited by the feminized testes (Morris and Mahesh 1963).

TUMOR DEVELOPMENT IN THE FEMINIZED TESTES. As in the case of dysgenetic gonads, the feminized testes have a high incidence of malignancy (Morris and Mahesh 1963; Sternberg and Barclay 1967; Gordon et al. 1969; Zourglas and Jones 1965). Morris and Mahesh (1963) reported that of fifty patients over the age of thirty years, eleven had malignant germinomas, fifteen had tubular ademonas, and ten had developed cysts. Malignant tumors are rare in the younger cases of testicular feminization in contrast to that of XY gonadal dysgenesis (Morris and Mahesh 1963; Sternberg and Barclay 1967). In testicular feminization, the testes are generally removed only after puberty as they have definite hormonal function and will produce normal feminization (Boczkowski 1968). By contrast, the rudimentary gonads are not active in hormone production and in some cases they develop tumors at a very early age. Therefore, they are usually removed as soon as the case is confirmed as gonadal dysgenesis and not a condition of testicular feminization.

STEROID HORMONE LEVELS IN XY FEMALES WITH TESTICULAR FEMINIZATION. In testicular feminization, plasma testosterone levels are either within the range characteristic of normal males or considerably above the range of normal males (Ismail et al. 1972; Rivarola et al. 1967; Saez et al. 1971; Rosenfield et al. 1971; Southern et al. 1965; French et al. 1965), indicating normal secretion of testosterone by the feminized testes. By comparison, in pure XY gonadal dysgenesis where the streak gonads are inactive in hormone production, the testosterone is most likely produced by the adrenal cortex. The complete form of testicular feminization exhibits a normal or elevated male pattern of dehydroepiandrosterone excretion (Wade et al. 1968; Ismail et al. 1972; Morris and Mahesh 1963). Testosterone is normally converted by an intracellular 5-α-reductase to dehydrotestosterone (see Figures 3 and 4) in which form it binds to the cell nuclei within male secondary sex organs before it exerts its characteristic influence (Rosenfield et al. 1971). Testosterone concentrations in the testicular vein plasma and testosterone glucuronide levels in the urine are within the normal male pattern (Wade et al. 1968; Southern et al. 1965).

In these XY females, the free testosterone index is also within the normal male range (Rosenfield et al. 1971). Testicular binding globulin is elevated and is within the female range in individuals exhibiting the complete form of testicular feminization (Rosenfield et al. 1971). Urinary ketosteroids are elevated or within the normal range in both the incomplete and complete forms of testicular feminization (Morris 1953; Morris and Mahesh 1963; French et al. 1965; Rivarola et al. 1967) and the urinary estrogens are within the normal range of men or women (Morris and Mahesh 1963; Griffiths et al. 1963; French et al. 1965; Wade et al. 1968; Ismail et al. 1972). In the incomplete form of testicular feminization, the level of plasma estrogens is within the normal range for either female or males, and estradiol-17β levels are in the female range. By contrast, the complete form exhibits estradiol-17β levels that are characteristically similar to that of normal males (Wade et al. 1968). In the complete form, the level of gonadotrophins is slightly elevated or within normal limits (Morris and Mahesh 1963; Rivarola et al. 1967; Rosenfield et al. 1971) and the ratio of sulfated dehydroepiandrosterone to dehydroandrosterone corresponds to the male pattern (Laurian et al. 1971).

ETIOLOGY OF THE XX MALE

The development of a male phenotype in the absence of Y chromosome material must be taken into account in any proposal concerning the etiology of the XX male syndrome. Several theories have been advanced and will be discussed below.

The Gene Theory

The gene theory postulates an autosomal mutant gene causing human XX zygotes to develop into phenotypic males. In *Drosophila melanogaster,* Sturtevant (1945) found a recessive autosomal mutant, *tra,* that transforms females into males. Under these conditions, he found that the sex ratios of the offspring change from the normal, expected ratio resulting in a higher number of males than expected. If there is a similar autosomal mutant causing a comparable effect in man, we would expect that the sex ratio of close relations of XX males to deviate significantly from that of the normal population. To test this hypothesis de la Chapelle (1972) tabulated the number and sex of close relatives of XX males and calculated the sex ratios of their sibs, uncles and aunts, and cousins. He found no significant difference between the calculated sex ratios and that expected among the offspring of random matings.

Under these same conditions, consanguinity among the parents of XX males would be expected and a high incidence of XX males would be found among sibs, cousins, or uncles. Consanguinity has been specifically ruled out in a number of cases of XX males and familial occurrence among sibs, cousins, and uncles have not been reported (de la Chapelle 1972).

In some species, such as *Drosophila,* sex is determined by the balance of sex chromosomes to the number of autosomal sets of chromosomes. In contrast to *Drosophila,* the Y chromosome in humans is a very strong male determining element (Abdel-Hameed and Shoffner 1971). As a result, it is probably unlikely that a single autosomal mutant can change XX human zygotes into phenotypic males.

Mosaicism

A second theory postulates the existence of undetected mosaicism in the XX male. The presence of sex chromatin in tissue cells has been confirmed, indicating the existence of two X chromosomes in cell nuclei of the examined tissues (de la Chapelle 1972; Andersson et al. 1972; Lisker et al. 1970; Lindsten et al. 1966; Boczkowski et al. 1968; de la Chapelle et al. 1965; de la Chapelle et al. 1971; Michalova et al. 1972; Powers et al. 1970; Therkelsen 1964). Karyotyping reveals a modal count of forty-six chromosomes. Drumsticks are observed on the polymorphonuclear leukocytes, also indicating the presence of two X chromosomes (Lindsten et al. 1966). Labelling with tritiated thymide reveals a late labelling X chromosome. This evidence further indicates the presence of two X chromosomes (de la Chapelle 1972; de la Chapelle et al. 1965, 1971; Powers et al. 1970; Lindsten et al. 1966; Boczkowski et al. 1968; Paris Conference 1971). Mosaicism can be present, even if a normal karyotype is detected in some tissues. To eliminate the possibility of mosaicism, as many types of body tissues as possible should be cultured, examined, and karyotyped. The XX male, then, may have Y chromosome material in some cell lines, though this is unlikely. In some studies, over 1,000 cells from several different tissues, such as leukocytes, skin, fascia, bone marrow, and testes, have been examined and no evidence of mosaicism has been detected (de la Chapelle et al. 1965).

Several attempts to detect either the presence of Y chromosome fluorescent material in the tissues of XX males or the presence of Y fluorescent material incorporated in one of the X chromosomes have been performed. Up to now these studies have failed to detect the presence of a Y chromosome material even though several types of tissues and several thousand cells have been examined (Breg et al. 1971; George and Polani 1970; Michalova et al. 1972; Andersson et al. 1972; Caspersson et al. 1971). If the Y chromosome were present in only 0.5

percent of the cells, it would be detected by this method (de la Chapelle 1972). The frequency of sex chromatin positive Sertoli cells in sex chromatin positive males is low and sometimes the cells are all chromatin negative. Therkelsen (1964) has reported one such case of XX male in which sex chromatin was absent in the Sertoli cells. Certain factors may be present in these individuals affecting the expression of sex chromatin in testicular cells (de la Chapelle 1972).

On the basis of the available evidence, the possibility that the XX male is an example of undetected mosaicism in some tissues is therefore unlikely (see discussion of X–Y interchange hypothesis below).

Eliminated Y Chromosome Theory

This theory is similar to the theory of mosaicism. Both postulate the existence of a stemline containing a Y chromosome which triggers testicular differentiation in early fetal development. As the name implies, the eliminated Y chromosome theory assumes that the Y chromosome became completely lost from all fetal cells, probably at a very early stage of its development, rather than a case of actual mosaicism where some tissues would still contain the Y chromosome in their nuclei.

There are several possible ways in which this could occur:

1. Upon the formation of an XY zygote, mitotic nondisjunction could have produced: (a) an XX cell line and a YY cell line; and (b) an XXY cell line and a Y cell line or all four cell lines could have been produced (Figure 5). Later on, the YY and Y cell lines would be eliminated as they

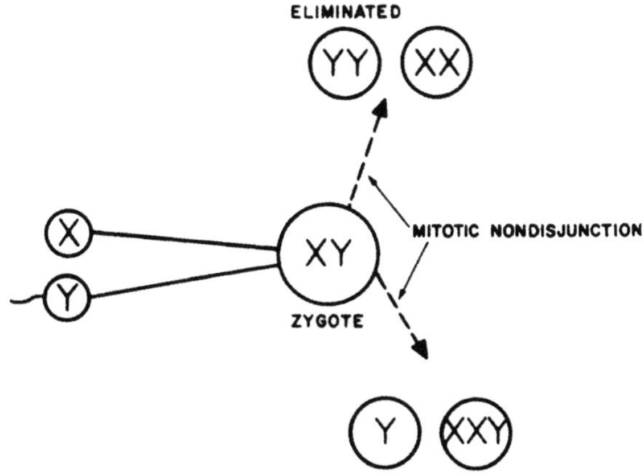

Figure 5. Eliminated Y chromosome theory which requires one nondisjunctional event. Either of these two possible mitotic nondisjunctional events are assumed to be followed by the elimination of all cell lines except those with XX nuclei

are lethal (de la Chapelle 1972; Jones 1969). The XY or XXY lines would also be lost, leaving the remaining XX line as the only one left thereafter.

2. An XX ovum could also have been produced by meiotic nondisjunction and fertilized by a normal Y containing sperm. The result would be an XXY zygote. In a later stage, after testicular differentiation had been initiated, a mitotic error could produce an XX and an XXYY line. The XXYY and the XXY lines are then postulated as subsequently eliminated (Figure 6).

The first proposal is more probable, as only one nondisjunctional event would be required. The second proposal requires the occurrence of two nondisjunction events or other divisional errors and the elimination of the XXY and XXYY lines which are viable (de la Chapelle 1972). The occurrence of nondisjunction is usually associated with elevated maternal age. The mean age of the mothers of XX males is 27.8 years and does not differ significantly from that of the total population (de la Chapelle 1972). By contrast, mothers of XXY Klinefelter males are generally older, the mean age being 32.6 years (de la Chapelle 1972). Thus a maternal age effect is not present in the XX male syndrome to account for nondisjunctional events.

Race and Sanger (1969) in a study of the X_g locus and sex chromosome abnormalities submitted data on X_g^a inheritance in XX males. The blood antigen X_g^a is an X-linked dominant character (Mittwoch 1967). On the basis of the twenty XX males tested, the results indicate that the distribution of X_g^a blood groups in XX males is ten times more likely to represent a male than a female distribution, and that the distribution is 2.4 times more likely to represent a normal male distribution than that of an XXY distribution. If the distribution in XX males was representative of a female distribution or of an XXY Klinefelter male distribution, it would support the eliminated Y theory. By 1972, thirty-two cases of XX males had been reported with X_g blood group data. With the addition of these

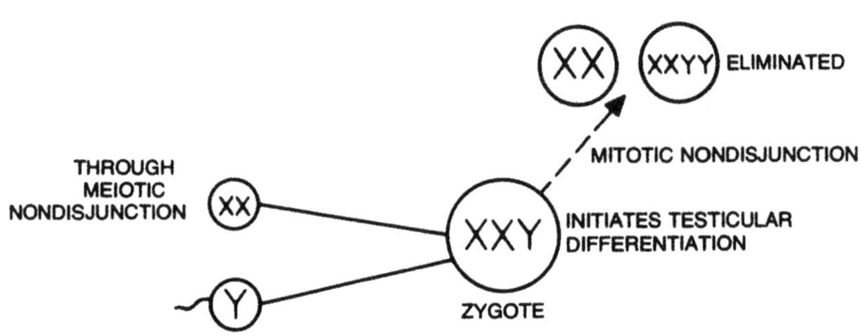

Figure 6. Eliminated Y chromosome theory which requires two nondisjunctional events. Mitotic nondisjunction event is assumed to be followed by the elimination of all cell lines except those with XX nuclei

figures, the XX male distribution appears to differ insignificantly from those of XX females, XY males, and XXY males, and is farthest from that expected of XY males (de la Chapelle 1972). An accurate interpretation of XX male inheritance of the X_g locus will not be possible until more cases are reported. At the present time, the number of XX males reported is small and cannot be used to give reliable estimates.

Available X_g blood group by itself neither confirms nor disproves either the eliminated Y hypothesis or the mosaicism theory as the cause of XX male syndrome. However, both of these hypotheses are unlikely due to: (1) the failure to detect mosaicism in several body tissues of the XX males; (2) the above-mentioned observation that maternal age effect is lacking.

X–Y Interchange Hypothesis

The X–Y interchange hypothesis was proposed by Ferguson-Smith (1966). During the terminal association of the X and Y chromosomes in paternal meiosis, an opportunity exists in which crossing over could occur resulting in interchange of genetic material between the X and Y chromosomes. The translocation of material could include the male determining genes on the Y chromosome resulting in an X chromosome containing Y chromosome material. An X^y sperm would result from this interchange. If this X^y sperm were to fertilize a normal X ovum, an XX^y zygote would result possessing the genes necessary for male differentiation.

Although the Y chromosome is known to vary in size from individual to individual, these variations can be accounted for wholly by variations in the length of the long arm of the Y chromosome (Bobrow 1971). Exceptionally large Y chromosomes are a rather common occurrence among normal males with oligospermia and in some syndromes, but has not been definitely associated with any particular disease state (Sperling and Lackman 1971). The long arm of the Y chromosome would genetically appear inert (Bobrow 1971). The male determining genes are then more likely to exist on the short arm of the Y chromosome. In addition, the short arm of the Y chromosome is known to be relatively stable in size (Bobrow 1971). Nielsen et al. (1972) have reported cytogenetic studies of a phenotypic male possessing testes of normal size, but with only the short arm of the Y chromosome detectable. A male phenotype in the presence of only the short arm of the Y chromosome provides support for the male determining loci being located on the short arm of the Y chromosome. Furthermore, Ferguson-Smith et al. (1969) also reported a phenotypic female in which the long arm of the Y chromosome was present.

The short arm of the X chromosome and the short arm of the Y chromosome associate terminally during paternal meiosis, and an opportunity exists for exchange of material. If Y chromosome material is located on one of the X chromosomes in the XX male, the characteristic bright fluorescence of the Y chromosome material might be detectable. As mentioned earlier, studies of this nature have been performed, but failed to detect the presence of Y translocated material onto an X chromosome (Andersson et al. 1972; Michalova et al. 1972; Casperson et al. 1971; Breg et al. 1971; George and Polani 1970). If the genes that determine male differentiation are located on the short nonfluorescent arm of the Y chromosome and if the genetic exchange is small, bright fluorescing Y chromosome material would not be detectable on the X chromosome. Also the fluorescent pattern of the X chromosome would probably remain unaltered (Andersson et al. 1972; Breg et al. 1971; George and Polani 1970).

The Y chromosome material has a characteristic bright fluorescence in *Drosophila* that is even more pronounced than that of humans (H. J. Barr, personal communication). When a portion of the Y chromosome in *Drosophila* is translocated onto an autosome, the fluorescence of this material dims (Jacobs and Ross 1966; Fraccaro et al. 1971). In humans, if dimming of fluorescence is characteristic of translocated chromosomal material, Y translocated material in XX males would be difficult to detect on the X chromosome. Such dimming of fluorescence of translocated material, however, has not as yet been observed in humans (Buhler et al. 1971).

ETIOLOGY OF THE XY FEMALE

The presence of a normal appearing Y chromosome and a female phenotype must also be accounted for in any proposal to explain XY gonadal dysgenesis or testicular feminization. The XY female condition is estimated to occur in one out of 20,000 to 62,400 births (Stenchever et al. 1969). The syndrome is a familial disorder transmitted through the maternal line (Park and Jones 1971; Laurian et al. 1971; Morris and Mahesh 1963; Gordon et al. 1969; Saez et al. 1971; Sternberg and Barclay 1967; French et al. 1965; Griffiths et al. 1963; Boczkowski et al. 1972; Mukerjee et al. 1972; Espiner et al. 1970). The carriers are normal females, but in some instances, decreased sexual hair has been reported (Morris and Mahesh 1963; Laurian et al. 1971; Boczkowski 1968) indicating some penetrance of the gene in carriers. The ratio of female to male to affected male in testicular feminization is 2:1:1, suggesting that an X-linked recessive gene or a male limited dominant gene is responsible for the disorder (Morris and Mahesh 1963; Laurian et al. 1971; Race and Sanger 1969).

The Gene Theory

The study by Race and Sanger (1969) using X_g blood group data from females with testicular feminization indicates that the distribution of their blood groups is of the male type and differs significantly from that of females. If a gene causing testicular feminization is located on the X chromosome, it is not within measurable distance of the X_g locus (Race and Sanger 1969; Boczkowski 1968) or within measurable distance of the genes causing color blindness and hemophilia (Morris and Mahesh 1963; Boczkowski 1968). In genetic studies, the frequency of crossing over is high between the X_g locus and the gene responsible for testicular feminization suggesting that the gene is probably autosomal (Sanger et al. 1969). Such autosomal inheritance of the syndrome is also suggested from data on consanguinity among parents and the familial occurrence of the XY female with gonadal dysgenesis and testicular feminization syndrome.

The X–Y Interchange Hypothesis

Ferguson-Smith (1966) further postulated that XY females with streak gonads could be accounted for within the framework of the X–Y interchange hypothesis. If an X–Y interchange occurred during paternal meiosis, the Y chromosome resulting would lack male determining genes. An XY zygote would result if this Y sperm fertilized a normal X ovum. Under these conditions, an XY female would result without male determining genes and with total testicular failure (Ferguson-Smith 1966; Dewhurst 1971).

The transmittance of XY gonadal dysgenesis is probably through the maternal line. The syndrome has been reported in monozygous twins (Frasier et al. 1964) and in siblings (Boczkowski et al. 1972; Broggen 1964; Chemke et al. 1970). The possibility that an X–Y interchange during paternal meiosis is the cause of XY gonadal dysgenesis would necessitate the condition being transmitted through the paternal line. Indications are that it is not. The probability that this type of translocation would occur several times in the same parent and in several offspring also appears unlikely.

ANDROGEN INSENSITIVITY

Androgen insensitivity is not evident in XY females with gonadal dysgenesis. Administered androgens cause masculinization and administered estrogens will cause breast development and pubic hair growth. The syndrome is due to a lack of hormone production rather than a deficiency in response to androgens. By comparison, both variants of testicular feminization syndrome exhibit a definite lack of response to androgens. However, some degree of response to estrogens must be present, as normal breast development occurs (Sternberg and Barclay 1967; Morris 1953). The cause of androgen insensitivity is uncertain, but may include impaired production of androgens, impaired utilization of these androgens once produced, or the production of antiandrogenic substances.

If a lack of response to androgens or a failure of testicular secretion existed early in fetal development, male differentiation would fail (Sternberg and Barclay 1967). At this early stage of development, Wolffian system development and phallus masculinization would not occur.

Testicular hormone production in testicular feminization is reasonably normal, and deficient hormone production by itself would fail to account for the phenomenon of androgen insensitivity. Studies indicate that the testes are capable of producing significant levels of both androgens and estrogens (Griffiths et al. 1963; French et al. 1966; Morris and Mahesh 1963), and that these hormones are of high biological activity (Morris and Mahesh 1963; Cardiff et al. 1972). Following gonadectomy, testosterone levels fall (Rosenfield et al. 1971; Ismail et al. 1972; French et al. 1965), urinary ketosteroid and estrogen levels decrease (Morris and Mahesh 1963; Laurian et al. 1971; French et al. 1965; Morris 1953), urinary gonadotrophins increase (Morris and Mahesh 1963; Morris 1953) and oxosteroid secretion falls to female levels (Wade et al. 1968; Ismail et al. 1972). This indicates that the source of the estrogens and androgens is primarily from the feminized testes rather than the adrenal glands (Wade et al. 1968; French et al. 1965; Saez et al. 1971). Further, the patient usually develops flushes after castration (Laurian et al. 1971; Morris 1953; Saez et al. 1971), although in some cases this does not occur (Wade et al. 1968). In vitro studies of testicular tissue from patients with testicular feminization have shown that the biosynthesis of testosterone from androstenedione is normal and that androstenedione is the major metabolite of the feminized testes. The testes also possess the necessary enzymatic systems for metabolism of estrone (Wade et al. 1968; Cardiff et al. 1972; Rosenfield et al. 1971). On the basis of these studies, the production of androgens and estrogens by the feminized testes does not appear to be defective (see Figures 3 and 4).

In the incomplete form of testicular feminization, if androgen insensitivity exists, it is partial. Saez and co-workers (1971) have observed a sharp increase in nitrogen retention in a patient with incomplete testicular feminization under testosterone therapy, suggesting that the response to androgens was normal. In another study by Rosenfield and collaborators (1971), however, testosterone and dehydrotestosterone were administered to a patient with incomplete testicular feminization and plasma concentrations of testosterone rose to over twice that observed in normal males suggesting that the patient had a partial androgen unresponsiveness. Saez et al. (1971) suggest an enzymatic defect in 17-ketosteroid reductase as the cause of androgen insensitivity. This type of defect would create a partial enzyme block and subsequently the build-up of several precursor steroids in the plasma. If a defect of this type exists it may only be localized to the testes. Their study indicates that steroid biosynthetic pathways are normal in the adrenal glands and in the peripheral tissues. The cause of the incomplete form of testicular feminization could then be an enzyme defect, or possibly it is a variant of the complete form and the differences exhibited between the two syndromes is only one of degree.

Dihydrotestosterone is possibly a biologically active form of testosterone in humans (Strickland and Smith 1969). A decrease in the conversion of testosterone to dihydrotestosterone by the testes has been observed in the complete form of testicular feminization (Rosenfield et al. 1971; Wade et al. 1968; Strickland and Smith 1969). There are also indications that this difficulty in conversion is present in other body tissues as well. Studies comparing testosterone conversion in the abdominal skin of normal men and women with that of individuals with testicular feminization have shown that the conversion of testosterone to 5-α-dihydrotestosterone is more efficient in normal men and women (Northcutt et al. 1969; Laurian et al. 1971). Wilson and Walker (1969) were not able to duplicate these results, but they did demonstrate that the skin from the labia majora of normal females and the scrotum of normal males were more efficient in this conversion than that of individuals with testicular feminization. Mauvais-Jarvis and co-workers (1969, 1970) have also shown that patients with testicular feminization have a very low capacity to reduce both testosterone and dihydrotestosterone to androstanediols and that the conversion of androsterone and isoandrosterone is characteristic of normal males. End organ resistance to the effects of androgens has been proposed as a likely cause of the clinical symptoms associated with testicular feminization. In a study by Rosenfield et al. (1971) of the complete form of testicular feminization, administration of high concentrations of dihydrotestosterone failed to completely suppress plasma testosterone. Hypothalamic-pituitary resistance to the effects of androgens is suggested and further supported by this patient's elevated gonadotrophin excretion.

The end organ resistance in testicular feminization is highly specific as indicated by the normal response to estrogens. This is indicated by closure of epiphyses and the protein anabolic influence of administered estrogens (Rosenfield et al. 1971). Estrogen responsiveness is also indicated by the degree of breast development present in the syndrome. The secretion of estrogens by the feminized testes would then be expected to be at a level comparable to normal females, but the excretion of estrone, estradiol-17β, and estriol is within male limits (Wade et al. 1968). It is known that androgens are capable of suppressing the action of estrogens (Morris and Mahesh 1963; French et al. 1965). The normal breast development exhibited in the complete form of testicular feminization may then be caused by the nonaction of androgens and the subsequent efficient action of estrogens. Breast development could also be present as a result of high levels of estrogens in the blood without subsequent excretion in the urine to enable their detection (Wade et al. 1968).

The poor conversion of testosterone to dihydrotestosterone then does not appear to be the primary defect in androgen resistance, and decreased reductase activity is presumed to be a secondary phenomenon (Rosenfield et al. 1971). In general, it is possible that any abnormality in the intracellular metabolism or mode of action of testosterone could lead to the clinical picture of androgen resistance and to the variants of the feminizing testis syndrome (Rosenfield et al. 1971).

The production of antiandrogenic substances, such as 16-α-hydroxyprogesterone, is another possible explanation of androgen insensitivity. If these substances are produced, they must come from other than the testes since the androgen insensitivity exists even after gonadectomy (French et al. 1965). Other intracellular pathways of testosterone metabolism might exist and lead to the formation of inhibitory compounds (Rosenfield et al. 1971). In vivo studies of extrahepatic 5-α-reductase in patients with the syndrome show that this activity more closely resembles that of females than of males (Mauvais-Jarvis et al. 1970). The activity of this enzyme could possibly be decreased as a consequence of androgen ineffectiveness.

Several possible mechanisms could be responsible for androgen insensitivity. Abnormal and rapid degradation of testosterone is also a possible cause, but rapid degradation must be balanced by excessive testosterone production to account for the hormone levels observed. Further, in vitro studies do not indicate excessive testosterone production. At present, the exact defect in androgen insensitivity is not certain and further studies will be necessary before a definite cause can be established.

LEYDIG CELL HYPERPLASIA

Besides the usual connective tissue elements, the stroma of the testis contains characteristic cells known as the interstitial cells or cells of Leydig. These cells occur in groups of various sizes and are quite distinct in the human testis. Histochemical studies have shown that compounds with the chemical properties of testosterone are present in the interstitial cells but not elsewhere in the testis. Furthermore, the evidence is quite conclusive that the interstitial (Leydig) cells of the normal testes secrete the male sex hormone. The female sex hormone also occurs in the male, about 80 percent of the estrogen arises from the Leydig cells and about 20 percent from the adrenal gland. In undescended or partially descended testes (cryptorchid) of otherwise normal males, the epithelium of the seminiferous tubules, but not the interstitial tissue, atrophies.

The term hyperplasia refers to the increase in amount of tissue produced by an increase in the number of cells. Hyperplasia often accompanies the regeneration of damaged tissues. During different stages of normal male development, there is a correlation between the amount of interstitial tissue and androgen content.

Pronounced Leydig cell hyperplasia has been observed in the testes of intersex chickens (Abdel-Hameed and Shoffner 1971). Accurate determination of low levels of steroid hormones became feasible through the new and highly sensitive technique of radioimmunoassay. Radioimmunoassay determinations of blood plasma levels for two estrogens, i.e., estrone and estradiol, in addition to two androgens, i.e., testosterone and dihydrotestosterone, were compared in nine adult chicken intersexes against that of five laying hens and another five adult males. In intersex chickens, estrone and estradiol plasma levels are similar or slightly higher than the roosters but are much lower than that of laying hens. Their testosterone and dihydrotestosterone levels, on the other hand, are similar or slightly higher than the laying hens but are markedly lower than that of the rooster. It seems as if the hyperplasia of Leydig cells observed on the light microscope level in these intersexes (Abdel-Hameed and Shoffner 1971) are not associated with active production of steroid hormones. The presence of such hyperplasia of Leydig cells is intriguing, and data on their ultrastructural organization may help in understanding their function.

It is possible that the Leydig cells of intersex chickens are potentially capable of normal steroidogenic activity but do not receive adequate hormonal stimulation from the pituitary gland due to certain malfunction of either the pituitary or the hypothalamus. To test the validity of this assumption, the effects of administering chorionic gonadotrophin hormones (PLH-PFSH) and hypothalamic hormones (LH-FSH-RH) to the intersexes will be examined in some detail.

Similarly, the observed hyperplasia of Leydig cells in human XX male testes (de la Chapelle et al. 1965; Boczkowski et al. 1968; Therkelsen 1964; Lindsten et al. 1966) is associated with plasma testosterone levels that are much lower than that of XY normal males (Neuwirth et al. 1972). Furthermore, the response of human XX males to administered pituitary gonadotrophins is poor, indicating the possibility that Leydig cells are deficient in hormone production (de la Chapelle 1972).

By comparison, XY human females with testicular feminization also exhibit Leydig cell hyperplasia and show testicular ultrastructure similar to that of fetal testis (Morris and Mahesh 1963); however, their plasma testosterone levels are similar to, or even higher than those of normal males. In the complete form of XY testicular feminization, the level of gonadotrophins is similar to that of normal males (Morris and Mahesh 1963; Rivarola et al. 1967; Rosenfield et al. 1971), thus confirming the normal hormone secretion of their Leydig cells.

From this comparison of Leydig cell hyperplasia in human XX males and XY females with testicular feminization, it is obvious that Leydig cell hyperplasia by itself does not result in lower production of testicular hormones. In this case, the decisive factor in affecting Leydig cell function seems to be the presence or absence of the Y chromosome. In chickens, the presence of the W chromosome does not result in a similar effect to that of the Y chromosome in man. This also indicates that the W chromosome in birds is not a strong element in sex determination as is the case with Y chromosome in mammals.

In some cases, however, the presence of a Y chromosome in man fails to direct embryonic testicular differentiation and results in complete failure of testicular function as is the case in the XY female with pure gonadal dysgenesis. By comparison, the embryonic testis of the XY female with testicular feminization seems to undergo normal development and subsequent function, producing normal male androgen levels. Therefore, the feminized testis appears to represent a case of androgen unresponsiveness. In this case, end organ resistance to the effects of androgens is the likely cause of testicular feminization. The effects of androgens are intimately correlated with hypothalamic-pituitary function, especially in later stages of development (Ismail et al. 1972). Therefore, hypothalamic-pituitary resistance to androgens may be the case in testicular feminization, and we suggest that extensive experimental work in this area should be conducted in the future.

DETECTION

The detection of the XX male syndrome and that of the XY female should preferably occur immediately after birth, but in most instances, this is not the case. The XX male is usually normal in appearance; somatic abnormalities are rare. Consequently, the XX male is externally indistinguishable from normal male infants (Stewart et al. 1969). The condition is usually not recognized until adulthood, with the most frequent complaint being sterility (de la Chapelle 1972). In studies of newborn male infants, XX males have been detected (Stewart et al. 1969; Ratcliffe et al. 1970; de la Chapelle 1972). It is possible that the condition could be identified successfully at birth. Buccal smear examination of newborns could be carried out as a routine practice. In males, the presence of sex chromatin would indicate the necessity of further investigation into the nature of the disorder. The fluorescent method, however, is a faster and more reliable method in the detection of the Y chromosome. If Y chromosome material is present in only 0.5 percent of the cells examined, it would be detectable (de la Chapelle 1972). A male infant failing to exhibit a Y chromosome has a higher probability of being an XX male. The sex-chromatin technique would fail to distinguish between an XXY and an XX sex chromosome complement. The inconveniences of the sex-chromatin technique are time delay, occasional failure to produce adequate preparations, and the unreliability of the test (Kaplan et al. 1969; Khudr and Benirschke 1971). Furthermore, the fluorescent technique is fairly simple and requires only the preparation of smears (Khudr and Benirschke 1971).

This method could also be employed in the identification of XY females at birth. A buccal smear for sex chromatin masses would not enable us to distinguish between XO Turner's syndrome and the XY female, while the fluorescent method would make identification of the disorder faster and more reliable.

As with the XX male, the XY female with testicular feminization is frequently detected only after puberty. They are raised as females and their sexual development at puberty is reasonably normal. The first indication that any abnormality is present occurs when they fail to menstruate (Kaplan et al. 1969). The XY female with gonadal dysgenesis, on the other hand, is usually detected because of failure of development of secondary sexual characteristics, although sometimes the condition is recognized in childhood as a result of gonadoblastomas (Sternberg and Barclay 1967; Dewhurst 1971).

The incidence of inguinal hernias in females with testicular feminization occurs in about 1.6 percent of all observed cases of female inguinal hernias (Kaplan et al. 1969). During operations for inguinal hernias, the gonads can be examined and the presence of tubes or uterus noted (Kaplan et al. 1969). Not all XY females develop hernias; however, the occurrence of a hernia presents an additional opportunity to detect the disorder in childhood, secondary in importance to detection at birth. As an additional means of detection, the female child with inguinal hernia or masses, whether visible or palpable, in the inguinal canals should have a buccal smear examination (Sternberg and Barclay 1967; Kaplan et al. 1969).

Detection in infancy is preferable to chance detection later in childhood, adolescence, or adulthood. Detection is especially important in the case of XY gonadal dysgenesis, where the development of gonadoblastomas is common in early childhood. Early detection also would aid in the proper handling of these disorders and in the detection of carriers and other affected family members in the cases of testicular feminization and XY gonadal dysgenesis (Dewhurst 1971; Stenchever et al. 1969; Kaplan et al. 1969).

THERAPY AND COUNSELING

In the therapy and treatment of testicular feminization and XY gonadal dysgenesis, the question of whether the gonads should be removed in infancy or when puberty is completed is a main consideration (Sternberg and Barclay 1967; Dewhurst 1971; Kaplan et al. 1969; Morris 1953). The argument for immediate gonadectomy is the danger of malignancy. In testicular feminization, tumors seldom arise before puberty, the risk of malignancy becoming greater after puberty (Kaplan et al. 1969; Dewhurst 1971; Sternberg and Barclay 1967). Once the gonads are removed, replacement estrogen therapy can be initiated to prevent the hot flashes and other characteristic menopausal symptoms such as breast atrophy, loss of libido, weight gain, etc. (Kaplan et al. 1969; Dewhurst 1971). Removal of the gonads before puberty prevents the natural occurrence of secondary sexual development (Kaplan et al. 1969; Dewhurst 1971; Morris 1953), and estrogen-induced puberty may not be as satisfactory as the natural event (Kaplan et al. 1969). Therefore, the treatment of testicular feminization would then have a more satisfactory outcome if the testes were removed after natural puberty. In incomplete testicular feminization, the presence of some virilization at birth makes further virilization at puberty likely (Dewhurst 1971; Kaplan et al. 1969; Sternberg and Barclay 1967). In this case, the gonads are removed to prevent further masculinization such as further clitoral enlargement, deepening of voice, and development of facial hair. Feminization and sexual development can be invoked by replacement estrogen therapy (Dewhurst 1971). By contrast, the gonads have no hormonal function in XY gonadal dysgenesis, and the risk of malignancy is extremely high. Streak gonads should be removed and estrogen therapy initiated to induce puberty. The XX male, however, does not present the clinical problem of malignancy common to the XY female condition.

The XY female is clearly and unmistakably feminine in behavior and outlook. Attitudes toward homemaking, marital role, pregnancy, and raising a family are female (Money et al. 1967). An important consideration in the clinical management of the XY female is one of helping the patient in psychological adjustment to the condition without adversely affecting sexual identity. The recommended policy is to describe the condition as one of diseased ovaries that will possibly become malignant at some future time necessitating removal (Dewhurst 1971; Morris 1953). Emphasis should be made that the patient is a normal female.

Mention of the presence of the testes or the male karyotype should be avoided (Stenchever et al. 1969; Kaplan et al. 1969; Dewhurst 1971; Morris 1953). Cooperation of the family of the patient in the detection of other family members has been reported as a result of this type of approach (Stenchever et al. 1969). The parents should also be informed of the probability that subsequent children may be similarly affected.

The attitude of the parents and the patient is an important consideration, as is the clinical consideration of gonadectomy. Each should be handled properly to prevent unnecessary problems (Dewhurst 1971; Kaplan et al. 1969; Stenchever et al. 1969; Money et al. 1967; Morris 1953). Early detection is then important, not only for detection of neoplasms and prescription of the most satisfactory treatment, but also for counseling to psychologically prepare the patient (Money and Ehrhardt 1973).

SUMMARY AND CONCLUSIONS

The XX male and XY female in man represent cases in which the phenotypic sex is opposite to that which the individual should have had in accordance with his chromosomal constitution. The occurrence of these two phenomena are not clearly understood and appear to represent a genetic paradox. The XX male and the XY female are normal in appearance but sterile. In all cases, some degree of gonadal dysgenesis, from partial to total, is characteristic. In XY females with testicular feminization, the production of hormones is normal or near that which is normal for males. The XX male is characteristically a case of hypogonadism, while the streak gonads in XY females with gonadal dysgenesis have no hormonal function. Histologically, the testes in both the testicular feminization syndrome and the XX male syndrome exhibit a lack of spermatogenesis, and the histology is usually distinctive from other forms of gonadal dysgenesis. The frequency of XY female condition is 0.002 to 0.005 percent (1/20,000–62,400) and the XX male, 0.006 percent.

The failure of normal testicular differentiation in early fetal stages is directly responsible for the phenomena of XX males and XY females. In humans and mammals the hormonal secretions of the fetal testis induce male development (Mittwoch 1967). Failure of the testis at any time during these critical stages of development results in some form of sexual abnormality. The XY female with complete gonadal dysgenesis represents a case of total testicular failure, while the XY female with testicular feminization and the XX male represent cases of partial testicular failure. The Y chromosome in XY females with gonadal dysgenesis failed to direct testicular differentiation and resulted in the total failure of testicular function. The embryonic testis in the XY female with testicular feminization, on the other hand, did not fail in either function and appears to represent a case of androgen unresponsiveness. It is now recognized that the effects of androgens are intimately bound up with hypothalamic-pituitary function at a later stage of existence (Ismail et al. 1972). Therefore, end organ resistance to the effects of androgens seems to be the likely cause of testicular feminization, suggesting hypothalamic-pituitary resistance to androgens.

Several proposals have been forwarded to explain the development of male primary and secondary sexual characteristics in the presence of a normal appearing female karyotype. The possibility that a mutant gene is responsible for the XX male syndrome in man is unlikely, although this phenomena does occur in other species. Familial occurrence and consanguinity are not characteristic of the XX male syndrome. Although the presence of mosaicism in the XX male is a distinct possibility and cannot be totally disregarded in the etiology of the syndrome, elimination of this possibility would entail examination of all body tissues. Very detailed studies have been performed and the presence of Y chromosome material has not been detected. Further, maternal age effect is lacking in the syndrome and makes the probability of nondisjunction unlikely. The interchange hypothesis can adequately explain the XX male syndrome, and the available evidence does not provide evidence against this mechanism. The male-determining loci are likely to be located on the short arm of the Y chromosome, and during the terminal association of the X and Y chromosomes in paternal meiosis, an exchange of genetic material on the short arms of X and Y chromosomes could occur.

Evidence indicates that testicular feminization and XY gonadal dysgenesis are transmitted through the maternal line and that an autosomal gene is responsible. The X–Y interchange hypothesis, however, can explain the presence of a Y chromosome and the development of a female phenotype. The XY female with gonadal dysgenesis may result from the loss of male determining genes on the Y chromosome. Under these conditions testicular development and secretions would be lacking, and a female phenotype would develop.

REFERENCES

ABDEL-HAMEED, F., R. N. SHOFFNER
1971 Intersexes and sex determination in chickens. *Science* 172:962–964.

ANDERSSON, L., S. BERGMAN, J. REITALU, S. ANSEHN
1972 A case of XX male-cytogenetic findings by autoradiography and fluorescence microscopy. *Hereditas* 70:311–314.

BACHMAN, R., F. F. DE LA CRUZ, M. AL-AISH, F. SANTELL
1971 Short arm deletion of an X chromosome in a 19-year-old girl. *American Journal of Mental Deficiency* 75:435–441.

BISHUN, N. P.
1968 Karyotype-phenotype correlations in a series of females with amenorrhea. *Acta Genetica Basel.* 18:479–486.
BOBROW, M.
1971 Length variation in the quinacrine-binding segment of human Y chromosomes of different sizes. *Cytogenetics* 10:190–198.
BOCZKOWSKI, K.
1968 Genetical studies in testicular feminization syndrome. *Journal of Medical Genetics* 5:181–188.
BOCZKOWSKI, K., Z. JANCZEWSKI, J. PHILLIP, M. MIKKELSEN
1968 A clinical and cytological study of XX males. *Hereditas* 62:285–292.
BOCZKOWSKI, K., J. TETER
1969 Intersexuality characterized by female external genitals and intra-abdominal rudimentary testes. *American Journal of Obstetrics and Gynecology* 103:1167–1168.
BOCZKOWSKI, K., J. TETER, J. STEINNADEL
1972 Sibship occurrence of XY gonadal dysgenesis with disgerminoma. *American Journal of Obstetrics and Gynecology* 113:952–955.
BREG, W. R., G. M. DILLARD, E. M. SHORT, P. W. ALLDERDICE, D. A. MILLER, O. J. MILLER
1971 Fluorescence and Y translocation in XX males. *Lancet* 1:1241–1242.
BROGGEN, A.
1964 Two sisters with primary amenorrhea and male chromosome pattern. *Human Chromosome Newsletter* 12:5.
BUHLER, E. M., H. I. MULLER, G. R. STRADLER
1971 A strongly fluorescent abnormal chromosome in a malformed child. *Humangenetik* 12:64–66.
CARDIFF, C., V. UDEN, D. W. KILLINGER
1972 Steroid biosynthesis in gonadal tissue from patients with testicular feminization. *Canadian Journal of Biochemistry* 50:849–855.
CASPERSSON, T.
1971 Translocations causing non-fluorescent Y chromosomes in human XO/XY mosaics. *Hereditas* 68:317–324.
CASPERSSON, T., A. DE LA CHAPELLE, F. LINDSTEN
1971 Absence of brightly fluorescent Y material in XX men. *Annals of Genetics* 14:173–176.
CHEMKE, J., R. CARMICHAEL, J. M. STEWART, T. H. GEER, A. ROBINSON
1970 Familial XY gonadal dysgenesis. *Journal of Medical Genetics* 7:105–111.
CHRISTAKOS, A. C., J. L. SIMPSON, J. B. YOUNGER, C. D. CHRISTIAN
1969 Gonadal dysgenesis as an autosomal recessive condition. *American Journal of Obstetrics and Gynecology* 104:1027–1030.
COHEN, M. M., M. W. SHAW
1965 Two XY siblings with gonadal dysgenesis and female phenotype. *The New England Journal of Medicine* 272:1083–1088.
DE LA CHAPELLE, A.
1972 Nature and origin of males with XX sex chromosomes. *American Journal of Human Genetics* 24:71–105.
DE LA CHAPELLE, A., H. HORTLING, J. WENNSTROM, M. NIEMI, C. J. JOHANSSON
1965 Two males with female chromosomes. *Acta Endocrinological Supplement* 100:90.
DE LA CHAPELLE, A., S. SIMILA, M. LANNING, M. KONTTURI, C. JOHANSSON
1971 Two further males with female karyotypes. *Humangenetik* 11:286–294.

DEWHURST, C. J.
1971 The XY female. *American Journal of Obstetrics and Gynecology* 109:564–588.

EGGEN, R. R.
1969 "Cytogenetics," in *Clinical Diagnosis by Laboratory Methods*. Edited by I. Davidson and J. B. Henry, 1208–1240. Philadelphia: W. B. Saunders.

ESPINER, E. A., A. M. O. VEALE, V. E. SANDS, P. H. FITZGERALD
1970 Familial syndrome of streak gonads and normal male karyotype in five phenotypic females. *The New England Journal of Medicine* 283:6–11.

FERENCZY, A., R. M. RICHART, O. J. MILLER
1971 Gonadoblastoma occuring in a female with XO/XY fragment gonadal dysgenesis. *American Journal of Obstetrics and Gynecology* 109:564–569.

FERGUSON-SMITH, M. A.
1966 "Sex chromatin, Klinefelters syndrome and mental deficiency," in *The Sex Chromatin*. Edited by K. L. Moore, 277–315. Philadelphia: W. B. Saunders.
1966 X–Y chromosomal interchange in the aetiology of true hermaphroditism and of XX Klinefelters syndrome. *Lancet* 2:475–476.

FERGUSON-SMITH, M. A., E. BOYD, M. E. FERGUSON-SMITH, J. G. PRITCHARD, A. F. M. YUSUF, G. BRAY
1969 Isochromosome for long arm of Y chromosome in patient with Turner's syndrome and sex chromosome mosaicism (45, X/46, XY qi). *Journal of Medical Genetics* 6:422–425.

FORD, C. E., J. L. HAMERTON
1956 The chromosomes of man. *Nature* 178:1020–1023.

FRACCARO, M., L. TIEPOLO, O. ZUFFARDI, C. BARIGOZZI, S. DOLFINI
1971 Fluorescence and Y translocation in XX males. *Lancet* 1:858.

FRASIER, S. D., R. A. BASHORE, H. D. MOSIER
1964 Gonadoblastoma associated with pure gonadal dysgenesis in monozygous twins. *The Journal of Pediatrics* 64:740–745.

FRENCH, F. S., B. BAGGETT, J. J. VAN WYK, L. M. TALBERT, W. R. HUBBARD, F. R. JOHNSTON, R. P. WEAVER
1965 Testicular feminization: clinical morphological and biochemical studies. *Journal of Clinical Endocrinology* 25:661–677.

FRENCH, F. S., J. J. VAN WYK, B. BAGGETH, W. E. EASTERLING, L. M. TALBERT, F. R. JOHNSTON
1966 Further evidence of a target organ defect in the syndrome of testicular feminization. *Journal of Clinical Endocrinology* 26:493–503.

GEORGE, K. P., P. E. POLANI
1970 Heterochromatin and XX males. *Nature* 228:1215–1216.

GOLDBERG, M. B., A. L. SCULLY, M. D., I. L. SOLOMON, M. D., H. L. STEINBACH
1968 Gonadal dysgenesis in phenotypic female subjects. *American Journal of Medicine* 45:529–543.

GORDON, D. L., E. KRMPOTIC, B. B. ZELINGER, L. E. ROSENBLUM
1969 Sex chromosome mosaicism in the testicular feminization syndrome. *Obstetrics and Gynecology* 34:779–782.

GRIFFITHS, K., J. K. GRANT, W. G. WHYTE
1963 Steroid biosynthesis in vitro by cryptorchid testes from a case of testicular feminization. *Journal of Clinical Endocrinology* 23:1044–1055.

HENRY, J. B., A. F. KRIEG
1969 "Endocrine measurements," in *Clinical diagnosis by laboratory*

methods. Edited by I. Davidson and J. B. Henry, 601–644. Philadelphia: W. B. Saunders.

HOFFENBERG, R., W. P. U. JACKSON
1957 Gonadal dysgenesis in normal looking females. A genetic theory to explain the variability of the syndrome. *British Medical Journal* 1:1281–1284.

ISMAIL, A. A. A., D. W. DAVIDSON, A. R. SONKA, J. A. LORAINE, G. L. FOSS
1972 Hormonal studies in a patient with the testicular feminization syndrome. *Acta Endocrinologica* 71:409–416.

JACOBS, P. A., A. ROSS
1966 Structural abnormalities of the Y chromosome in man. *Nature* 210:352–354.

JONES, T. C.
1969 "Anomalies of sex chromosomes in tortoiseshell male cats," in *Comparative Mammalian Cytogenetics.* Edited by K. Benerschke, 414–433. New York: Springer-Verlag.

JOST, A.
1947 Recherches sur la differentiation sexuelle de l'embryon de Lapin III. Rôle des gonades foetalis dans la differentiation sexuelle somatique. *Archives d'Anatomie Microscopique et de Morphologie Expérimentale* 36:271.

KAPLAN, S. A., W. H. SNYDER, JR., S. LITTLE
1969 Inguinal hernias in females and the testicular feminization syndrome. *American Journal of Diseases of Children* 117:243–251.

KARIMINEJAD, M. H., M. A. MOVLAVI, M. A. NASSERGHODOSSI, D. J. GHAFOORZADEH, Y. BEHHATNIA
1972 Gonadoblastoma associated with mixed gonadal dysgenesis. *American Journal of Obstetrics and Gynecology* 113:410–414.

KHUDR, G., K. BENIRSCHKE
1971 Fluorescence of the Y chromosome: a rapid test to determine fetal sex. *American Journal of Obstetrics and Gynecology* 110:1091–1095.

LAURIAN, LYDIA, M. D., J. O. FINKELSTEIN, N. SOFERMAN, M. LOEWENTHAL, L. DELEGDISCH
1971 Testicular feminization syndrome. *Israel Journal of Medical Science* 7:251–254.

LAURITSEN, J. G., J. JONASSON, A. J. THERKELSEN, F. LASS, J. LINDSTEN, G. B. PETERSEN
1972 Studies on spontaneous abortions. Fluorescence analysis of abnormal karyotypes. *Hereditas* 71:160–163.

LINDSTEN, J., C. G. BERGSTRAND, K. G. TILLINSER, H. G. SCHWARZASHER, L. TREPOLO, S. MULDAL, B. HOKFELT
1966 A clinical and cytogenetical study of three patients with male phenotype and apparent XX sex chromosome constitution. *Acta Endocrinologica* 52:91–112.

LISKER, R., F. FLORES, A. COBO, F. G. ROJAS
1970 A case of XX male syndrome. *Journal of Medical Genetics* 7:394–398.

MANOLOV, G., Y MANOLOVA, G. FISKESJO, A. LEVAN
1971 The complexity of the fluorescent pattern of the human Y chromosome. *Hereditas* 68:328–331.

MAUVAIS-JARVIS, P., J. P. BERCOVICI, F. GAUTHIER
1969 In vivo studies on testosterone metabolism by skin of normal males and patients with the syndrome of testicular feminization. *Journal of Clinical Endocrinology* 29:417–420.

MAUVAIS-JARVIS, P., J. P. BERCOVICI, O. CREPY, F. GAUTHIER
1970 Studies on testosterone metabolism in subjects with testicular feminization syndrome. *Journal of Clinical Investigation* 49:31–40.
MAUVAIS-JARVIS, P., H. H. FLOCH, J. P. BERCOVICI
1968 Studies on testosterone metabolism in human subjects with normal and pathological sexual differentiation. *Journal of Clinical Endocrinology* 28:460–471.
MICHALOVA, K., J. NEUWIRTH, J. RABOCH
1972 Evidence of the absence of Y fluorescent material in XX men. *Humangenetik* 14:251–254.
MILLER, O. J.
1964 The sex chromosome anomalies. *American Journal of Obstetrics and Gynecology* 90:1078–1139.
MITTWOCH, U.
1967 *Sex chromosomes.* New York: Academic.
MONEY, J., A. A. EHRHARDT
1973 *Man and woman, boy and girl.* Baltimore: Johns Hopkins University Press.
MONEY, J., ANKE A. EHRHARDT, DANIEL N. MASICA
1967 Fetal feminization induced by androgen insensitivity in the testicular feminizing syndrome: effect on marriage and maternalism. *The Johns Hopkins Medical Journal* 123:105–114.
MORRIS, J. M.
1953 The syndrome of testicular feminization in male pseudohermaphrodites. *American Journal of Obstetrics and Gynecology* 65:1192–1211.
MORRIS, J. M., V. B. MAHESH
1963 Further observations on the syndrome, testicular feminization. *American Journal of Obstetrics and Gynecology* 87:731–748.
MUKERJEE, D., J. M. BOWEN, J. M. TRUJILLO, A. CORK
1972 Increased susceptibility of cells from cancer patients with XY gonadal dysgenesis to simian papovavirus 40 transformation. *Cancer Research* 32:1518–1520.
NEUWIRTH, J., L. SLARKA, J. RABOCH
1972 Different chromosome variants of Klinefelters syndrome and plasma testosterone. *Humangenetik* 15:93–95.
NIELSEN, J., V. FRIEDRICH, A. CHRISTENSEN, H. H. GODT, L. S. STROMGREN
1972 A phenotypic male with karyotype 45, X/45, X, ace + (?Yq-). *Humangenetik* 15:319–326.
NORA, J. J., A. K. SINHA
1968 Direct transmission of the Turner phenotype. *American Journal of Diseases of Children* 116:343–350.
NORTHCUTT, R. C., D. P. ISLAND, G. W. LIDDLE
1969 An explanation for the target organ unresponsiveness to testosterone in the testicular feminization syndrome. *Journal of Clinical Endocrinology and Metabolism* 29:422–425.
PANTELAKIS, S. N., O. M. CHRYSSOSTOMIDOU, D. ALEXIOU, T. VALAES, S. A. DOXIADIS
1970 Sex chromatin and chromosome abnormalities among 10,412 liveborn babies. *Archives of Disease in Childhood* 45:87–92.
PARIS CONFERENCE (1971)
1972 Standardization in human cytogenetics. *Birth Defects: Original Article Series* 8(7). White Plains, New York: The National Foundation, March of Dimes.

PARK, I. J., H. W. JONES
1971 Familial male hermaphroditism with ambiguous external genitalia. *American Journal of Obstetrics and Gynecology* 108:1197–1205.
PEARSON, P. L., M. BOBROW
1970 Definitive evidence for the short arm of the Y chromosome associating with the X chromosome during meiosis in the human male. *Nature* 226:959–961.
PEREZ-BALLISTER, B., R. G. GREENBLATT, J. R. BYRD
1970 Familial gonadal dysgenesis. *American Journal of Obsterics and Gynecology* 107:1262–1263.
POWERS, H. O., R. L. NEU, H. SMULYAN, L. I. GARDNER
1970 An adult phenotypic male with 46, XX chromosome complement. *Journal of Clinical Endocrinology and Metabolism* 31:576–579.
RACE, R. R., R. SANGER
1969 X_g and sex chromosome abnormalities. *British Medical Bulletin* 25:99–103.
RATCLIFFE, S. G., A. L. STEWART, M. M. MELVILLE, P. A. JACOBS
1970 Chromosome studies on 3,500 newborn male infants. *Lancet* 1:121–122.
RIVAROLA, M. A., J. M. SAEZ, W. J. MEYER, F. M. KENNY, C. J. MEGLON
1967 Studies of androgens in the syndrome of male pseudohermaphroditism with testicular feminization. *Journal of Clinical Endocrinology* 27:371–378.
ROSENFIELD, A. L., A. M. LAURENCE, S. LIAO, R. L. LANDAU
1971 Androgens and androgen responsiveness in the feminizing testes syndrome. Comparison of complete and incomplete forms. *Journal of Clinical Endocrinology* 32:625–632.
SAEZ, J. M., D. PERETTI, A. M. MORERA, M. DAVID, J. BERTRAND
1971 Familial male pseudohermaphroditism with gynecomastia due to testicular 17–ketosteroid reductase defect, volume I. Studies in vivo. *Journal of Clinical Endocrinology and Metabolism* 32:604–610.
SANGER, R., P. TIPPETT, J. GAVIN, A. GOOCH, R. R. RACE
1969 Inheritance of testicular feminization: some negative linkage findings. *Journal of Medical Genetics* 6:26–27.
SCHWINGER, E., F. PERA
1972 On splitting of the Y fluorescent body in man. *Humangenetik* 14:107–111.
SMITH, MOYRA, W. J. PEPLER
1969 Female phenotype associated with XY karyotype: report of a case with gonadal dysgenesis. *South African Medical Journal* 43:1414–1415.
SOLARI, A. J., L. L. TRES
1970 The three dimensional reconstruction of the XY chromosomal pair in human spermatocytes. *Journal of Cell Biology* 45:43–53.
SOLTAN, H. C.
1968 Genetic characteristics of families of XO and XXY patients including evidence of source of X chromosomes in 7 aneuploid patients. *Journal of Medical Genetics* 5:173–180.
SOUTHERN, A. L., S. TOCHIMOTO, N. C. CARMODY, K. ISURUI
1965 Plasma production rates of testosterone in normal adult men and women and in patients with the syndrome of feminized testes. *Journal of Clinical Endocrinology* 25:1441–1450.

SPERLING, K., I. LACKMAN
1971 Large human Y chromosome with two fluorescent bands. *Clinical Genetics* 2:352–355.
STENCHEVER, M. A., A. B. P. N., G. K. JONES, J. A. JARVIS
1969 Testicular feminization syndrome: chromosomal histologic, and genetic studies in a large kindred. *Obstetrics and Gynecology* 33:649–657.
STERNBERG, W. H., D. L. BARCLAY
1967 Women with male sex chromosomes: the syndromes of testicular feminization and XY gonadal dysgenesis. *Birth Defects*, reprint series 141. White Plains, New York: The National Foundation, March of Dimes.
STEWART, A. L., A. J. KEAY, P. A. JACOBS, M. M. MELVILLE
1969 A chromosome survey of unselected liveborn children with congenital abnormalities. *The Journal of Pediatrics* 74:449–458.
STONOVA, N. S., T. G. SELEZNIOVA
1968 Chromosome aberrations in cases of human spontaneous abortions. *Genetika* (Moscow) 4:126–144.
STRICKLAND, A. L., F. S. SMITH
1969 Absence of response to dihydrotestosterone in the syndrome of testicular feminization. *Journal of Clinical Endocrinology and Metabolism* 29:1284–1286.
STURTEVANT, A. H.
1945 A gene in *Drosophila melanogaster* that transforms females into males. *Genetics* 30:297–299.
TETSUJI, K., K. OHAMA, T. SOFUNI, T. IKEDA
1971 An XY human female with ovaries. *American Journal of Obstetrics and Gynecology* 110:98–102.
THERKELSEN, A. J.
1964 Sterile male with the chromosome constitution 46, XX. *Cytogenetics* 3:207–218.
WACHSTEIN, M.
1951 Male hermaphroditism: a type showing female habitus, absence of uterus, and male gonads, often associated with testicular tubular adenoma: Report of a case and review of literature. *American Journal of Clinical Pathology* 21:10.
WADE, A. D., G. S. WILKINSON, J. C. DAVIS, T. N. A. JEFFCOATE
1968 The metabolism of testosterone, androstenedione and oestrone by testes from a case of testicular feminization. *Journal of Endocrinology* 42:391–403.
WILSON, J. D., J. D. WALKER
1969 The conversion of testosterone to 5α- and androstan-17β-ol-3-one (dihydrotestosterone) by skin slices of man. *Journal of Clinical Investigations* 48:371–379.
ZOURGLAS, P. A., H. W. JONES
1965 Clinical, histologic, and cytogenetic findings in male hermaphroditism II. Male hermaphrodites with feminine external genitalia (testicular feminization). *Obstetrics and Gynecology* 25:768–778.

Hormonal Factors in Formative Processes

E. N. KHRISANFOVA

Among the many pressing tasks modern anthropology faces, one of the most important is to study the factors that affect the development of the human morphotype. Like many other natural sciences human morphology has long developed through descriptive methods, but nowadays it shows a tendency to come to grips with the causes of the observed somatic variations. Of course, at the highest level of integration one finds it particularly difficult to specify the role played by the numerous endogenous and exogenous factors affecting the individual somatic features and the constitution of a person as a whole. From the methodical point of view, it is essential that in such a study the possible determinants of morphological variability should be chosen correctly and scientifically.

Naturally, a researcher attempting to tackle the problem of the formation of the human morphotype from the standpoint of unity and integrity of the organism will be tempted to look into specific features of the basic system providing for such integrity and, above all, those elements of this system that have a marked influence on the formative processes. The sex steroids of gonadal and adrenal origin whose functions are associated with activation of the genetic apparatus of the cell occupy a position of special importance in the system of neurohormonal regulation governing the translation of genotype into phenotype. These steroids exhibit a high level of morphogenetic and metabolic activity and a broad spectrum of action, influencing structures of particular interest for the morphologist.

Although the effects of sex hormones in vertebrates are well known, the relationship of morphology and gonadal activity in healthy human beings is still obscure. In fact, there have been only a few initial attempts to analyze this relationship at the level of modern hormonal biochemistry.

For several years now, a group of researchers and post-graduates of the

Chair of Anthropology of the Moscow State University has been carrying out a project of study of morphohormonal relationships.[1] At various times over the years the following scientists have made their contributions: L. V. Zhelnina, L. V. Bets, Ye. B. Savostyanova, R. G. Sedova, G. M. Elgurt, Ye. P. Titova, Ye. S. Sayapina, T. A. Blagoveshchenskaya, and others (Khrisanfova 1971: 37–38). All in all, over 500 persons have been examined and classified into fifteen age and sex groups, each separated from the next by an interval of one year.[2]

This paper sets forth some of the results of the "statical" research program. The program of studies of dynamic relationships of morphological and hormonal factors is still in progress.

The study is based on classical endocrinological methods of determining quantitative secretion, which have been applied to the study of healthy human populations comparatively recently, since the advent of sufficiently accurate methods of hormone determination. It is well known that the results of some experiments and clinical observations indicate that the end effect of a formative reaction may depend to a considerable degree on the concentration of hormones present (Gallien 1962; Elger et al. 1970), although the ratio of active to inactive hormone components in peripheral blood also seems to be of great importance. In this respect, individual and intraindividual variation in the quantity of excretion of sex hormones is of interest, since excretion is a measure of secretion. Here it should be borne in mind that the level of secretion of sex hormones itself may depend on a number of endogenous and exogenous factors, such as the functional state of other elements of the neuroendocrine system, sex and age of the subject, biological rhythms, motor activity, nutrition, etc. Some of these factors (primarily exogenous) can be controlled for season, nutritional regime, motor activity, sex and age, while others can at this stage be merely abstracted (excepting, of course, pathology — both endocrine and nonendocrine). The quantity of secretion of the "working hormone" is the result of all these influences.

When studying the levels of secretion of sex hormones, one problem is estimating the role of genetic determination. The first results of our experiment in comparing the excretion of sex steroids (estrogens and 17-Keto-steroids [17-KS]) in mono- and dizygotic twins support the hypothesis that the genetic factor is important (Holzinger index for estrogens and 17-KS: 0.8–0.9). Some facts supporting the belief that the genetic factor is important are obtained from clinical data as well (Zhukovskiy 1971; Bordzhadze and Starkova 1971:4).

Another way of indirectly estimating the degree of genetic control

[1] The research program is carried out in cooperation with the Pediatric Clinic (Chief Physician, Professor M. A. Zhukovsky) of the Institute of Endocrinology and Hormonal Chemistry, Academy of Medical Sciences of the Soviet Union.

[2] An average group included from forty to fifty subjects.

involves the determination of an individual hormonal profile. Some literary sources point to the relation stability of excretion types (Tanner et al. 1964; Tanner and Gupta 1968). The data at our disposal (both literary and our own) indicate that throughout a period of "microlongitudinal" observations (lasting from two to six weeks) the interindividual differences in the 17-KS, estrogens, and testosterone excretion show consistent differences. This means in effect that a higher or lower level of hormone excretion is a feature of biochemical individuality. Therefore, in the majority of cases the individual hormonal profile tends toward relative stability. This trend is the most pronounced in extreme variants, making it feasible to establish morpho-hormonal correlations even on the basis of a single examination. At the same time, in different periods of ontogenesis, under different physiological conditions, there probably exists some specificity of intraindividual variability. Thus, in the absence of menstrual cycles (eight-year old girls), estrogens are more stable than 17-KS, whereas in girls with an established cycle, 17-KS were more stable than estrogens. 17-KS stability has been observed in girls over long periods. Thus, each of ten girls aged twenty-one to twenty-three, observed for three years, maintained her own level of excretion of these hormones. Again estrogens have proved more variable, with increases that are not always strictly proportional.[3]

The relative stability of an individual hormonal profile may be maintained in certain cases for a long period of life. This is suggested by the data from a group of twenty-three-year-old women featuring a high negative correlation between the level of estrogens and body length. At the same time, in a group comprising 300 pubertal (fifteen-year-old) girls, accelerated sexual development accompanied reduction in body length. This finding suggests that in this case the group, or at least its extreme variants, had a relative stability of the hormonal profile throughout a whole ontogenetic period. It is true, however, that the morphologic effect of intensified sex hormone secretion cannot always be defined clearly enough.

The quantitative factor, however, characterizes but one facet of the morpho-hormonal relationship. Another important aspect concerns individual (and even local) variations in the threshold sensitivity of peripheral tissues. It is highly probable that this factor depends on the age, sex, and innervation of the subject, as well as on the nutritional status and other exogenous factors. The problem of hormone receptors and the mechanisms of hormonal effects on the genes is beyond the scope of anthropology, being in the realm of molecular biology. Hence, we must

[3] It should be noted, however, that since the variability of hormonal indices is effected to a considerable degree by the methods of investigation (see above), including the accuracy of the quantitative determination technique used, it is reasonable to assume that further improvement and standardization of these techniques will modify the variability estimate.

study this aspect of morpho-hormonal relationships indirectly, studying their specificity in individuals with regard to sex, age, and other factors.

In morpho-hormonal investigations it is important to study the sphere and degree of influence of the hormonal factor in various sex-age groups. Naturally, clinical and experimental data must be utilized as widely as possible, but with one reservation — they cannot always be automatically extended to healthy populations. In each case the specific group should be studied on the basis of its specific conditions with due regard for the possible effect of factors generally associated with the hormonal status of man. In such investigations the individualizing technique seems to be the most effective. Another possible way of studying morpho-hormonal relationships is to match hormonal indices with constitutional types reflecting genetically determined morphological variation.

Table 1 presents the distribution of estrogens and 17-KS along the astheno-pycnomorphic coordinate in young women and pubertal girls. It shows a clear trend, all indices being considerably higher in the pycnomorphic variant than in the asthenomorphic one. This distribution agrees with correlations of hormonal indices with some somatic measurements obtained in these groups. For example, estrogens (r of the order of $+4-0.5$) correlate with the thigh girth and fat deposition in the women's group, the correlation with estriol being particularly significant. These findings are in agreement with the current concept of the estriol in the development of other derivatives of the "skin organ" as well. Similar correlations have been observed between 17-KS and chest circumference and sagittal diameter. In other words, the pycnomorphic tendency in female constitution is reliably associated with an increased quantitative secretion of estrogens and adrenal androgens. A similar distribution of hormonal indices is found in the group of fifteen-year-old girls at an advanced stage of puberty. Such a definitive tendency stresses the typological importance of certain somatic secondary sexual indices in women.

Table 1. Distribution of estrogens and neutral 17–KS in women's groups

Type of constitution	Estrogens (mg./24 hr)				Neutral 17–KS (mg./24 hr)
	Total	estradiol	estron	estriole	
A. Fifteen-year-old girls					
Asthenomorphic	19.82	4.29	6.64	8.89	4.57
Pycnomorphic	30.37	7.23	10.09	13.05	6.90
Group mean	25.64	5.68	8.36	11.60	5.57
B. Young women					
Asthenomorphic	32.82	4.33	10.66	17.83	5.44
Pycnomorphic	48.37	6.11	16.94	25.32	9.02
Group mean	42.89	5.10	13.56	24.23	8.01

In our previous works we posed a question as to the exact phase of ontogenesis in which the morpho-hormonal relationships found in adults are manifested. According to our preliminary data, as early as the prepuberty stage (nine to eleven years) the girls of digestive and muscular types are ahead of the asthenoid girls of the same age in terms of excretion of adrenal androgens and estrogens. Thus, eleven-year-old girls featuring higher levels of 17-KS and estrogen excretion not only had, on the average, larger total somatic measurements (height, weight, chest circumference), but also a higher level of fat deposition, a larger anterior–posterior-chest diameter relative to the transverse diameter, and a larger thigh girth relative to chest circumference. In other words, a definite trend toward pycnomorphicity begins as early as the beginning of the pubertal period. It is associated with a high level of androgenic activity of the adrenal cortex, which may in part be ascribed to the effect of endogenous estrogens providing for a higher activity of the adrenal cortex in females. In this age group the "accelerated type" is consistently associated with increased secretion of adrenal androgens and an increased androsterone-to-etiocholanolone ratio (1:2.2 and 1:3.3 for the accelerated and retarded types, respectively). Significant correlations are observed between 17-KS and estrogen secretion and some bone age indices (hand–wrist ossification index and cortical index of II metacarpal) (Table 2). These data are in agreement with clinical observations: children with an intensified secretion of sex hormones feature a very high level of cortical bone growth exceeding, in terms of the cortical index of II metacarpal, the level of development in children of the same age by three, four, or even more years.

Among males, nineteen-year-old boys of a muscular type were definitely ahead of asthenomorphic boys of the same age in terms of the 17-KS index. This is in agreement with Tanner's data showing a positive correlation of 17-KS with mesomorphy and a negative correlation of 17-KS with ectomorphy (as Sheldon described it). Of course, the 17-KS

Table 2. Correlation of neutral 17–KS with some somatic indices in nine-year-old girls

Indices correlated with neutral 17–KS	Correlation coefficients
Body length	+0.458**
Body weight	+0.438**
Chest sagittal diameter	+0.414*
Chest circumference	+0.409*
Thigh girth	+0.422**
Cortical index of II metacarpal	+0.453**
Index of ossification of hand-wrist	+0.469**

Key: * Reliable with $\beta_1 = 0.95$.
** Reliable with $\beta_2 = 0.99$.

index is far from reflecting the level of androgenization of the organism. Yet even this index shows a reliable correlation with the muscular development, anterior–posterior-chest diameter, and chest circumference (r of the order of +0.4–0.5), which agrees with the above data on the women's group as well as with the results of other authors (Tanner et al. 1964; De Moor et al. 1963, 1964; Cristescu and Lazăr 1967; Kreze and Galvánek 1972). In this connection, the role of chest circumference in the clinical picture of the syndrome of *pubertas precox*, including cases of hyperandrogeny of adrenal origin, is especially interesting. Children with excessive secretion of sex steroids exhibit very high values (on the basis of the regression equation in body length) when compared not only with children of the same chronological age, but also somewhat with children of the same biological age as indicated by bone age.

The lead of the muscular type compared with the asthenoid in terms of 17-KS secretion is manifest at an early stage of puberty, but before gonad activation (nine-year-old boys) there seems to be no clearly defined 17-KS distribution by constitutional types. Thus, for example, there existed only a trend toward the correlation of this hormonal index with muscular development. In boys aged ten to eleven, however, the role of androgens as stimulants of anabolic processes seems quite pronounced (Viru and Kyrge 1969).

We also have some tentative results of morpho-hormonal comparisons of a comparative primatological nature. An experiment was conducted in a small group of adult female Old World monkeys (*Papio hamadryas*) which are known to have some features of similarity with man in hormonal spectrum, secretion rhythm, and physiological mechanisms of sex hormones (Alekseeva 1961; Shubert et al. 1972). Characteristically, the monkey with the highest level of estrogen and 17-KS secretion had the largest weight, the largest chest and abdomen circumference, and thigh girth. This monkey also exhibited pronounced brachymorphy, as well as an absolute and relative limb shortening. Of course, a few nonsystematic observations are insufficient for any serious generalization, yet the very fact of a certain similarity between men and a catarrhine nonhominid primate in morpho-hormonal relationships is interesting.

The above-described tentative results of morpho-hormonal comparisons naturally fall far short of exhausting the problem; in fact they merely touch on one of its aspects. It is also clear that sex hormones do not account for all of the hormonal influences on constitutional development. Comprehensive neuro-endocrinological investigations are needed, particularly concerning the state of the hypothalamus-hypophysis-adrenals-gonads system, the functional activity of the thyroid gland, and the somatotropic function of the hypophysis. Obviously, only thorough exploration of the individual specificity of each functional system (and above all, the endocrine system, with its decisive role in the formation and maintenance of morphotypes), followed by synthesis and theoretical development of the data will make possible integrated studies of functional constitution. Such studies will call for the joint efforts of many researchers. The study of sex steroid levels provides a good start for research into one of the most important functional systems.

REFERENCES

ALEKSEEVA, L. V.
1961 "Izuchenie ekskretsii estrogenov u obez'yan" ["The study of estrogen excretion in monkeys"]. Mezhdunarodnyy biokhimicheskiy kongress. Referaty sektsioniykh soobshcheniy 1.

BORDZHADZE, V. K., N. T. STARKOVA
1971 Dlinnaya u khromosoma pri semeynoy forme gipogonadizma [Chromosome length in the familial form of hypogonadism]. *Problemy endokrinologii* 4.

CRISTESCU, M., A. LAZĂR
1967 Quelque aspects biochimiques en corrélation avec le dévelopment morpho-fonctionnel des enfants et des adolescents. *Annuaire roumain anthropologie* 4.

DE MOOR, P., O. STEENO, E. MEULEPAS, A. HENDRIX, M. OSTYN
1963 Influence of body size and of sex on urinary corticoid excretion in a group of normal young males and females. *Journal of Clinical Endocrinology* 23.
1964 Anthropomorphology and corticosteroid excretion in normal young adult males and females. *Acta Clin. Belg.* 19(5).

ELGER, W., H. STEINBECK, B. CUPCEANCU, F. NEUMANN
1970 Influence of methyltestosterone and cyproterone acetate on Wolffian duct differentiation in female rat fetuses. *Journal of Endocrinology* 47(4).

GALLIEN, L.
1962 Comparative activity of sexual steroids and genetic constitution in sexual differentiation of amphibian embryos. *General and Comparative Endocrinology* 2.

KHRISANFOVA, E. N.
1971 Nekotorye aspekty gormonal'nykh issledovaniy v antropologii [Some aspects of hormone research in anthropology]. *Voprosy antropologii.*

KREZE, A., M. GALVÁNEK
1972 Kontrolné hodnoty individuálných 17-keto-steroidoy a ich antropometrické korelácie [Control values of individual 17-ketosteroids and their anthropometric correlations]. *Bratislavské lekarské listy* 57(2).

SHUBERT, K., G. KHOBE, N. P. GONCHAROV, G. V. KATSIYA, V. I. VORONTSOV
1972 Obmen steroidov u primatov. O vydelenii testosterona, spitestosterona i androstendiona iz mochi pavianov gamadrilov [The exchange of steroids in primates: on the separation of testosterone, epitestosterone, and androsterone from the urine of Old World monkeys]. *Problemy endokrinologii* 2.

TANNER, J. M., D. GUPTA
1968 A longitudinal study of the urinary excretion individual steroids in

children from eight to eighteen years old. *Journal of Endocrinology* 41(2).

TANNER, J. M., G. A. HARRISON, J. S. WEINER, N. A. BARNICOT
1964 *Human biology*. New York/Oxford: Oxford University Press.

TZATCHEVA, L., B. PANDOVA
1971 Quelques résultats de l'application de la methode jumeaux dans les recherches médicoanthropologiques. *Bulletin of the Association of Anatomy* 146.

VIRU, A. A., P. K. KYRGE
1969 "Izuchenie osobennostey adrenokortikotropnoy regulyatsii obmena veshchestv, proizvedennoe u 15 mal'chikov 10–11 let" ["A study of the characteristics of adrenocorticotropal regulation in the exchange of materials, conducted among 15 boys aged 10-11 years"]. Materialy 9 nauchnoy konferentsii po vozrastnoj morfologii, fiziologii i biokhimii.

ZHUKOVSKIY, M. A.
1971 *Detskaya endokrinologiya.*

Polyploidy in Humans and Other Vertebrates

FATHI ABDEL-HAMEED

Each species has a characteristic chromosomal number and morphology. In higher plants and animals, the nuclei of the somatic tissues contain two similar sets, or a diploid number ($2n$) of chromosomes. One set is donated by a haploid male gamete and represents the paternal genome, while the other represents the maternal genome. A degree of variability in the karyotypes of members of a given species is common in natural populations. If variability involves deviation from normal structure for certain chromosomes, it is described as a structural aberration, while changes in chromosome number are described as numerical aberrations.

Under normal conditions, various somatic tissues show the same ploidy level, i.e., homogeneous for diploidy. Infrequently, however, an individual may possess more than one cell, an occurrence known as mosaicism. In tissue cultures, especially long-term cultures, a progressive alteration in chromosome constitution over time can result in heteroploid cell lines, i.e., cell lines showing great variation in chromosomal number. The chromosome number of a single somatic cell can differ from diploidy in two ways: (1) in the case of aneuploidy, the chromosome number is anything other than an exact multiple of the basic number (x) or of the haploid number ($1n$) of a true diploid; and (2) in the case of true polyploidy or euploidy, the chromosome number is exactly three or more times the gametic number of a true diploid or that of a hypothetical diploid ($3n$, $4n$, . . . etc.). Various forms of aneuploidy, including monosomy ($2n - 1$), nullisomy ($2n - 2$), trisomy ($2n + 1$), double trisomy ($2n + 1 + 1$), and tetrasomy ($2n + 2$), have been observed in several plant and animal groups.

Since the discovery of the first known aneuploid condition in man, trisomy 21 (Lejeune et al. 1959), several other autosomal and sex-chromosome aneuploidy conditions have been described, and it has become obvious that many clinical syndromes and pathological conditions in man are due to specific structural or numerical changes in the chromosomes. A large body of information concerning the incidence of aneuploidy in man has accumulated over the past fifteen years. On the other hand, very little is known about true polyploidy and its effect in higher vertebrates, including man. This paper is an attempt to discuss this area in some detail.

FORMS OF EUPLOIDY AND THEIR ORIGIN

In plants, where polyploidy is so common, there are numerous expressions of it. Stebbins (1950) described autopolyploidy, allopolyploidy, segmental allopolyploidy, autoallopolyploidy, and genomic allopolyploidy.

For all practical purposes, however, I believe that limiting classification to three types is adequate and simpler: the two extremes, typical autopolyploidy and true or genomic allopolyploidy, with all intermediate conditions designated as segmental allopolyploidy.

Autopolyploidy

When the chromosome number of a single diploid is doubled, the result is four sets of chromosomes. Each chromosome in the haploid complement is represented by four identical (homologous) chromosomes, i.e., homologs. Due to their homology, they may pair during the first meiotic prophase, forming quadrivalents instead of bivalents, a characteristic feature of meiosis in autotetraploids. Autopolyploids may result from hybridization between two very closely related races followed by chromosome doubling.

Typical autopolyploidy seems to be the primary form of euploidy in animals species, and may be the only possible form in higher vertebrates, particularly primates. In plants, autoploidy is the least common, and alloploidy is quite frequent. About one-fourth to one-third of flowering plant species are alloploids (Stebbins 1950), most of them classified as segmental allopolyploids.

Allopolyploidy

The prefix "allo" indicates that two or more well-differentiated and genetically isolated species are involved. In the case of allopolyploidy, interspecific hybridization is followed by a doubling of the chromosome number. Chromosome doubling is a necessary step to restore fertility since the interspecific hybrids are, in most cases, extremely sterile (Abdel-Hameed 1970). Interspecific hybridization is quite rare among animals, especially higher mammals. When it occurs, as is the case of the mule, consequent doubling of the chromosome number is almost impossible. For these two reasons, true allopolyploidy is virtually absent in higher vertebrates.

In certain groups of lower vertebrates, interspecific or even intergeneric hybridization is possible. Greenfield and colleagues (1973) have recently described the occurrence of natural hybrids between two different genera of fresh water cyprinids. Several other examples of spontaneous interspecific hybridization have been described and appear to be common in other fishes, most likely due to the absence of strong genetic barriers or isolating mechanisms. In cases of gene flow between distinct species or genera, we would expect that pre- and/or postmating isolating mechanisms are weak and inefficient. We believe that ethological and hybrid inviability are not well-developed isolating mechanisms in these fishes, due partially to their unique external fertilization and to simultaneous spawning and fertilization of several species using the same nest.

The prefix "allo' is occasionally used in animal studies to specify a diploid hybrid, such as an allodiploid. While alloploidy is the commonest type of polyploidy in plants, it is indeed the rarest in animals. However, Stebbins (1950) predicted that alloploidy will be found more commonly in animals as more data become available. He explains the scarcity of identified animal hybrids in terms of the small percentage of interspecific hybrids in animals as compared to plants.

Segmental Allopolyploidy

Stebbins (1950) originated the term segmental allopolyploidy and defined it as a polyploid containing two pairs of genomes which possess in common a considerable number of homologous chromosomal segments or even whole chromosomes, but differ from each other in respect to a sufficiently large number of genes or chromosomal segments, so that the different genomes produce sterility when present together at the diploid level. Segmental allopolyploids represent an intermediate condition in a continuous spectrum where completely homozygous autopolyploidy is at one extreme and allopolyploidy, derived from parental species so distinct that they barely form a hybrid, is at the other. Upon closer cytogenetical examination, many cases that were previously designated as allopolyploids proved to be segmental allopolyploids, and it is believed that the majority of plant allopolyploids should be reclassified as segmental allopolyploids (Stebbins 1950).

POLYPLOIDY IN MAN

In higher vertebrates, polyploidy is usually restricted to certain metabolically active tissues, such as the liver, and homogeneous polyploidy throughout body tissues is extremely rare. Many biologists, therefore, assumed that whole-body polyploidy does not occur in man. However, recent observations indicate that such polyploidy seems to occur spontaneously in higher vertebrates at a much higher frequency than previously thought (Abdel-Hameed and Shoffner 1971; Abdel-Hameed 1972). It is safe to assume that polyploidy is a lethal condition in man and other mammals. Evidence for whole-body tetraploidy or triploidy has been found in man, but the effects of tetraploidy are more severe and less frequently observed than triploidy (Carr 1971a, 1971b). Both triploidy and tetraploidy in man represent cases of typical autopolyploidy.

Autotetraploidy in Man

Since the haploid or gametic chromosome number (1*n*) in man is twenty-three, the tetraploid nuclei contain ninety-two chromosomes. Tetraploid cells seem to be common in certain mammalian tissues, such as liver, where they are about twice the size and contain more nucleoli than the diploid cells. They are also quite common in cultures from normal tissues. For example, studies on amniocentesis show a high percentage of tetraploid nuclei in some cultures of amniotic fluid cells (Kohn and Robinson 1970; Milunsky et al. 1970; Sperling and Saling 1971). In at least one case, Kohn and Robinson (1970) found that all mitotic cells examined from the amniotic fluid culture were tetraploid, but the infant, who was born at term, proved to be a normal diploid. For unknown reasons, a high percentage of tetraploid cells is characteristic of certain amniotic fluid cultures. One might suspect that this form of polyploidy is due to an artifact in the *in vitro* culture technique, especially when observations on duplicated sex chromatin in interphase nuclei of human amnion epithelium indicate that polyploid cells make up less than 10 percent (Klinger and Schwarzacker 1958). It is also known that tetraploid cells arise quite commonly in normal cultures (Carr 1971a), but the spontaneous appearance of balanced triploid cells in these cultures is exceedingly rare (Pawlowitzki and Cenani 1967; Sellyei et al. 1971). We would expect to see a random pattern of occurrence if tetraploidy is indeed culture artifact, but this does not seem to be the case. Therefore, such tetraploidy is considered to be representative of the *in vivo* condition of otherwise a normally developing diploid fetus. The biological and clinical significance of amniotic tetraploidy is not known and certainly requires further investigation.

Carr (1965) reported the first known case of human tetraploidy, which were found among several specimens of spontaneous abortion.[1] In twenty studies of presumably unselected spontaneous abortions in humans published in the last ten years, fifteen tetraploid abortuses were reported among the 1,485 abortions examined (Carr 1971a). The frequency of tetraploidy would therefore be 1.01 percent of the abortions studied (Table 1). This estimate is believed to be representative as this survey probably includes the majority of reported observations on unselected spontaneous abortions. Interestingly, the majority of the 1,485 fetuses were normal in appearance, and only 361 cases showed developmental abnormalities (24.3 percent). All tetraploid specimens were abnormally developed. Thus, of the abnormally developed embryos, 4.2 percent were tetraploid (Table 1). In a recent study, Carr (1971b) carefully examined seven tetraploid specimens. Karyotype analysis was possible in six of them and revealed that four were 92, XXYY, and two were 92, XXXX. None of these specimens showed an embryo or embryonic remnant, five were intact empty sacs, and the other two were incomplete, ruptured sacs. Carr (1971a) reported that if the tetraploid specimen is not macerated and is complete, it usually consists of an intact sac that is empty in most cases, but occasionally contains a nondescript embryonic remnant.

Table 1. A summary of twenty unselected cytogenetic studies of spontaneous abortions showing the frequency of tetraploid, triploid, and abnormally developed abortuses

Number of abortuses		Percentage of total	Percentage of abnormal development
Abnormally developed	361	24.3	100
Tetraploid	15	1.01	4.2
Triploid	56	3.77	15.5
Total	1,485		

Autotetraploidy may result from several possibilities including the following: (1) fertilization of an unreduced, i.e., diploid ovum ($2n$) by an unreduced spermatozoa ($2n$); (2) fertilization of a normal haploid ovum ($1n$) by three normal sperms; (3) fertilization of a normal ovum ($1n$) by two sperms one of which is normal ($1n$) while the other is unreduced ($2n$); (4) endomitosis during first zygotic division; the centromeres of metaphase chromosomes split but the products do not segregate into

[1] There has been only one example of tetraploidy in a live infant (Kohn and Robinson 1967), which proved to be a mosaic for tetraploid and normal diploid cell lines ($4n/2n$).

separate nuclei, thus resulting in endotetraploidy; (5) failure of cytokinesis following the completion of karyokinesis at the first zygotic division.

All tetraploid abortuses examined so far are known to have either XXXX or XXYY sex chromosome constitution. Therefore, Carr (1971b) concluded that there is no reason to suspect any mechanism other than failure of cytokinesis following karyokinesis. However, this conclusion is based on examination of a very small number of tetraploid specimens and it is necessary to obtain more data before drawing any conclusion.

The mean gestation age at the time of abortion for the seven tetraploid specimens examined by Carr (1971b) was estimated at seventy-five days, and subsequently modified to seventy-seven days (Carr 1971a). Tetraploid abortuses with either XXXX or XXYY sex chromosomes do not seem to differ in either their mean maternal or gestation age, the mean maternal age being 26.9 years. Information concerning tetraploid conception is quite limited, reflecting the degree to which tetraploidy is lethal.

Judging by the literature, a fully-formed embryo in a purely tetraploid abortus has not yet been found (Carr 1965, 1971a, 1971b; Boué et al. 1967; Gropp 1967; Mikamo 1970; Stonova and Slezniova 1968; Thiede and Metcalfe 1966; Waxman et al. 1967). This suggests that tetraploidy is serious enough to interfere with early stages of fetal development; however, some believe that the formation of a well-developed tetraploid embryo is not impossible (Carr 1971b).

Autotriploidy in Humans

In contrast to tetraploidy, autotriploidy is quite common in human conception. It is estimated that about 20 percent of the cases of spontaneous abortion examined involved abnormal karyotypes; one-fifth of these were triploid (Butler et al. 1969; Schindler and Mikamo 1970). In evaluating the incidence of triploidy in humans, we must consider spontaneous abortions separately from conceptuses reaching term.

The percentage of triploidy reported in abortion surveys varies widely. Combining data from six surveys (Table 2), we estimate the percentage of triploidy in spontaneous abortions at 4.42 percent, similar to that reported by other investigators (Schindler and Mikamo 1970; Carr 1971a). Assuming a spontaneous abortion rate of about 20 percent of all pregnancies, it can be inferred that triploidy occurs in about 1 percent of all conceptions (Edwards et al. 1967). Using Carr's data (1965), the survival rate of triploid embryos is estimated at much less than 0.01 percent of live births. Therefore, the embryonic mortality of triploid conceptuses must exceed 99 percent (Edwards et al. 1967).

Table 2. Triploidy in spontaneous abortion surveys

Survey	Number of specimens analyzed	Number of triploid specimens	Percent of triploidy specimens
Carr (1965)	200	9	4.50
Schlegel et al. (1966)	30	3	10.00
Singh and Carr (1967)	146	5	3.42
Stenchever et al. (1967)	101	3	2.97
Szulman (1965)	25	5	20.00
Waxman, Arkaki, and Smith (1967)	85	1	1.18
Combined	587	26	4.42

The majority of cases of triploidy in humans have been detected in early spontaneous abortion (Penrose 1961; Schlegel et al. 1966; Stonova and Selezniova 1968; Szulman 1965; see Table 3). Almost all triploid conceptuses are aborted by the end of the third month of gestation; however, a few cases are known to have been alive while others died shortly before birth. Only fourteen cases of triploidy surviving to term or near term are known in man (Böök and Santesson 1960; Ellis et al. 1963; Ferrier et al. 1964; Gropp, Jussen, and Odunjo 1964; Bernard et al.

Table 3. Some reported triploid early human spontaneous abortuses

Reference	Karyotype[a]	Sex chromatin
Mosaic triploids		
Penrose (1961)	(1) XX/XXY	
Atkin (1962)	(1) ?/XXX	+ some cells ++ some cells
Aspillaga et al (1964)	(1) XY/XXY	–
Homogeneous triploids		
Carr (1963)	(1) XXY	–
Szulman (1965)	(4) XXY	
Carr (1967a)	(6) XXY	
	(2) XXX	
	(1) XYY	
Carr (1967b)	(1) XXX	
	(1) XXY	
	(1) triploid unable to karyotype	+
Edwards et al. (1967)	(2) XXY	–
Carr (1971b)	(13) XXY	+ some cells – some cells
	(9) XXX	++ some cells
	(5) triploid unable to karyotype	

Key: – no sex chromatin observed in interphase nuclei; + presence of one sex chromatin mass; ++ presence of two sex chromatin masses

[a] Numbers in parentheses preceeding karyotype indicate number of triploids in the study reported.

1967; Edwards et al. 1967; Sacrez et al. 1967; Schmid and Vischer 1967; Lejeune et al. 1967; Papiernik-Berkhauer 1968; Butler et al. 1969; Schindler and Mikamo 1970; Schmickel, Floyd, and Pooley 1971; Uchida and Lin 1972). Available data on twelve of them are summarized in Table 4. Of these twelve, six are $2n/3n$ mosaics, five of which are reported alive. The remaining six represent cases in which only a single triploid ($3n$) cell line was detected, i.e., pure or homogeneous triploids. All were born alive but died from two to twenty-three hours after delivery (Table 4).

It is clear that the triploid condition, lethal in humans, is also lethal in other mammals. Edwards (1958) induced triploidy in mice by using colchicine to suppress the extrusion of the second polar body of the eggs. In early embryonic stages, he found a high proportion of triploids, but found a decrease in fertility and no triploids in the offspring at term. This indicates an early death of triploid conceptuses.

In mammals, sexual dimorphism is well defined and therefore normal individuals are diploid with either XX or YY sex chromosome constitution. Under polyploid conditions, such clear-cut dimorphism is lacking since there are more than two possible sex chromosome complexes as shown in Table 5. The three possible triploid genotypes would be expected to occur in 1:2:1 ratio (Table 5) if they were produced at random and had equal viability, but this seems not to be the case. Among ninety-three triploid abortuses with known sex chromosome compliments, fifty-five had XXY, as compared to only four with XYY genotype (Table 5). The reasons for such imbalance of gonosomal types are not clearly understood at present, and there is a possiblity that it may reflect a more lethal condition in 69, XYY triploids leading to very early and therefore undetected abortion. The XXY and XXX genotypes do not seem to differ in their mean gestational age at the time of abortion, and their combined mean of 105.8 days is close to the 106.7 days estimated for abortuses with normal karyotypes. The mean maternal age for women aborting triploid embryos (27.2 years) is similar to that for both tetraploids and chromosomally normal abortuses (Table 5). Therefore, evidence for maternal age effect is lacking in human polyploidy, in contrast to the well-known relationship between trisomy and increased maternal age (Penrose 1961). On the other hand, the mean age at the time of abortion in tetraploidy is only seventy-seven days, which is strikingly shorter than for any other class of abortuses, normal or abnormal. Furthermore, the physical characteristics of tetraploid specimens show a much greater retardation of development than in triploid cases. By all measurable criteria, triploidy is less lethal than tetraploidy (Carr 1971a); while pure triploidy is compatible with live birth in humans, pure tetraploidy is apparently incompatible with embryonic development to term, or even close to term.

Table 4. Human triploids carried to term or past the usual stage of abortion

Reference	Phenotype	Gestation (weeks)	Postnatal age at death	Karyotype	Sex chromatin
Mosaic triploids					
Böök and Santesson (1960)	M	40	Alive	46, XY/ 69, XXY	?
Ellis et al. (1963)	F	?	Alive	46, XX/ 69, XXX	?
Ferrier et al. (1964)	M	?	Alive	46, XY/ 69, XXY	–
Lejeune et al. (1967)	?	40	Alive	46, XX/ 69, XXY	+
Sacrez et al. (1967)	F	36	21 hr.	46, XY/ 69, XXY	+
Schmid and Vischer (1967)	M	43	Alive	48, XXYY/ 71, XXXYY	–
Pure triploids					
Bernard et al. (1967)	?	32	15 hr.	69, XXY	?
Edwards et al. (1967)	?	32	17 hr.	69, XXY	–
Butler et al. (1969)	F	35	23 hr.	69, XXX	+
Schindler and Mikamo (1970)	M	39	6 hr.	69, XXY	–
Schmickel, Floyd, and Pooley (1971)	M	39	2 hr.	69, XXY	+
Uchida and Lin (1972)	M	approximately 29	4 hr.	69, XXY	

Table 5. Sex chromosome complements in triploid and tetraploid conditions in man

Theoritically possible genotypes	Expected ratio		Observed[a] cases	Mean gestation age in days	Mean maternal age in years
	Diandry	Digyney			
Triploidy (3n):					
69, XXX	1	2	34	104.9	27.0
69, XXY	2	2	55	106.7	27.3
69, XYY	1	0	4	?	?
Mean				105.8	27.2
Tetraploidy (4n):					
92, XXXX	1		2	86	23.5
92, XXXY	2		0	–	–
92, XXYY	1		4	67.5	29.5
92, XYYY	0		0	–	–
Mean				77.0	26.5
Mean for chromosomally normal abortuses				106.7	27.1

[a] Triploidy figures represent data collected by several investigators; tetraploidy figures represent the most recent and extensive work of Carr (1971b). For further details, see Carr 1965, 1971a, 1971b.

While triploid/diploid mosaicism ($3n/2n$) is relatively rare among abortuses (Carr 1971b), it seems to be common in living infants (see Table 4). Occasionally, triploidy may be associated with aneuploidy, a condition known as hypertriploidy. For example, Carr (1971a, 1971b) described three hypertriploid specimens all having the modal seventy chromosomes which indicates the presence of an extra chromosome in addition to the three genomes ($3n + 1$). Their mean age at abortion was eighty-three days, and their mean material age was 40.7 years.

It is difficult to say whether the multitude of developmental abnormalities described recently in the literature can be designated as a new clinical syndrome. Numerous external and internal abnormalities have been found in human triploids. Some of the more common external abnormalities are abnormal skull shape, low set and malformed ears, syndactyly of hands and feet, single palmar creases, abnormal dermatoglyphs, and body asymmetry. The most common internal deficiencies are generally cardiovascular and urogenital in nature. Mental and physical retardation are common (see Edwards et al. 1967; Butler et al. 1969; Schindler and Mikamo 1970; Carr 1971a, 1971b).

None of these abnormalities, however, seem to be characteristic. Furthermore, the wide variation in their expression makes it difficult to define a readily recognizable syndrome for human triploidy, in contrast to the many chromosomal abnormalities that are known to have clear-cut, clinical manifestations. The following abnormalities are commonly, but not always, associated with triploidy in humans:

1. *Anomaly of the chorionic villi*, probably the most consistent phenotypic feature of triploidy, involves the placenta and not the embryo. Marked cystic appearance of the chorionic villi, variously called hydatidiform degeneration or hydatidiform mole depending on the degree of villous swelling, has been observed in 85 percent of triploid abortuses (Carr 1971b).

2. *Placenta size* in many triploids is about twice the normal size for comparable states of embryonic development. Histologically, the placenta is usually immature and shows hydatidiform-like changes in the stroma of the stem villi.

3. *Syndactyly* of the third and fourth fingers is a rather consistent characteristic of most pure triploids and some mosaics.

4. *Coloboma* of the choroid and iris sometimes appears as a large cleft in the nasal edge of each iris (Edwards et al. 1967; Schmid and Vischer 1967), but occasionally involves all structures of the eye ball (Schindler and Mikamo 1970). Although "fusion failure" is a fairly common defect in triploid embryos, there is no characteristic syndrome (Carr 1971a).

5. *Low set, malformed ears.*

6. *Large posterior frontanelle*, sometimes with saw-edged parietal bones (Edwards et al. 1967).

7. *Hypertelorism of the eyes.*
8. *Single palmar creases.*
9. *Abnormal external genitalia*: The 69, XXX female has no labia minora or vaginal orifice, a barely visible urethral orifice (Edwards et al. 1967), and hypoplastic ovaries, containing very few ova (Butler et al. 1969). The 69, XXY male has bilateral intra-abdominal testes, exhibiting marked Leydig-cell hyperplasia and a low number of seminiferous tubules, i.e., tubular hypoplasia, with severe hypospadias and a hypoplastic penis (Schindler and Mikamo 1970). In both pure and mosaic triploids with at least one cell line containing a Y chromosome, some degree of intersexuality is evident (Butler et al. 1969).
10. *Cryptorchidism*, i.e., failure of the testes to descend into the scrotum.

Table 6 summarizes the abnormalities found in two pure triploids (3*n*) and five diploid-triploid mosaics (2*n*/3*n*). Syndactyly was present in five cases, and coloboma and cryptorchidism were present in three of the seven cases. Pure and mosaic triploids differ notably in their survival rate: neither pure triploid survived the postnatal period, while the mosaic cases were all living, despite abnormalities. One mosaic (Gropp, Jussen, and Odunjo, 1964) was a three and one-half-month-old boy whose only abnormalities were a harelip and a cleft palate. In this case, triploid cells were found only in the palatal area. A ten-year-old mosaic was found to have body asymmetry, physical and mental retardation, colomba, syndactyly, and cryptorchidism (Ferrier et al. 1964).

The seven cases studied comprise a rather heterogenous group. The lack of similarities is especially noticeable among mosaics. However, if we consider only the pure triploid infants (3*n*) known so far in the literature (see Table 4), a certain homogeneity appears: they have a tendency toward premature birth, developmental retardation, and reduced viability. The pure triploid infants born near term show striking similarities in their low birth weight (about 1,500 grams); small, hypospadic, and sometimes ambiguously developed external genitalia; severe respiratory distress; and syndactyly of the third and fourth fingers. These similarities may represent features of the "pure triploidy syndrome". While some authors feel that such homogeneity is quite characteristic in pure triploidy (Schindler and Mikamo 1970), others argue that the wide variation observed in both mosaic and pure triploids makes it rather difficult to designate a well-defined syndrome (Carr 1971a, 1971b).

It is important here to re-examine our findings and the methods used for investigation. Most embryos from spontaneous abortions are dead for some time before being aborted, and retention in the uterus would certainly lead to some breakdown of tissue known as maceration (Carr 1971b). In the presence of severe maceration, moderate abnormalities may escape detection, and it is easy to confuse the effects of maceration

Table 6. A sample of surviving triploids to term showing some of the most common abnormalities observed in human triploidy

	Pure triploids		Diploid-triploid mosaics				
	Edwards et al. (1967)	Schindler and Mikamo (1970)	Böök and Santlesson (1960)	Ellis et al. (1963)	Ferrier et a (1964)	Gropp, Jussen, and Odunjo (1964)	Schmid and Vischer (1967)
Karyotype	69, XXY	69, XXY	46, XY 69, XXY	46, XY 69, XXX	46, XY 69, XXY	46, XY 72, XXY	48, XXYY 71, XXXYY
Hemopoietic system	3*n*	3*n*	2*n*	2*n*	2*n*		2*n*
Age	Dead at premature birth (32 weeks)	Expired 6 hours after birth	3-year-old	6-year-old	10-year-old	3½-month-old	11-month-old
Sex chromatin	negative	negative			negative	negative	positive (IX)
Coloboma	yes	yes					yes
Syndactyly		yes	yes	yes	yes		yes
Cryptorchidism	yes	yes			yes		no
Physical retardation		yes			yes		yes
Malformed ears	yes	yes					yes
Hydatidiform placenta	yes	yes					
Severe respiratory distress	yes	yes					
Hyprospadias	yes	yes					
Microphthalmia	yes	yes					
Low umbilical cord		yes					yes
Micrognatia			yes				yes
Other			brain lesion	body asymmetry	body asymmetry	harelip and cleft palate	

with defects in development. For this reason, Carr (1971b) reported that developmental abnormalities innate in spontaneous abortuses tend to be over-diagnosed. Furthermore, only a very small number of pure triploid abortuses have been found. I believe that data obtained from additional cases may make it possible to define a readily recognizable syndrome for pure triploidy.

It can also be noted that the five mosaics (Table 6) all have a diploid hemopoietic system (peripheral leucocytes and, where analyzed, the bone marrow). Böök and co-workers (1962) observed that *in vitro*, triploid cells show severe mitotic disturbances. Thus, he postulated that a triploid hemopoietic system could not have functioned normally. Ferrier et al. (1964) suggest that diploid blood-forming tissue could either be derived from a diploid stemline or from competition between diploid and triploid cells, resulting in the elimination of the 3*n* cell line. Evidence that diploid cells have a selective advantage over triploid cells is provided by the observation (Böök, Masterson, and Santesson 1962) that the relative numbers of diploid and triploid cells in skin cultures can change with time, resulting in a higher proportion of the diploid type. Yet Aspillaga et al. (1964) found no evidence that triploid cells were at a disadvantage in cultures from a mosaic fetus. The fact that known pure triploid cases (see Tables 4, 6) developed to term with 3*n* hemopoietic systems may be evidence against Böök's postulation of the inability of such a triploid system to function.

It would be tempting to propose that the severity of the malformations in the mosaics is related to the relative numbers of triploid and diploid cells in the individual, but evidence does not support this assumption (Schmid and Vischer 1967). Cases with as high as 50 percent triploid cells (Ellis et al. 1963) show mild manifestations of the syndrome, while those with a lower proportion of triploid cells (Ferrier et al. 1964; Schmid and Vischer 1967) have many severe abnormalities.

A striking feature of the reported triploid cases is the negative findings in the analysis of sex chromatin in several triploid individuals with 69, XXY complements. Although some investigators have reported some sex chromatin positive cells, the positive ones are usually found in a very low proportion (Edwards et al. 1967). This is in contrast to diploid cells with two or more X chromosomes, which are usually positive for sex chromatin. The absence of an allocyclic X indicates that the two X's are active. Schmid and Vischer (1967) found sex chromatin positive cells, but their patient had a 71, XXXYY complement. Since they found only one heteropycnotic body per cell, it appears that two X's were active in this case also. It has been suggested that the activity of the X chromosomes may be correlated with the presence of three haploid sets of chromosomes (Schmid and Vischer 1967). This abnormal ratio of autosomes to active X chromosomes may be the cause of many of the abnormalities

(Schmid and Vischer 1967). Considering the sex chromosome imbalance, it is not surprising that many cases show abnormal development of sex structures (Edwards et al. 1967; Ferrier et al. 1964; Schindler and Mikamo 1970).

The abnormalities involving sexual differentiation in human triploids are of special interest. Most mosaic triploids with at least one cell line containing a Y chromosome showed some degree of intersexuality, and complete XXY triploids (69, XXY males) were intersexual but with testes showing marked Leydig cell hyperplasia (Butler et al. 1969).

Studies of the nuclear area of triploid cells have proven to be interesting. Aspillaga and co-workers (1964) demonstrated two types of cells in tissue from a triploid–diploid mosaic fetus. Cells with large nuclei showed a karyotype of 69, XXY. The other cell line, with smaller nuclei, had a diploid (46, XY) karyotype. Schimd and Vischer (1967) also found cells with larger nuclei in fibroblast cultures in a mosaic case.

Edwards and his collaborators (1967) compared the nuclear area of liver cells from two triploid cases as well as liver diploid cells from controls of both sexes. Triploid cells were shown to have an average nuclear area roughly 50 percent larger than diploid nuclei, which had an average area of 8.5 square units. It was suggested by Edwards that the difference in nuclear size between diploid and triploid cells might be used in diagnosing triploidy when cultures cannot be obtained.

ETIOLOGY OF TRIPLOIDY IN MAN AND OTHER VERTEBRATES

In normal development of bisexual species, a haploid male gamete fuses with a haploid female gamete, resulting in a diploid zygote. In triploidy, however, three haploid sets of chromosomes are incorporated in the zygote. Basically, there are three anomalies of fertilization which can result in triploidy: (1) diandry, in which two haploid sperm fertilize a haploid egg; (2) digyny, in which formation of either the first or the second polar body of the fertilized egg is suppressed; and (3) aneugamy, in which a diploid gamete of one sex fuses with a haploid gamete of the opposite sex (Beatty 1957; Austin 1960).

Diandry arises from the entry of two sperms into the vitellus of the egg, with both participating fully in the process of fertilization (Austin 1960). The entrance of more than one sperm into the egg is normally prevented by the fact that *in vivo*, only a small number of sperms reach the egg, as well as by reactions causing the egg to be impermeable after the first sperm enters. Nonetheless, the fact that diandry does occur in mammals is shown by the presence of two male pronuclei (identified by sperm tails) in fertilized eggs of rats, mice, and hamsters. The incidence of this phenomena varies within and between different species and seems to be under genetic control (Austin 1960), but normally occurs in about 1 or 2 percent of fertilizations. Large variations in incidence can be induced experimentally. Delayed fertilization after ovulation causes an increased incidence in rats, rabbits (Austin 1960), and in pigs (Hunter 1967). Induction of hyperthermia increases the frequency of polyspermy in rats (Pikó and Bomsel-Helmreich 1960). There is evidence for polyspermy in the chick (Pikó 1961). Katzberg and Hendricky (1966) observed changes in the zona pellucida of baboon ova as a result of treatment with human chorionic gonadotrophin. They postulated that these changes would lead to a decrease in the function of sperm selectivity. In his study of human triploid abortuses from women who had been using oral contraceptives, Carr (1967b) proposed malfunction of the mechanism preventing second sperm entrance, due to the presence of excess gonadotrophin, as the cause for triploidy.

Polygyny, in its triploid form, i.e., digyny, derives from the fertilization of a binuclear oocyte or of an egg in which either the first or second polar body has been suppressed after fertilization. The commonest pathway is through second polar body suppression. Polygyny apparently is under some type of genetic control as it occurs spontaneously in silver mice but is rare in other strains (Beatty 1957). Polygyny is seen as the result of delayed mating or insemination in some groups of rabbits and pigs (Austin 1960) and might be the mechanism involved in the formation of triploid frogs which were fertilized after induction of overripeness in the egg (Witschi and Laguens 1963). Cold treatment after fertilization suppresses formation of the second polar body in axolotls (Fankhauser and Humphrey 1942), whereas heat treatment of mouse eggs suppresses second polar body formation (Beatty 1957). That polyandry and polygyny are both induced by heat treatment and aging of the egg is not unreasonable, for both are functionally dependent on properties of the vitelline surface.

Carr (1967b) suggests that a hormone imbalance affecting tubal transport of the egg might cause delayed fertilization with subsequent second polar body suppression, thereby causing triploidy in human abortuses. Miller and co-workers (1971) also suggest that a hormone imbalance in chicks can cause aging in the ovum resulting in second polar body suppression. The suppression of the second polar body seems to be the most common of the mechanisms (Austin 1960). This can be induced experimentally by colchicine treatment of the gametes of mice (Edwards 1958). Second polar body suppression is also the proposed cause of triploidy in the newt (Böök 1940) and has been demonstrated unequivocally as the causal agent in a case of human diploid–triploid mosaicism (Ellis et al. 1963). Frequency distribution studies on twenty cases of human abortuses led Edwards et al. (1967) to conclude that polygyny (more specifically digyny) is the commoner mechanism in human triploidy.

Aneugamy is attributable to: (1) fertilization of an egg in which the first or second meiotic division has been suppressed, resulting in a uninuclear diploid egg (whereas suppression of the first or second polar body in polygyny results in the formation of a binuclear egg); or (2) mitotic error in the germ cell precursors causing formation of a diploid gamete which then fertilizes, or is fertilized by, a haploid gamete of the opposite sex. Experimental evidence for the induction of aneugamy is meager; however, suppression of the second meiotic division after delayed fertilization has been induced in rabbits (Shaver and Carr 1969). Aneugamy is postulated as the mechanism responsible for triploidy in a few naturally occurring cases. Fertilization of a diploid (2A-WW) egg by a normal (A-Z) sperm is postulated as the cause of triploidy in a 3A-ZWW chick embryo (Bloom 1969). There is evidence that diploid sperm may be produced in the chicken (Ray-Chaudhuri 1967), which might account for both the production of a ZZZ triploid chick embryo (Donner, Chyle, and Sainerova 1969) and a triploid 3A-ZZZW chick embryo in which nondisjunction of sex chromosomes in meiosis in the egg preceded fertilization (Bloom 1970). Fertilization by a diploid sperm could also be responsible for the triploid 3A-ZZW chickens studied by Bloom (1969). However, they could also have been the result of fertilization by two A-Z sperm (polyspermy) of a normal A-W egg (Bloom 1969).

Schindler and Mikamo (1970) suggest that comparing the frequency of the three possible sex chromosome complements with the ratios expected from diandry might indicate which mechanism is playing the greatest role in the production of human triploidy. The expected ratio for diandry would be 1XXX:2XXY:1XYY; for digyny it would be 1XXX:1XXY. The literature shows a ratio of 17XXX:30XXY:4XYY. If we assume that the XYY type has reduced viability and consider only the XXX:XXY ratio, we see that it most closely approximates the ratio expected for diandry (Schindler and Mikamo 1970). Recent observations of Uchida and Lin (1972) support this finding. A male triploid infant (3A-XXY) who lived only four hours after six and one-half months gestation, provided cytological evidence for the parental origin of the extra genome (Uchida and Lin 1972). Fluorescent microscopy of fluorochrome-stained preparations from this infant and his parents revealed that the extra genome was donated by the father, possibly through fertilization by a 2A-XY diploid sperm (Uchida and Lin 1972).

In vertebrates other than humans, digyny appears to be the major mechanism by which triploidy is induced experimentally. This may be because the second meiotic division of the ovum can be influenced by various experimental methods (Edwards et al. 1967).

The cause of triploidy in humans is still undetermined but it is probably either diandry or digyny, or a combination of the two.

Carr (1967b) has found an interesting correlation between triploidy and oral contraceptives. Selecting for abortuses from women who had become pregnant soon after discontinuance of oral contraceptives, he found that four out of eight abortuses were either triploid or diploid–triploid mosaics. He proposes that a hormonal imbalance caused by the contraceptives might lead to delayed ovulation and hence, delayed fertilization. Delayed fertilization in frogs (Witschi and Laguens 1963) and delayed ovulation in rats (Butcher and Fugo 1967) have been shown to increase the incidence of chromosomal abnormalities including triploidy.

It can be seen that triploidy is rarely found in human births because of its extreme lethality during embryonic development. When this chromosomal abnormality is found at term, it is expressed in a variety of malformations. The major significance of triploidy, therefore, is to increase the incidence of spontaneous abortions.

POLYPLOIDY IN OTHER VERTEBRATES

Triploidy, in sexually reproducing species, is usually accompanied by segregational sterility. This is due to defective gamete formation because meiotic pairing is usually limited to only two homologues at a time. Thus, the third homologue of a triploid (especially if it is an acrocentric chromosome where chiasma may not be formed in the short arm) may easily fail to pair with either of the other two, thereby becoming a univalent. Even if it does pair to form a trivalent, it may still not be incorporated in either nuclei or be randomly distributed to the gametes, so that some gametes have two homologues and others only one. Also, the presence of a third homologue may interfere with pairing and even division between homologues, the result being three univalents which are then randomly distributed to the gametes (Strickberger 1968). Since normal development depends upon the proper kinds and amounts of genetic materials (i.e., the presence of whole genomes), the presence of unbalanced gametes forming unbalanced zygotes easily brings about impaired development.

Triploid species can persist without sexual reproduction only if they have some sort of natural form of parthenogenic or gynogenetic reproduction which allows them to overcome the segregational sterility characteristic of bisexual triploids. Several triploid unisexual vertebrate populations have been discovered. Pennock (1965) reported the occurrence of three parthenogenic, all-female triploid species of Teiid lizard from the genus *Cnemidophorus* (*C. velox, C. tesselatus,* and *C. exsanguis*). This was the first demonstration of parthenogenesis and polyploidy occurring together throughout a natural, self-reproducing population of vertebrates. Uzzell (1964) reported the occurrence of two triploid, all-female gynogenetic species of salamander from the genus *Ambystoma* (*A. platineum*, and *A. tremblayi*), and Uzzell (1970) reported another triploid species of *Cnemidophorus* (*C. uniparens*). Triploidy is known to be maintained through the mechanism of premeiotic mitosis without cytokinesis in six species: *A. tremblayi, A. platineum, C. uniparens* (Uzzell 1970) and in the three all-female gynogenetic species of fish from the genus *Poeciliopsis* (Cimino 1972). Prior to meiosis in these species, the chromosome number of the nucleus of the triploid oogonium is raised endomitotically to a hexaploid, followed by two conventional meiotic divisions producing a triploid ovum with a genetic complement identical to that of the mother (Cimino 1972). Ploidy and inheritance data suggest that all other triploid *Cnemidophorus* species utilize this pathway of premeiotic mitosis, and that *Poecillia* triploid populations do also (Uzzell 1970). These unisexual populations of vertebrates have been thought to arise as a result of hybridization between distinct bisexual species (Uzzell 1970).

Organisms with even-numbered sets of chromosomes (e.g., tetraploids) have a better likelihood of achieving an equal separation of chromosomes during meiosis, since univalents and trivalents may be absent. Segregational sterility in recently formed tetraploids, (auto- or segmental-allotetraploids), can be overcome through the process of diploidization, whereby chromosome pairing is reduced to bivalents with the characteristic precise separation and disjunction of homologous chromosomes at meiosis in diploids. Such chromosomal diploidization would result in increased fertility, and, therefore, increased viability, as in the case of cultivated hexaploid wheats (Riley 1960). Most likely this has also been the case in several bisexual species of tetraploid fish, for example, the Catostomid fish (Uyeno and Smith 1972), *Misgurmus follilis* (Raicu and Taisescu 1972), and eleven species of the family *Cyprinidae* (Wolf et al. 1969). The enormous diversity in genome size found among living fishes must reflect the fact that in this class of vertebrates, nature is still experimenting with various means of gene duplication (Ohno 1970). The result is the absence of a well-defined sex chromosome system in most species.

Triploid amphibian embryos, in general, show an increase in both cell size and number in proportion to the level of ploidy. This increase in cell size is compensated by a corresponding decrease in cell number in various organs and tissues. During embryonic development, the general course of differentiation, as well as its rate of progress, appear normal up to the pentaploid level as long as whole sets of genes are present, i.e., all genes are present in the same number (Fankhauser 1939, 1952).

Induced and spontaneous triploidy has been observed in the larvae of the common newt, *Triturus viridescens*, and larvae of the lungless salamanders of *Eurycea bislineata* (Fankhauser 1939). Naturally occurring triploid populations of salamanders have been reported (Uzzell 1964; Uzzell 1970) as well as triploid reptile populations of Gekkonid (Kluge and Eckardt 1969) and Teiid (Fankhauser 1952; Fankhauser and Humphrey 1942; Pennock 1965).

In triploidy, in animals with a definite sex chromosome mechanism, the homogametic sex is normal, and the heterogametic sex is changed to an intersexual condition. Triploidy has a striking effect on sex differentiation in salamanders. At the time of metamorphosis, triploid testes appear normal, while triploid ovaries are much reduced in differentiation (Fankhauser 1952). This observation led Fankhauser and Humphrey (1942) to postulate that, in amphibians, the female is the heterogametic sex while the male is homogametic. Earlier observations from Böök (1940) support this theory. He reported finding a uniformly triploid male newt which showed no signs of intersexuality, but which, on the basis of histological examination of the testes, appeared to be at least partially fertile. As a rule, these triploids are sterile (Beçak et al. 1966). However, fertility has been reported in triploid and tetraploid axolotl females and in both sexes of triploid *Triturus pyrhogaster* (Kawamura 1951). A reduction of life span seems typical of triploid axolotls despite their general vigor (Fankhauser 1952).

Variations in the hemopoietic system have been found in triploid amphibians. Böök (1940) observed that the nuclei of the erythrocytes of the newt examined were definitely larger than normal. Uzzell (1964) reported that the triploid female salamanders he studied had a larger erythrocyte mean cell size than any other group in the *Ambystoma jeffersonianum* complex. He also found a positive correlation between mean cell and nuclear size. Areas of erythrocytes of triploid salamanders were, on the average, 1.5 times the area of the erythrocytes of diploid salamanders.

Beçak and co-workers (1966) reported the existence of the first known bisexual tetraploid amphibian species, a South American frog of the family *Ceratophrydidae*. There is no indication of abnormalities in male and female gonadal development. Apparently tetraploidy does not affect the fertility of these animals. Beçak's further investigations (Beçak 1969) demonstrated these South American frogs are the first proven example of a series of polyploidizations occurring during speciation in vertebrates. In the genus *Odontophrynus*, two diploid (a tetraploid and an octaploid) species have been identified. Proportional increases in nuclear volume were found and the relative DNA values of the species conformed to a 1:2:4 ratio respectively. Artificial hybrids of diploid males and tetraploid females yielded viable triploids. However, progeny from diploid females and tetraploid males did not survive gastrulation. Erythrocyte measurements show that while cell volume in the diploid and tetraploid species is different, the amount of hemoglobin per cell in both species is about the same.

In the class Aves, triploidy can be compatible with both embryonic and postnatal life (Table 7). In fowl, the male is the homogametic (ZZ) sex, whereas the female is the heterogametic (ZW) sex. With the exception of two adult male triploids (3A-ZZZ), one produced experimentally and the other spontaneously (Shoffner et al. 1972), all adult triploids reported have been intersexes. The all-triploid embryo reported by Donner and collaborators (1969) and the mosaic reported by Bloom and Buss (1966) showed retarded development and physical malformation with little mitotic activity. However, the embryo reported by Bloom (1970) appeared to be developing normally at the four-day stage; he speculated that development to term would have been probable. Bloom (1969) reported that triploid cell nuclei were larger than haploid and diploid nuclei and contained three nucleoli (modal number). In recent studies of the hemopoietic system of triploid chickens (Abdel-Hameed, Neat, and Briles 1971a, 1971b; Abdel-Hameed 1972) the triploids showed higher mean erythrocyte (RBC) whole cell and nuclear width and length values when compared with diploids. RBC size in triploids was approximately 1.5 times that in diploids. This ratio corresponds to the number of

Table 7. Reported cases of triploidy in *Gallus domesticus*

Chromosome variation[a]		Reference
Adult pure triploids		
(1) Adult triploid	3A–ZZ	Ohno et al. (1963)
(1) Adult triploid	3A–ZZW	Shoffner (personal communication)
(1) Adult triploid	3A–ZZW	Sarvella (1970)
(5) Adult triploid	3A–ZZW	Abdel-Hameed, Neat, and Briles (1971a)
(17) Adult triploid	3A–ZZW	Abdel-Hameed (1972)
Adult mosaic triploids		
(1) Adult triploid/diploid	3A–ZZZ/2A–ZW	Abdel-Hameed, Neat, and Briles (1971a)
(1) Adult triploid/diploid	3A–ZZZ/2A–ZW	Abdel-Hameed (1972)
Embryonic triploids		
(3) Embryonic triploid	3A–ZZZ	Donner, Chyle, and Sainerova (1969)
(2) Embryonic triploid	3A–ZZW	Bloom (1969)
(1) Embryonic triploid	3A–ZWW	Bloom (1969)
(1) Embryonic triploid	3A–ZZZW	Bloom (1970)
(4) Embryonic triploid	3A–ZWW	Miller, Fechheimer, and Japp (1971)
(1) Embryonic triploid/diploid	3A–ZZZ/2AZW	Bloom and Buss (1966)

[a] Numbers in parentheses preceeding chromosome variation indicate number of triploids in the study reported.

chromosome sets in the somatic nuclei. Thus, it appears that RBC size is an accurate indicator of ploidy level. In triploids, DNA and RNA content per RBC was found to be 1.4 and about 2.0 times normal, respectively.

RBC size or whole RBC length, which can be determined relatively easily, can possibly be used in screening for triploidy and intersexes in chickens (Abdel-Hameed, Neat, and Briles 1971b; Abdel-Hameed and Neat 1972). The increase of triploid RBC volume would be expected to result in a higher hemoglobin level than in diploids, but surprisingly, it is very similar to that in triploids (Abdel-Hameed 1972). This similarity in hemoglobin concentration must be the result of a mechanism that maintains a constant amount despite the obvious difference in RBC size. This suggests the possibility of either genetic inactivation of one set of chromosomes in the triploid cells, or the presence of a homeostatic system which maintains a metabolically desirable hemoglobin concentration despite triploidy and its effect on cell size (Abdel-Hameed 1972). The first possibility has been ruled out, and our recent observations support the presence of a homeostatic mechanism. It has been found that RBC count in the blood of triploid animals is significantly smaller than that in the diploids, but not significantly different from (actually about equal to) that of two-thirds of the diploid animals. Thus, even though RBC size in triploid animals is about one and one-half times that of the diploids, the number of RBC's in triploid animals is only two-thirds that of the diploids. Therefore, the increase in volume in the triploid RBC's is accompanied by a reciprocal decrease in the number of RBC's (Figure 1). As a result, the packed volume of RBC's and the concentration of hemoglobin in blood were, as expected in the light of these recent observations, very similar (Figure 1).

A similar relationship exists in macrocytic anemia in humans where the number of RBC's is much lower than normal, but the packed red cell volume and hemoglobin concentration are normal (Abdel-Hameed 1972). In humans, however, macrocytic anemia is either megaloblastic (due to deficiency of vitamin B_{12} or folic acid), or is caused by a variety of diseases. In either case, whole-body triploidy is definitely not involved. The megaloblastic macrocytic anemias result in certain disorders of hemopoiesis, including megaloblastic changes in bone marrow, but they usually respond to the administration of vitamin B_{12} and folic acid which restore normal RBC size and number. It is unlikely that nutritional therapy would change the RBC number and size of healthy triploid chickens into that of normal diploid ones.

Developmental homeostasis (canalization) leads to a standard phenotype in spite of genetic or environmental changes. This homeostasis is a dynamic stability brought about by changes in processes which result in the stability of blood composition, for example, in vertebrates (Cannon 1932; Rendel 1967). In a vital trait such as hemoglobin content, an

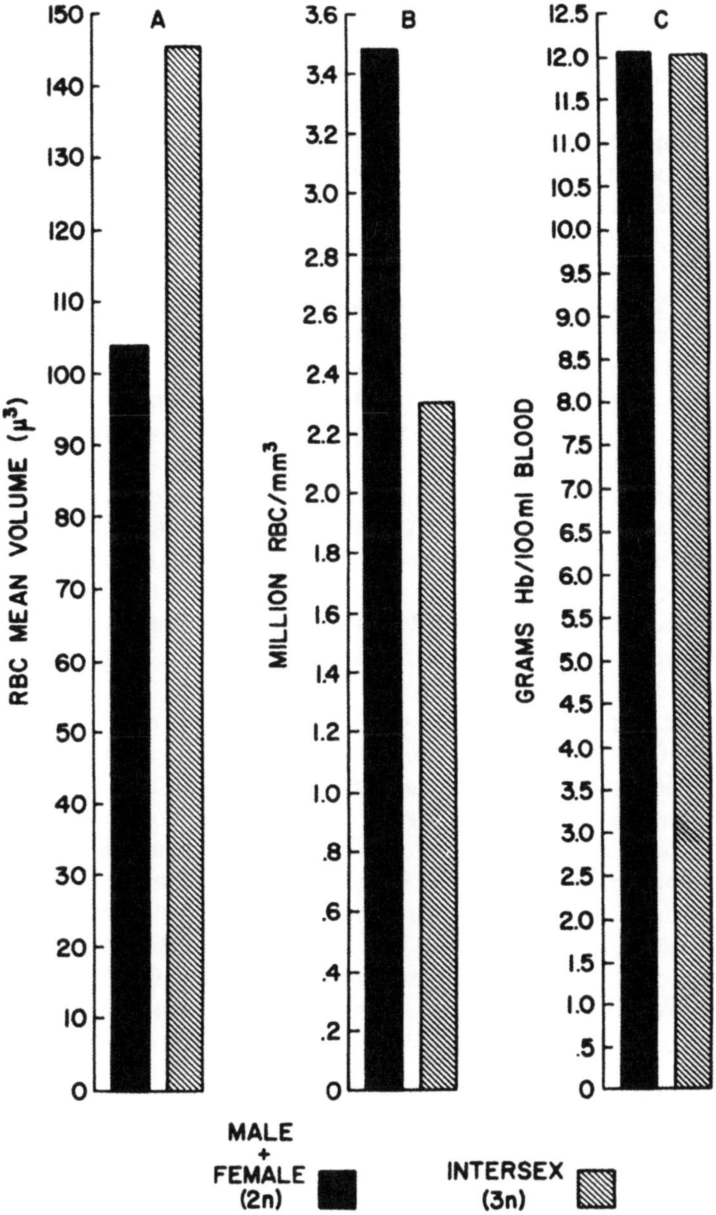

Figure 1. Differences between mean RBC volume, mean RBC number, and mean hemoglobin concentration of triploid intersex chickens and combined male and female chickens: $2n$, diploid, $3n$, triploid

amount essential for normal metabolic activity has most likely been already achieved and stabilized in diploid animals as a result of long-term natural selection. This optimal concentration is also maintained in triploid animals by a reduction in their RBC number to compensate for the increase in RBC size resulting from triploidy. Processes that control RBC number seem to change in order to maintain a constant concentration of hemoglobin for the species. Reduction in RBC number is probably mediated by erythropoietin, which induces erythropoiesis and controls hemoglobin synthesis (Marks and Rifkind 1972).

In higher plants, the most immediate and universal effect of polyploidy is an increase in cell size. This does not always increase the size of the plant as a whole, or even its individual organs, since a common effect of polyploidy is a reduction in the number of cell divisions which take place during development (Stebbins 1971). A similar process may take place in triploid chickens, as all of them encountered so far possess a general body size that is intermediate between normal males and females. These triploid animals have a packed RBC volume that is similar to that in normal animals and, most likely, they possess comparable total blood volume and total RBC mass. The total mass of living material produced during development is apparently fixed by the genetic constitution of the species, and diploidy seems to be the most optimum degree of ploidy for gene expression in higher organisms.

Changes in ploidy result in changes in gene dosage and, therefore, must have a definite effect on nucleocytoplasmic interactions. Detailed examination of nucleocytoplasmic relations in polyploid animals of several amphibian species reveals a normal course of morphogenesis during the development of triploid, tetraploid, and pentaploid embryos; thus, whole sets of genes may be multiplied in these amphibians up to the pentaploid level without disturbing the harmonious activity of the whole genotype, as long as all genes are present in the same dosage (Fankhauser 1952).

My studies on triploid chickens (Abdel-Hameed 1972, 1973) present the only known cases of a normally functioning homeostasic hemopoietic system in adult triploid higher vertebrates. Such homeostasis may be characteristic in cases of naturally occurring polyploidy in lower vertebrates, such as fishes and amphibians. Beçak's (1969) hemoglobin determinations in polyploid frog species are indeed suggestive of this.

Triploidy is incompatible with life past mid-term in the mouse, as ascertained through the extensive observations of Fischberg and Beatty (1951). They demonstrated that triploid embryos can implant and develop to advanced embryonic stages with morphogenesis and organogenesis proceeding normally through nine and one-half days gestation. However, at this point they discovered an important difference in development between triploid and diploid embryos. Despite the expected reverse proportion between all size and number in triploid embryos, all of them were smaller than their average diploid litter mates. A possible explanation is that increased cell and nucleus size combined with decreased cell number causes changes in the physiology of the embryo, reducing the efficiency of the physiological system and slowing growth. This places triploid embryos at a disadvantage during intrauterine competition and most, therefore, die during the latter part of gestation (Fischberg and Beatty 1951). Pikó and Bomsel-Helmreich (1960) induced triploidy in the rat and at eight to fifteen days gestation found eight embryos that were highly retarded and clearly abortive. However, two triploid embryos, both ten days old, were of normal size and appearance.

It is interesting to note that polyploidy, up to the pentaploid level, in the presence of complete genomes, is compatible with amphibian embryonic development (Fankhauser 1952), whereas it markedly retards development past mid-term in mice (Fischberg and Beatty 1951). Perhaps the increase in whole cell and nuclear size found in triploids causes a corresponding decrease in mitotic activity compared with diploid embryos. This decreased mitotic activity would be accompanied by a decrease in the total cell surface area. Since all biochemical processes take place at the cellular level, it is feasible that cell surface area is a crucial parameter in embryonic development. If this is the case, a decrease in cell surface area may have a deleterious effect on development, placing triploid embryos at a selective disadvantage. This disadvantage may be more critical in the more developmentally complex embryo.

Mosaic triploidy has been reported in the cat (Chu, Thuline, and Norby 1964; Pyle et al. 1971), in mink (Nes 1966) and in cattle (Dunn, McEntree, and Hansel 1970). However, no sperm were found in cattle testes or epididymis, thereby making them functionally sterile.

POLYPLOIDY AND SEX DIFFERENTIATION

In animals with well-established chromosomal sex determining mechanisms, polyploids suffer developmental abnormalities. Several theories have been proposed to explain this phenomenon. The theory of "genic balance" proposes that sex is due to the ratio of autosomes to the number of sex chromosomes, and is based on observed sex differences in the staining of many somatic nuclei. In female somatic nuclei, dark staining spots called sex chromatin bodies or masses (Barr bodies) are found which are absent in normal males. The number of sex chromatin bodies usually found is one less than the number of X chromosomes (Strickberger 1968). It has been postulated that in females one of the X chromosomes is somatically inactive, so that the dosage relationship between her sex chromosomes and autosomes will be the same as in the somatic tissues of the male (Lyon 1962). Harnden (1961) postulated that the number of contracted X chromosomes responsible for the formation of the sex chromatin body is related to the number of sets of autosomes in the cell and can be predicted on the basis of the following formula:

$$S = X - A/2$$

where S is number of sex chromatin bodies; X is the number of X chromosomes; and A is the number of chromosome sets.

Mittwoch and collaborators (1963) postulated that all X chromosomes are potentially capable of giving rise to a sex chromatin body, but that each set of autosomes suppresses the heterochromatization of one X chromosome, thus establishing a 1:2 ratio of active sex chromosomes to autosomal sets of chromosomes. Inactivation of the X chromosome(s) in triploids is irregular. The XXY triploids reported by Edwards and co-workers (1967) and Schindler and Mikamo (1970) were sex chromatin negative rather than the expected sex chromatin positive, as were the two mosaics reported by Ferrier and co-workers (1964) and Schmid and Visher (1967), each having XXY and XXXYY triploid cell lines respectively (Table 4). In other cases reported, the expected sex chromatin constitution was found.

If we apply Harnden's formula to the triploids, the result is one and one-half sex chromatin bodies for 69, XXX and one-half for 69, XXY. Since "halves" do not exist biologically Carr (1971b) interpreted the results to mean an unstable effect of the autosome complement. Under these conditions, the number of Barr bodies for 69, XXX would be one or two; for 69, XXY it would be zero or one. This is exactly what is found in practice. Therefore, the problem of understanding the irregularity and the obvious difference between diploids and triploids in their patterns of X chromosome condensation seems to have been resolved. Furthermore, Harnden's formula holds for tetraploid abortuses, since 92, XXXX specimens exhibit a maximum of two Barr bodies per cell and 92, XXYY is chromatin negative (Carr 1971a, 1971b).

Hamerton (1968) postulated that structural genes for sex development are the same in both sexes, and that sex determination is brought about by the heterochromatic sex chromosome. He suggests that in females, the euchromatic X carries the structural and operator gene for ovarian determination, whereas, the heterochromatic X may carry a gene-regulating ovarian atresia. In the male, the heterochromatic Y could be a carrier of male-determining regulator gene switching on a structural gene, located on another chromosome. Any imbalance in the amount of heterochromatic sex chromosome material, as in several reported triploid cases, would therefore lead to abnormal development, such as a disturbance in previously sex determined gonads. In *Drosophila* triploids, Maroni and Plaut (1973) proposed that dosage-compensated sex-linked genes are controlled by a positively acting regulatory factor of autosomal origin. They claim that with this hypothesis it is possible to explain dosage compensation as a consequence of general regulatory mechanisms without invoking a special device which applies only to the X chromosomes.

Our histological data indicate that both right and left gonads of the majority of the examined intersex chickens were composed only of testicular tissues, i.e., their gonads were not ovotestes but rather malformed testes (Abdel-Hameed and Shoffner 1971). However, the gonads of some seem to contain a few primary follicles arrested at an early stage of development, in addition to the more obvious and abundant seminiferous tubules. These observations seem to contradict the data of Ohno et al. (1963) who described a single adult triploid intersex chicken with a left gonad that is, histologically, an ovotestis.

In addition to our data, only two cases of adult triploid chickens have been reported in the literature. The first was described by Ohno et al. (1963) as a 3A-ZZ intersex. However, it is likely that its karyotype was actually 3A-ZZW since Ohno and co-workers reported difficulty in identifying the W chromosome with certainty in the domestic fowl. The recent development of the feather-pulp technique in Shoffner's laboratory (Shoffner et al. 1967) has made it possible to ascertain the presence of the W chromosome in the females of many avian species (Krishan and Shoffner 1966; Krishan, Haiden, and Shoffner 1965) including chickens. The second case was a parthenogenetic 3A-ZZW chicken (Sarvella 1970). Even though Sarvella did not refer to it as an intersex, its external and internal morphology, as well as the histology of its gonads, was essentially identical to that of our intersex chickens.

It may be possible to explain the apparent contradiction between our data and that of Ohno et al. (1963) if we assume that intersex 3A-ZZW adult triploids may range in their gonadal structure between two extremes, i.e., genuine ovotestes and malformed testes.

Different cell types involved in steroid biosynthesis may possess a basically similar and characteristic ultrastructure, the morphological components of which seem to be essential for steroid production (Fawcett, Long, and Jones 1969; Christensen and Gillim 1969). Thus, as suggested by Fawcett, Long, and Jones (1969), in all steroidogenic tissues investigated so far, the presence of large amounts of smooth endoplasmic reticulum appears to be distinctive. Studies of the steroidogenic capabilities of the various cell fractions suggest that certain of the steroid dehydrogenases, hydroxylases, and other enzymes necessary for steroid biosynthesis may be bound to the membranes of the smooth endoplasmic reticulum, while others are present only in the mitochondria or in the cytoplasmic ground substance.

Fawcett (1963) has suggested that the membrane of the smooth reticulum may also be involved in the storage of cholesterol prior to its conversion into hormonally active steroids. Thus, one of the most characteristic features of steroid-producing tissues, the presence of smooth endoplasmic reticulum, may be explained in terms of function. Furthermore, the presence of such smooth ER in steroidogenic cells, i.e., testicular Leydig cells and ovarian theca interna cells, can be used as an indirect qualitative method to assess whether or not these cells in intersex chickens are active in steroid production.

Our data on steroid hormone radioimmunoassay of adult triploid chickens show that their levels of estrogen (estrone, estradiol) are similar or slightly higher than the normal roosters, but are much lower than those of laying hens. Their testosterone levels (testosterone, dihydrotestosterone), on the other hand, are similar or slightly higher than those of laying hens, but are markedly lower than those of the roosters. It seems as if the hyperplasia of Leydig cells observed on the light microscope level in these triploids (Abdel-Hameed and Shoffner 1971) is not associated with active production of steroid hormones. The presence of such hyperplasia of Leydig cells is intriguing, and more data on their ultrastructural organization may help in understanding their function.

It is possible that the Leydig cells of triploid chickens are potentially capable of normal steroidogenic activity but do not receive adequate hormonal stimulation from the pituitary gland due to certain malfunction of either the pituitary or the hypothalamus. To test the validity of this assumption, the effects of administering chorionic gonadotrophin hormones (PLH-PFSH), hypothalamic hormones (LH-FSH-RH), and gonadal hormones should be examined in detail.

So far, no serious attempt has been made to elucidate the problem of sex determination in chickens and other avian species. Such determination has been difficult because of the small size of the microchromosomes and the lack of adequate cytological techniques to identify both autosomes and sex chromosomes, as well as the lack of studies on sexual deviants. Therefore, the recent advances in karyotype analysis (Shoffner et al. 1967), in addition to the data obtained in our laboratory on intersex chickens, represent an exceptional opportunity to characterize the mechanisms of sex determination in this group.

If the W chromosome is strongly female-determining in chickens, we would expect 3A-ZZW individuals to be phenotypically female. Even though observations on the phenotypes of 2A-ZO and 2A-ZZW chickens are lacking, we can still conclude that the W chromosome is not a strong female-determining element. The available data indicate that sex determination in chickens and probably birds in general is unlike that in mammals, and most likely is similar to that operating in *Drosophila*, that is, dependent upon the ratio between the Z chromosomes and the number of autosomal sets of chromosomes (Abdel-Hameed and Shoffner 1971). Thus, the observed intersex phenotype and oviduct-type tissues in intersex chickens, especially those with 3A-ZZW complement, can be explained, at least partially, in the light of data from recent sex-reversion experiments. Male chicken embryo sex reversion following estradiol benzoate treatment occurs for a limited period of embryonic development. Afterward, the rapidly proliferating testicular tissues replace the degenerating ovarian-type tissues (Narbaitz and De Robertis 1970). Intersex chickens probably develop in a similar fashion under the effect of indigenous male–female hormones during early embryonic stages. The production of such hormones is probably determined by somatic-sex chromosome balance.

Mittwoch's recent studies (1969, 1971) provide new insight into the understanding of the genetics of human sex differentiation. Before six weeks of development, both male (Wolffian) and female (Mullerian) ducts are present in an embryo. In the presence of embryonic testes, the Wolffian duct develops and the Mullerian duct regresses. In the absence of testes, the Mullerian duct develops and the Wolffian duct regresses. Therefore, Mittwoch (1969) proposed that the Y chromosome carries a gene(s) responsible for the development of the male phenotype. Furthermore, she reports (1971) that it appears that the embryonic gonadal rudiment must reach a critical size by a given stage of development in order to become a testis. If it fails to reach the required size, the gonad develops into an ovary. The incipient mammalian testis is larger than the incipient ovary. Therefore, it is possible that the mammalian Y chromosome is responsible for the increased growth rate of the gonadal rudiment. This situation is opposite that in birds. In embryonic gonadal development, the ovary grows faster than the testis. However, interestingly, the female fowl is the heterogametic (ZW) sex as is the mammalian (XY) male. Therefore, the presence of a Y (or W) chromosome may be responsible for the resulting increased growth of the dominant gonad of the heterogametic sex (Mittwoch 1971).

The presence of an additional haploid set of autosomes leads to masculinization of the otherwise 2A-XXY female fruitflies (*Drosophila melanogaster*), whereas in man, an additional set of autosomes causes feminization of the phenotype. In human triploid infants with XXY sex chromosomes, there is a definite tendency toward intersexuality, whereas 47, XXY karyotype is associated with an undoubted male phenotype (Mittwoch 1973). Such differences are indeed intriguing, but they indicate the presence of a basically different mechanism of sex determination in man from that of fruit flies, even though both have an XX/XY scheme of sex determination.

Experimental evidence shows that the length of time required for DNA replication depends upon the amount of DNA per nucleus. In several plant species, the mitotic cell cycle time span, especially the period spent in DNA synthesis (the S period), was found to be proportionate to the amount of DNA present (Van't Hof 1965). A general tendency for this relationship seems to exist, even though it may not be found in every individual case examined (Mittwoch 1971). Studies on the cell cycle of tumors led Goldfeder (1965) to suggest the existence of such a relationship in mammalian cells. His suggestion was later supported by the observation that ascites tumor cells with tetraploid nuclei require about twice as long to complete DNA synthesis as diploid ascites cells (Lennartz, Schummelfeder, and Maurer 1966). Lengthening of the mitotic cycle time would be expected to lead to a slowing down of the rate of cell proliferation and, therefore, to a retardation of growth and development. Growth retardation has been observed as a characteristic feature of many clinical syndromes involving an increase in chromosome number, and, as a consequence, DNA content per nucleus, as in the case of trisomy or polyploidy. Hall (1965, 1966) suggested that the similarities in clinical features seen in the three human trisomy syndromes (D, E, and G) might be a result of a lower rate of cell multiplication. Furthermore, the degree of growth disorder is apparently directly proportional to the size of the extra chromosome. A similar situation seems to be evident in polyploidy where the degree of developmental retardation (and lethality) is greater in tetraploidy than triploidy (Carr 1971a, 1971b). It also follows that developmental retardation is, as expected, much greater in polyploidy than in trisomy, since the increase in relative DNA content (per nucleus) corresponds to the increase in the number of whole sets of chromosomes.

Interestingly, the mitotic cell cycle under monosomic condition ($2n - 1$) of Turner's 45, X is apparently shorter (as expected) than in normal male and female cells (Polani 1970). Thus, a similar trend seems to be found in sex chromosomes despite their characteristic heterochromatic nature and their very different role in sex determination. The second X chromosome of normal females and the Y chromosome of normal males are both late replicating (German 1964) and are considered to be largely heterochromatic. Schmid and Leppert (1969) found that heterochromatin in rodents requires less time for DNA synthesis than euchromatin. This may explain why differences in heterochromatin seem to be less disruptive to the cycle than differences in euchromatin (Mittwoch 1969, 1971).

It is believed that the sex chromosomes have originated from an ordinary pair of homologous chromosomes, but diverged gradually during the course of evolution, thus resulting in two very distinct chromosomes in mammals, i.e., X and Y. However, when the chromosome number of the diploid 2A-XX female and 2A-XY male is doubled, the clear cut sexual dimorphism observed at the diploid level will no longer exist at the tetraploid level. The immediate products of chromosome doubling are 4A-XXXX females and 4A-XXYY males which would produce 2A-XX ova and 2A-XY spermatozoa, thus resulting in 4A-XXXY progeny. The 4A-XXXY genotype may develop into an all-male progeny if the Y chromosome is a strong male determining element; otherwise it will develop into an all sterile intersex progeny. Thus, tetraploidy would be incompatible with sexual reproduction. Therefore, the possibility of polyploid evolution is denied to mammals, birds, and even to reptiles due to their well-established chromosomal sex-determining mechanism (Ohno 1970). Although triploidy is compatible with postnatal life in chickens, the 3A-ZZZ are sterile males and the 3A-ZZW are sterile intersexes (Abdel-Hameed and Shoffner 1971; Abdel-Hameed 1972). Here again, we find that triploidy results in intersexual development of the heterogametic sex, but not of the homogametic sex.

At least in contemporary higher vertebrates, polyploidy is either extremely lethal or else, as in the case of chickens, it produces infertility. Under such conditions, we may consider polyploidy a genetic and evolutionary dead end. However, it is believed that polyploidy may have played an important role in vertebrate evolution and higher vertebrate diversity (for references and further details, see Ohno 1970).

SUMMARY AND CONCLUSIONS

Polyploidy is very common in plants as evidenced by the fact that about one-third of all flowering plant species are polyploid. In groups such as the wheat family, polyploidy occurs in two-thirds of its representative species (Stebbins 1950, 1971). By comparison, polyploidy seems to be extremely rare in animals, especially so in higher vertebrates where it is usually incompatible with postnatal life.

In view of the common belief that polyploidy causes embryonic mortality and is therefore absent in higher vertebrates, the recent finding of spontaneously occurring triploid chickens represents an important exception. Triploid chickens not only survive postnatally, but carry out nearly normal functions, except for reproduction, and develop into healthy adult sterile males (3A-ZZZ) or intersexes (3A-ZZW). The intersexes appear in a very low frequency; about 1:2000 of the commercially-sexed females in egg producing flocks are intersexes (Abdel-Hameed and Shoffner 1971). Apparently, these intersex individuals are largely indistinguishable from their female sibs either when sexed at hatching or at the time of transference to laying cages. They are detectable at sexual maturity when their secondary sexual characteristics are obviously intermediate between that of normal adult males and females.

The reasons for the drastic differences observed between plants and animals in terms of relative success of polyploidy can be summarized as follows:

1. The highly developed mechanisms of sex determination, well-

differentiated sex chromosomes, and almost universal bisexuality in higher vertebrates appear to be related to their incompatibility with polyploidy. On the other hand, the lack of sex chromosomes and the hermaphroditic reproductive mode in most flowering plant species are perfectly compatible with polyploid evolution. Of course, plants with sex chromosomes (such as *Rumex*) have become tetraploid, at least for the autosomes, by simply keeping the sex chromosomes at the diploid level, i.e., 4A-XX and 4A-XY. Animal polyploids have rarely exhibited such a mechanism to escape the conflict between polyploidy and bisexuality.

2. Natural hybridization in animals seems to be extremely rare under natural conditions due to the multitude of well-developed isolating mechanisms acting as efficient barriers against gene flow, even between individuals from closely-related species. Conversely, natural hybridization between plant species is so common that almost all plant polyploids are allopolyploids. Due to lack of hybridization in animals, the majority of animal polyploids are autopolyploids. Interestingly, though, allopolyploidy has been demonstrated in certain lower vertebrates, i.e., salamander, lizard, and fish populations of the *Ambystoma, Cnemidophorus, Poecilia*, and *Poeciliopsis* genera (Uzzell 1970). Therefore, reports of spontaneously occurring polyploid populations and individuals as well as lower vertebrate models of polyploid evolution lends credence to the hypothesis that polyploidization did, in fact, play an important role in higher vertebrate evolutionary development (Ohno 1970).

3. The much lower level of developmental and physiological complexity in plants as compared to animals, enables plants to tolerate the effects of polyploidy.

Ohno (1970) postulated that polyploid evolution became impossible as soon as vertebrates evolved to the reptilian stage. At this point, further increase in genome size would have to occur through tandem duplication, not polyploidy. He argued that polyploid evolution has been possible only in fish and amphibians, and that the enormous diversity in genome size among living fishes is a reflection of the fact that, in this group of vertebrates, nature is still experimenting with various means of gene duplication. Furthermore, most of nature's experiments with gene duplication (through polyploidy and tandem duplication) must have occurred at the fish and amphibian stages of vertebrate evolution (Ohno 1970). This experimentation has apparently been possible due to a rather primitive mechanism of sex determination in which the chromosomal determiners (X/Y or Z/W) are still in their initial state of differentiation. Under such conditions, it seems feasible that two distinct genome lineages became stabilized at the reptilian stage, and further evolution of one to mammals and the other to birds was accomplished without substantial increase in genome size (Ohno 1970).

In primates, there is a wide range of possible chromosome numbers (between $2n = 20$ to $2n = 80$). A number of theories have been suggested to explain such wide variation, including polyploidy and chromosome fission. Polyploidy was rejected as being incompatible with the sex-determination system of primates (XX/XY type). Furthermore, in most Catarrhini, only one pair of "marked" (Chiarelli 1966) chromosomes exhibits a long achromatic region. If polyploidy had occurred, one would expect different numbers of such chromosomes. Recent cytological data on primate chromosomes seem to rule out the polyploidy hypothesis and favor the "chromosome fission" explanation (Egozcue 1969).

It seems obvious that even though polyploidy is far less frequent in animals than plants, it does occur at a higher frequency than earlier estimates indicated, and has probably played an important role in the evolution of present-day higher vertebrates.

Carr (1965) reported the first known human tetraploid embryos as well as seven more tetraploid specimens from human spontaneous abortions (Carr 1971b). There have been no reported cases of a formed embryo in purely tetraploid abortuses. This does not mean that formation of an embryo is impossible in tetraploidy; however, it does suggest that tetraploidy has a more serious effect on development than triploidy (Carr 1971a, 1971b). It is interesting to note Carr's apparent change in approach to polyploidy's compatibility with human post-natal life over the past eight years. There appears to be a decided de-emphasis on polyploid scarcity, and a growing emphasis on the feasibility of the existence of viable polyploid individuals as more cases of human polyploidy were discovered.

The deficiencies of human triploid cells are particularly obvious when they are subjected to an environment which induces an increased mitotic turnover, such as during cell culture *in vitro*. The *in vivo* conditions which are most similar to cultivation *in vitro* occur in the blood-forming organs during post-natal life (Böök, Masterson, and Santesson 1962). Because earlier reports showed a diploid blood cell line (Böök, Masterson, and Santesson 1962; Carr 1967; Ellis et al. 1963; Schmid and Vischer 1967), for many years it was believed that triploidy involving the blood-forming tissues possibly led to intrauterine death and abortion in humans. However, all homogeneous triploid chickens have hemopoietic systems with triploid cells, and consistent triploidy has been reported in peripheral blood analyses performed on six live born triploid infants (Butler et al. 1969; Edwards et al. 1967; Schindler and Mikamo 1970; Schmickel, Floyd, and Pooley 1971; Uchida and Lin 1972). Thus, it appears that a triploid hemopoietic system may function normally.

However, under physiological stress conditions such as vacuum-simulated high altitude conditions of 15,000 feet, triploid chickens show a different hemopoietic response than that of the diploids. The data seem to indicate that the triploid hemopoietic system in chickens is slower in its response to polycythemia conditions than the diploid, probably due to the slower rate of erythropoiesis and longer cell cycle in triploid cells. Such a lowering of the rate of cell proliferation has been observed with higher amounts of DNA.

Experimental evidence shows that the time required for DNA synthesis is proportional to the amount of DNA per nucleus (Mittwoch 1969, 1971). An increased DNA content has been reported in triploid chicken cells (Abdel-Hameed 1972). One could speculate that an imbalance caused by the conflicting processes of Y chromosome-directed embryonic cell proliferation, as well as reduced mitotic rate due to increased DNA, could be responsible for abnormal triploid development.

In triploid animals with a definite sex chromosome mechanism, the homogametic sex is normal and the heterogametic sex is changed to an intersexual condition.

Problems concerning control mechanisms of cell growth and cell proliferation in embryogenesis, especially those dealing with patterns of normal development and sex differentiation, seem formidable. Yet we certainly must come to grips with them in order to attain a basic understanding of the multitude of interacting biochemical pathways induced by the genotype, as well as the ultimate end product, the phenotype.

REFERENCES

ABDEL-HAMEED, F.

1970 Reproductive and developmental differentiation within section *Primigenia* of the genus *Clarkia. Transactions Illinois Academy of Science* 63:347–358.

1972 Hemoglobin concentration in normal diploid and intersex triploid chickens: Genetic inactivation or canalization? *Science* 178:864–865.

1973 Triploidy in higher vertebrates. *Genetics* 74 (Supplement):s2.

ABDEL-HAMEED, F., H. J. NEAT

1972 A study on erythrocyte cell volume and number in adult normal and triploid intersex chickens. *Poultry Science* 51:1376–1382.

ABDEL-HAMEED, F., H. J. NEAT, W. E. BRILES

1971a Erythrocyte size as a diagnostic aid in identifying triploidy in intersex chickens. *Poultry Science* 50:1542.

1971b Differences in erythrocyte measurement of intersex and normal adult chickens. *Poultry Science* 50:1847–1854.

ABDEL-HAMEED, F., R. N. SHOFFNER

1971 Intersexes and sex determination in chickens. *Science* 172:962–964.

ASPILLAGA, M. J., R. J. SCHLEGEL, R. NEU, L. I. GARDNER

1964 Triploid/diploid mosaicism in tissue from an early human fetus. *Journal of Pediatrics* 65:1098–1099.

AUSTIN, C. R.

1960 Anomalies of fertilization leading to triploidy. *Journal of Cellular and Comparative Physiology* 56 (Supplement 1):1–15.

BEATTY, R. A.

1957 *Parthenogenesis and polyploidy in mammalian development.* Cambridge: Cambridge University Press.

BEÇAK, M., W. BEÇAK, M. N. RABELLO

1966 Cytological evidence of constant tetraploidy in the bisexual South American frog. *Chromosoma* 19:188–193.

BEÇAK, W.

1969 Gene action and polymorphism in polyploid species of amphibians. *Genetics* 61:183–190.

BERNARD, R., K. STAHL, J. COIGNET, F. GIRAUD, M. HARTUNG, Y. BRUSQUET, P. PASSERON

1967 Triploidie chromosomique chez un nouveauné polymalformé. *Annales de Génétique* 10:70–74.

BLOOM, S.

1969 Chromosome abnormalities in early chicken (*Gallus domesticus*) embryos. *Chromosoma* 28:357–369.

1970 Trisomy — 3, 4 and triploidy (3A-ZZZW) in chick embryos: autosomal and sex chromosomal nondisjunction in meiosis. *Science* 170:457–458.

BLOOM, S., E. G. BUSS

1966 Triploid/diploid mosaic chicken embryo. *Science* 153:759–760.

BÖÖK, J. A.

1940 Triploidy in *Triton taeniatus* LAURA. *Hereditas* 26:107–114.

BÖÖK, J. A., J. G. MASTERSON, B. SANTESSON

1962 Malformation syndrome associated with triplody — further chromosome studies of the patient and his family. *Acta Genetica* 12:193–201.

BÖÖK, J. A., B. SANTESSON

1960 Malformation syndrome in man associated with triploidy (69 chromosomes). *Lancet* i:858–859.

BOUÉ, J. G., A. BOUÉ, P. LAZAR

1967 Les aberrations chromosomiques dans les avortements. *Annales de Génétique* 10:179–187.

BUTCHER, R. L., N. W. FUGO

1967 Overripeness and the mammalian ova. II. Delayed ovulation and chromosome anomalies. *Fertility and Sterility* 18:297–302.

BUTLER, L. J., C. CHANTLER, N. E. FRANCE, C. G. KEITH

1969 A live-born infant with complete triploidy (69, XXX). *Journal of Medical Genetics* 6:413–421.

CANNON, W. B.

1932 *The wisdom of the body.* New York: Norton.

CARR, D. H.

1963 Chromosome studies in abortuses and still born infants. *Lancet* 2:603–606.

1965 Chromosome studies in spontaneous abortions. *Obstetrics and Gynecology* 26:308–326.

1967a Chromosome anomalies as a cause of spontaneous abortion. *American Journal of Obstetrics and Gynecology* 97:283–293.

1967b Chromosomes after oral contraceptives. *Lancet* 2:830–831.

1971a Genetic basis of abortion. *Annual Review of Genetics* 5:65–80.

1971b Chromosome studies in selected spontaneous abortions: polyploidy in man. *Journal of Medical Genetics* 8:164–174.

CHIARELLI, B.

1966 Marked chromosomes in Catarrhine monkeys. *Folia Primatalogica.* Basel 4:74–80.

CHRISTENSEN, A. K., S. W. GILLIM

1969 "The correlation of fine structure and function in steroid-secreting cells, with emphasis on those of the gonads," in *The gonads*. Edited by K. W. McKerns. New York: Appleton-Century-Crofts.

CHU, E. H., H. C. THULINE, D. E. NORBY

1964 Triploid-diploid chimerism in a male tortoiseshell cat. *Cytogenetics* 3:1–18.

CIMINO, M.

1972 Meiosis in triploid all-female fish. *Science* 175:1484–1486.

DONNER, L., P. CHYLE, H. SAINEROVA

1969 Malformation syndrome in *Gallus domesticus* associated with triploidy. *Journal of Heredity* 60:113–115.

DUNN, H. O., K. MCENTREE, W. HANSEL

1970 Diploid-triploid chimerism in a bovine true hermaphrodite. *Cytogenetics* 9:245–259.

EDWARDS, J. H., C. YUNCKEN, D. I. RUSHTON, S. RICHARDS, U. MITTWOCH

1967 Three cases of triploidy in man. *Cytogenetics* 6:81–104.

EDWARDS, R. G.

1958 Colchicine-induced heteroploidy in the mouse. 1. The induction of triploidy by treatment of the gametes. *Journal of Experimental Zoology* 137:317–384.

EGOZCUE, J.

1969 "Primates," in *Comparative mammalian cytogenetics*. Edited by K. Benirschke, 357–389. New York: Springer-Verlag.

ELLIS, J. R., R. MARSHALL, I. C. S. NORMAND, L. S. PENROSE

1963 A girl with triploid cells. *Nature* 198:411.

FANKHAUSER, G.

1939 Polyploidy in the salamander *Eurycea bislineata*. *Journal of Heredity* 30:379–388.

1952 Nucleo-cytoplasmic relations in amphibian development. *Intern. Rev. Cytology* 1:165–193.

FANKHAUSER, G., R. R. HUMPHREY

1942 Induction of triploidy and haploidy in axolotl eggs by cold treatment. *Biological Bulletin* 83:367–374.

FAWCETT, D. W.

1963 "Structural and functional variations in the membranes of the cytoplasm," in *Intracellular membraneous structure*. Edited by S. Steno and E. V. Cowdry. *Symposia of the Society for Cellular Chemistry* 14:15–40.

FAWCETT, D. W., J. A. LONG, A. L. JONES

1969 The ultrastructure of endocrine glands. *Recent Progress in Hormone Research* 25:315–380.

FERRIER, P., S. FERRIER, G. STALDER, E. BUHLER, F. BASATTER, D. KLEIN

1964 Congenital asymmetry associated with diploid-triploid mosaicism and large satellites. *Lancet* 1:80–82.

FISCHBERG, M., R. A. BEATTY

1951 Spontaneous heteroploidy in mouse embryos up to mid-term. *Journal of Experimental Zoology* 118:321–332.

GERMAN, J.

1964 The pattern of DNA synthesis in the chromosomes of human blood cells. *Journal of Cell Biology* 20:37–55.

GOLDFEDER, A.

1965 Biological properties and radiosensitivity of tumors: determination of the cell-cycle and time of synthesis of deoxyribonucleic acid using tritiated thymidine and autoradiography. *Nature* 207:612–614.

GREENFIELD, DAVID W., F. ABDEL-HAMEED, G. DECKERT, R. FLINN
1973 Hybridization between *Chrosomus erythrogaster* and *Notropis cornutus* (Pisces: Cyprinidae). *Copeia* 1:54–60.

GROPP, A.
1967 Chromosomenuntersuchungen bei Spontanabortus. *Verhandlungen der Deutschen Gesellschaft für Pathologie* 51:278–282.

GROPP, A., A. JUSSEN, F. ODUNJO
1964 Near-triploid chromosome constitution in epithelial-cell cultures of palatal mucosa from a case of cleft palate. *Lancet* 1:1167.

HALL, B.
1965 Delayed ontogenesis in human trisomy syndromes. *Hereditas, Genetiskt Arkiv* 52:334–344.
1966 Follow up investigation of new born mongoloids with respect to growth retardation. *Hereditas, Genetiskt Arkiv* 56:99–108.

HAMERTON, J. L.
1968 Significance of sex chromosome derived heterochromatin in mammals. *Nature* 219:910–914.

HARNDEN, D. G.
1961 Nuclear sex in triploid XXY human cells. *Lancet* ii:488.

HUNTER, R. H. F.
1967 The effects of delayed insemination on fertilization and early cleavage in the pig. *Journal of Reproduction and Fertility* 13:133–147.

KATZBERG, A. A., A. G. HENDRICKY
1966 Gonadotropin-induced anomalies of the zona pellucida of the baboon ovum. *Science* 151:1225.

KAWAMURA, T.
1951 Reproductive ability of triploid newts, with remarks on their offspring. *Journal of Science of Hiroshima University*, Series B, Division 1, 12:1–10.

KLINGER, H. P., H. G. SCHWARZACKER
1958 Sex chromatin in polyploid nuclei of human amnion epithelium. *Nature* 181:1150–1152.

KLUGE, A. G., M. J. ECKARDT
1969 *Hemidactylus garnotii* Dumeril and Bibron, a triploid all-female species of gekkonid lizard. *Copeia* 4:651–664.

KOHN, G., B. H. MAYALL, M. E. MILLER, W. J. MELLMAN
1967 Tetraploid-diploid mosaicism in a surviving infant. *Pediatric Research* 1:461–469.

KOHN, G., A. ROBINSON
1970 Tetraploidy in cells cultured from amniotic fluid. *Lancet* 2:778–779.

KRISHAN, A., G. J. HAIDEN, R. N. SHOFFNER
1965 Mitotic chromosomes and the W-sex chromosome of the great horned owl (*Bubo U. virginianus*). *Chromosoma* 17:258–263.

KRISHAN, A., R. N. SHOFFNER
1966 Sex chromosomes in the domestic fowl (*Gallus domesticus*), Turkey (*Meleagris fallopavo*), and the Chinese pheasant (*Phasianus colchicus*). *Cytogenetics* 5:53–63.

LEJEUNE, J., C. SALMON, R. BERGER, M. O. RETHORE, A. ROSSIER, J. C. JOB
1967 Chimère 46, XX/69, XXY. *Annales de Génétique* 10:188–192.

LEJEUNE, J., R. TURPIN, M. GAUTHIER
1959 Le Mongolism, premier example d'aberration autosomique humaine. *Annales de Génétique* 1:41–49.

LENNARTZ, K. J., N. SCHUMMELFEDER, W. MAURER
1966 Dauer der DNA-Synthesephase bei Ascitestumoren der Maus unterschiedlicher Ploidie. *Naturwissenschaften* 53:21–22.

LYON, M. F.
1962 Sex chromatin and gene action in the mammalian X-chromosome. *American Journal of Human Genetics* 14:135–148.

MARKS, P. A., R. A. RIFKIND
1972 Protein synthesis: its control in erythropoiesis. *Science* 175:955–961.

MARONI, G., W. PLAUT
1973 Dosage compensation in *Drosophila melanogaster* triploids. 11. Glucose-6-phosphate dehydrogenase activity. *Genetics* 74:331–342.

MIKAMO, K.
1968 Interfollicular overripeness and teratologic development. *Cytogenetics* 7:212–233.
1970 Anatomic and chromosomal anomalies in spontaneous abortion. Possible correlation with overripeness of oocytes. *American Journal of Obstetrics and Gynecology* 106:243–254.

MILLER, R. C., N. S. FECHHEIMER, R. G. JAPP
1971 Chromosome abnormalities in 16–18 hour chick embryos. *Cytogenetics* 10:121–136.

MILUNSKY, A., J. W. LITTLEFIELD, L. ATKINS
1970 Tetraploidy in amniotic fluid cells. *Lancet* 2:979.

MITTWOCH, U.
1969 Do genes determine sex? *Nature* 221:446–448.
1971 Sex determination in birds and mammals. *Nature* 231:432–434.
1973 Triploidy and intersexuality in *Drosophila* and man. *Genetics* 74 (Supplement):183.

MITTWOCH, U., N. B. ATKIN, J. R. ELLIS
1963 Barr bodies in triploid cells. *Cytogenetics* 2:323–330.

NARBAITZ, R., E. M. DEROBERTIS, JR
1970 Steroid producing cells in chick intersexual gonads. *General and Comparative Endocrinology* 14:164–169.

NES, N.
1966 Diploid-triploid mosaicism in a true hermaphrodite mink. *Hereditas* 56:159–170.

OHNO, S.
1970 The enormous diversity in genome sizes of fish as a reflection of nature's extensive experiments with gene duplication. *Transactions American Fisheries Society* 99:120–130.

OHNO, S., W. A. KITTRELL, L. C. CHRISTIAN, C. STENIUS, G. A. WITT
1963 An adult triploid chicken (*Gallus domesticus*) with a left ovotestis. *Cytogenetics* 2:42–49.

OHNO, S., U. TETTENBORN, R. DOFUKU
1971 Molecular biology of sex differentiation. *Hereditas* 69:107–124.

PAPIERNIK-BERKHAUER, E.
1968 Enfant triploide à terme et therapeutique hormonale. *Bulletin de la Fédération des Sociétés de Gynécologie et d'Obstétrique de Langue Française* 20:248–249.

PAWLOWITZKI, I. H., A. CENANI
1967 Sporadic triploid cells in human blood and fibroblast cultures. *Humangenetik* 5:65–69.

PENNOCK, L. A.
1965 Triploidy in parthenogenic species of teiid lizard genus *Chemiodphorus*. *Science* 149:539–540.

PENROSE, L. S.
1961 Mongolism. *British Medical Bulletin* 17:184–189

PENROSE, L. S., J. D. A. DELHANTY
1961 Triploid cell cultures from a macerated fetus. *Lancet* i:1261–1262.

PIKÓ, L.
1961 La polyspermie chez les animaux. *Annales de Biologie Animale, Biochemie, Biophysique*. Paris 1:323–383.

PIKÓ, L., O. BOMSEL-HELMREICH
1960 Triploid rat embryos and other chromosomal deviants after colchicine treatment and polyspermy. *Nature* 186:737–739.

POLANI, P. E.
1970 Growing points in cell biology: cytogenetic aspects. *Proceedings of the Royal Society of Medicine* 63:955–959.

PYLE, R. L., D. F. PATTERSON, W. C. D. HARE, D. F. KELLY, T. DIGIULIO
1971 XXY sex chromosome constitution in a Himalyan cat with tortoise-shell points. *Journal of Heredity* 62:220–222.

RAICU, P., E. TAISESCU
1972 *Misgurmus fossilis*, a tetraploid fish species. *Journal of Heredity* 63:92–93.

RAY-CHAUDHURI, R.
1967 Mitotic and meiotic chromosomes of the Koel, *Eudynamys scolopacea*. *Nucleus* 10:179–189.

RENDEL, J. M.
1967 *Canalisation and gene control.* London: Logos.

SACREZ, R., J. CLAVERT, D. WILLARD, Y. RUMPTER, J. M'BEDE, R. MEYER, R. KORN
1967 La triploidie chez l'enfant. *Pediatrie* 22:267–275.

SARVELLA, P.
1970 Sporadic occurrence of parthenogenesis in poultry. *Journal of Heredity* 61:215–219.

SCHINDLER, A. M., K. MIKAMO
1970 Triploidy in man: report of a case and a discussion on etiology. *Cytogenetics* 9:116–130.

SCHLEGEL, R. J., R. L. NEU, J. CARNEIRO LEAO, E. FARIAS, M. J. ASPILLAGA, L. I. GARDNER
1966 Observations on the chromosomal, cytological and anatomical characteristics of 75 human conceptuses. *Cytogenetics* 5:430–446.

SCHMICKEL, R. D., A. D. FLOYD, J. M. POOLEY
1971 A live born infant with 69 chromosomes. *Journal of Pediatrics* 79:97–103.

SCHMID, W., M. F. LEPPERT
1969 Rates of DNA synthesis in heterochromatic and euchromatic segments of the chromosome complements of two rodents. *Cytogenetics* 8:125–135.

SCHMID, W., D. VISCHER
1967 A malformed boy with double aneuploidy and diploid-triploid mosaicism 48, XXYY/71, XXXYY. *Cytogenetics* 6:145–155.

SELLYEI, M., S. DERZSY, L. VASS
1971 69, XYY sporadic triploidy in an adult. *Humangenetik* 11:244–246.

SHAVER, E. L., D. H. CARR
1969 The chromosome complement of rabbit blastocysts in relation to the time of mating and ovulation. *Canadian Journal of Genetics and Cytology* 11:287–293.
SHOFFNER, R. N., A. KRISHAN, G. J. HAIDEN, R. K. BAMMI, J. S. OTIS
1967 Avian chromosome methodology. *Poultry Science* 46:333–344.
SHOFFNER, R. N., N. WANG, J. S. OTIS
1972 Gametogenesis in triploid chickens. *Genetics* 71 (Supplement):59.
SPERLING, K., E. SALING
1971 Pränatale Chromosomenanalyse mit Mosaikbefund 46, XX/92, XXXX. *Humangenetik* 11:139–145.
STEBBINS, G. L.
1950 *Variation and evolution in plants.* New York: Columbia University Press.
1971 *Chromosomal evolution in higher plants.* Reading: Addison-Wesley.
STONOVA, N. S., T. G. SELEZNIOVA.
1968 Chromosome aberrations in cases of human spontaneous abortions. *Genetica* 4:126–144.
STRICKBERGER, M. W.
1968 *Genetics.* New York: Macmillan.
SZULMAN, A. E.
1965 Chromosomal aberration in spontaneous human abortions. *New England Journal of Medicine* 272:811–818.
THIEDE, M. D., S. METCALFE
1966 Chromosomes and human pregnancy wastage. *American Journal of Obstetrics and Gynecology* 96:1132–1138.
UCHIDA, I., C. C. LIN
1972 Identification of triploid genome by flourescence microscopy. *Science* 176:304–305.
UYENO, T., G. R. SMITH
1972 Tetraploid origin of the karyotype of catostomid fishes. *Science* 175:644–646.
UZZELL, T. M.
1964 Relations of the diploid and triploid species of the *Ambystoma jeffersonianum* complex. (*Amphibia, Caudata*). *Copeia* 2:257–306.
1970 Meiotic mechanisms of naturally occurring unisexual vertebrates. *American Naturalist* 104:433–445.
VAN'T HOF, J.
1965 Relationships between mitotic cycle duration, S period duration and the average rate of DNA synthesis in the root meristem cells of several plants. *Experimental Cell Research* 39:48–58.
WAXMAN, S. H., D. T. ARKAKI, J. B. SMITH
1967 Cytogenetics of fetal abortions. *Pediatrics* 39:425–432.
WHITE, M. J. D.
1954 *Animal cytology in evolution.* Cambridge: Cambridge University Press.
WITSCHI, F., R. LAGUENS
1963 Chromosomal aberrations in embryos from overripe eggs. *Developmental Biology* 7:605–616.
WOLF, U., H. RITTER, N. B. ATKIN, S. OHNO
1969 Polyploidization in the fish family *Cyprinidae* order *Cypriniformes*. *Humangenetik* 7:240–244.

Biographical Notes

M. Fathi Abdel-Hameed (1938–   ) was born in Tanta, Egypt. He studied agricultural sciences and poultry production at Alexandria University where he received his B.S. and M.S. in 1958 and 1961, respectively. He was awarded a Fulbright Fellowship to study environmental horticulture and genetics at the University of California at Davis where he received his M.S. and Ph.D. in 1963 and 1967, respectively. He was engaged in postdoctoral research on the ultrastructure and function of chromosomes at Argonne National Laboratory and in medical genetics and cancer cytogenetics at the University of Kentucky Medical Center. He is now an Associate Professor of biological sciences at Northern Illinois University and has published many articles in *Science* and other reputed journals.

Malkhaz G. Abdushelishvili (1926–   ) is Doctor of History, Head of the Department of Anthropology of the Institute of History, Academy of Sciences, Georgian SSR. He was educated at the Institute of Medicine of Tbilisi (Georgia), and received his anthropological education as a postgraduate student at the Institute of Anthropology of the State University of Moscow. He has conducted craniological and somatological expeditions in Georgia, Armenia, Ossetia, Chechen-Ingusheti, Kabardino-Balkaria, Karachaevo-Cherkessia, Adigei, and also in North-West India and North-East India. He is the author of twelve books and 130 other papers. His publications include: "The materials of the expedition in 1950 on the anthropology of the modern population of the Caucasus" (1953); "The paleoanthropology of the Samtavro grave" (1954); "The anthropological structure of the modern population of Armenia" (1963); "The anthropology of the ancient and modern population of Georgia" (1964); "Craniology of the ancient and modern population of the

Caucasus" (1966); "The genesis of the aboriginal population of the Caucusus in the light of anthropological data" (Tokyo 1969); "Certain problems of ethnic anthropology in South-West Asia in the light of the latest research" (1973); "The correlation of the morpho-physiological and racial characters" (1976); "Craniology" (1976); "The method of the categorical calculations for intra and intergroup analysis of the large number observations" (1977).

T. I. ALEKSEEVA. No biographical data available.

V. P. ALEXEEV. No biographical data available.

E. CAPUCCI. No biographical data available.

EDMUND CHURCHILL received his B.S. from the University of Maryland and M.A. from Columbia University. He was Professor of Mathematics, Antioch College, 1947 to 1970, Senior Associate, Webb Associates, Yellow Springs, Ohio, 1970–1977, and is Senior Consultant, Anthropology Research Project, Inc., Box 498, Yellow Springs, Ohio and sometime Consultant to the Advisory Group for Aeronautical Research and Development (NATO); the Office of Community Standards, U.S. National Bureau of Standards; the Peruvian Institute of Fashion; and the National Nutrition Survey, U.S. Department of Health, Education and Welfare.

V. CORRENTI. No biographical data available.

M. CRESTA. No biographical data available.

EMILE CROGNIER. (1941–) received his doctorate in sciences from the University of Paris in 1972. He has done research work at the CNRS and taught courses at the University of Paris. He has coauthored a book (in press), with four other authors, about a world-wide survey of human diversity. His contribution is the result of a three-year study of human biological diversity in the European and Mediterranean area.

M. E. DE CASTRO E ALMEIDA. No biographical data available.

HERTHA DE VILLIERS (1926–), M.Sc. (Wits), Ph.D. (Wits), is presently Associate Professor, Department of General Anatomy, University of the Witwatersrand, Johannesburg, RSA. Her fields of research endeavor and interests are: modern southern African populations, their morphology, dermatoglyphics and genetics, and the evolution and morphology of Stone and Iron Age southern African populations.

ALBERT DUCROS (1934–) was born in Beni-Abbès, Algeria. He studied biology, physical and cultural anthropology at the University of Paris, receiving his D.Sc. in 1971 with a thesis in physical anthropology on micro-evolution factors in a Basque village (he was awarded the "prix Broca" by the Société d'Anthropologie de Paris). He is currently Maître de Recherche at the CNRS (National Center of Scientific Research) and a member of the National Committee of this Institution. His chief interests and publications have been first in osteological anthropology (methodology, sexual dimorphism) then, on physical variation, evolution, and adaptation of living populations. He conducted field research in France, East Greenland, Comoro Islands and French Polynesia. He is co-secretary of the "Société de Biométrie Humaine".

JACQUELINE DUCROS (1939–) was born in Tananarive, Madagascar. She received her Licence ès Sciences Naturelles at the University of Paris La Sorbonne, then a "Doctorat de 3e Cycle" in physical anthropology in 1965. At present she is Chargé de Recherche at the CNRS (National Center of Scientific Research). She participates at multidisciplinary field research concerning physical variation in France, the Indian Ocean, and Polynesia and she has published dermatoglyphic analysis about Eskimo, Basques, Somali, and Danakil.

OTTÓ G. EIBEN (1931–) received his Ph.D from the University of Debrecen and his C.Sc from the University of Budapest. He is an Associate Professor of Physical Anthropology and Head of the Department of Anthropology at Eötvös University, Budapest. His main research fields are: study of growth and development of children, variation of physique, and population genetics. The results of his scientific activity have been published in three books and over eighty articles, as well as several chapters which he has contributed to other works. He is the Editor of *Anthropologiai Közlemények* [Anthropological Communications], and his editing and science organizing activity is significant, too.

PEGGY ERSKINE. No biographical data available.

DZIGBODI K. FIAWOO was born in Ghana. He received his undergraduate degree from Cornell University, his M.A. from Columbia University, and his Ph.D. from Edinburgh University. He was a postgraduate Research Associate at the Laboratory of Human Development at Harvard University (1961–1962) and a Senior Commonwealth Fellow at Cambridge University's Centre of African Studies (1970–1971). He was also a visiting lecturer at the Centre for West African Studies, Birmingham University, Birmingham, England. He is Senior Lecturer (Acting Head) of the Department of Sociology, University of Ghane Legon, Ghana. He has done research on child development in Ghana, as well as on traditional religion and social change, with special reference to the Ewe-speaking people, and is currently National Director of a Ghana Government Project on the Young Child, sponsored by UNICEF.

PRANAB GANGULY (1929–) was born in Varanasi City, Uttar Pradesh, and was educated in Calcutta. He received his B.Sc. in anthropology in 1949 from Calcutta University, where he also obtained his M.Sc. in 1951, Ph.D. in 1966, and D.Sc. in 1973. Dr. Ganguly is currently Superintending Anthropologist in the Anthropological Survey of India, Calcutta. He has done much original research in the Andaman and Nicobar Islands, Assam, Maharashtra, and Uttar Pradesh. His research interests include physical anthropology, social biology, and population studies, and he has published widely in these fields. The Société d'Anthropologie of Paris awarded the Prix Bertillon to Dr. Ganguly in 1973 for his study of human physique in relation to socio-economic environment. In 1975 the Calcutta University awarded him the Sir Asutosh Mookerjee Gold Medal for his work on Andaman Negritos, and in the same year he also received the Bengal Immunity Research Prize and Gold Medal for his contribution on the secular trend in body height of the Indian population. In 1976 Dr. Ganguly received the Griffith Memorial Prize for his research on human micro-variation in the Nicobar Islands. He was elected President of the Anthropology and Archaeology Section, Indian Science Congress Association, for the year 1976–1977.

LINDA M. GLUESING. No biographical data available.

ELENA Z. GODINA (1941–) was educated at the Department of Anthropology, Moscow State University. Since 1963 she has been on the staff of the Institute and Museum of Anthropology as a research worker. Her main interest is growth and development of children. She has done field work in the mountains of Tien Shan and Pamirs and her publications include some twenty articles published in different editions.

JEAN HIERNAUX (1921–) is an anthropobiologist, now Director of Research at the CNRS (France) and head of a team of research and teaching in human ecology at Paris VII University. He received a doctorate of medicine in Brussels and one of science in Paris. He spent many years of research and teaching in Rwanda and Zaire, where his main interests concerned the human response to climate and food needs, and Iron Age archaeology. His present research deals mainly with migration from rural to urban environments. He is the author of *The people of Africa* (1974) and a number of monographs and papers on anthropobiology of Central Africa, African archaeology, and human taxonomy.

SEIICHIRO INOKUCHI (1926–) is Professor of Anatomy at Showa University School of Medicine in Tokyo, Japan. His interests lie in myofibrous organization of skeletal muscles, subcutaneous adipose tissue, and blood vessel wall.

SANAE KANDA (1925–) is Professor of Anatomy at Hyogo College of Medicine (Japan). He was educated and graduated M.D. at Osaka University and has done fieldwork in India, Pakistan, South East Asia, and Japan. His publications include: *Physical anthropology of minorities in India and Pakistan; Physical anthropology of Thailand; Growth of Japanese children;* and *Japanese prehistorical skulls.*

JUDIT KATONA-APTE (1943–), a native of Hungary, received her B.A. in anthropology and is currently a doctoral candidate in the Department of Nutrition at the University of North Carolina School of Public Health. She is specializing in nutrition and anthropology and has done field work in India and in the United States. Her interests and publications focus on medical and nutritional anthropology, methodology, acculturation, and nutritional adaptation.

E. N. KHRISANFOVA. No biographical data available.

KUNIHIKO KIMURA (1927–) was born in Tokyo, Japan. He received his B.S. from University of Tokyo in 1951; his Ph.D. (D.M.) from Osaka City Medical School in 1957; and his D.Sc. from University of Tokyo in 1962. He successively held Assistant or Associate Professor of Anatomy at Toho University, Tokyo, and University of Tokyo; and Professor of Anatomy at Kanagawa Dental College, Yokosuka, and Tokyo University of Education. He has been Professor of Anatomy at National Defense Medical College since 1976. He has authored and coauthored nine books, and eighty articles on physical anthropological and gross anatomical studies, mainly on growth of children.

FRITZI KOHN DE BRIEF. No biographical data available.

GABRIEL WARD LASKER was born in York, England, and educated in the United States. After two years in China he entered graduate study at Harvard where he received his Ph.D. in 1945. Since then he has been on the faculty of the Department of Anatomy at Wayne State University with brief leaves for field research (in Mexico and Peru) and to teach anthropology elsewhere. He has been President of the American Association of Physical Anthropologists, Chairman of the Council of Biology Editors, and is Editor-in-Chief of *Human Biology*. He is author of numerous journal articles and a textbook in physical anthropology.

LAURENCE MALCOLM (1929–) is a New Zealander and graduated from the Otago Medical School in Dunedin in 1953. He has held posts in public health administration in Papua New Guinea from 1957–1974 and more recently was head of the Health Planning Unit of the Department of Public Health involved in the preparation of the first National Health Plan for the country. His research interests include child growth and development, nutrition, community biology, community health, and planning and evaluation of health services. He is a Fellow of the Royal College of Physicians, Edinburgh, a Fellow of the Australian College of Health Administrators, and a Member of the Faculty of Commonwealth Medicine.

NATALIA N. MIKLASHEVSKAYA (1929–) graduated from the Moscow State University (Ph.D. in anthropology). Since 1952 she has taught anthropology at the Department of Anthropology, Moscow State University, and worked in the field of ethnic anthropology. She has conducted fieldwork in Middle Asia, Caucasus, and Siberia. Since 1962 she has been a research worker at the Institute and Museum of Anthropology and has studied head and face growth in different ethnic groups of the USSR. Now her main interest is growth and development of children in different climatic zones. She has published more than sixty scientific articles, and edited a book: *Growth and development of a child.*

LAURA NEWELL MORRIS (1933–) has an undergraduate degree from the University of New Mexico, an A.M. from Northwestern University and a Ph.D. from the University of Washington. Her special areas of research include primate biology, genetics and development, and human demography. She has done fieldwork in Latin America and Jamaica under a PHS postdoctoral award. Her research has also involved rural Philippine people and she is continuing a study of the Japanese-American population of Seattle. She is an Associate Editor of the *American Journal of Physical Anthropology*. She has edited works on population genetics and on primate population biology. At present she is an Associate Professor in the Departments of Anthropology and Orthodontics and a Research Affiliate of the Regional Primate Research Center, University of Washington, Seattle.

PETER E. NUTE (1938–) received his B.S. (zoology) from Yale University in 1960, studied Medicine at Yale for the following two years, and was awarded his Ph.D. (anatomy and physiology) by Duke University in 1969. He has pursued research interests in molecular genetics and evolution as a Postdoctoral Fellow in the Division of Medical Genetics at the University of Washington, and is now an Assistant Professor of Anthropology and a Research Assistant Professor of Medicine at that

University. A member of the Center for Inherited Diseases, he is also on the Consulting Staff of the University Hospital and a Research Affiliate of the Regional Primate Research Center at the University of Washington.

GEORGES OLIVIER (1912–) was born in Paris. He has been a colonial medical doctor, an anatomy "agrégé", and Professor at the Paris School of Medicine and at the Paris Faculty of Sciences, as well as President of the Paris Society of Anthropology. He is now Professor of Biological Anthropology at the University Paris VII, and President of the European Association of Anthropology. His field of research is anthropological anatomy and variability of human present-day populations.

JANA PAŘÍZKOVÁ. No biographical data available.

MIROSLAV PROKOPEC (1923–) is a leading scientist in the Institute of Hygiene and Epidemiology, Prague, Czechoslovakia. He received his M.A., Ph.D. and D.S. from Charles University, Prague and also studied at U.C.L., London. His main interests are growth studies and physique in relation to hereditary and environmental factors. He has conducted several state-wide child growth studies and anthropometrical surveys of population samples in Czechoslovakia and is following up semi-annually a group of 300 Prague children from birth to maturity (started in 1956). During his field work in India (1963–1964) he studied social influence on growth and in Australia (1969) the Rembarranga tribe in the Northern Territory. He was Visiting Curator in the South Australian Museum in 1974–1975 and studied human osteological material from Roonka, S.A. He has published over 100 scientific papers and articles, and is author or coauthor of several books and text-books e.g. *Tracing man's evolution* (1957), *Physical and mental development of the present generation of children* (1969), *Antropologie* (1967).

DAVID RABINOW. No biographical data available.

PETER RÂMNEANTZU (1902–) was born near Timişoara, Romania. He received his Doctor's degree in medicine from the University of Cluj, Romania, and his M.P.H. from Johns Hopkins University of Baltimore (1939). He was Professor of Hygiene and Preventive Medicine at the University, and the Director of the Institute of Hygiene and Public Health of Timişoara, since 1946. He has conducted many field and scientific works in public health problems. He has authored four books and 220 articles in Romanian, and three books and thirty-two articles (some coauthored) in other languages (English, German, French, Russian). He has been a member of the Academy of Medicine from Bucharest (1942), of the Superior Sanitary Council (1942), of the American Statistical Association (1934), and of other societies. He is the author of *Medical biometry and vital statistics* (1939). He has been awarded four prizes by the Romanian Academy for different works (1936, 1937, 1940, 1947) and he was a Fellow of the Rockefeller Foundation (1932–1933).

John R. K. Robson is a graduate of Kings College Medical School, Durham University in England. Following training as a surgeon, he specialized in clinical nutrition at London, Edinburgh, and Newcastle universities where he obtained his doctorate in nutrition research. He worked for the British Government for ten years, in Tanganyika (now Tanzania) in Africa, followed by five years as a Regional Advisor in the Western Pacific and Eastern Mediterranean Regions for the World Health Organization. In 1967 he was appointed Director of the Nutrition Program in the University of Michigan where he became increasingly involved in the anthropological aspects of nutrition. Since 1975 he has been Chief of the Nutrition Division in the Medical University of South Carolina. He is Editor of *Ecology of Food and Nutrition* — an international journal.

J. B. Romain. No biographical data available.

Raghbir Singh (1933–   ) received his M.Sc. and Ph.D. from Delhi University. He has taught anthropology at the University of Delhi and was Assistant Research Officer (Anthropologist) at the National Institute of Nutrition, Hyderabad from 1958 to 1963. He is currently guiding and conducting research in human growth, physique and body composition. He is a member of the Editorial Board of *Annals of Human Biology* (London) and is also an Associate and a referee of scientific papers for the journal *Current Anthropology* (Chicago). His publications include: "Nutritional somatometric study of adult males with special reference to skinfold measurements and inter-correlations" (1964); "A study of fat, muscle, bone and comparison between roentgenogrammetric and somatometric methods for assessing subcutaneous fat and muscle thickness in the upper arm of adult males" (1967); and "Somatometric study of Punjabi Hindu Khatri (male) with special emphasis on growth (cross sectional) and relationship between arm girth and body weight measurements" (1972), Miami, Florida.

Nina S. Smirnova (1933–   ) was educated at the Department of Anthropology Moscow State University (Ph.D. in anthropology). Since 1955 she has been on the staff of the Institute of Anthropology Moscow State University as a research worker where she studies regularities of morphological variability in man. Her publications include some thirty articles published in different editions.

VALERIA S. SOLOVYEVA (1932–   ) graduated from the Moscow State University (Ph.D. in anthropology). Since 1955 she has been working as research worker at the Institute and Museum of Anthropology, studying peculiarities of growth and sexual maturation of adolescents of different ethnoterritorial groups. She has published more than forty articles mainly in the periodical issue of *Voprosi Antropologii* [Problems of Anthropology] as well as in other editions.

G. SPEDINI. No biographical data available.

WILLIAM ARTHUR STINI (1930–   ) received his undergraduate and graduate education at the University of Wisconsin where he earned his Ph.D. degree in 1969. He is currently Professor of Anthropology at the University of Arizona. His main area of interest is human adaptation to environmental stress, especially the effects of nutritional deprivation on the growth process. He has done research on the growth of poorly nourished children in Latin America and experimental work on the effects of malnutrition in laboratory animals.

JAROSLAV SUCHÝ (1926–1975) was Professor of Anthropology at Charles University, Prague (Pedagogical Faculty). He studied physical anthropology in Prague and received his M.A. in 1949 and Ph.D. in 1956. He became an Assistant Professor in 1959 and Professor in 1972. He worked in various branches of physical anthropology and in methodology. His main scientific interest was devoted to child growth studies, to problems of the Gypsies, and to gerontology. He published well over 150 scientific works and contributions and published, edited or co-edited several textbooks and books. He edited *Anthropological Archive*, issued by the National Museum Society in Prague, and was Scientific Secretary of the Czechoslovak Anthropological Association at the Czechoslovak Academy of Sciences, and President of its Prague Section. As a university teacher he trained the present generation of anthropologically-oriented biologists — teachers and scientists. His books include: *How is man evolving?* (1972); *Developmental anthropology of Czechoslovakian inhabitants* (1972); *Antropologie* (1967).

ERIC SUNDERLAND (1930–   ) graduated in 1950 at the University of Wales, being awarded the degree of B.A. in Anthropology and Geography. In 1951 he was awarded the degree of M.A. by the University of Wales and in 1954 the degree of Ph.D. by the University of London, both in anthropology. Since 1958 he has been a member of staff at the University of Durham, England, and is currently both Professor and Head of the Department of Anthropology there. He has published extensively in human population genetic studies and he has carried out several periods of fieldwork, mainly in Middle Eastern countries.

CHARLES SUSANNE (1943–) is a physical anthropologist and human genetist at the Free University of Brussels. He was educated at the Free University of Brussels where he received his doctorate in 1966 and where he started his teaching career. He is now Professor of Anthropology and Human Genetics, President of the Société Royale Belge d'Anthropologie et de Préhistoire, and Editor of the Newsletters of the European Anthropological Association. His research interests include the heritability of physical characters, growth characters, and cytogenetical aspects of the human heredity. He has published about eighty articles in the fields on interaction of environment and heredity.

H. TISSIER. No biographical data available.

F. VECCHI. No biographical data available.

A. A. VORONOV. No biographical data available.

HENRY C. WOMACK (1938–) did his undergraduate work at Alcorn A and M and holds an M.S. degree in biology and a Ph.D. in anatomy from Wayne State University. He is currently Assistant Professor of Physiology and Health Science at Ball State University. His major research interest is in human body composition, especially variations in body fat distribution.

Index of Names

Abbie, A. A., 56
Abdel-Hameed, Fathi, 411–414, 417–444, 459–492, 499
Abdushelishvili, M. B., 45–52, 98, 115, 116, 499
Abel, W., 62, 64
Abramovici, Lydia, 349
Abramson, J. H., 56, 293, 294
Acheson, Roy M., 208, 293, 316, 326
Agadzhanian, N. A., 81
Agarwala, S. N., 364
Ager, J. A. M., 116
Aguero, O., 351
Aitken, F. C., 319
Al-Aish, M., 420
Alegria, Cogerino, 349
Alekseeva, T. I., 81–86, 131, 500
Alexeev, V. P., 49, 97–107, 137, 500
Alexiou, D., 423, 424
Allderdice, P. W., 430, 434
Allen, J. A., 69
Alleyne, E. A. O., 396, 398
Allison, A. C., 113
Amirhakimi, G., 375
Andersson, L., 421, 430, 434
Ansehn, S., 421, 430, 434
Antonov, A. M., 345
Arguedas, H. A., 387
Arkaki, D. T., 464, 465
Arroyare, G. A., 392
Aschkenasy, A., 399
Aschkenasy-Lelu, P., 399
Ascoli, W., 326
Ashcroft, M. T., 387
Aspillaga, M. J., 465, 472, 473
Astrand, P. O., 201
Atkin, –., 465
Austin, C. R., 473, 474
Aykroyd, W. R., 329

Babakov, O. B., 100
Bablet, J., 391
Bachman, R., 420
Baggett, B., 427, 428, 429, 434, 436, 438
Bailey, K. V., 237
Baker, P. T., 81, 243, 244, 246
Bakwin, H., 317
Bannister, W. H., 142
Barclay, D. L., 425, 426, 427, 428, 434, 436, 441, 442
Barigozzi, C., 434
Barin-Baum, D. E., 379
Barnes, R. H., 393
Barnicot, N. A., 7, 142, 143, 144, 146
Bartley, M. H., Jr., 396
Basatter, F., 467
Bashore, R. A., 426, 435
Basu, M. N., 315
Beatty, R. A., 473, 474, 482, 483
Beçak, M., 478
Beçak, W., 482
Behnke, A. R., Jr., 202
Behrhorst, C., 387
Belekova, R. B., 246
Benirschke, K., 441
Benoist, Jean, 255, 256, 258
Bentzon, A. W., 317, 319
Bercovici, J. P., 437, 438
Berg, B. N., 399
Bergmann, C., 69
Bergher, E., 349, 355
Bergman, S., 421, 430, 434
Bergstrand, C. G., 421, 422, 430, 439
Bernard, R., 465, 467

Bernini, L., 120
Bernstein, M., 296
Bertrand, J., 427, 428, 434, 436, 437
Bets, L. V., 452
Biasutti, R., 97
Bielicki, T., 279
Billewicz, W. Z., 347
Billy, G., 284
Birdsell, J. B., 139, 140, 149
Biswas, P. C., 148
Bishun, N. P., 420
Blagoveshchenskaya, T. A., 452
Bloom, S., 475, 479
Bobrow, M., 418, 419, 433
Boczkowski, K., 421, 422, 426, 427, 428, 430, 434, 438, 439
Bojlen, K., 350
Bolerig, L., 392
Bomsel-Helmreich, –., 474, 483
Bompar, F., 391
Bordzhadze, V. K., 452
Bose, C., 339
Bose, K. O., 366
Böök, J. A., 465, 467, 471, 472, 474, 478, 491
Bowles, G., 316, 317
Boué, J. G., 464
Boué, A., 464
Bowen, J. M., 426, 434
Bourliere, F., 56
Boyd, E., 433
Boyd, W., 122
Boyne, A. W., 317, 319
Boyer, S. H., 142, 143
Bramwell, F., 391
Brasel, J., 391
Bray, G., 433
Breg, W. R., 430, 434
Brehme, H., 149
Briles, W. E., 479, 480
Broggen, A., 435
Brooke, R. T., 243, 246
Brozek, A., 294
Brozek, L. J., 294
Bruch, H. A., 235
Brusquet, Y., 465, 467
Buettner-Janusch, J., 142
Buettner-Janusch, V., 142
Buhler, E. M., 434, 465, 467
Bunak, V. V., 47, 49, 115, 131
Burrell, R. J. W., 350, 351
Buss, E. G., 479
Butcher, R. L., 476
Butler, L. J., 464, 467, 469, 473, 491
Byrd, J. R., 425

Cannon, W. B., 480
Cantalini, C., 319
Capucci, E., 111–113, 500
Cardiff, C., 436
Carfagna, M., 355
Carmichael, R., 425, 435
Carmody, N. C., 428
Carneiroheao, J., 465
Carr, D. H., 462, 463, 464, 465, 466, 468, 469, 470, 472, 474, 475, 476, 488, 491
Carter, J. E., 137
Caspersson, T., 430, 434
Castellanos, H., 392
Castellanos, R., 351
Cavalli-Sforza, L. L., 329
Cenani, A., 462
Chai, C. K., 149
Chamla, M. C., 187, 293
Chandrasekhar, S., 329, 364, 366
Chantler, C., 464, 467
Chase, H. P., 345
Cheek, D. B.; 391
Chemke, J., 425, 435
Cheboksarov, N. N., 50
Chepurkovskii, E. M., 97
Chiarelli, B., 491
Choi, Sing C., 20
Choremis, C., 120
Chow, –., 239
Christakos, A. C., 420, 425
Christensen, A. K., 433, 485
Christian, C. D., 420, 424
Chryssostomidou, O. M., 423, 424
Chu, E. H., 483
Churchill, Edmund, 19–43, 500
Churchill, Thomas, 20, 30
Chyle, P., 474
Cimino, M., 477
Clark, I., 392
Clauser, C. E., 30
Clavert, J., 467
Cobo, A., 421, 422, 430
Cohen, I. A., 81
Cohen, M. M., 425
Cohen, P., 142, 143, 144
Coignet, J., 465, 467
Cole, L., 304
Consolazio, C. F., 389, 390
Conterio, F., 329
Coope, E. M., 149
Cork, A., 426, 434
Correnti, V., 111–113, 500
Cravioto, Joaquin, 237
Crema, M., 350
Crenier, E., 59
Crepy, U., 437, 438
Cresta, M., 111–113, 500
Cristescu, M., 456
Crognier, Emile, 56, 69–79, 500
Crooke, W., 321

Crosby, E. F., 142
Csébfalvi, K., 167
Cummins, H., 62, 149, 151
Cupceanu, B., 452
Curtain, C. C., 120

Dabiere, C S., 345
Da Fonseca Sacarrao, G., 279
Damon, Albert, 21, 56, 317, 372
Dandekar, V. M., 339
Daniels, G. S., 21
David, M., 427, 428, 434, 437
Davidson, D. W., 427, 428, 429, 436, 440, 444
Davila, E. H., 208
Davis, J. C., 428, 429, 436, 437, 438
Dean, R. F. A., 392
Deane, R., 301
Debets, E. E., 48, 49, 50, 101
De Castro e Almeida, M.E., 279–285, 500
Deckert, G., 461
De Díaz Ungría, Adelaida, 349
De George, F. V., 58
De Jong, W. W. W., 142
De la Chapelle, A., 420, 421, 422, 423, 424, 429, 430, 431, 432, 433, 434, 439, 440, 441
De la Cruz, F. F., 420
Delegdisch, L., 419, 427, 429, 434, 436, 437
Delicardie, Elsa R., 237
Dellaportas, G. J., 387
De Macedo, F., 279
De Moor, P., 456
Deniker, J., 97
Derzsy, S., 462
Despeignes, Mimose, 259
De Uifalvy, Charles, 321
De Villiers, Hertha, 287–298, 500
Dewhurst, C. J., 420, 425, 426, 427, 435, 441, 442
Dillard, G. M., 430, 434
Dixon, W. J., 19, 325
Dobzhansky, T., 153, 155
Dolfini, S., 434
Donner, L., 475, 479
Dorken, U., 120
Doxiadis, S. A. 423, 424
Drabik, J., 122
Drillien, C. M., 400
Dronamraju, K. R., 326
Ducros, Albert, 55–65, 501
Ducros, Jacqueline, 55–65, 501
Dunn, H. O., 483
Duperuis, C. W., 30, 98, 187
Du Plessis, J. P., 294
Durante, G., 391
Dutta, P. C., 322

Dýcka, Z., 203

Easterling, W. E., 436
Eckardt, M. J., 478
Edwards, J. H., 464, 465, 467, 471, 474, 484
Edwards, R. G., 466, 469, 475
Eggen, R. R., 420
Egozcue, J., 491
Ehrhardt, A. A., 419, 420, 442, 443
Eiben, O. G., 165–176, 501
Eiselt, E., 201, 203, 204
Elger, W., 452
Elgurt, G. M., 452
Ellis, J. R., 465, 467, 471, 474, 491
Elsner, P. W., 131
Erskine, Peggy, 19–43, 501
Espiner, E. A., 425, 426, 434
Etemadi, –., 379
Eyman, S. L., 208
Eysenck, H. J., 131

Fankhauser, G., 474, 477, 478, 482, 483
Fanas, E., 465
Fawcett, D. W., 485
Fechheimer, N. S., 479
Ferenczy, A., 426
Ferguson-Smith, M. A., 420, 433, 435
Ferguson-Smith, M. E., 433
Ferrier, P., 465, 467, 471
Ferrier, S., 465, 467, 470, 472, 473, 484
Ferro-Lyzzi, G. E., 347
Fessas, P. H., 120
Fetter, V., 219
Fiawoo, D. K., 301–312, 501–502
Figurelli, E., 355
Filippi, G., 120
Finkelstein, J. O., 419, 427, 429, 434, 436, 437
Finney, D. J., 351
Fischer, A., 202
Fischberg, –., 482, 483
Fisher, R. A., 192, 195, 198
Fiskesjo, G., 417, 418
Fitzgerald, P. H., 425, 426, 434
Flinn, R., 461
Flores, F., 421, 422, 430
Fluchmann, F., 350
Floyd, A. D., 467, 491
Forbes, G., 375
Ford, C. E., 418
Foss, G. L., 427, 428, 429, 436, 440, 444
Foster, Phillip, 311
Fowler, G. B., 293
Fraccaro, M., 434
France, N. E., 464, 467
Franco, G., 351
Frankel, T., 59

Frasier, S. D., 426, 435
French, F. S., 427, 428, 429, 434, 436, 438
Friedlaender, J. S., 318
Friedrich, V., 433
Frisancho, A. R., 243, 245, 326, 369, 372, 400
Frisancho, R. B., 243, 246
Frisch, R., 351
Froehlich, J. W., 317
Fugo, N. W., 476
Fujinaga, D., 122

Gabel, –., 318
Gait, E. A., 363
Gajdusek, D. C., 120
Gallien, L., 452
Galton, Sir Francis, 326
Galvánek, M., 456
Gampel, B., 56, 293, 294
Ganguly, Pranab, 315–333, 502
Gardner, L. I., 421, 422, 430, 465
Garn, S. M., 326, 342, 369, 372, 375
Gatto, J., 120
Gauthier, F., 437, 438
Gauthier, M., 459, 467
Geber, M., 301
Gavin, J., 435
Geertz, Clifford, 400
Geer, T. H., 425, 435
George, K. P., 430, 434
German, J., 488
Gessain, R., 55, 62, 64, 65
Ghafoorzaden, D. J., 426
Gillim, S. W., 485
Ginzburg, B. B., 131
Ginzburg, V. V., 81, 100
Giraud, F., 465, 467
Gluesing, Linda M., 414, 417–444, 502
Godina, E. Z., 243–247, 248–254, 502
Godt, H. H., 433
Goldberg, A. L., 392
Goldberg, M. B., 420
Goldfeder, A., 488
Goldman, R. F., 56
Goldsmith, G., 351
Goldspink, G., 389
Goldstein, Marcus, 369, 371, 372
Goncharov, N. P., 456
Gonzalez Orihuela, Gisela, 349
Gooch, A., 435
Goodman, M., 148, 153
Gopalan, C., 86
Gordon, D. L., 428, 434
Gordon, J. E., 233, 235
Gormah, J. G., 120
Gower, J. C., 91, 92, 95
Grab, B., 192, 195, 198
Grant, J. K., 429, 434, 436
Grande, F., 376
Gray, H., 192, 195, 198
Graystone, J. E., 391
Grech, J. L., 142
Green, D., 122
Greenblatt, R. G., 425
Greenfield, David W., 461
Greengard, P., 399
Greulich, W. W., 208, 295
Greyson, J. S., 293
Griffiths, K., 429, 434, 436
Grigorova, O. P., 81
Grobbelaar, C. S., 293, 295
Grossman, M. I., 59
Grunnofer, H. J., 31
Gropp, A., 464, 465, 470, 471
Gruska, S., 355
Guha, B. S., 50
Gupta, P., 322, 339, 453
Gusinde, M., 91
Guzman, M. A., 235

Haiden, G. J., 485
Halban, J., 382
Haldane, J. B. S., 411
Hall, B., 488
Hall, S. K. P., 294
Hamerton, J. L., 418, 484
Hamill, P., 371
Hamilton, Buchanan, 328, 329
Hanuvi, J. J., 392
Harnden, D. G., 484
Harris, H., 147, 294, 295
Harris, R., 7, 62
Harrison, G. A., 7, 62, 243, 244
Hartung, M., 465, 467
Hautvast, J., 355
Hayern, M. G., 391
Headings, V. E., 148
Healy, M. J. R., 193, 208, 351
Heath, B. H., 137
Hempel, J. M., 415
Hendrix, A., 456
Hendricky, G., 474
Heneage, P., 387
Henrotta, I. E., 86
Henry, J. B., 422
Hertzberg, H. T. E., 21, 30
Hiernaux, Jean, 69, 89–95, 293, 294, 502
Higginson, J., 350
Hoffenberg, R., 425
Hokfelt, B., 421, 422, 430, 439
Holliday, R. M., 347
Hollingsworth, M. J., 400
Holm, –., 55
Holt, S. B., 297
Hooton, E., 98, 187
Hortling, H., 421, 422, 430, 439

Hosseini, E., 379
Howells, W. W., 187
Hubbard, W. R., 427, 428, 429, 434, 436, 438
Huehns, E. R., 143, 144, 146
Hughes, D. R., 151
Husiman, T. H. J., 142
Huizinga, J., 62, 279
Hulse, F. S., 7, 62
Humphrey, R. R., 474, 478
Hunt, E. E., 154, 316, 375
Hunter, R. H. F., 474
Hytten, F. E., 347

Ikeda, T., 425
Inokuchi, S., 379–384, 503
Ishimoto, G., 141, 143, 144, 146, 148
Island, D. P., 437
Ismail, A. A. A., 427, 428, 429, 436, 440, 444
Isuri, K., 428
Ivanovsky, A., 316
Iwamoto, M., 149

Jackson, F. S., 149
Jackson, W. P. U., 425
Jacobs, P. A., 423, 424, 434, 440
Jalavisto, Eeva, 184
Janczewski, Z., 421, 422, 430, 439
Japp, R. G., 479
Jardine, N., 94, 95
Jarosz, E., 326
Jarvis, J. A., 420, 434, 442, 443
Jayant, K., 400
Jeffcoate, T. N. A., 428, 429, 436, 437, 438
Jelliffe, D. B., 233, 235, 346, 400
Jenkins, T., 295, 298
Johansson, C. J., 421, 422, 430, 439
Johnson, R. E., 389, 390
Johnston, F. E., 369, 371
Johnston, F. R., 427, 428, 429, 434, 436, 438
Jolly, C. J., 143, 144, 146
Jonasson, J., 424
Jones, A. L., 485
Jones, G. K., 420, 434, 442, 443
Jones, H. W., 428, 434
Jones, T. C., 432
Jopp, W., 7, 62
Jost, A., 419
Jussen, A., 465, 470, 471

Kalmus, H., 295
Kalyuzhny, I. T., 246
Kanda, S., 177–186, 503
Kaplan, B. A., 326, 369, 371
Kaplan, S. A., 441, 442, 443
Kariminejad, M. H., 426
Karn, M. N., 70, 400
Karve, I., 339
Katona-Apte, Judit, 363–368, 503
Katsberg, A. A., 474
Katsiya, G. V., 456
Kattamis, C., 120
Kawamura, T., 478
Kay, Won Chuel, 31
Keay, A. J., 423, 424, 440
Keith, C. G., 464, 467
Kenntner, G., 316
Kenny, F. M., 425, 428, 429, 440
Kerr, B. A., 149
Keys, A., 294
Khan, P., 120
Khobe, G., 456
Khosla, T., 294
Khrisanfova, E. N., 451–457, 503
Khudr, G., 441
Kidson, Chev., 120
Kikta, Paul, 19, 30
Killinger, D. W., 436
Kimura, Kunihiko, 207–212, 317, 319, 503
Kinsell, L. W., 392
Kinzey, W., 56, 399
Kitchen, H., 141
Klein, D., 465, 467
Klinger, H. B., 414
Kluge, A. G., 478
Kmietowics-Zukorskaya, A., 355
Knussman, R., 100, 125, 273
Kohashi, Y., 382
Kohn, G., 462, 463
Kohn de Brief, Fritzi, 349–361, 503
Kondik, V. M., 245
Konlande, J. E., 347
Kontturi, M., 421, 422, 430
Korn, R., 467
Korner, A., 392
Kouba, E. O., 141
Kpedekpp, –., 302
Kramer, G., 153
Krehl, W., 351
Kretschmer, E., 166
Kreze, A., 456
Kreig, A. F., 422
Krishan, A., 485
Krmpotic, E., 428, 434
Krogman, W. M., 319
Kroh, G., 31
Kucheman, C. F., 243, 244
Kulkarni, A., 153
Kwong, E., 393
Kyrge, P. K., 456

Lackman, I., 433
Laguens, R., 474, 476
Lai, L. Y. C., 148

Lamprecht, De V., 293, 294
Landau, R. L., 427, 428, 429, 436, 437, 438, 440
Lanning, M., 421, 422, 430
Larkin, F. A., 347
Lasker, Gabriel W., 369–377, 503
Lass, F., 424
Latte, B., 120
Laubscher, N. F., 294
Laughlin, W. S., 60, 131
Laurence, A. M., 427, 428, 429, 436, 437, 438, 440
Laurian, Lydia, 419, 427, 429, 436, 437
Lauritsen, J. G., 424
Lazar, P., 456, 464
Leon, H. A., 392
Le Bideau, G., 58
Lebret, L. J., 259
Lee, M., 375
Leguèbe, A., 60
Lehmann, H., 116
Leitch, I., 319
Lejeune, J., 415, 459, 467
Lemeshav, S., 371
Lennartz, K. J., 488
Leppert, M. F., 488
Levan, A., 412, 417, 418
Levin, M. G., 98
Levine, V. E., 350
Levitsky, D., 393
Lewis, H. E., 316
Liao, S., 427, 428, 429, 436, 440, 444
Liao, Shu-Chung, 347
Liddle, S., 441, 442, 443
Lin, C. C., 467, 475, 491
Lindsten, J., 421, 422, 424, 430, 434, 439
Lisker, R., 421, 422, 430
Little, S., 441, 442, 443
Livson, N., 351
Loewenthal, M., 419, 427, 429, 434, 436, 437
Long, J. A., 485
Loots, J. M., 293, 294
Loraine, J. A., 427, 428, 429, 436, 440, 444
Lourie, –., 7
Louw, M. E. J., 294
Lovell, H. G., 387
Lowe, C. H., Jr., 153
Lowe, C. R., 294
Lubbe, A. M., 294
Lundman, B., 97
Luschan, F., 115
Lyon, M. F., 484

McCloy, C. H., 185
McConville, John T., 30
MacCullum, J. B., 391
McEntree, K., 483
McFarland, Ross, 21, 372
McFie, J., 392
Macintyre, M. N., 415
McLaughlin, S. D., 317
McMahon, T., 389
McNeil, D., 351
Mackovà, E., 201, 202
Mahalanobis, P. C., 339
Mahesh, V. B., 426, 427, 428, 434, 435, 436, 438, 440
Mair, E., 131
Majumdar, D. N., 339
Makkonen, T., 185
Malá, H., 265, 266
Malcolm, L. A., 233–240, 504
Malhotra, M. S., 81, 327
Manalova, Y., 417, 418
Manolov, G., 417, 418
Margen, S., 392
Marks, P. A., 480
Maroni, G., 484
Marquer, P., 187, 293
Marshall, R., 465
Marshall, W. A., 316
Martin, R., 166, 187, 340
Masica, Daniel N., 442, 443
Masterson, J. G., 472, 491
Materese, G., 355
Materese, S., 355
Matiegka, J., 221
Maurais-Jarvis, P., 437, 438
Maurer, W., 488
Mayr, E., 139, 154, 155
M'Bede, J., 467
Meglon, C. J., 427, 428, 429, 440
Melville, M. M., 423, 424, 440
Mendez, J., 387
Meredith, H., 315, 371, 372
Merhauslovà, J., 398
Mertens, R., 153
Metcalfe, S., 464
Meyer, R., 467
Meyer, W. J., 427, 428, 429, 440
Michelova, K., 430, 434
Midlo, C., 62, 149, 151
Mikamo, K., 464, 467, 469, 470, 471, 473, 475, 484
Mikkelson, M., 421, 422, 430, 439
Miklashevskaya, N. N., 243–247, 248–254, 504
Milicer, H., 351
Miller, D. A., 430, 434
Miller, O. J., 426, 430, 434
Miller, Peter S., 318
Miller, R. C., 474, 479
Millot, Jacques, 255
Mirrakhimov, M. M., 81, 246

Mittwoch, U., 419, 432, 443, 464, 465, 467, 484, 487, 488
Money, J., 419, 420, 442, 443
Montero, Luis, 349
Montgomery, R. D., 392
Mood, A. M., 325
Moore, M. A. S., 243, 244
Moral, Paul, 259
Morales, S., 387
Moreno, L. F., 351
Morera, A. M., 427, 428, 434, 436, 437
Morgan, J. J. B., 304
Moroney, M. J., 325, 330
Morris, J. M., 426, 427, 428, 429, 434, 435, 436, 442, 443
Morris, Laura Newell, 139–156, 504
Morrissey, L., 393
Mosier, H. D., 426, 435
Mourant, A., 97
Movlavd, M. A., 426
Muir, A., 391
Mukherjee, D. P., 315, 327, 328, 426, 434
Muldal, S., 421, 422, 430, 439
Mulden, –., 185
Muller, H. I., 434

Nagamine, S., 56, 59
Napier, J. R., 155
Napier, P. H., 155
Nasserghodossi, M. A., 426
Natishvili, A. I., 116
Neat, H. J., 479, 480
Nes, N., 483
Neu, R. L., 421, 422, 430, 465
Neumann, F., 452
Neuwirth, J., 430, 434, 439
Newman, M. T., 149, 392, 400
Newman, R., 375
Nielsen, J., 433
Niemi, M., 421, 422, 430, 439
Nigi, H., 141, 143, 144, 146
Noble, A., 345
Nora, J. J., 420
Norby, D. E., 483,
Normand, I. C. S., 465
Normet, L., 391
Northcutt, R. C., 437
Noyes, A. N., 143
Nute, Peter E., 139–156, 504–505

O'Brien, D., 345
O'Brien, Ruth, 21
Odunjo, F., 465, 470, 471
Oettlé, A. G., 350
Ohama, K., 425
Ohno, S., 477, 479, 485, 489, 490
O'Keefe, B., 361
Olivier, A., 391
Olivier, G., 279–285, 505
Oracca-Tetten, –., 310
Osborne, J. A., 387
Osborne, R. H., 58
Oshanin, L. V., 48
Osofsky, H. J., 345
Oster, H., 319
Ostyn, M., 456

Pal, A., 315
Pallardel, D., 400
Pantelakis, S. N., 423, 424
Pařízková, J., 201–204, 398, 505
Park, I. J., 434
Parot, S., 187
Parpaćen, Joel Valencia, 349
Partridge, J. W., 392
Pascale, L. R., 59
Passeron, P., 465, 467
Pataryas, H. A., 147
Pattats, M., 120
Pavlovskii, O. M., 82, 131
Pawlowitzki, I. H., 462
Payne, F. J., 387
Pearson, P. L., 418, 419
Pecora, L. J., 389, 390
Pende, Nicola, 273
Pennock, L. A., 476, 478
Penrose, L. S., 70, 400, 465, 466
Pepler, W. J., 425
Pera, F., 417
Peretti, D., 427, 428, 434, 436, 437
Petersen, G. B., 424
Phillip, J., 421, 422, 430, 439
Pikó, L., 474, 483
Plant, W., 484
Polani, P. E., 430, 434, 488
Pollitzer, W. S., 140
Pomelli, S., 120
Pooley, J. M., 467, 491
Potapov, M. I., 122
Poulik, E., 148, 153
Powers, H. O., 421, 422, 430
Pritchard, J. G., 433
Prokopec, Miroslav, 213–231, 505
Prychodko, W., 141, 143, 144, 148, 153
Pyle, R. L., 483
Pyle, S. I., 208
Pyzuk, M., 326

Rabinow, David, 19–43, 505
Raboch, J., 430, 434, 439
Race, R. R., 432, 434, 435
Raicu, P., 477
Rakshit, H. K., 315, 321
Ramnathan, K. S., 86
Râmneantzu, Peter, 273–277, 505–506
Rao, C. R., 339

Ratcliffe, S. G., 423, 424, 440
Ray-Chaudhri, R., 475
Reitalu, J., 421, 430, 434
Reklus, E., 153
Rendel, J. M., 480
Revelle, R., 351
Rhoads, John, G., 20
Richards, S., 464, 465, 467
Richardson, Margaret A., 316
Richart, R. M., 426
Rife, D. C., 149
Rifkind, R. A., 480
Rijk, J., 293
Riley, –., 477
Ring, A., 238, 239
Riopelle, A. J., 399
Ripley, W., 97
Risley, Sir Herbert H., 320, 321, 363
Robert, J., 62
Roberts, D. F., 82, 86, 149
Roberts, Jean, 21
Robinson, A., 425, 435, 462, 463
Robson, John R.K., 345–348, 506
Roche, A. F., 208
Rodahl, K., 60
Rogers, G. G., 293
Roginskii, Ia., 131
Rojas, F. G., 421, 422, 430
Romain, J. B., 255–260, 506
Romer, M. A., 120
Rosenblum, L. E., 428, 434
Rosenfield, A. L., 427, 428, 429, 436, 437, 438, 440
Ross, A., 434
Rosso, P., 345
Rumpter, Y., 467
Rushton, D. I., 460, 465, 467
Rychkov, Ju. G., 98

St. Surin, Jacques, 255
Sabharwal, K. P., 387
Sacrez, –., 467
Saez, J. M., 427, 428, 429, 434, 436, 437, 440
Sainerova, H., 475
Saller, 166, 187
Sanchez, J., 400
Sanretto, A. M., 347
Sands, V. E., 425, 426, 434
Sanger, R., 432, 434, 435
Sanka, A. R., 427, 428, 436, 440, 444
Santell, F., 420
Santesson, B., 465, 471, 472, 491
Sarvella, P., 479, 485
Savostyanova, Ye. B., 452
Sayapina, Ye. S., 452
Schindler, A. M., 464, 467, 469, 470, 471, 473, 475, 484
Schlaginhaufen, O., 293
Schlegel, R. J., 465
Schmickel, R. D., 467, 491
Schmid, W., 467, 469, 472, 484, 491
Schreider, Eugene, 69, 293, 295, 326
Schroeder, W. A., 142
Schull, W. J., 295
Schummelfader, N., 488
Schwinger, E., 417
Scotch, N., 56
Scragg, R. F. R., 238, 239
Scrimshaw, N. S., 233, 235, 389
Scully, A. L., 420
Sedova, R. G., 452
Selezniova, T. G., 424, 464, 465
Sellyei, M., 462
Selye, H., 392
Sen, D., 315, 339
Severtsov, A. N., 82
Shakir, A., 351, 355
Shaver, E. L., 475
Shaw, M. W., 425
Shchekochikhuna, L. K., 82
Sheldon, W. H., 172, 273
Shelly, R., 349, 355
Shelton, William C., 21
Shmal'gauzen, I. I., 131
Shoffner, R. N., 417, 430, 439, 462, 479, 485, 486, 489
Short, E. M., 430, 434
Shubert, K., 456
Sidransky, H., 392, 393
Siggs, E. B., 399
Simila, S., 421, 422, 430
Simons, E. L., 155
Simpson, J. L., 420, 425
Singal, B. M., 148, 153
Singer, R., 7, 62
Singh, Raghbir, 339–343, 465, 506
Sinha, A. K., 420
Siniskalko, M., 120
Sirotnin, N. N., 202
Skerlj, B., 375
Slarka, L., 439
Sloane, H. S., 59
Slob, A., 279
Slobodkin, L. B., 387
Slome, C., 56, 293, 294
Smirnova, N. S., 83, 131–137, 506
Smith, C. A., 345
Smith, F. S., 437
Smith, G. R., 477
Smith, J. B., 464, 465
Smith, Moyra, 425
Smulyan, H., 421, 422, 430
Snyder, W. H., Jr., 441, 442, 443
Soferman, N., 419, 427, 429, 434, 436, 437
Sofuni, T., 425

Solari, A. J., 419
Solevyeva, V. S., 243–247, 248–254, 507
Solomon, I. L., 420
Soltan, H. C., 420
Sorensen, –., 235
Southern, A. L., 428
Spedini, G., 111–113, 507
Sperling, K., 433
Sporcq, J., 90, 91, 94
Sprynarova, Š., 201, 202, 203
Sribnik, H. H., 399
Stahl, K. A., 465, 467
Stalder, G., 465, 467
Stamatoyannopoulos, G., 142
Starkova, N. T., 452
Stebbins, G. L., 460, 461, 482, 489
Steeno, O., 456
Steinbach, H. L., 420
Steinnadel, J., 434, 435
Stenchever, M. A., 415, 420, 434, 442, 443, 465
Steinbeck, H., 452
Sternberg, W. H., 425, 426, 427, 428, 434, 436, 441, 442
Stevens, S. S., 172
Stewart, A. L., 423, 424, 440
Stewart, J. M., 425, 435
Stini, William A., 1–4, 163–164, 387–403, 507
Stonova, N. S., 424, 464, 465
Stoudt, Howard W., 21, 372
Stradler, G. R., 434
Strickberger, M. W., 476, 483
Strickland, A. L., 437
Stromgren, L. S., 433
Strydom, N. B., 293
Sturt, A. E., 238, 239
Sturt, R. J., 238, 239
Sturtevant, A. H., 429
Suchý, Jaroslav, 263–272, 507
Suda, A., 207
Sunderland, E., 7–17, 507
Susanne, C., 187–198, 508
Sutherland, I., 351, 361
Suzuki, S., 56, 58, 316
Szulman, A. E., 465

Tadayyon, G., 347
Taggart, N. R., 347
Taisescu, E., 477
Takahashi, E., 326
Talbert, L. M., 427, 428, 429, 434, 436, 438
Tanaka, T., 141, 143, 144, 146
Tangen, L., 414
Tanner, J. M., 185, 193, 195, 208, 209, 210, 213, 246, 307, 316, 342, 349, 350, 351, 355, 361, 387, 453, 456
Tashian, R. E., 148
Taylor, C. E., 233
Teter, J., 426, 434, 435
Tetsuji, K., 425
Therkelson, A. J., 421, 422, 423, 430, 431, 439
Thiede, M. D., 464
Thieme, F. P., 295
Thomas, R. B., 391
Thuline, H. C., 483
Thurman, T. F., 142
Thurston, Edgar, 321
Thurstone, L. L., 22
Tiepolo, L., 434
Tillinser, K. G., 421, 422, 430, 439
Timmons, C. F., 143
Tippett, P., 435
Tissier, H., 279–285, 508
Titlbacková, S., 219
Titova, Ye. P., 452
Tjio, J. H., 412
Tobias, P. V., 86, 287
Tochimoto, S., 428
Townsend, P. K., 347
Toyamasu, T., 143, 148
Trepolo, L., 421, 422, 430, 439
Tres, L. L., 419
Trompiz, Gabriel, 349
Trotter, Mildred, 20
Truett, Bruce, 23
Trujillo, J. M., 426, 434
Tucker, Pearl E., 30
Tucker, W. B., 172
Turpin, R., 414, 459, 467
Turusbekov, B. T., 243, 246
Twiesselmann, F., 187
Tzagournia, M., 392

Vacher, J., 293
Valaes, T., 293
Valsik, J. A., 350, 351
Vandenberg, Steven G., 20
Van Graan, C. H., 293
Van't Hoff, J., 487
Van Valen, L., 154
Van Wyk, J. J., 427, 428, 429, 434, 436, 438
Vass, L., 462
Veale, A. M. O., 425, 426, 434
Vecchi, F., 111–113, 508
Velez, H., 396
Verney, E., 392, 393
Vilhjalmdottir, L., 393
Viljoen, J. M., 293
Villarejos, V. M., 387
Vinnikov, Ya. F., 244
Viola, Giacinto, 273
Viru, A. A., 456
Vischer, D., 467, 469, 471, 472, 484, 491

Vitale, J. J., 396
Vlastovsky, V. G., 318
Voronov, A. A., 115–127, 508
Volkov-Dubrovin, V. P., 83, 131
Vorontsov, V. I., 456
Vorster, D. J. M., 288

Wachstein, M., 425
Wachtlova, M., 203
Waddell, –., 321
Waddington, Conrad, 400, 415
Waddy, B. B., 301
Wade, A. D., 428, 429, 436, 437, 438
Wade, P. T., 142, 143, 144, 146
Wagle, D. S., 392
Walker, A. R. P., 112
Walker, J. D., 437
Ward, N., 293
Waszczenko-Zacharenki, E., 122
Waterlow, J. C., 392, 396, 398
Watts, E. S., 399
Waxman, S. H., 464, 465
Weaver, R. P., 427, 428, 429, 434, 436, 438
Weiner, J. S., 7, 62, 69, 77, 81, 390
Welch, N. Noreen, 345
Welbourne, H. F., 301, 312, 392
Welon, Z., 279
Wennstrom, J., 421, 422, 430, 439
White, Robert M., 31
Whitehouse, R. H., 208, 209, 210
Whyte, W. G., 429, 434, 436
Wilkinson, G. S., 428, 429, 436, 437, 438
Willard, D., 467
Wilson, D. C., 351, 361
Wilson, J. D., 437
Winick, M., 345
Witschi, F., 474, 476
Wolanski, Napoleon, 239, 326
Wolf, U., 477
Wolferman, Beatriz, 349
Womack, Henry, 369–377, 508
Womble, W. H., 91
Wright, S., 400
Wyndham, C. H., 86, 293

Yamazaki, C., 317
Yanez, L., 400
Young, R. A., 143
Younger, J. B., 420, 425
Yuncken, C., 464, 465, 467
Yusuf, A. F. M., 433

Zaino, E. C., 120
Zelinger, B. B., 428, 434
Zenkevich, P. I., 131
Zhelnina, L. V., 452
Zhukovskiy, M. A., 452
Zourglas, P. A., 428
Zubov, A. A., 50
Zuffardi, O., 434
Zukowski, W., 355

Index of Subjects

Abdomen circumference. *See* Anthropometry
Abkhazians. *See* Transcaucasus
Abortions. *See* Pregnancy
Adductor longus muscle, 379, 380, 381
Adighe Caucasians, 46
Adiposity. *See* Bodyweight; Japanese children; Obesity; Vendans
Adrenal cortex: hormone development factors, 455; malnutrition effects, 400; protein effects, 392
Aerobic capacity, 201, 202, 203, 293
Aerospace Medical Research Laboratory, 20
Afghanistan. *See* Skin color
Age effects. *See* Body frame variability; Brachycephalization; Eskimos; Ghanaian children; Japanese children; Papua New Guineans; Pregnancy; Sex hormones
Airforce personnel, 21
Allele variation. *See Macaca fascicularis*; Pedi; Vendi
Allopolyploidy. *See* Polyploidy
Altitude effects on growth, 243–258, 412
Amenorrhea, 346, 420
Aminoacids in muscle development, 392
Amphibians, 477, 478
Androgens, 454, 455
Anemia, 480
Aneuploidy, 420, 459
Angina pectoris, 274
Anterior-Asiatic subrace. *See* Transcaucasus
Anthropology. *See* Anthropometry; Caucasians; Sara Madjingay
Anthropometry: abdomen circumference, 291; acromial height, 290; airforce personnel, 21; arm measurements, 10, 11, 12, 13, 15, 17, 291, 292, 339, 340, 341; biacromial breadth, 177, 178, 179; bicrystal diameter, 248–251, 290; biiliac breadth, 177, 178, 179; buttock circumference, 22, 134; calf circumference, 22, 31; cephalic index, 270, 271; chest measurements, 132, 133, 135, 177, 248–251, 269, 290, 291, 455; cormic index, 292; cristale height, 290; epiphyses widths, 133; facelength factor, 23; facial index, 270, 271; factor analysis, 19–43; foot length, 22; hand–foot factor, 23, 31; head measurements, 23, 30, 31; leg length, 132, 177, 178, 179, 291, 292; longbone factor, 22, 31, 310; lower leg circumference factor, 23; maps, 89–95; nasal index, 270, 271; neck circumference, 22; pelvis width, 133; pregnant women, 346; radial height, 290; shin circumference, 134; shoulder measurements, 132, 134, 135, 248–251, 292; sitting height, 2, 290, 341; stylion height, 290; suprasternal height, 290; tibinale height, 290; torso length, 132; tragion height, 290; waist–back factor, 30; weight–circumferential factor, 22, 31. *See also* Body frame variability; Burundi; Eskimos; Female physique; Gypsies; Kirginians; Rwanda; Zaire
Arabian peninsula. *See* Skin color
Arm measurements. *See* Anthropometry; Skin color
Armenians. *See* Transcaucasians
Armenoid Caucasians, 46

Asia Minor Caucasians, 46, 50
Assyroid Caucasians, 46, 50
Autochthonic character, 47
Autopolyploidy. *See* Polyploidy
Azerbaijanians. *See* Transcaucasus
Azraq oasis, 7

Barr bodies, 483–484
Belgians: body measurements, 187, 190, 194; family inheritance, 189–193; genotype/phenotype factors, 187–198; skin reflectance measurements, 60; standardization of measurements, 187–188, 190
Biceps brachii, 379, 380, 381, 382
Bicrystal diameter. *See* Anthropometry
Biological adaptability: arctic type, 60, 84, 85; climatic factors, 69–79, 89; continental type, 84, 85; environmental factors, 84, 225–227, 302, 309, 310, 311; Europeoids, 83; foothills type, 84, 85; genetic factors, 2, 9, 10, 17, 81, 89, 90, 93, 95, 118, 119, 125–127, 139–127, 139–156, 189–198, 295, 391, 411, 412; geographical factors, 9–10, 15, 69–79, 84, 97–113, 214–218, 225–231; maps, 97; metabolic factors, 81, 82, 84; Mongoloids, 83; skeletal factors, 69, 75, 77, 83, 207–212; Soviet populations, 82, 83; biomass, 374, 375. *See also* Altitude; Anthropometry; Body frame variability; Caucasians; Chromosomal variations; Nutrition
Biomedical Computer Programs, 19
Bira tribes (C. Africa), 89–95
Blood group frequencies: 2, 111–113, 120–124, 244
Blood levels, 111–113, 115–127
Body frame variability: adiposity, 134, 135, 370–371; aerobic capacity, 201, 202, 203, 293; age factors, 201–204; asthenic, 135, 137, 455; athletic, 135, 137, 165–176; body surface, 132; brachymorphic, 135, 137; Buryats, 132, 133, 134, 135, 136–137; Chukchi, 132, 133, 134, 135, 136–137; height, 202, 203, 264, 265, 267, 301, 308, 309, 360; hormone effects, 451–456; male/female, 132–135, 201–204; muscle fiber, 379–384, 395; span, 290; stature, 132, 135, 177, 290, 397; Tajiks, 132, 133, 134, 135, 136–137; weight, 132, 135, 201–204, 268, 290, 301, 305, 306, 310, 340–342, 347, 360, 370–373, 389, 455; Wood Nenets, 132, 133, 134, 135, 136–137. *See also* Anthropometry; Belgians; Chromosomal variations; Czechoslovakians; Genetic factors; Japanese children; Nutrition; Pedi; Skinfold measurements; Venda
Body mass, lean: calorific deficiency factors, 392; Colombian population, 395; Mexican populations, 369–377, 395; U.S. population, 369–377. *See also* Body frame variability
Body size vs. race, 2–3, 131–137
Brachycephalization: age factors, 281; genetic factors, 284–285; history, 284; male/female factors, 280; relation to tuberculosis deaths, 279–285
Brachymorphic body frames, 135–137
Breast feeding: Indian mothers, 366; nutrition factors, 366, 367, 398; Papua New Guineans, 237–238
Burundi population: adaptability vs. climate, 89–95
Buryats. *See* Body frame variability
Buttock circumference. *See* Anthropometry

Calcium blood levels, 111,112
Calf measurements. *See* Anthropometry
Canalization, 412, 413, 480
Cancer. *See* Skin cancer
Cardiovascular systems, 244
Caucasians: Adighe, 46; anthropological genesis, 45–52; Armenoid type, 46; Asian Minor subrace, 46, 50; Assyroid type, 46; autochthonic character, 47; brachycephalic forms, 48; Caspian types, 46, 49, 50, 51, 115–125; contemporary population, 46; cranial features, 99, 100; doliocephalic forms, 48; doliocranic forms, 48; Europeoid types, 46, 48, 49; Iberian type, 46; Indian groups, 51; Indo-Pamir subrace, 46, 47; Kòlkhic type, 46; Mediterranean-Balkan subrace, 46, 47; paleoanthropological factors, 48, 49; Persian Azerbaidjan group, 46; physical features, 46, 47, 48, 51; Pontic type, 46, 48, 49, 50
Center of Anthropological Research, 55
Chad. *See* Sara Madjingay
Chest measurements. *See* Anthropometry
Children. *See* Czechoslovakian males; Gypsies; Japanese children; Papua New Guineans
Chileans, 316
Cholesterol blood levels, 111, 112
Christians. *See* Skin color
Chromatin-positive testicular dysgenesis, 420
Chromosomal variations: aneuploidy, 420; autosomal mutants, 429; detection at birth, 440–442; diploidy, 411; effects on fetal loss, 412, 413; effects on new-born, 413; eunuchoid, 420; hormone factors,

422–423; mosaicism, 430–431, 459, 469; sex, 420, 421–444; tetraploidy, 411; translocation, 413; trisomy, 21, 413; tumor developments, 428; XX-male, 414, 421–424, 429–434, 443–444; XY-female, 414, 417, 425–429, 434–442, 443–444
Chromosome Y: structure and function, 417–420
Chromosome 21; 413
Climatic effects on adaptation. *See* Biological adaptability
Clines, 2, 93, 94, 97
Colombian population nutrition, 389
Color blindness, 296
Contraceptives, 476
Cormic index. *See* Anthropometry
Craniological characteristics, 2, 100, 101, 282, 283
Cyrenaica. *See* Skin color
Czechoslovakian males: children, 291, 220; environmental factors, 225–227; geographical variations, 214–218, 225–231; growth factors, 213, 219–231; heights (1921–1970), 213–223; weights (children), 223–225

Dahomeian males: blood groups, 111, 112; blood levels, 111, 112; genetic factors, 111; hemoglobin gene, 111; survival, 113
De-brachycephalization, 284
Demographic data, 369–377
Dermatoglyphics: Kyzyl-Djar males, 244; *Macaca fascicularis*, 148–153; Pedi males, 296–298; Venda males, 296–298
Diet. *See* Nutrition
Diploidy, 411, 459, 461. *See also* Chromosomal variations
Doliocephalic subjects: Caucasian, 48; non-resistance to epidemics, 274
Doliocranic subjects, 48
Down's syndrome, 413
Drosophila willistonii, 153
Dyad separation. *See* Female physique

Ectomorphy, 455
Education factors: in adaptation, 227, 302–304, 309; in menarche, 350, 356, 359
Endocrinal deviations, 274
Environmental factors in adaptation. *See* Biological adaptation
Epidemics, 285
Eskimos (Ammassalimiut) males: adaptation to cold, 60; age/skinfold factors, 55–56; arm reflectance, 60–62; finger patterns, 62–65; forehead reflections, 60–62; panniculus adiposus, 56, 58; ridge counts, 62–65; skin reflectance, 60–62; subscapular skin fold, 55; suprailiac skinfold, 55; tricipital skin fold, 55; weight/height factors, 57–58
Estrogens: distribution in females, 454
Ethnogenesis, 86
Eunuchs. *See* Chromosomal variations
Euploidy. *See* Polyploidy
European female athletes, 165–176
Europeoid Caucasians, 46, 48, 49, 102
Evans Electroselenium Ltd., 8, 60

Facelength. *See* Anthropometry
Factor analysis. *See* Anthropometry
Female physique: athletes, 165–176; body measurements, 166, 169–176; dyad separation, 167–168; hormone factors, 425, 451–455; morphological differences, 166, 168; sterility, 414
Fertility. *See* Papua New Guinea; Pregnancy; Sara Madjingay
Fetal abnormalities, 346, 412, 413
Fezzan. *See* Skin color
Fingerprints. *See* Eskimos; Dermatoglyphics
Food Composition Tables, 347
Foot length. *See* Anthropometry
Forehead measurements. *See* Eskimos; Skin color

Gall bladder lithiasis, 274
Gastroduodenal ulcer, 274
Genetic factors in adaptation, 2, 10, 17, 111, 118, 119, 125–127, 139–156, 189–198, 295, 391, 411, 412
Genogeography of Transcaucasus, 115–127
Genotypes, 141–142, 187–198, 411, 412. *See also* Chromosomal variations
Geographical effects on morphology, 214–218, 225–231
Georgians. *See* Transcaucasus
Ghanaian children: age/height factors, 304; age/weight factors, 304, 305; boys/girls growth 308; environment/growth, 311; heights, 301, 308, 309; nutrition, 307; "privileged", 302, 309; rural, 303; Secondary school, 302, 309, 310, 311; suburban, 303; weights, 301, 305, 306, 310
Glucocorticoids determination, 392
Gonadal dysgenesis, 414, 420, 425, 440, 443–444
Growth: altitude effects, 243–254, 412; altitude/sex factors, 243; environmental effects, 301–312; Ghanaian children, 301–312; gypsy children, 263–271;

Growth *continued*
Japanese children, 185; Kirgizians, 243, 244, 248-253; Kyzyl-Djar, 244-247; Quechuans, 243, 244. *See also* Biological adaptation
Gynandromorphism, 274
Gynecomastia, 420
Gypsies: anthropometric measurements, 264-271; children's growth, 263-271; history 263-264

Habbanite Jews, 7, 17
Haiti, 255-260
Harpenden caliper, 55, 177
Head measurements. *See* Anthropometry
Height. *See* Body frame variability; Indians
Hemoglobin. *See* Blood groups; Blood levels; Dahomeian males; Genetic factors; *Macaca* spp.
Hepatitis (viral), 274
Hereditary characteristics. *See* Chromosomal variability; Genetic factors
Heterosis, 75
Homeostasis, 412, 480
Hormones. *See* Chromosomal variations; Female physique; Male physique; Sex hormones; Thyroid hormone
Humu tribes (C. Africa), 89-95
Hutu tribes (C. Africa), 89-95
Hyperplasia, 391, 438-440
Hypertension, 274
Hypertrophy, 301, 382

Iberian Caucasians, 46
Immunological specifications, 2
Indians: anthropometric measurements, 320; female mortality, 363, 364, 365; food preparation, 364; height decrease, 315, 319-330; life expectancy, 367; marriages, 364; nutrition/height factors, 326, 327; nutrition/reproductive cycle factors, 326, 327; survival/height factors, 329. *See also* Caucasians; Punjabis
Indo-Pamir Caucasians, 46, 47, 115
Influenza, 274
Iran. *See* Skin color
Iraq. *See* Menarche; Skin color
Isogenes, 97
Isogenic lines, 97, 98, 102
Italian students' menarche, 355

Japanese-American hybrids (white): Kimurs, 208, 209, 211; skeletal maturation, 207-212; Tanner-Whitehouse, 208, 209, 210
Japanese children: adiposity, 178, 186; age variations, 178, 179, 181-184; arm circumference, 177, 178, 179; biacromial breadth, 177, 178, 179; biiliac breadth, 177, 178, 179; body weight, 177, 178, 179; chest circumference, 177, 178, 179; growth/body composition factor, 185; leg length, 177, 178, 179; sex variations, 178, 179, 180; skinfold thicknesses, 177-181; stature, 177, 178, 179
Jordan. *See* Skin color

Kakhatians. *See* Transcaucasus
Kazakhs, 102
Kenya. *See* Skin color
Kirghizs, 102, 243-254
Klinefelter's syndrome, 413, 420, 421
Kolkhic Caucasians, 46
Kurdish Jews, 7, 17
Kyzyl-Djar males: altitude effect on growth, 243-254; anthropometric measurements, 244, 245; blood factors, 244; cardiovascular system, 246; dermatoglyphics, 244; growth factors, 243-254; nutrition factors, 247; puberty, 244; thyroid function, 246

Lebanon. *See* Skin color
Leg measurements. *See* Anthropometry
Lega tribes (C. Africa), 92
Leydig cell hyperplasia, 438-440
Leydig cells, 420, 421
Longbone factor. *See* Anthropometry
Longevity: Indian women, 367; relation to somatotype, 273-277

Macaca arctoides: genotype/phenotype correlations, 141-142; hemoglobin variations, 141-142
Macaca fascicularis: allelic variability, 147-148, 153; dermatoglyphic data, 148-153; genotype/geographical variation, 143-148, 153-156; hemoglobin variations, 143-147
Macaca nemestrina: dermatoglyphic data, 148-153
Magnesium blood levels, 111, 112
Male physique: chromosome factors, 414, 421-424, 429-434; hormone factors, 422-423, 455-456; sterility, 414, 421
Malnutrition. *See* Nutrition
Marriage. *See* Indians
Martinique, 255, 258
Mbuti pygmies, 91
Mbuti tribes (C. africa), 89-95
Mediterranean-Balkan Caucasians, 46, 47, 115
Menarche: education factors, 302, 309, 350, 356, 359; environmental factors,

256–257, 259, 349, 355, 356, 359, 361; Iraqui students, 355; Italian students, 355; nutrition factors, 258, 259, 351, 357, 358; Polish students, 355; protein intake, 359; Tanzanian students, 355; U.S. students, 355. *See also* Puberty
Menstrual factors, 238, 239, 256, 257–260. *See also* Menarche
Mesomorphy, 455
Mexican population bodymass, 369–377
Micronesians, 315
Mongoloid Caucasians, 46, 48, 49, 202
Monkeys. *See Macaca* spp.; Muscle fiber composition
Monosomy, 459
Morocco. *See* Skin color
Mortality, 238, 363, 364, 365, 466
Mosaicism, 430–431, 459, 469
Müllerian ducts, 419
Muscle fiber composition (crabeating monkey): analysis, 380, 381, 382; biceps brachii, 379, 380, 381, 382; counting, 380, 381, 382
Muscle fiber composition (human); adductor longus, 379, 380, 381; analysis, 380, 381, 382; aminoacid exchange, 392; belly, 383, 384; biceps brachii, 379, 380, 381, 382; counting, 380, 381, 382; hyperplasia, 391; hypertrophy, 382, 391; sartorius, 379; sternocleidomastoid, 379, 381, 382
Muscle fiber replacement, 391, 392
Muslims. *See* Skin color
Mutation. *See* Chromosomal variations
Myocardial infarction, 274

Neck measurements. *See* Anthropometry
Negroids. *See* Dahomeians; Pedi; Venda
Neuritis, 274
Normotypes, 275–277
Nullisomy, 459
Nursing mothers. *See* Breastfeeding
Nutrition: adaptation factor, 307, 387–403; breastfeeding factor, 398, 399, 400; Colombian population, 389; effects on adrenal cortex, 400; effects on fetuses, 400; effect on fertility, 398–399; effect on menarche, 258, 259, 351, 357, 358; effect on pregnancy, 399; gene variation, 400–401; Kyzyl-Djar males, 247; metabolism factors, 388–392; Papua New Guineans, 235; Pedi, 294; protein factors, 392–394; puberty, 258–259; reproductive cycle factor, 363-368; Vendans, 294

Obesity, 274. *See also* Body weight; Japanese children; Vendans

Papua New Guinea Children: age/weight factor, 234; breastfeeding, 237–238; feeding, 235; growth patterns, 233–235; mortality, 238; skinfold measurements, 236
Papua New Guinea females: fertility, 238, 239; menstrual factors, 238, 239; protein intake, 239; sexual factors, 238–239
Pakistan. *See* Skin color
Palestine. *See* Skin color
Palm patterns: *Macaca* spp., 150–153
Pedi group (South Africa): adiposity/age factor, 294; anthropometric measurements, 289–292; dermatoglyphics, 296–298; rural, 288, 293; nutrition, 294; sero-genetic marking, 295; skinfold measurements, 294; urban, 288, 293
Pelvis width. *See* Anthropometric measurements
Persian-Azerbaidjan Caucasians, 46
Phenotypes, 2, 142, 187–198, 411, 412. *See also* Chromosomal variations; Genetic factors
Phylogenetic developments, 82
Pigmentation, 2. *See also* Skin color
Pigs: protein intake vs. body weight, 393–394
Plague, 284
Plants: polyploidy, 482
Pneumonia, 274
Polish students' menarche, 355
Polypeptides, 2
Polyploidy: abortion factors, 463–466, 476; amniotic fluid, 462; amphibians, 477, 478; aneugamy, 475; associated deformities, 469–473, 483; definition, 460–461; diandry, 473–474; digyny, 473–474, 475; human, 462–476, 480, 483; in chromosome variation, 411, 414; lethal, 466, 476; mosaicism, 469; oral contraceptive factor, 476; origin, 463; plants, 482
Pontic Caucasians, 46, 48, 49, 50
Populations: adaptation/nutrition, 388; characteristics, 98, 371–377; metabolism, 388–391. *See also* Biologocal adaptability
Potassium blood levels, 111, 112
Pregnancy: abortions, 463–466, 476; age factor, 346; amniotic fluid, 462; anthropometric measurements, 346; fertility, 346; infection factors, 346–347; menstrual factor, 346; nutrition factors, 345–348, 366–367, 398–399, 400; spontaneous abortions, 463; tetraploid cells, 462
Protein: bloodlevels, 113; deficiency, 391–392, 393; effect on adrenal cortex,

Protein *continued*
392, intake, 239, 347, 359, 365, 390, 393-394
Puberty: class vs. menarche, 256–257, 259; in Haiti, 255–260; hormone distribution, 454; income effects, 258; Kyzyl-Djar males, 244; menstruation, 256, 257–260; nutrition effects, 258, 259; race vs. menarche, 256–257
Punjabi males: body weight, 340, 341, 342; upper arm girth, 339, 340, 341; sitting height, 341; wrist girth, 341

Races: analysis, 97–113; body size, 2–3, 131–137; Caucasian, 45–52; dermatoglyphics, 244, 296; distribution, 104–105; formation, 99; formative centers, 106–107; geography, 97–113
Reflectance spectrometer, 8
Republic of Chad. *See* Sara Madjingay
Rheumatoid polyarthritis, 274
Ridge count. *See* Dermatoglyphics
Rwandans: adaptation/climate factors, 89–95; biometric maps, 94; forest populations, 89, 93; gene flow, 89, 90, 92, 93, 95; morphology, 91; savanna populations, 89, 92, 93

Sagyndyk (Russia), 244
Sara Madjingay (Republic of Chad): anthropometric measurements, 69–76; climatic adaptation, 69–79; differential survival, 75–76; endogamy/exogamy factor, 77–79; fecundity factors, 71–76, 79; height, 69; heterosis, 75; leg length, 69; parental couple studies, 71–76; selection factors, 70, 72, 73; skeletal frame, 69, 75, 79; trunk length, 69; weight factors, 73
Sartorius muscle, 379
Sero-genetic marking, 295
Sertoli cells, 422
Serum proteins, 2
Sex hormones: age factors, 453–454; excretion 452; genetic factors, 452; morphological relationships, 451–457; profile, 453; twins, 452
Shin measurements. *See* Anthropometry
Shoulder measurements. *See* Anthropometry
Sidorov (Russia), 244
Sitting indices. *See* Anthropometry
Skeletal factors, 48, 69, 75, 77, 82, 83, 84, 207–212
Skeletal muscle. *See* Muscle fiber composition
Skin cancer, 7
Skin color: Afghanistan/Iran males, 8, 10, 12, 13, 14; Arabian peninsular males, 8, 10, 12, 13; arm measurements, 10, 11, 12, 13, 15, 17; Cyrenaicans, 8–14; Fezzan males, 8–13; forehead measurements, 9, 11, 12, 13, 16; genetic effects, 9, 10, 17; geographical effects, 9–10, 15; Habbanite females/males, 7, 17; Iraq/-Syrians, 8, 10, 12, 14; Jordanian males, 8, 10, 12, 13; Kenyan males, 8–13; Kurds, 7, 17; Lebanese, 8, 10, 12, 13; male/female variations, 11, 12, 13, 14, 15; measurement, 8; Moroccan males, 8, 10, 13; Muslim/Christian variations, 14–16; Pakistanis, 8, 10, 12, 14; Palestinians, 8, 10, 12, 13; Sudanese males, 8, 10, 12, 13; Tripolitanian males, 8, 10, 12, 14; Turks, 8, 10, 12, 13; ultraviolet light effects, 7, 17; Yemenites, 7, 17
Skin folds, 55, 134, 177–181, 291, 371, 372, 373
Skin reflectance, 9, 10–17, 60–62
Smallpox, 284, 285
Sodium blood levels, 111, 112
Somatotypes: male/female, 275, 276; normotypes, 275–276; obesity, 274; relation to age, 273–277; sthenic brevitype, 274; sthenic longitype, 274
South Siberian race, 102, 103
Stature measurements, 132, 135, 177. *See also* Body frame variability
Sterility, 414, 421
Sternocleidomastoid muscle, 379, 381, 382
Sthenic somatotypes, 274
Sudan. *See* Skin color
Syria. *See* Skin color

Tajiks. *See* Body frame variability
Talyshes. *See* Transcaucasus
Tanzanian students menarche, 355
Taste blindness, 295
Testicular dysgenesis, 420
Testicular feminization, 426, 428, 436–438, 442
Testis development, 413, 419, 443, 484
Tetraploidy, 411, 462, 466
Tetrasomy, 459
Thalassemia, 116, 119, 120
Thyroid function vs. growth, 246
Torso measurements. *See* Anthropometry
Transcaucasians: Abkhazians, 115, 117, 121; abnormal hemoglobins, 115–127; Armenians, 115, 117; Azerbaidjanians, 115, 117, 121, 124, 126; bloodgroup distribution, 120, 121–123; B-thalassemia, 116, 119, 120; gene frequencies, 118, 119, 125, 127; genogeography, 115–127; Kakhetians, 115, 117, 121, 124, 126; Talyshes, 115, 117, 120; tasting factors, 124–127

Translocation, 413
Tripolitania. *See* Skin color
Triploidy, 462, 464, 466
Trisomy, 21, 413, 459
Tristan da Cunhans, 316
Tuberculosis, 274, 279–280, 284–285
Tumors, 428
Turkey. *See* Skin color
Turkmen: anthropological factors, 100–101
Turners' syndrome, 413, 320
Tutsi tribes (C. Africa), 89–95
Twins, 453

United States: population body mass, 371–377; students' menarche, 355
Urea blood levels, 111, 112

Venezuelan students' menarche, 349–361
Vendan males (South Africa): adiposity/age factor, 294; anthropometric measurements, 289–292; dermatoglyphics, 296–298; nutrition, 294; rural, 288, 293; serogenetic marking, 295; skinfold measurements, 291, 294; urban, 288, 293.

Weight measurement. *See* Anthropometry; Body frame variability
Weight somatotype, 274
Wolffian ducts, 419
Wood Nenets. *See* Body frame variability
Wrist size. *See* Anthropometry

Yap Islanders, 316
Yemenite Jews, 7, 17

Zaire populations: adaptability/climate factors, 89–95

Bei Fragen zur Produktsicherheit wenden Sie sich bitte an:
If you have any questions regarding product safety,
please contact:

Walter de Gruyter GmbH
Genthiner Straße 13
10785 Berlin
productsafety@degruyterbrill.com